Cambridge Studies in Biological and Evolutionary Anthropology 61

The Monkeys of Stormy Mountain: 60 Years of Primatological Research on the Japanese Macaques of Arashiyama

The Arashiyama group of Japanese macaques holds a distinguished place in primatology as one of the longest continuously studied non-human primate populations in the world. The resulting long-term data provide a unique resource for researchers, allowing them to move beyond cross-sectional studies to tackle larger issues involving individual, matrilineal and group histories.

This book presents an overview of the scope and magnitude of research topics and management efforts that have been conducted on this population for several decades, covering not only the original troop living around Kyoto, Japan, but also the two subgroups that were translocated to Texas, USA and to Montréal, Canada. The chapters encompass topics including life history, sexual, social and cultural behaviour and ecology, giving an insight into the range of current primatological research. The contributors underscore the historic value of the Arashiyama macaques and showcase new and significant research findings that highlight their continuing importance to primatology.

JEAN-BAPTISTE LECA is a post-doctoral research fellow and Lecturer in the Department of Psychology, University of Lethbridge, Canada. His research explores the determinants of behavioural innovations and traditions, and the evolution of non-conceptive sexuality, including the motivational mechanisms underlying female-to-male mounting in Arashiyama Japanese macaques.

MICHAEL A. HUFFMAN is an Associate Professor in the Department of Social Behaviour and Ecology at the Primate Research Institute, Kyoto University, Japan. His research on free-ranging and captive Japanese macaques encompasses sexual behaviour, reproductive physiology and energetics, enrichment, social learning, cultural behaviour, self-medication and parasite ecology.

PAUL L. VASEY is a Professor in the Department of Psychology at the University of Lethbridge, Canada. His research focuses on the development and evolution of non-conceptive sexuality from a cross-species and cross-cultural perspective. He has conducted research on sexual behaviour in free-ranging Japanese macaques at Arashiyama and on the captive subgroup in Montréal.

Cambridge Studies in Biological and Evolutionary Anthropology

Series editors

HUMAN ECOLOGY
C. G. Nicholas Mascie-Taylor, University of Cambridge
Michael A. Little, State University of New York, Binghamton
GENETICS
Kenneth M. Weiss, Pennsylvania State University
HUMAN EVOLUTION
Robert A. Foley, University of Cambridge
Nina G. Jablonski, California Academy of Science
PRIMATOLOGY
Karen B. Strier, University of Wisconsin, Madison

The Monkeys of Stormy Mountain

60 Years of Primatological Research on the Japanese Macaques of Arashiyama

Edited by

JEAN-BAPTISTE LECA

University of Lethbridge, Canada

MICHAEL A. HUFFMAN

Kyoto University, Japan

PAUL L. VASEY

University of Lethbridge, Canada

CAMBRIDGE
UNIVERSITY PRESS

CAMBRIDGE
UNIVERSITY PRESS

University Printing House, Cambridge CB2 8BS, United Kingdom

One Liberty Plaza, 20th Floor, New York, NY 10006, USA

477 Williamstown Road, Port Melbourne, VIC 3207, Australia

314-321, 3rd Floor, Plot 3, Splendor Forum, Jasola District Centre, New Delhi - 110025, India

79 Anson Road, #06-04/06, Singapore 079906

Cambridge University Press is part of the University of Cambridge.

It furthers the University's mission by disseminating knowledge in the pursuit of
education, learning and research at the highest international levels of excellence.

www.cambridge.org
Information on this title: www.cambridge.org/9781108823920

© Cambridge University Press 2012

First published 2012
First paperback edition 2020

A catalogue record for this publication is available from the British Library

Library of Congress Cataloging in Publication data
The monkeys of Stormy Mountain : 60 years of primatological research on the Japanese
    macaques of Arashiyama / [edited by] Jean-Baptiste Leca, Michael A. Huffman, Paul L. Vasey.
        p.   cm. – (Cambridge studies in biological and evolutionary anthropology; 61)
    Includes bibliographical references and index.
    ISBN 978-0-521-76185-7
    1. Japanese macaque–Behavior–Japan–Kyoto.   2. Social behavior in animals–Japan–
    Kyoto.   3. Primates–Research–Japan–Kyoto.   I. Leca, Jean-Baptiste.
    II. Huffman, Michael A.   III. Vasey, Paul L.
    QL737.P93M653 2012
    599.8´644–dc23   2011038243

ISBN 978-0-521-76185-7 Hardback
ISBN 978-1-108-82392-0 Paperback

We dedicate this book to the memory of the late Nobuo Asaba, Nancy Collinge, Larry Fedigan, Carole Gauthier, Baldev Grewal, Naonosuke Hazama, Junichiro Itani, Shunzo Kawamura and Eiji Ohta, as well as all the other people who directly or indirectly contributed to the development of research on the Arashiyama monkeys and to our professional development under their mentorship. Their examples live on in us, and those we mentor.

Jean-Baptiste Leca, Michael A. Huffman and Paul L. Vasey

Contents

Contributors

Patrick Bélisle, Collège Lionel-Groulx, 100 rue Duquet, Ste-Therèse, Québec, J7E 3G6, Canada

Alisa Chalmers, 698 Fulton St. #3F, Brooklyn, NY 11217, USA

Constance Dubuc, Département d'Anthropologie, Université de Montréal, C.P. 6128, Succursale Centre-ville, Montréal, Québec, H3C 3J7, Canada

Linda M. Fedigan, Department of Anthropology, #830 Social Sciences Bldg, University of Calgary, 2500 University Drive N.W., Calgary, Alberta T2N 1N4, Canada

Mariko Fujimoto, Graduate School of Asian and African Area Studies, Kyoto University, 46 Shimoadachi-cho, Yoshida Sakyo-ku, Kyoto 606–8501, Japan

Noëlle Gunst, Department of Psychology, University of Lethbridge, 4401 University Drive, Lethbridge, Alberta T1K 3M4, Canada

Michael A. Huffman, Section of Social Systems Evolution, Department of Ecology and Social Behavior, Primate Research Institute, Kyoto University, 41–2 Kanrin, Inuyama, Aichi 484–8506, Japan

Eiji Inoue, Laboratory of Human Evolution Studies, Graduate School of Science, Kyoto University, Kitashirakawa, Oiwake-cho Sakyo-Ku, Kyoto 606–8502, Japan

Katharine M. Jack, Department of Anthropology, Tulane University, 7041 Freret Street, New Orleans, Louisiana 70118, USA

Masao Kawai, Professor Emeritus, Kyoto University, Kyoto, Japan

Naoki Koyama, Professor (retired), The Center for African Area Studies, Kyoto University, Kyoto, Japan

Jean-Baptiste Leca, Department of Psychology, University of Lethbridge, 4401 University Drive, Lethbridge, Alberta T1K 3M4, Canada

Andrew J. J. MacIntosh, Center for International Collaboration and Advanced Studies in Primatology, Primate Research Institute, Kyoto University, 41–2 Kanrin, Inuyama, Aichi 484–8506, Japan

Yukiyo Maekawa, Nanzan Girls' High School, Nanzan University, 18 Yamazato-cho, Showa-ku, Nagoya, Aichi 466–8673, Japan

Eiji Ohta, Teacher (deceased)

Ann C. O'Neill, 1402 Kona Kai Lane, Escondido, CA 92029, USA

Mary S. McDonald, Pavelka Department of Anthropology, University of Calgary, 2500 University Drive NW, Calgary, Alberta T2N 1N4, Canada

Sergio M. Pellis, Canadian Centre for Behavioural Neuroscience, University of Lethbridge, 4401 University Drive, Lethbridge, Alberta T1K 3M4, Canada

Vivien C. Pellis, Canadian Centre for Behavioural Neuroscience, University of Lethbridge, 4401 University Drive, Lethbridge, Alberta T1K 3M4, Canada

Jean Prud'homme, Département d'Anthropologie, Université de Montréal, C.P. 6128, Succursale Centre-ville, Montréal, P.Q., H3C 3 J7, Canada

Anne E. Russon, Department of Psychology, Glendon College, York University, 2275 Bayview Ave., Toronto, Ontario M4N 3M6, Canada

Masaki Shimada, Department of Animal Sciences, Teikyo University of Science and Technology, 2525 Yatsusawa, Uenohara, Yamanashi 409–0193, Japan

Keiko Shimizu, Department of Zoology, Faculty of Science, Okayama University of Science, 1–1 Ridaicho, Kita-ku, Okayama, 700–0005, Japan

Yukimaru Sugiyama, Primate Research Institute, Kyoto University, 41–2 Kanrin, Inuyama, Aichi 484–8506, Japan

Yukio Takahata, School of Policy Studies, Kwansei Gakuin University, 2–1 Gakuen, Sanda City 669–1337, Japan

Yuji Takenoshita, Faculty of Children Studies, Chubu-Gakuin University, 30–1 Naka Oida-cho, Kagamigahara, Gifu 504–0837, Japan

Doug P. VanderLaan, Department of Psychology, University of Lethbridge, 4401 University Drive, Lethbridge, Alberta T1K 3M4, Canada

Paul L. Vasey, Department of Psychology, University of Lethbridge, 4401 University Drive, Lethbridge, Alberta T1K 3M4, Canada

Foreword

MASAO KAWAI

Kyoto University Professor Emeritus

In 1951, the Primate Research Group was created in the Department of Zoology at Kyoto University, and it was from this point that systematic research on Japanese macaques began. In 1952, the troop on the island of Koshima (Miyazaki Prefecture) was successfully provisioned, followed shortly thereafter by successful provisioning of the Takasakiyama troop ('yama' means mountain in Japanese) under a plan proposed by Tamotsu Ueda, the mayor of Oita City (Oita Prefecture). Junichiro Itani[1] had begun to study the Takasakiyama troop in April 1950, but after provisioning began, he was officially put in charge of research on the provisioned troop. Based on individual identification of the entire troop, he was the first to document this species' hierarchical social structure and described their vocal communication system, among other important findings.

Ueda's intention was to provision and habituate the monkeys in order to create a wild monkey park to attract tourists. For this, it was necessary to have the cooperation of researchers, and thus this joint effort of science and business made the establishment of the wild monkey park at Takasakiyama possible. At that time, the Takasakiyama troop was quite large, containing 180 members. Being located close to the natural hot springs of Beppu, many tourists visited the park and it became a very profitable venture. Based on the success of Takasakiyama, other wild monkey parks were soon established across the country.[2]

---

[1] Junichiro Itani (1926–2001, Professor Emeritus of Kyoto University), Shunzo Kawamura (1924–2003, Professor Emeritus of Kyoto University) and Masao Kawai (born 1924, Professor Emeritus of Kyoto University) are the three most prominent pioneers of primatology in Japan who started their research together under the direction of Kinji Imanishi (1902–1992, Professor Emeritus of Kyoto University), going on to make major contributions to the discipline and to nurture generations of primatologists in Japan and from abroad who came to work with them.

[2] In 1957, there were 19 monkey parks in operation across the country. In order to promote adequate protection, understanding and respect for wild monkeys and to facilitate information exchange, all wild monkey parks became members of the League for the Protection of Wild Japanese Monkeys (日本野猿愛連盟), established in September 1957. Yaen (Wild Monkey), a journal dedicated to the dissemination of information about Japanese monkeys, was established and published four times a year in Japanese. The administrative office was set up within the Japan Monkey Centre located in Inuyama with M. Kawai in charge. Koshima was the only troop that did not become part of the tourist parks, and maintained its status as a research site.

Within the larger Kyoto City area, Arashiyama, a small town located on the western edge of the city, has been a tourist destination famous for its scenic beauty, from ancient times. Sonosuke Iwata, a wealthy entrepreneur and owner of Iwatayama – the mountain was named after his family – planned to lure the monkeys to a location on the steep slopes of the mountain and provision them there to make another wild monkey park. The Arashiyama-Kyoto troop lived some distance away from Iwatayama and so it would be a very difficult task to induce them to come to the feeding site. In 1954, this ambitious project was entrusted to Naonosuke Hazama, a researcher in the Department of Zoology at Kyoto University. Hazama undertook this project and was eventually success-ful in luring and habituating the group to settle on the slopes of Iwatayama. The Iwatayama Nature Park was established in March 1957.

The two principal researchers in charge of provisioning at the time were Hazama and Kinya Nakajima. They were supported by the Primate Research Group. Thus the system of management and operation of the Iwatayama Nature Park was established under the collaborative efforts of business and research. The recording of individual identification and maintenance of the troop's genealogies from 1954 up to the present have been made possible because of this business and research collaboration. Hazama's systematic and thorough work on individual body weight and development, and the troop's natural plant food diet, were significant contributions to our fundamental knowledge of the Japanese macaque.

A noteworthy point to mention about research on the Japanese macaques of Arashiyama is the continued collaborative relationships between Japanese and foreign researchers, in particular from the USA. In 1966, a joint Japan–America scientific collaboration was established between Shunzo Kawamura, then at Osaka City University, and John Emlen of the University of Wisconsin-Madison. In 1972, after the troop fission of 1966, the Arashiyama A-troop was transported to Texas. This was the first instance for an entire Japanese macaque troop to be relocated overseas, and the last. The US research team carried out independent studies on this troop.

From this time onward, a number of researchers from the USA, Canada, UK, France, India, Tanzania, Indonesia and other countries in the world, came to Arashiyama to conduct research, making many valuable scientific contri-butions. Among these, the unique longitudinal research on stone handling, a cultural behaviour in Japanese monkeys documented by M. A. Huffman, has gathered wide attention. Huffman's enthusiasm for primatological research and his organisational abilities have resulted in many other researchers from abroad coming to Arashiyama over the years, since he started his work there in the late 1970s.

Masao Kawai at his home in Sasayama, Hyogo Prefecture, Winter 2011
(photo by G. Ohashi).

It is indeed my great pleasure to welcome the publication of this multi-authored book, a collaboration between researchers from Japan and around the world, presenting the 60-year history of research on the Arashiyama Japanese macaques and its major contributions, as well as the most recent findings from work on the Arashiyama-Kyoto troop and on its sister troop members in North America. In the long history of Japanese monkey research, this is an epic endeavour. The genealogies of the Arashiyama troops are complete for the 60-year period of research to date. A number of new research topics may arise from this unique demographic dataset. Based on this book, I am anticipating many new and valuable research results will spring forth from the Arashiyama monkeys in the future.

Preface

Many researchers may have heard of the Arashiyama macaques, but are unaware of the scope and magnitude of research and management efforts that have been conducted on this population, as well as its overall significance for wider issues in the area of evolutionary studies and educational outreach. In 1991, Linda M. Fedigan and Pamela J. Asquith co-edited the volume, *The Monkeys of Arashiyama: Thirty-five Years of Research in Japan and the West*. Since the completion of this book, a number of interesting developments have taken place in the fields of research, management and education related to the Arashiyama macaques. These developments prompted us to publish a second edited volume that updates the first one in a number of areas, covers research on the Arashiyama macaque troops since the publication of the Fedigan and Asquith volume, and also takes advantage of the long-term records available from years of collaborative research between observers working at the site.

Research at Arashiyama has now passed the half-century benchmark – an impressive feat that characterises few other field sites. In June 2008, a symposium on the Arashiyama macaques, entitled *Half a Century of Research on the Behavior & Evolution of the Arashiyama Macaques*, was organized for the Human Behavior and Evolution Society Meetings in Kyoto, Japan, by Paul Vasey. In September 2010, another symposium, entitled *The Japanese Macaques of Arashiyama: Demographic Studies, Behavioral Research, and Management Efforts*, was organized for the International Primatological Society XXIII Congress in Kyoto by Jean-Baptiste Leca. These symposia highlighted the active and on-going primatological research at the site. Some of the participants were invited to contribute to this edited volume, and they produced a series of chapters which represent the current state of research on the Arashiyama macaques. We feel privileged that they accepted to participate in this project.

Our overarching goal in editing this new book is to underscore the truly historic and, indeed, the value and diversity of primatological research activities that have been carried out on the different populations of Arashiyama macaques studied for several decades, namely the original troop still living around Kyoto, Japan, as well as the two subgroups translocated to south Texas, USA, and to Montréal, Canada. The Arashiyama macaques continue to yield new and significant research findings and these are showcased in this book.

We are very grateful, first and foremost, to Professor Masao Kawai, one of the great names in Japanese primatology, for agreeing to write the Foreword to this book. We also wish to thank all the contributors to this edited volume: Patrick Bélisle, Alisa Chalmers, Constance Dubuc, Linda M. Fedigan, Mariko Fujimoto, Noëlle Gunst, Eiji Inoue, Katharine M. Jack, Masao Kawai, Naoki Koyama, Andrew J. MacIntosh, Yukiyo Maekawa, Eiji Ohta, Ann C. O'Neill, Mary S.M. Pavelka, Sergio M. Pellis, Vivien C. Pellis, Jean Prud'homme, Anne Russon, Masaki Shimada, Keiko Shimizu, Yukimaru Sugiyama, Yukio Takahata, Yuji Takenoshita, and Doug P. VanderLaan. It is extremely gratifying that so many researchers continue to find value in Arashiyama as a study site.

We extend our sincere appreciation to the researchers, students, staff and friends who provided permission to work, assistance and valuable specific information about the Arashiyama macaques. We are particularly grateful to the following people in Japan: the late Nobuo Asaba, his son, Shinsuke Asaba, Jun Hashiguchi, Takeya Kawashima, Shuhei Kobatake, Nanako Kunugi, Koichiro Ota, Ryoko Sugano, Yaeko Sugano and Shinya Tamada (Iwatayama Monkey Park, Arashiyama). We also thank Tetsuro Matsuzawa (Kyoto University Primate Research Institute, Japan), Shunkichi Hanamura (Kyoto University Laboratory of Human Evolution Studies, Japan), Charmalie A.D. Nahallage (University of Sri Jayewardenepura, Sri Lanka), Duane Quiatt (University of Colorado, USA), Klara J. Petrzelkova (Academy of Sciences of the Czech Republic), and Natasha Tworoski (University of Minnesota, USA). At the Université de Montréal in Canada, we thank Bernard Chapais, Carole Gauthier and Jean Prud'homme. At the University of Lethbridge in Alberta, Canada, we thank Afra Foroud, Hester Jiskoot, Nadine Duckworth, Derek Rains and Christine Reinhart. In the USA, we thank Stefani Kovacovsky. We are deeply indebted to the Enomoto family (Eiji Enomoto, Yoshimasa Enomoto, Keiko Enomoto and Ryō Enomoto) for providing us with (more than) logistic assistance at Arashiyama.

We thank Fred B. Bercovitch, Christine A. Caldwell, Alan Dixson, Noëlle Gunst, Peter Henzi and Sergio M. Pellis for critically reading several draft chapters. We thank Noëlle Gunst for preparing the index. Thanks are due to Martin Griffiths, Stacey Meade, Jonathan Ratcliffe, Lynette Talbot, Zewdi Tsegai and their colleagues at Cambridge University Press for an outstanding job in producing this volume.

The editors' work was supported by the following funding sources. J.-B. Leca was supported by a Lavoisier postdoctoral Grant (Ministère des Affaires Etrangères, France), a JSPS (Japan Society for the Promotion of Science) postdoctoral fellowship complemented with a Grant-In-Aid for scientific research (sponsored by Ministry of Education, Science, Sports and Culture, Japan, to M. A. Huffman), travel funds from the HOPE Project, a core-to-core programme sponsored by JSPS, and a Natural Science & Engineering Research Council (NSERC) of Canada Discovery Grant to Paul L. Vasey. Paul L. Vasey was supported by research grants from NSERC, LSB Leakey Foundation, and research funds from the University of Lethbridge. We thank the Global COE (Centers of Excellence) Program.

We believe the work presented in this book will spark the interest of a new generation of primatologists who might make the journey to Arashiyama and play their own role in the long history of research at this site. Let us hope so, in which case, some of us might look forward to reading the centenary research volume on the Arashiyama monkeys!

Jean-Baptiste Leca, Michael A. Huffman and Paul L. Vasey

Lethbridge, Canada, and Inuyama, Japan

Introduction

MICHAEL A. HUFFMAN, PAUL L. VASEY AND JEAN-BAPTISTE LECA

Bordering the western side of Kyoto, the ancient capital of Japan, is a pictur-esque little town called Arashiyama (嵐山: stormy mountain). Nestled at the foot of the mountains bordering Kyoto city, it has been the subject of many artists' brush and pen. Perhaps it is most famous for Togetsukyo bridge, which crosses over the Hozu river at the point where it flows out of the mountain range. This scenic area is often depicted on contemporary postcards and many Edo period woodblock prints. It is, and always has been, a popular area for local and international tourists.

Less known, is the world that exists in the mountains above, with its near vertical slopes densely forested with towering hardwoods and pines. In this world there is a group of inhabitants living in a very complex society bound by rules. They are colloquially referred to simply as 'saru' by most Japanese, but are known technically as the 'Nihonzaru' or Japanese monkeys (*Macaca fuscata fuscata*). Their human-like qualities and emotions have been the sub-ject of many anthropomorphic legends, folktales and paintings. However, it was not until after the end of World War II that attempts were made to system-atically observe these animals and to unravel the mysteries concerning their human-like attributes, or as some may argue, our own monkey-like behaviour. The monkeys of Arashiyama were one of the first primate groups to receive scientific attention by primatologists in Japan, and indeed in the world.

The Arashiyama group of Japanese macaques holds a distinguished place in primatology as one of the longest continuously studied non-human primate populations in the world. Habituation of the Arashiyama-Kyoto macaques and informal observations began in 1948 by school teacher Eiji Ohta and later by Junichiro Itani and others in 1951. Systematic data collection and habituation with provisioning by Japanese primatologist, Naonosuke Hazama, followed in 1954 and the first scientific publications pertaining to the Arashiyama macaques appeared in the early 1960s. Non-Japanese researchers began working at the

The Monkeys of Stormy Mountain: 60 Years of Primatological Research on the Japanese Macaques of Arashiyama, eds. Jean-Baptiste Leca, Michael A. Huffman and Paul L. Vasey. Published by Cambridge University Press. © Cambridge University Press 2012.

site in the late 1960s and since that time Arashiyama has been the focus of numerous international collaborative research efforts.

In 1972, a subgroup of the Arashiyama population was transferred to south Texas in the USA where they live under semi-free ranging conditions and where research by non-Japanese primatologists continued until 1999. They became known as the 'Arashiyama West' population of Japanese macaques. In 1984, a subgroup of the Texas population was, in turn, translocated to the University of Montréal, Canada, where they were studied until 1998. Since demographic data have been continuously collected for the Arashiyama-Kyoto population, these monkeys represent unique resources for behavioural studies that are contingent on individual identification and known genealogical relationships. The value of this longitudinal dataset cannot be overstated as it has allowed researchers to move far beyond cross-sectional studies and tackle larger issues pertaining to individual, matrilineal and group histories. Important scientific findings have been derived from this longitudinal collaborative research on such topics as kinship, group fission, behavioural traditions and male transfer, as well as reproductive and non-reproductive sexual behaviour. Taken together, research at Arashiyama has made a significant and enduring contribution to the discipline of primatology, and more broadly, to the study of animal behaviour and human evolution.

The present edited volume covers research on the Arashiyama troops since the publication of Fedigan and Asquith's 1991 volume, *The Monkeys of Arashiyama: Thirty-five Years of Research in Japan and the West* (State University of New York Press). Like Fedigan and Asquith's book, this volume takes advantage of the long-term data available from years of collaborative research between observers working at Arashiyama. It includes 19 chapters by a variety of Japanese and international researchers. Three chapters include short box essays that relate to the chapter's topic. Contributions fall into four main categories: (1) Historical perspectives; (2) Sexual behaviour; (3) Cultural behaviours, social interactions and ecology; and (4) Management and education.

At the beginning of the 'Historical perspectives' section, Huffman, Fedigan, Vasey and Leca (chapter 1) provide a brief timeline of six decades of research activities on the Kyoto, Texas and Montréal populations of Arashiyama macaques. This chapter is then followed by a few English translations of Japanese essays written by pioneers involved in the early efforts to study the Arashiyama-Kyoto troop of Japanese monkeys. These essays provide the reader with a sense of the human drama behind early research at Arashiyama and a glimpse of the early days of primatological research in Japan. This is the first time any of these accounts have appeared in English. Chapter 2 and chapter 4 were both originally written in Japanese for a more general audience to showcase the pioneering efforts that contributed to making this site an important starting point for primatology.

The author of the first essay (chapter 2), Mr Eiji Ohta (born *c*. 1928) was an educator and a founding member of the Arashiyama Natural History Society, a group of local naturalists, primatologists and community members who promoted research and education at Arashiyama for many years. Ohta provides a detailed account of his first encounters with the monkeys of Arashiyama and the accidental destruction of the first research station at this field site. Chapter 4 was written by Kyoto University Professor Emeritus Naoki Koyama (born 1941), who elucidated, in collaboration with Sophia University Professor Emeritus Koshi Norikoshi (born 1941), the influence of matrilineal kin relationships on incest avoidance and group restructure after troop fissioning. Koyama's essay is based on the first detailed studies of primate troop fission and underscores the importance of individual recognition in conducting such research. The third essay (chapter 3) is an account of research activities at Arashiyama during the late 1950s and was written especially for this volume by Kyoto University Professor Emeritus Yukimaru Sugiyama. He spent one season at Arashiyama, before moving on to study Japanese monkeys at Takasakiyama, on the island of Kyushu, langurs in India and chimpanzees in East and West Africa.

The following chapters are based on quantitative and qualitative research. Chalmers, Huffman, Koyama and Takahata (chapter 5) present an unprecedented 50-year record of life-history traits (e.g. age-specific reproductive parameters, infant survival and mortality) among 200 female Japanese macaques from 14 matrilines, all members of the Arashiyama-Kyoto troop. Their chapter emphasises the characteristics of long-lived individuals (i.e. >25 years). Extending from 1954 to 2007, this dataset complements and enhances Koyama *et al.*'s (1992) report, which analysed the demographic parameters of all troops feeding at the Arashiyama site from 1954 to 1983. The authors urge for a continuation of such long-term studies at this field site to better understand the various female life-history traits, and more particularly those of the very old individuals. In most of the subsequent chapters, researchers and educators capitalise on these valuable longitudinal demographic data to address various questions related to a wide range of behavioural research, as well as population management, and public education. Key features of these longitudinal demographic data include the systematic identification of all group members and detailed knowledge of genealogical relationships in the Arashiyama-Kyoto population, but also in the subgroups translocated to the USA (Arashiyama West troop) and Canada (Montréal troop).

In the second section of the volume, entitled 'Sexual behaviour', Huffman and Takahata (chapter 6) use records collected at Arashiyama by several researchers, including themselves, to investigate the possible existence of long-term mating preferences in male and female Japanese macaques over their reproductive lifetimes. They argue that the best mating strategy for high-ranking males is to mate preferentially with high-ranking and mid-aged females because this may ultimately result in higher reproductive success. However, after considering various factors, such as age, dominance rank and maternal kinship, they show that both males and females change mate selection patterns over the course of their life, as a consequence of their life-history trajectories and variation in their social status. While females tend to seek novelty, males, in contrast, seek familiarity and perhaps, in turn, social stability within the troop through their associations with adult females. Huffman and Takahata's findings shed further light upon our understanding of the relative roles of both sexes in the maintenance, and ultimate evolution, of mating systems.

In chapter 7, O'Neill examines the relationships between behavioural and endocrine changes during the ovarian cycles of female Japanese macaques in the Arashiyama West colony. She provides a detailed analysis of female pre-conceptive and post-conceptive reproductive behaviour, as well as same-sex sexual behaviour, in relation to hormonal variation over the course of the female reproductive cycle. Her study reveals cyclic patterns in reproductive sexual behaviour by females, with significant increases in attractivity (e.g. mounting and holding behaviour received from males) and proceptive behaviours (e.g. mounting, holding and other sexual behaviours directed toward males) during the follicular and periovulatory phases of the cycle, followed by a complete absence of these behaviours during the luteal phase. Functionally, these behavioural changes may serve to locate and attract males who will then be in place for mating when ovulation does occur. Her demonstration of a similar relationship between homosexual behaviour and endocrine condition supports the argument that, although non-reproductive, same-sex mounting in Japanese macaque females is primarily sexual in motivation. O'Neill's research also provides evidence for the reliability of faecal analysis and enzyme immuno-assay techniques for evaluating the hormonal status in Japanese macaques.

In chapter 8, Jack describes and explores the mating strategies and behaviours of male Japanese macaques living in the Arashiyama West colony, in order to determine how these affect their ability to attract and gain access to female mates. Using a stepwise regression analysis, she tests the influence of male rank, age, spatial distribution, affiliative and aggressive interactions with females, and display performance on male mating strategies. Her results show that male age, affiliation and aggression are significant variables, and together explain a substantial proportion of the observed variation in males' mating frequency score (MFS). The negative correlation between male age and MFS may reflect female aversion to mates from preceding years and is likely the result of limitations on male dispersal at the Texas site. High rates of affiliation and aggression may be the by-product of increased proximity between males and females during the mating season. However, given the rarity of male to female aggression during the non-mating season, Jack argues that the significant correlation between male aggression and MFS may indicate that males are employing a form of sexual coercion to acquire mating opportunities.

In chapter 9, Pavelka and Fedigan tackle the hypothesis that natural selection has favoured the cessation of reproduction in female monkeys and apes who are grandmothers because the benefits of caring for grandchildren outweigh the benefits of continuing to produce more babies. To do so, they employ a large dataset on Arashiyama West female Japanese macaques and review their work on the costs and benefits of reproduction in old age versus post-reproductive grandmothering. Their results show that post-reproductive grandmothers are extremely rare in this population, but when present, they increase their grandchild's chance of surviving to age one. They suggest that the cessation of reproduction in these females is a by-product of selection favouring longevity rather than as a result of direct selection for reproductive termination. The new data they present here further support the view that reproductive costs do not appear to increase with age. They argue that continuing to reproduce until death is beneficial, not costly, to the fitness of Arashiyama West female Japanese macaques. Overall, Pavelka and Fedigan's findings provide considerable insight into the evolution of menopause in humans.

The last three chapters of this section focus on non-conceptive sexual behaviour. First, Vasey and VanderLaan (chapter 10) review some of their long-term research on female homosexual behaviour in the Arashiyama-Kyoto macaques. In this troop, a substantial proportion of females routinely engage in same-sex mounting and courtship within the context of temporary, but exclusive, sexual relationships that are virtually identical to those found in heterosexual consortships. Although numerous sociosexual hypotheses have been tested to account for such female–female sexual behaviour in Japanese macaques, none has been supported. Vasey and VanderLaan review evidence pertaining to female–female courtship, mount postures, pelvic thrusting and genital stimulation and conclude that these behaviours can be primarily, and objectively, described as 'sexual' in character. They argue that female homosexual behaviour in the Arashiyama macaques has broad implications for mating system dynamics because males must not simply compete intra-sexually for access to female mates, but they must do so *inter*-sexually as well. In a related essay, Takenoshita (box 10) provides a descriptive account of male Japanese macaque homosexual behaviour at Arashiyama, and discusses a series of possible proximate and ultimate explanations for this seldom observed behaviour.

In chapter 11, VanderLaan, Pellis and Vasey address the question of why same-sex sexual behaviour is so widespread among members of the Arashiyama-Kyoto macaque population. In addition to adult homosexual behaviour, juvenile males frequently mount one another. The authors first describe how juvenile male–male mounting is often associated with social play and more generally with tension reduction. They go on to propose a developmental model of how male–male mounting progresses through the juvenile stage and gives rise to adulthood patterns of non-conceptive mounting. The chapter closes with consideration of the biogeography and evolution of adult female–male and female–female mounting in Japanese macaques. In essence, the authors provide a springboard for further study into these puzzling phenomena and how they may vary in frequency and form among populations across the Japanese archipelago.

To conclude this set of chapters about non-conceptive patterns of sexual behaviour, Inoue (chapter 12) examines the contexts of male masturbation in Japanese macaques at Arashiyama, and investigates possible mechanisms and functional aspects of the behaviour. His results are consistent with the hypothesis that males masturbate more frequently when opportunities to copulate are rare. Inoue speculates that male masturbation does not negatively affect reproductive success and may actually function to increase sperm quality because the next ejaculate may contain fewer, but quicker-moving sperm. Taken together, this work underscores the important point that masturbation, when it is adaptive, likely serves different functions depending on the species in question.

The third section of the book compiles comparative, longitudinal and experimental studies on cultural behaviour, social interactions and feeding ecology of the Arashiyama macaques. In chapter 13, Leca, Gunst and Huffman examine one of the most thoroughly documented behavioural traditions in non-human primates, namely, stone-handling behaviour in the Arashiyama-Kyoto macaques. The authors show how 30 years of research (1979–2009) on this behaviour has elucidated the gradual transformation of stone-directed behavioural patterns that could be regarded as tool-use precursors and, as such, contributed to the understanding of cumulative culture in animals. They emphasise that Arashiyama is the first field site where a combination of longitudinal, comparative and experimental approaches has provided sound evidence for the long-term maintenance, inter-troop variability and social transmission of a single cultural behaviour in Japanese macaques. Finally, the authors argue that even traditional behaviours with no obvious function and no apparent adaptive value can not only be practised on a daily basis, and maintained over several decades within a social group, but can also be modified through the transgenerational accumulation of behavioural diversity and complexity.

In chapter 14, Shimada first provides a detailed structural analysis of 'social object play' (SOP) behaviour (i.e. social play while holding a portable object) in juvenile macaques at Arashiyama. He further explores a subcategory of SOP where multiple individuals treat a single object as the target of play and the object holder escapes from the others, referred to as 'play-chasing with a target object' (PCT). He then goes on to compare SOP and PCT in the provisioned Arashiyama-Kyoto troop and the non-provisioned Kinkazan troop. His results show that PCT is an established play pattern in the former (where the objects held during play have a low nutritional value) but not in the latter (where the objects held are nutritionally valuable). To explain differences in PCT between the two troops, Shimada proposes a causal model about the effect of food condition on the dissemination of this particular play pattern. Finally, he discusses whether variation in the frequency and form of PCT between these two troops can be explained in terms of local traditions.

In their related essay, Pellis and Pellis (box 14.1) provide an overview on the form of play fighting across macaque species (e.g. number and age of participants, degree to which dominance and familial relationships influence the selection of play partners, and the type of attack and defence behavioural patterns). They argue that species differences in play relate to species differences in social systems and dominance styles, ranging from 'despotic' (e.g. Japanese macaques) to 'egalitarian' (e.g. Tonkean macaques). From this cross-species perspective, it is thus not surprising that in their play fighting, Japanese macaques tend to be more competitive, whereas Tonkean macaques tend to be more cooperative. Pellis and Pellis conclude by raising the question as to whether the mounting behaviour that pervades the play fighting of Japanese macaques is a by-product of the species' 'hyper-sexual' tendencies, or whether it is just an exaggeration of what occurs in other macaque species.

On a final note about play behaviour, Russon and Vasey (box 14.2) systematically compare the frequency, form and context of spontaneous eye-covering play (i.e. deliberately closing or covering one's eyes during a play sequence) in orangutans and in the colony of Arashiyama Japanese macaques housed at the Université de Montréal. Their findings suggest that eye-covering play in both Japanese macaques and orangutans does not involve pretending to be blind (i.e. acting as if one can't see, doesn't exist or exists in some altered form). However, it could involve relatively sophisticated cognition in the sense that actors may learn about the possibility of pretence by discovering the fallibility of mental representations when tested against the real world. The authors' observations demonstrate that Japanese macaques are more cognitively limited in their eye-covering play compared with orangutans, with only primary representations characterising the former, whereas secondary representations appeared to characterise the latter.

In chapter 15, Fujimoto explores the effects of the social style characteristic (i.e. strict dominance hierarchy and social asymmetry) on the form of grooming interactions among adult female Japanese macaques at Arashiyama. She does so by undertaking a systematic analysis of the behavioural sequences involved in adult female Japanese macaque allo-grooming episodes. Fujimoto shows that the participants adopt different behavioural patterns depending on whether the grooming partner is kin or non-kin. She also points out that the participants change grooming roles in a symmetrical way. Finally, Fujimoto discusses such symmetrical role reversals in grooming in terms of the turn-taking occurring in human conversation.

In a related essay on grooming behaviour, Leca (box 15) reports the first case of dental flossing behaviour in Japanese macaques and considers its idiosyncratic presence in the Arashiyama-Kyoto troop as a grooming-related innovation. Since this behaviour is always associated with self- or allo-grooming activity, he suggests that the dental flossing innovation is a by-product of grooming. This tool-use innovation could be a transformation of grooming patterns via the running of hair between the teeth to remove louse eggs. He then proposes a possible scenario for the emergence of the different dental flossing techniques observed in the innovator. To explain why this behaviour has not yet spread to other group members, Leca discusses the various constraints on the social transmission of behavioural innovations within a group.

In chapter 16, Bélisle, Prud'homme and Dubuc experimentally analyse the influence of kinship degree, cost of resource defence by dominant individuals, and priority of access to the resource, on co-feeding in the Arashiyama Japanese macaques housed at the Université de Montréal. In their experimental approach, two adult females with a clear dominance relationship had access to a feeding box containing a limited amount of highly prized food. Their results show that kinship degree, defence cost and priority access all have a significant impact on co-feeding. In the first experiment they conducted, they found that the rates of tolerated co-feeding increased abruptly with degree of kinship, underscoring the importance of kin selection in the evolution of altruism in primates beyond the mother–offspring bond. In the second experiment they conducted, they showed that co-feeding time was significantly lowered when the costs of food defence for the dominant female were reduced, suggesting that a substantial proportion of the time spent in co-feeding during the first experiment may simply reflect a selfish strategy on the part of the dominant female not to expend energy, rather than altruism. In a third experiment, the authors demonstrate that when the subordinate female is given prior access at the feeding box, it substantially increases the co-feeding time of that female and her dominant counterpart. This result indicates that priority of access positively affects respect of possession from dominant females towards subordinate ones. In a final experiment, the authors demonstrate that the latter effect is even more prevalent in Tonkean macaques, a species with a more egalitarian system of dominance than that of Japanese macaques.

In chapter 17, Huffman and MacIntosh first provide a historical overview of the few studies on feeding ecology in Japanese macaques at Arashiyama.

They then combine information from records collected over a 45-year period to create the first comprehensive list, that has been published in English, of plant-foods consumed by the Arashiyama-Kyoto macaques. They point out that most of the previous reports were published in Japanese, and thus, have remained largely inaccessible to the larger primatological community. In order to determine the extent to which the Japanese macaques living at Arashiyama consume plant items that are potential medicinal foods, they compile an additional database listing those plant items containing secondary metabolites, as well as their potential role in health maintenance. Huffman and MacIntosh's impressive list aims to facilitate future work on this topic at Arashiyama and to stimulate similar work at other study sites across Japan. Such research may shed light on the major potential of plants for different forms of self-medication, including health maintenance, disease prevention and healing.

In the fourth and final section of the book, we provide two examples of the value of the Arashiyama macaques in terms of population management and public education. In chapter 18, Shimizu describes the effects of different administration methods and different doses of two types of hormonal contraceptives on the ovulatory cycles of female Japanese macaques living at Arashiyama. She shows that this hormone-induced sterility is reversible, since females eventually return to normal cycles and experience normal pregnancy after the treatment is stopped. Finally, Shimizu discusses the costs and benefits of various approaches for controlling fertility in Japanese macaque troops and how hormonal contraception could solve some of the current problems that are experienced between monkey parks and local human populations.

In the final chapter of our volume, Takenoshita and Maekawa (chapter 19) emphasise the major role of the Arashiyama-Kyoto macaques in science and environmental education in Japan. They point out that the Arashiyama site has a long history of collaboration between researchers, naturalists and educators in research and educational activities. After reviewing the current situation of science education in Japan, they discuss the positive role of the Iwatayama Monkey Park (Arashiyama Monkey Park in Japanese) as a field site for science education where primatologists have worked as collaborative educators. Finally, they introduce their own educational practices as an example of a contemporary attempt to form a new type of collaboration between educators and researchers. By regarding Arashiyama as a field site for education, Takenoshita and Maekawa show that the scientific value of Arashiyama macaques goes far beyond research.

Following the presentations of these research papers, our edited volume closes with a comprehensive bibliography of the scientific publications derived from 60 years of primatological research on the Arashiyama Japanese macaques (including the Arashiyama-Kyoto, Arashiyama West and Montréal troops). We hope this extensive (and impressive) listing will help readers in their search through the massive literature produced from work on the Arashiyama macaques, and stimulate further studies on this extraordinary monkey population.

Part I

Historical perspectives

1 A brief historical timeline of research on the Arashiyama macaques

MICHAEL A. HUFFMAN,

LINDA M. FEDIGAN, PAUL L. VASEY

AND JEAN-BAPTISTE LECA

The history of research on the Japanese macaques, starting at Arashiyama-Kyoto in 1948 in Japan to one fissioned troop's relocation in Texas, has been described in detail by Huffman (1991) and Fedigan (1991), respectively, in *'The Monkeys of Arashiyama: Thirty-five Years of Research in Japan and the West'*. We refer the reader to these two chapters for historic details prior to 1991. Here we present a brief historical timeline of research on the three main study groups of Arashiyama macaques (respectively in Kyoto, Japan, Texas, USA and Montréal, Canada) up to the present, spanning more than 60 years, to put the work presented in this book into perspective.

Looking at the impressive list of 115 people who have played a role as researcher and/or advisor of research on the Arashiyama macaques since 1948, we can see the high degree of connectivity this site has had throughout the history of primatology (Table 1.1). Arashiyama has played the dual role of a training and study site for graduate students and many prominent Japanese and international researchers alike. From the 93 researchers who conducted research on one or more of these three main Arashiyama research groups, 75 academic degrees (24 PhD or DSc degrees, 46 MA or MSc degrees and five Bachelor degrees) have been awarded, in full or in part for their work. Regardless of the capacity of their participation in the activities, every individual has made an important contribution to the accumulation of knowledge that makes the Arashiyama macaques the valuable resource that they are.

The first overseas scientists, two graduate students from the USA, to conduct research at Arashiyama were Steven Green and Gordon Stephenson. They were part of a joint Japanese–American research project set up in 1968 by senior Kyoto University Professor Ryoji Motoyoshi and Professor John T. Emlen of the University of Wisconsin. From these humble beginnings,

The Monkeys of Stormy Mountain: 60 Years of Primatological Research on the Japanese Macaques of Arashiyama, eds. Jean-Baptiste Leca, Michael A. Huffman and Paul L. Vasey. Published by Cambridge University Press. © Cambridge University Press 2012.

Table 1.1. *Who's who in primatological research of Arashiyama monkeys*

Researcher (degrees obtained for this research)	Period of fieldwork	Current academic affiliation/ Direct involvement	Site	Main advisor(s)
Ohta, Eiji	1948–1970s	Deceased	Kyoto	
Itani, Junichiro	1951	Deceased	Kyoto	
Hazama, Naonosuke	1951, 1954–1961	Deceased	Kyoto	
Kawamura, Shunzo	1951–periodically to 1980s	Deceased	Kyoto	
Sugiyama, Yukimaru	1958	Prof. Emeritus, Kyoto University	Kyoto	
Koyama, Naoki (DSc)	1964–1967, 1972–1990s	Prof. Emeritus, Kyoto University	Kyoto, Texas	S. Kawamura
Norikoshi, Koshi (DSc)	1967–1972	Prof. Emeritus, Sophia University	Kyoto	S. Kawamura
Green, Steven (PhD)	1968–1969	University of Miami	Kyoto, Texas	P. Marler
Stephenson, Gordon (PhD)	1969, 1972	Retired, University of Wisconsin	Kyoto, Texas	J. Emlen
Mano, Tetsuzo	1969–1972	Retired	Kyoto, Texas	
Casey, Denise E.	1972	Northern Rockies Conservation Cooperative	Texas	
Clark, Tim W.	1972	Yale School of Forestry and Environmental Studies	Texas	
Fedigan, Larry (PhD)	1972–1974, 1978–1979	Deceased	Texas	O. L. Davis
Fedigan, Linda M. (PhD)	1972–1996	University of Calgary	Texas	C. Bramblett
Nishida, Kenichi (BSc, MA)	1973–1974	n/a	Kyoto	
Gouzoules, Harold (PhD)	1973–1975, 1977–1978	Emory University	Texas	J. C. Neese
Wolfe, Linda D. (PhD)	1973–1978	East Carolina University	Kyoto, Texas	P. E. Simonds
Koyama, Takamasa	1974	Japan Women's University	Kyoto, Texas	
Takaragawa, Norihisa	1975–1976	Chiba City Zoo	Kyoto	
Gouzoules, Sarah (MA, PhD)	1975, 1977–1978	Emory University	Texas	R. Tuttle
Grewal, Baldev S. (MSc, PhD)	1975–1978	Deceased	Kyoto	Y. Sugiyama
Takahata, Yukio (MSc, DSc)	1975–1978	Kwansai Gakuin University	Kyoto	J. Itani
Quick, Larry B.	1976	n/a	Texas	
Ehardt, Carolyn (PhD)	1978–1979	University of Texas at San Antonio	Texas	C. Bramblett
Allan, Brian (Honour's thesis)	1979	n/a	Texas	L. M. Fedigan

Name	Years	Institution	Location	Advisor
Blount, Ben G. (PhD)	1979, 1981	Social & Ecological Informatics, Helotes	Texas	J. Gumperz
Noyes, M. J. Sabra	1979–1981	n/a	Texas	
Huffman, Michael A. (BSc, MSc, DSc)	1979–1980, 1983–1989, periodically to the present	Kyoto University	Kyoto	J. Itani
Griffin, Lou	1980–2002	n/a	Texas	
Bullard, Jeff (MA)	1981	n/a	Texas	L. M. Fedigan
Matsuura, Nagako (BSc)	1981	n/a	Kyoto	T. Kawamichi
Mehrhof, Barbara (MA)	1981	n/a	Texas	J. F. Oates
Suzuki, Hisayo (BSc)	1981, 1983–1989	n/a	Kyoto	T. Kawamichi
Pavelka, Mary S.M. (MA, PhD)	1981–1987, 1991, 1993–1996	University of Calgary	Texas	L. M. Fedigan
Hauser, Marc	1982	Harvard University	Texas	
Masataka, Nobuo	1982	Kyoto University	Kyoto, Texas	
Platt, Meredith M.	1982	Retired	Texas	
Negayama, Koichi	1982–1983	Waseda University	Kyoto, Texas	
Ando, Akihito	1983	Mukogawa Women's University	Texas	
Collinge, Nancy (MA)	1983	Deceased	Texas	L. M. Fedigan
Kamada, Jiro	1983	Kansai University of Social Welfare	Texas	
Kondo-Ikemura, Kiyomi	1983	Health Sciences University of Hokkaido	Texas	
Nakamichi, Masayuki	1983	Osaka University	Kyoto, Texas	
Yoshida, Atsuya	1983	The University of Tokushima	Texas	
Quiatt, Duane	1984	Professor Emeritus University of Colorado	Kyoto	
Mignault, Christiane (MSc)	1984–1985	Collège Edouard-Montpetit	Montréal	B. Chapais
Sicotte, Pascale (MSc)	1984–1987	University of Calgary	Montréal	B. Chapais
Chapais, Bernard	1984–1998	University of Montréal	Montréal	
Larose, François (MSc)	1985–1986	Grant MacEwan University	Montréal	B. Chapais
Inoue-Nakamura, Miho (MSc)	1985–1987	Kyoto University	Kyoto	O. Takenaka
Girard, Michèle (MSc)	1986–1988	n/a	Montréal	B. Chapais
Primi, Ginette (MSc)	1987	n/a	Montréal	B. Chapais
Gauthier, Carole (MSc)	1987–1998	n/a	Montréal	B. Chapais

Table 1.1. (cont.)

Researcher (degrees obtained for this research)	Period of fieldwork	Current academic affiliation/Direct involvement	Site	Main advisor(s)
Ogawa, Hideshi	1987–1988, 2000–2002	Chukyo University	Kyoto	B. Chapais
Tejeiro, Shona (MSc)	1988	University of Montréal	Montréal	B. Chapais
Lecomte, Michel (MSc)	1989	Collège de Valleyfield	Montréal	M.S.M. Pavelka
Cunneyworth, Pamela (MA)	1990	n/a	Texas	P. Asquith
Giancarlo, Christine (MA)	1990–1991	Mount Royal University	Texas	B. Chapais
Prud'homme, Jean (MSc)	1991–1998	University of Montréal	Montréal	M.S.M. Pavelka
Paterson, James D.	1992	University of Calgary	Texas	T. Nishida (H. Ihobe)
Wyman, Tracy (MA)	1992–1999	University of Calgary	Texas	B. Chapais
Takenoshita, Yuji (MSc)	1992–1998, 2003	Chubu-Gakuin University	Kyoto	
Gates St-Pierre, Claude-Eric (MSc)	1993	n/a	Montréal	M.S.M. Pavelka
Norman, Sherry (MA)	1993	University of Victoria	Texas	M.S.M. Pavelka
Tillekeratne, Sasrika (MA)	1993		Texas	
Koyama, Nicola F. (PhD)	1993–1994	Liverpool John Moores University	Kyoto	R. Dunbar
Vasey, Paul L. (PhD)	1993–present	University of Lethbridge	Montréal, Kyoto	B. Chapais
Jack, Katherine (MA)	1994–1996	Tulane University	Texas	M.S.M. Pavelka
Lamarsh, Craig (MA)	1994–1996	n/a	Texas	M.S.M. Pavelka
Shimizu, Keiko	1994–present	Okayama University of Science	Kyoto	
Savard, Leanne (MSc)	1996	n/a	Montréal	B. Chapais
Zamma, Koichiro (MSc, PhD)	1996–1998	Hayashibara Great Ape Research Institute	Kyoto	T. Nishida
Kaneko, Makoto (MSc)	1996–1998	n/a	Kyoto	T. Nishida
O'Neill, Ann (MA)	1997–1998	n/a	Texas	L.M. Fedigan
Belisle, Patrick (MSc, PhD)	1997–1998	Collège Lionel-Groulx	Montréal	B. Chapais
Wakibara, James (MSc)	1998–1999	Tanzania National Parks	Kyoto	T. Nishida (M. A. Huffman)
Ueno, Ari (MSc)	1999	University of Shiga Prefecture	Kyoto	T. Kano
Barrett, Gordon (MSc, PhD)	1998–2001	Health Canada	Kyoto	A. Mori

			Site	Main advisor[a]
Shimada, Masaki (MSc, PhD)	1999–2000, 2004	Teikyo University of Science	Kyoto	T. Nishida
Kashiwabara, Sho (MSc)	2000	Kyoto University	Kyoto	A. Mori
Nishie, Hitonaru (MSc)	2000–2002	Kyoto University	Kyoto	T. Nishida
Kovacovsky, Stefani	2001	n/a	Kyoto	
Inoue, Eiji (MSc, PhD)	2001–2003	Kyoto University	Kyoto	T. Nishida, J. Yamagiwa
Fujimoto, Mariko (MSc, PhD)	2002–2004	Kyoto University	Kyoto	A. Nishimura, H. Takeshita
Nishimura, Hirohisa (MSc)	2003	Kyoto University	Kyoto	M. A. Huffman
Hanamura, Shunkichi (MSc)	2003–2004	Kyoto University	Kyoto	J. Yamagiwa, N. Nakagawa
Asai, Kenichiro (MSc)	2004–2005	Kyoto University	Kyoto	J. Yamagiwa, N. Nakagawa
Leca, Jean-Baptiste	2004–present	University of Lethbridge	Kyoto	P. L. Vasey
Duckworth, Nadine (BSc)	2005	n/a	Kyoto	J. Yamagiwa
Chalmers, Alisa (MSc)	2006–2009	Cornell Medical College	Kyoto	P. L. Vasey
VanderLaan, Doug (PhD)	2007	University of Lethbridge	Kyoto	J. Yamagiwa, N. Nakagawa
Inuma, Kenji (MSc)	2007–2008	Kyoto University	Kyoto	
Gunst, Noëlle	2008–present	University of Lethbridge	Kyoto	

*People included in this list either published at least one article/report/abstract from research on Arashiyama monkeys, obtained an academic degree from research on Arashiyama monkeys, or played a leading role in the supervision of students conducting research on Arashiyama monkeys. We excluded research on biological samples collected from captive housed Arashiyama (Kyoto) macaques, for which no field work was done.

Site: (1) Kyoto: Arashiyama-Kyoto troops (A, B, E), Japan; (2) Texas: Arashiyama-West (A-troop), USA; and (3) Montréal troop (artificially formed and captive group), Canada.

[a] 'Main advisor(s)' refers to the academic advisor(s) of researchers who conducted research on Arashiyama monkeys for their academic degrees. Despite extensive searching, some information could not be obtained or double-checked. Those cases are noted as n/a.

Arashiyama has taken on a very active role in the internationalisation of primatology in Japan, hosting many researchers from such countries as the USA, Canada, the UK, France, India and Tanzania. This activity has resulted in publications on a wide variety of topics ranging from basic ecology and biology to social behaviour, endocrinology, genetics and reproductive physiology (see Appendix).

1.1   Arashiyama troops in Japan

The beginning of research at Arashiyama was simultaneous with the beginning of primatological research in Japan. This is no coincidence given that the beginnings of primatology in Japan came from a group of scientists based at Kyoto University in the laboratory of Ecology led by Denzaburo Miyadi and Kinji Imanishi. The first members included a group of then young students, Junichiro Itani, Shunzo Kawamura, Naonosuke Hazama, Masao Kawai, Tadao Umesao and Kisaburo Tokuda. Unknown to them at the time, Eiji Ohta (see chapter 2) had already begun his own observations of the monkeys at Arashiyama. Ohta was an educator and later became one of the founding members of the Arashiyama Natural History Society, a gathering of local naturalists, primatologists and other members of the local community who supported and promoted research and education at the site for many years. Of the Kyoto University Primate Research Group members, Hazama, in collaboration with local entrepreneurs, agreed upon a plan to observe and habituate the monkeys for the joint purpose of tourism and scientific research in 1954. The land on which the provisioning grounds were set up and the tourist facilities constructed was donated by businessman Sonosuke Iwata, of which the family name is given to the mountain it sits upon (Mt. Iwata or Iwatayama, in Japanese). The Iwatayama Monkey Park in Arashiyama was officially opened to the public in March 1957. Iwata supplemented the financial operation of the tourist park until 1974, when due to financial problems it was sold to Kyoto City and was made a protected historical reserve (Huffman, 1991).

With the assistance of Eiji Furukawa, Shizuma Hirose, Kinya Nakajima and others, provisioning succeeded in 1955, two years after Hazama's ecological study began (see Huffman and MacIntosh: chapter 17). It was at this time that Hazama and his associates began to name each monkey, forming the basis of what were later to become the kin names of their progeny that laid the groundwork for some of the earliest, most detailed studies, on kinship-based social organisation and social networks in primates. In 1958 Hazama moved on to study monkeys living on Mt. Hiei, 18 km to the east of Arashiyama on the opposite side of Kyoto City (see Figure 1.1a). The maintenance of genealogical records and demographic data was continued by park staff and researchers, and formed the groundwork that allowed Naoki Koyama and Koshi Norikoshi, both graduate students of Kawamura, then based at Osaka City University, to document in detail the process of troop fission and male transfer between 1964 and 1966. After the troop fissioned into two stable social units, Arashiyama A and B troops, they coexisted on Mt. Iwata until 1972, B troop being dominant over A troop. During this time, the phenomena of male transfer and social re-organisation were also documented in detail (see Koyama, 1970; Norikoshi and Koyama, 1975; Norikoshi, 1977).

After this event, three troop fissions have been documented, albeit not in the detail that Koyama and Norikoshi did. The first fission was in early 1978. A 14-member splinter group (C troop) was formed by 12 members of the Kusha kin-group and two adult peripheral males who joined them. They were subsequently captured and sent to the Choju City Zoo in southern South Korea. In May 1978 a second splinter group of approximately 28 members was recognised by the park staff. Fifteen of these were selected and sent to the Kaibara Family Land, an amusement park in Hyogo Prefecture, Japan. In 1981, a small number of monkeys, not the product of natural troop fission, were transferred from Arashiyama to the Primate Research Institute of Kyoto University for breeding purposes. Members of this group have been the subjects of various behavioural and psychological investigations (e.g. Kubota, 1990, 1991; Leca *et al.*, 2007).

The last troop fission to date was completed by September 1986. B troop fissioned into two sister troops, E (n = 149) and F (n = 97). This event has been described by Huffman (1991). The two troops shared the same feeding grounds, but E troop was subordinate to F troop and eventually moved further and further away from the Iwatayama site into the mountains to the west, becoming dependent once again largely upon natural foods from the forest, but also occasionally from family garden plots at the edge of Arashiyama town. From the late 1990s up to a few years ago attempts were made to follow them in the forest and on the other side of the mountain range in the farm and residential areas on the outskirts of Kameoka City, by Akisato Nishimura and his students from Doshisha University in Kyoto. There has been no recent updated news about F troop's whereabouts since then. Of all the Arashiyama fissioned troops, F troop is the only one from Arashiyama to re-establish itself in a new home range away from Iwatayama, in part of the original Arashiyama troop's pre-1954 home range.

Management of the troops' genealogical records was first maintained by researchers with the assistance of the provisioning staff, with each successive generation training their juniors in individual identification of the monkeys.

From the beginning of his studies, Koyama was perhaps the most instrumental in maintaining the demographic database and genealogy of the troops well into the early 1990s. In the spring of 1976, Nobuo Asaba, a local entrepreneur, became the new park director. At this time, Yukio Takahata, then a graduate student of Itani's at Kyoto University, taught him to identify the monkeys. Asaba and his family took over the management of the park and Asaba breathed new life into it, just at the time when researchers in Japan were beginning to redirect their efforts towards wild, non-provisioned populations of macaques on the islands of Yakushima and Kinkazan. From this point on, the yearly birth, death, arrival and disappearance records were maintained and the genealogy updated yearly by Asaba and his hired staff, many trained biologists. When Asaba took on the duties of park director he relinquished the day-to-day duties of his previous business to his employees and made the maintenance of the park, provisioning of the monkeys, support of researchers and maintenance of the genealogical records his life's work, until his death in 2001 (Figure 1.1). His second son, Shinsuke Asaba, took over as director of the Iwatayama Monkey Park shortly after his father's death, and continues this work today, with the same dedication to protect the monkeys, educate the visitors and ensure the livelihood of his hardworking staff, without the financial support of universities or the government (Figure 1.1). In the tradition of the elder Asaba, this family-run business continues to generously support researchers who come to work here. Every year, undergraduate students from the Faculty of Science at Kyoto University conduct research at Arashiyama. The University of Osaka's Graduate School of Human Sciences and Osaka City University have also had a long record of research activity on the Arashiyama macaques (Table 1.1).

In other ways, fruitful collaborations between educators and scientists have also greatly utilised this invaluable resource for schoolchildren and the local community to educate them about the value of wildlife and protection of the environment. The Arashiyama E troop remains an important research resource in Japan (see Takenoshita and Maekawa: chapter 19). Without the Asaba family's continued dedication, this would not be possible.

1.2   Arashiyama West troop in Texas

In 1966, the original Arashiyama troop had grown to 163, at which time it fissioned. The resulting two groups were named the Arashiyama A and B troops. A by-product of the joint research project of Motoyoshi and Emlen was the relocation of Arashiyama A troop to the USA. A group of prominent American primatologists (including Clarence R. Carpenter, Carl B. Koford, John King, Eugene Sackett, William Mason, Peter Marler, Charles Southwick,

Figure 1.1. Iwatayama Monkey Park, Arashiyama-Kyoto site, Japan. (a) Field station and visitors centre in 1980; Mt. Hiei is the highest peak in the background on the other side of Kyoto City (photo by M. A. Huffman); (b) Nobuo Asaba (far right) with Masaki Shimada, Susumu Kashiwabara and Hitonaru Nishie (from left to right), three young researchers from Kyoto University in the feeding station office, October 2000 (photo by M. A. Huffman); (c) Shinsuke Asaba (front row, far right) with Yuusuke Kataoka, Shinya Tamada, Yuuto Kobayashi (back row from left to right), Nanako Kunugi and Yuuko Tomita (front row from left to right), five staff members (courtesy of S. Asaba); (d) First feeding site in the early 1950s (anonymous photographer, courtesy of the late N. Asaba).

Figure 1.1. (*cont.*)

John Vandenburg and Bruce Alexander) were gathered to investigate the logistics of relocating the troop (Huffman, 1991). Claud Bramblett, an anthropologist at the University of Texas in Austin, was also contacted by Emlen about the search for a new home for the Arashiyama A group. One day, Bramblett mentioned this to students in one of his primatology classes. By chance, one of the four daughters of a Texas rancher, Edward Dryden Jr. was taking that course. She mentioned to her father that a group of scientists were looking for a new home for these monkeys where they would be kept outdoors in a large enclosure with the core group intact. Intrigued and eager to help, he offered the use of his land. In February 1972, the A troop, with complete genealogies and individual recognition, was finally translocated to Dryden's ranch in La Moca, Texas. Arashiyama researchers, including Tetsuzo Mano along with their American counterparts, Tim Clark, Steven Green and Gordon Stephenson, who had monitored the group back at Arashiyama, accompanied the troop to Texas. The troop's adaptation to its new environment was one of the first topics of research (Figure 1.2).

Figure 1.2. Arashiyama West site at Dilley, Texas, USA. (a) Rheus-58, alpha female, with several of her offspring, in a prickly pear cactus (photo by Linda Fedigan); (b) Adult female foraging with her offspring on her back; (c) Lady Di (Betta-59-66-78) and her infant female Carmen (Betta-59-66-78-89) (photos b and c by the late K. Dickey, courtesy of L. Fedigan and T. Wyman).

Figure 1.2. (*cont.*)

The security of the troop and maintenance of the site were never easy and were dependent upon the generosity and goodwill of people like Dryden who provided land and facilities. The troop stayed at La Moca between 1972 and 1980. Even after Dryden's death in 1973, the group was financially sustained largely by the Dryden family who continued to run the site as a commercial operation, selling monkeys to zoos and research colonies, while maintaining the scientific value of the troop for research. In August 1980, due to financial considerations, the troop was moved to a new site on the Dilley homestead, 37 km to the north of La Moca (Fedigan, 1991).

Support of the troop at Dilley was taken over by a non-profit organisation of interested researchers called the Arashiyama West Institute under the leadership of Bramblett and Lou Griffin, with funds from an NSF grant received by the Arashiyama West Institute researchers. A core group of North American researchers oversaw a very productive 27-year period of research and educational activities at La Moca and Dilley that lasted up until 1999. During this time, the difficult task of site manager was taken on in succession by eight dedicated researchers (Tim Clark, 1972; Linda and Larry Fedigan, 1972–74, 78–79; Tim Johnston, 1972–73; Harold Gouzoules, 1973–75, 77–78; Sarah Gouzoules, 1977–78; Sabra Noyes, 1979–81; Lou Griffin, 1980–2002) (Fedigan, 1991). Of all the managers and researchers, Griffin in particular, dedicated many years of her personal time and resources at the site to ensure that the monkeys had a secure home. Faculty at the University of Alberta and the University of Calgary, Linda Fedigan, James Paterson and Mary Pavelka ran a successful field school at the site for many years. A new generation of Japanese researchers (Table 1.1), many from Osaka University, also periodically visited and conducted research on the troop and investigated a number of sociological topics, comparing these results with the sister troop back in Japan or other sites of free-ranging provisioned Japanese macaques. Though Bramblett never conducted research on the troop himself, he played an active and pillar role as student advisor, fundraiser and leader in the management of the facilities from 1972 to 1992.

In December 1999, research at the site ceased, when control of the Arashiyama West colony was taken over by the Animal Protection Institute (API). API (now known as 'Born Free, USA') is a national, non-profit animal advocacy organisation based in Sacramento, California. After taking over supervision and care of the Arashiyama West monkeys and purchasing the land on which they resided near San Antonio, Texas, API established the 'Born Free USA Primate Sanctuary'. Professor Linda Wolfe, a researcher with experience on both the Arashiyama-Kyoto and Arashiyama West troops, has been helping out at the Born Free Sanctuary for the last few years.

This sanctuary at Dilley (http://www.bornfreeusa.org/sanctuary/) is now home to several primate species (many of which were rescued from pet and entertainment venues), which are kept in separate outdoor enclosures. While research is no longer conducted on the Arashiyama West monkeys and visits to the sanctuary are not encouraged, video releases and requests for donations/ sponsors of individual monkeys do show that the Japanese monkeys continue to interact socially and range over native habitat in the large enclosures of the south Texas sanctuary.

1.3   The Montréal troop

In the early 1980s, Bernard Chapais contacted Lou Griffin, who was then manager at the Arashiyama West site, and he expressed an interest in creating a Japanese macaque colony at the University of Montréal. In 1984, a group of Japanese macaques was established there. This original group, derived from the Arashiyama West troop, was selected by Griffin and consisted of 15 individuals, namely an adult male and three matrilines with similar age/sex compositions. In each matriline, the mother [renamed A, B or C] was born in 1971 or 1973 and had three immature daughters born in 1981, 1982, and 1983, and one son born in 1984. The A and C families were members of the same genealogy (Betta) but were distantly related through the great-grandmother of A and C. The B family belonged to another genealogy (Rotte). The A family dominated the B and C families, and the B family dominated the C family.

In 1991, the troop was moved from its urban location near the University of Montréal campus to a larger rural location outside St-Hyacinthe, Québec, which was owned by the University of Montréal. Between the time when the Montréal lab (Laboratory of Behavioural Primatology) opened in 1984 until the time the St-Hyacinthe lab closed in 1998, Bernard Chapais and his graduate students conducted a series of elegant behavioural experiments and observations on Japanese macaque dominance systems. This research resulted in 12 Master's theses and two PhD dissertations. Of these students, Carole Gauthier, Jean Prud'homme and Paul Vasey continued to conduct research on the Arashiyama macaques. Until 1998, Carole Gauthier served as Bernard Chapais' research assistant and Jean Prud'homme served as the lab manager until 1998. After graduating from the University of Montréal, Paul Vasey conducted post-doctoral research at Concordia University in Montréal (under the supervision of Dr James Pfaus) on the neuroanatomy of the Arashiyama macaques using archived brain tissue. After obtaining a faculty position in 2000 at the University of Lethbridge, Canada, he then shifted his behavioural research to the Arashiyama E troop.

In 1993, an outbreak of encephalitis occurred in the Montréal colony (Lair et al., 1996). No major health problems had affected the colony since its establishment in 1984. Symptoms of the encephalitis included a variety of severe neurological deficits (e.g. paralysis, nystagmus, dyspnoea, dysphagia, etc.). Necropsied animals exhibited haemorrhagic cerebral infarcts with vasculitis. Mortality was virtually 100%. Despite intensive efforts to pinpoint the cause of the disease over a number of years, no definite cause was ever established. Owing to the significant level of health hazard involved, the acquisition of adequate veterinary services became highly problematic forcing the closure of the lab in 1998 in accordance with the regulations of the Canadian Council of Animal Care. The whole troop had to be euthanised.

References

Fedigan, L. M. (1991). History of the Arashiyama West Japanese macaques in Texas. In *The Monkeys of Arashiyama: Thirty-five Years of Research in Japan and the West*, eds. L. M. Fedigan and P. J. Asquith. Albany, NY: State University of New York Press, pp. 54–73.

Huffman, M. A. (1991). History of Arashiyama Japanese macaques in Kyoto, Japan. In *The Monkeys of Arashiyama: Thirty-five Years of Research in Japan and the West*, eds. L. M. Fedigan and P. J. Asquith. Albany, NY: State University of New York Press, pp. 21–53.

Koyama, N. (1970). Changes in dominance rank and division of wild Japanese monkey troop in Arashiyama. *Primates*, **11**, 335–390.

Kubota, K. (1990). Preferred hand use in the Japanese macaque troop, Arashiyama-R, during visually guided reaching for food pellets. *Primates*, **31**, 393–406.

(1991). Preferred hand use of the Japanese macaques during visually guided reachings. In *Primatology Today*, eds. A. Ehara, T. Kimura, O. Takenaka and M. Iwamoto. Amsterdam: Elsevier Science Publishers B. V. (Biomedical Division), pp. 269–270.

Lair, S., Chapais, B., Higgins, R., Mirkovic, R. and Martineau, D. (1996). Myeloencephalitis associated with a *viridans* group *Streptococcus* in a colony of Japanese macaques (*Macaca fuscata*). *Veterinary Pathology*, **33**, 99–103.

Leca, J.-B., Gunst, N., and Huffman, M. A. (2007). Japanese macaque cultures: Inter- and intra-troop behavioural variability of stone handling patterns across 10 troops. *Behaviour*, **144**, 251–281.

Norikoshi, K. (1977). Group transfer and social structure among male Japanese monkeys at Arashiyama. In *Keishiysu, Shinka, Reichōrui*, eds. T. Kato, S. Nakao and T. Umesao. Tokyo: Chukoron-Sha, pp. 335–370. (in Japanese)

Norikoshi, K. and Koyama, N. (1975). Group shifting and social organization among Japanese monkeys. In *Symposium of the 5th Congress of the International Primatological Society*, eds. S. Kondo, M. Kawai, A. Ehara and S. Kawamura. Tokyo: Japan Science Press, pp. 43–61.

2 *In search of the phantom monkeys*

EIJI OHTA
(Translation of the original and footnotes by Michael A. Huffman)

Togetsukyo bridge, Arashiyama. The Iwatayama Monkey Park is located in the background mountains (photo by M. A. Huffman).

2.1　My first encounter with monkeys

On 25 May 1948, hindered by the underbrush, constantly getting snagged on wild rose plants and frequently slipping on wet moss, I was climbing up the west slope of Mt. Karasu. I was searching for tree frogs that had not yet laid their eggs.

The Monkeys of Stormy Mountain: 60 Years of Primatological Research on the Japanese Macaques of Arashiyama, eds. Jean-Baptiste Leca, Michael A. Huffman and Paul L. Vasey. Published by Cambridge University Press. © Cambridge University Press 2012.

Just as I reached a gentle place in the slope, I heard an unfamiliar bark-like call coming from the bushes in front of me. In surprise, without thinking I pulled the sickle-shaped bush cutter out from my belt and stared into the bushes. I thought I saw a black object moving from one branch to another. The next moment a large bush rustled as something furry quickly moved past. Was it a squirrel or maybe a pack of wild dogs? With a feeling of tension and fear, I proceeded and gave a warning sound as I went. Hindered by the underbrush and stumbling all the way, I watched the object disappear off into the distance. Silence returned to the forest. Pausing on another gentle section of the steep slope I wiped the heavy sweat from my face and neck. Gazing into the forest my eyes met with a monkey's no more than 30 metres away up on the branch of a tree.

'Aa ha a monkey! I'm going to catch up with you.' That is what I thought as I quickly clambered up towards the tree. After getting only about 10 metres, the monkey jumped down from the tree into the understorey brush and disappeared. I searched around for the next 30 minutes but could not find it again. 'So these are the monkeys of Arashiyama!' While in Junior High School, I had often heard about them from the boatman of Hozu River, but this was the first time I had actually seen them with my own eyes. Judging from the movement in the brush, there seemed to be at least 100 of them. The figure of that monkey staring at me from the trees was very memorable and filled me with a mysterious feeling of wonder.

2.2   The beginning

On 28 November, the same year, I came upon two resting railroad track repairmen at a spot just to the side of tunnel no. 4 on the Sanin-line that winds up the steep valley above Hozu River. There were only a few scheduled round-trips a day on the line. Even in these hard times after the war, there were moments of relaxation. Munching on sweet potatoes, they had been waiting for the last two hours for the next train to pass by. When I asked what they were doing, one of them laughingly remarked: 'Today there is an inspection at the train station and someone will probably throw out some contraband rice along the tracks before getting there, we hope to be the lucky recipients.'[1]

One of them also said that just the day before he had seen a large white monkey on top of Byobu-rock, a small cliff in the area. That was all I needed to hear. I stood up abruptly, leaving them behind with their mouths open, as I headed up-stream for Kamome valley. They called out as I left: 'Be careful,

---

[1] In post-war Japan, times were hard and food was still scarce. It was common practice for government officials to confiscate rice from people getting off the trains from countryside destinations. City people would buy the rice from farmers.

it is dangerous up there.' I felt as if I must hurry, but progress was slow. Grabbing onto tree trunks and clumps of grass, I gradually climbed up the steep slope. There were areas here and there washed out by heavy rains from the 20 November typhoon. I climbed and traversed for about an hour. A sticky substance attached to the trunk of a tree stuck to my hands. 'Monkey faeces!' Looking carefully around the area, there was more on the leaves of a bush. There was also a distinctive scat on the ground and a small torn-off branch and some leaves unmistakably chewed on by a monkey. From the beginning of May, I had searched the forest, catching a single glimpse of one or hearing their calls, but I had never seen their faeces or feeding remains before. As time passed, their feeding remains became quite useful to find them. I also found their trails. These were the beginnings of a reunion with the monkeys. On that day, very carefully listening for their calls, searching for scat or signs of them having disturbed the vegetation, I walked the ridge from Karasugadake to Arashiyama, finally reaching the remains of Arashiyama castle, perched high up on the top of Arashiyama mountain.

2.3   Reunion day

On 24 February 1949, wearing high-top riding boots, with a sickle in my belt and towel around my neck, I headed towards Arashiyama on my bike. Leaving my bicycle near Daihikaku, I climbed the trail to Mt. Hatogasu. I rested for about 10 minutes at an elevation of about 300 metres. Quietly standing up, I pulled the sickle from my belt and swung it around, thrashing the understorey vegetation, and yelling at the top of my lungs. Suddenly I heard an eerie sound. It raised the hair on my neck! I peered in the direction of that sound and after a moment I discovered the monkey sitting on a branch looking at me. For the last nine months, I had often come across monkey scat, footprints, even half-eaten vegetation on the trail, but nothing more. This was only the second time that I had actually seen one. In order not to lose sight of him, I very carefully moved closer. After advancing a few metres, the monkey suddenly shook some branches and jumped down to the ground. Baring teeth, it threatened me with a husky bark. I stepped forward. At the same time he moved several steps forward as if he were going to jump on me. Immediately without thinking, I stepped back two or three paces. He quickly turned around and disappeared in the underbrush. Carefully moving around to the side, I edged closer. Feeling a presence, I looked to the left and saw a monkey peering out from behind some leaves in a tree, as if smiling mockingly at me. He jumped down to the ground barking and moved towards me. Postured as if he would lunge out toward me, he actually moved back slowly.

I realised that it was not the same monkey as before. Another monkey had taken his place and was threatening me. This one was bigger and huskier than the first monkey. Just as I was thinking that there might be others in the area, another one jumped out from the side and stood right in front of me. He was only metres away. He seemed to be a very strong adversary, not to be messed with. His intrepid expression was that of a boss who had endured many rough snow blizzards looking after his companions. This individual was different from the two previous monkeys. He was more confident and quietly stood his ground looking at me, after which he slowly backed up, never attempting to threaten me. When I attempted to move forward, he pushed off with all four legs and took off. It was the perfect silent attack and retreat.

I followed behind him just close enough not to lose sight. I was overwhelmed with excitement. He took off faster than my eyes could keep up with, jumping back into the vegetation, disappearing from sight. I didn't have the courage to enter the deep forest of Kamome valley that lay before me. I had had enough for one day. I gave up and climbed down the slope heading for the tracks and tunnel number 2. Partway down, I saw the shaking of branches and heard the calls of several monkeys off in a different direction from where I had just been.

I had been duped by the tactics of these males. They apparently had kept me busy while the rest of the group moved away to safety. I was at a loss for words over their brilliant strategy. I had been out looking for them since last year without so much as a glimpse of them. However, there is no doubt in my mind that they had been observing my every movement. After I found out they regularly looped around between Mt. Hatogasu and Mt. Karasu, with Kamome valley as their core area, I began to make contact with them more frequently. I was surrounded and threatened by a group of over-confident juveniles three times after that and stumbled off the edge of their trails in surprise many times. In the beginning, I thought I would be able to catch one, but I was proven wrong every time. Eventually they came to accept my presence and seemed to accept me as the lowest ranking monkey of their group!

2.4   The research station burns down

On 15 August 1950, five years after the war, it was so hot that you were bathed in sweat just quietly sitting. I was resting under the shade of a tree near the ruins of Arashiyama castle. I spotted three large stones, probably remains of the castle that were strewn about in the brush, probably unnoticed by anybody for all this time. Compared to the sweltering heat of the city below, it was cool and quiet up here. Puffing on a cigarette, I was enjoying the coolness of this place. Thinking that I would probably soon come across the troop, my heart was filled with a great feeling of accomplishment in having begun to understand their society. With all that had been happening lately, the expulsion of the communist party leader on 6 June, the Korean uprising on 25 June, and the burning of the Golden Pavilion in Kyoto on 2 July, I felt as I were in a different world up here.

I was gradually getting a handle on the troop's travel route and becoming a part of the troop, being tolerated to walk amongst them. I counted 42 monkeys in the troop. I named the top ranking male Mars, after the god of war in Greek mythology and the number two male in the group, Minerva, after the goddess of the arts. Mars♂ was a bit smaller than Minerva♂ but his expression was very stern and he appeared to be full of ambition. He was rather short tempered, aggressive and nervous acting. His build and expression reminded me of Takezo Miyamoto, the communist party leader in Japan. Minerva, on the other hand, had a very gentle disposition and was kind to females and young individuals. He was a very relaxed individual and had the best physical appearance of all the monkeys in the troop. He would sometimes fight fiercely with another adult male, apparently trying to challenge him for the position as leader.

Drifting off in thought thinking about them, I was suddenly distracted by the sounds of bells and sirens from down below. The piercing sound drew closer and closer, drawing me from the isolation of this place. I saw dark black smoke rising from the mountains on the other side of Horinji Temple. Was it a forest fire? I dug a small hole in the ground and buried my cigarette butt. I then quickly ran down the slope toward the temple. The trail being on the north side of the ridge made it impossible to know what was going on over there. Being careful not to slip, I finally reached a point where I could see down below. I shouted out 'Saikoji Temple is on fire, the research station is in danger!' Dazed, I ran the remaining 500 metres. I arrived to find that nothing was left of the research station. There were people running around in frenzy. Flames were rising up from the few remaining beams. Struck dumb, I approached the burning remains. Someone quickly pulled me away and scolded me for getting dangerously close. A fireman took my arm and led me away from the area. Worn out, I turned around to see the face of Yamaguchi Gamagurui (a nickname which means, one who is crazy about frogs).

I had first met 'Gamagurui' three years previously in March 1947. One day, I happened to pass by his house while he was out in the front making little frog figurines out of clay. At the time, I too was quite interested in frogs and was enthusiastically carrying out a study on them. We quickly became friends. He started up a conversation as if we had been friends for 10 years. By the end of that day, we had plans to build a small one-room research station large enough to put my research papers, books, maps, newspaper clippings and Gamagurui's clay frogs. We had a great time building the station. Later we brought a micro-scope and dissecting instruments. Here, the 60-year-old Gamagurui and I a 20-year-old ran experiments together, exchanged ideas and debated on many topics. In 1948, we began to compile data on the monkeys as well. In 1949, I was transferred from Matsuo Elementary School to Meirin Elementary School. It was about that time that I began taking pictures of the frogs and monkeys. Gamagurui would often laughingly say: 'No matter how hard I look at this photo with my pocket lens, I can't see anything that looks like a monkey. You've taken nothing but pictures of trees.' It was not until the beginning of 1950 that I was able to get pictures of Mars♂, Minerva♂, the aspiring young male, and the adult females.

Our frog research station had completely burned down. All the records we had collected concerning our research on the frogs and monkeys went with it. We lost all our pictures, books and research papers. Only one book was salvaged, *An Introduction to the Native Frogs of Japan* by Yoichiro Okada. Gamagurui had it clutched in his hands.

After that, we both wrote and visited each other several times a year. After about 10 years, correspondence from Gamagurui abruptly stopped. Several of my notes, concerning observations before August 1950 in the form of note-books and miscellaneous jottings were scattered around my home. I pulled together all of that material to write this essay. Minerva♂ was renamed Zao♂ by researchers and he became the alpha male by the time the troop was suc-cessfully provisioned at the Iwatayama feeding station in October 1954.[2]

The article was originally published as:

Ohta, E. (1975). Maboroshi no saru wo motomete. *Arashiyama Shizenkenkyūkai Kaihō (Reports of the Arashiyama Nature Research Society)*, **3**, 13–17.

---

[2] Mr Eiji Ohta later became principal of a local elementary school and frequently visited the Iwatayama Monkey Park, actively involving himself in educating young people about the ecol-ogy and behaviour of Nihonzaru for many years. Unknown to me, Eiji san passed away, before I could tell him about our plans for this book and the translation of his essay. He took me under his wing when I too was just a young man of 20, starting out in my first study of the Arashiyama macaques. He was a great source of information, inspiration and friendship.

3   *Arashiyama monkeys in the late 1950s*

YUKIMARU SUGIYAMA

Figure 3.1. Zao, the alpha male of the Arashiyama-Kyoto troop in 1958 (photo by Y. Sugiyama).

3.1   Early stages of provisioning and a change in the home range

From 1949, Naonosuke Hazama[1] and Eichi Furukawa carried out natural history studies of the wild Japanese macaques (*Macaca fuscata*) at Arashiyama. Stimulated by the success of provisioning the monkeys at Koshima in 1952 and Takasakiyama in 1953 (Kyushu, southern Japan), they began to feed the

[1] Naonosuke Hazama was an unsalaried researcher along with Kinji Imanishi in the Laboratory of Ecology; both from well-to-do families.

The Monkeys of Stormy Mountain: 60 Years of Primatological Research on the Japanese Macaques of Arashiyama, eds. Jean-Baptiste Leca, Michael A. Huffman and Paul. L. Vasey. Published by Cambridge University Press. © Cambridge University Press 2012.

34

monkeys in the forest and gradually enticed them to Iwatayama, a mountain property owned by Mr Sonosuke Iwata, located in the south-western corner of Arashiyama. Iwatayama was located on the eastern periphery of the troop's home range and the nearest location where tourists could easily approach to see the monkeys, and a panoramic sight of Kyoto City. The monkeys of Arashiyama gradually began to eat the provisioned food from around October 1954, settling down in the Iwatayama area.[2] However, it took several months for the monkeys to become accustomed to the new site and people, as the feeding ground was at the extreme periphery of their home range.

In April 1958, I began to study their behaviour as a graduate student of Professor Denzaburo Miyadi in the Laboratory of Ecology, Department of Zoology at Kyoto University. By that time, three and a half years had passed since provisioning succeeded and the site was officially open to the public as the 'Iwatayama Monkey Park', Arashiyama. The group frequently travelled far away from the park, sometimes for as much as one week at a time and did not come frequently to the feeding ground because it was located at the limits of their traditional home range. I sometimes tried to search for the monkeys in the centre of their home range, around Hatogasuyama and Kamomedani valley. The core area of the group was located in the northern part of Arashiyama, isolated from the surrounding northern and eastern forests by the Hozu River. There are forests adjacent to the south and south-west of this area too, but they were not a suitable habitat for monkeys and they only visited these locations occasionally. For this reason, their home range was basically isolated from other monkey populations in the area.

The main forest making up Mt. Arashiyama was owned by a certain private paper manufacturing company and most of the forest had already been deforested a few years before I visited the place. This was perhaps the reason why provisioning succeeded at the extreme periphery of the group's home range. Along with the progress of provisioning, the home range area of the group, about 5–8 km^2, was reduced to less than 2 km^2. The group infrequently visited Kamomedani valley, but mostly stayed in and around the forests of Iwatayama, particularly during winter when there was little natural food available for them in the forest.

3.2   Change in population size

In July 1958, the group contained 69 individuals, consisting of three adult males (at least seven years of age), five sub-adult males (five to six years), 14 adult females (at least five years), five sub-adult females (four years),

---

[2] See Huffman (1991) in the bibliography of this volume for details of the habituation and provisioning processes.

32 juvenile males and females (between one and four years) and ten infants (born in 1958). There were also five adult solitary males that appeared in the feeding grounds when the main group was absent, usually in the evening.

In 1954, it is reported that there were about 34 monkeys in the group at feeding time (Huffman, 1991). According to these figures, the rate of population growth was almost 20% per year. Even if there were 42 individuals, according to the estimate of Mr Eiji Ohta before artificial feeding was begun (see chapter 2), the growth rate would be 9%. This is tremendously high. I was surprised that nobody at the park in the early days had any apprehensions about this unusually high growth rate, and considered what the future consequences would be.

Thirteen infants were born in 1957 from 14 adult females (92.9%) and 11 in 1958 (78.6%), the following year from these same females. The birth rate per reproductive female was 85.7% on average. For wild Japanese macaques the birth rate is usually between 30–35%, a single birth every three years for reproductive females (Sugiyama and Ohsawa, 1982). In addition, most infants survived the vulnerable period until the next birth season, while in the wild only about 63–80% survive (Sugiyama and Ohsawa, 1982; Izawa, 2009). We can understand how it was unusual at Arashiyama after artificial feeding succeeded.

Hazama measured the body weights of all the monkeys at the Iwatayama feeding grounds (adult males: 14 kg, and adult females: 11 kg on average in June–July, a time when monkeys are lighter than in other seasons). They were heavier than wild monkeys though they fell within the standard range of variation (Sugiyama, 1996; Delson *et al.*, 2000). In those days, everybody at the park was happy that the number of monkeys was rapidly increasing and that the monkeys were becoming larger, because it pleased the visitors to see so many habituated monkeys near them.

At this stage, we should have recognised the warning signs of increasing population size, because compared with non-provisioned monkeys, the age difference between siblings was too short and the family size was growing too large. If family members cooperate with each other to compete against other family groups to monopolise a high-calorie food resource, dominance relationships may become more obvious and exaggerated. I began to doubt the existence of a strict dominance hierarchy and rank order in wild groups of monkeys, particularly among females and their offspring. I began to question whether the dominance hierarchy and ranking order among individuals was produced or exaggerated by artificial provisioning.

3.3   Personality

Zao, the alpha male, was estimated to be less than 15 years of age and weighed 13.5 kg (Hazama, 1964). He was always in the centre of the group when the monkeys were on the feeding ground (Figure 3.1). He was quite habituated to human observers. The central females – Tokiwa, Nose, Kojiwa, Cooper – and their families stayed around Zao but kept their distance from him on the feeding grounds. When infants came near to Zao, their mothers immediately pulled them back. The second-ranking male – Gongen – had already disappeared before I started my intensive study at Arashiyama.

In contrast, Lincoln, now the second-ranking male, was nearly 18 kg (Hazama, 1964) and older than Zao (15–20 years). He was often seen at the periphery of the group, usually between the park's entrance and the feeding grounds. Peripheral females – Rakushi, Shirayuki, Momo, Ai – and their offspring tended to stay near him, but not always as they sometimes scattered out away from each other. On many occasions, juveniles and infants could be seen playing with and grooming Lincoln. The mothers would allow their offspring to do so (Figure 3.2). The park staff would jokingly say that Lincoln was the headmaster of a kindergarten. He was aggressive towards observers, particularly when infants were near him.

Shan, the third-ranking male, estimated to be about 10 years of age, was slender and also very habituated to human observers. He was active and usually followed Zao around. Similar clear differences in the personality of individuals in other monkey groups could be seen. I became interested in the differences between the two males, Zao and Lincoln, and tried to clarify the characteristic

Figure 3.2. Lincoln, the second-ranking male, was always surrounded by juveniles who rarely hesitated to stay near him (photo by Y. Sugiyama).

features of these individuals. I followed each of them for some hours every day, recorded their behaviours every five minutes, and wrote down those individuals near them when the group was in or around the feeding ground of Iwatayama. This method resembles what later came to be known as 'focal animal sampling' (Altmann, 1974). I intended to elucidate the 'personality' of monkeys with quantitative data.

During those days, there was no funding for graduate students. When I was ordered by Imanishi to go to the Takasakiyama Natural Zoo as a scientific fellow with a salary, I decided to give up my study on the personality of Arashiyama monkeys. For the last 50 years, we have accumulated much information on the social structure of Japanese macaques, but even now, individual differences or the 'personality' of monkeys is a little studied subject.

3.4   Life history of a solitary male

Chikusha was born in 1954 just before the artificial feeding began. He had characteristic features and everybody easily recognised him. He disappeared from Arashiyama in September 1959. By this time, Hazama had begun to observe a wild monkey group on Mt. Hiei (Hieizan group). This site is located 18 km away from Arashiyama on the opposite, eastern side of Kyoto City. Making a detour through the forest north of Kyoto city, it is easily more than 20 km away. Hazama found Chikusha the following year, in 1960, in the periphery of the Hieizan group. Chikusha was following the group but did not yet appear to be a member. However, he did later join the group. He must have travelled around the city to the north by himself, but the exact route he took, and whether he was always alone throughout the whole year after he left Arashiyama is unknown. Moreover, we do not know if he directly came to Mt. Hiei or visited other areas along the way.

During the 1950s, male offspring were thought to develop in their natal group and only some of them went on to become central members of the group (Itani, 1954). Nobody knew why many juvenile and sub-adult, as well as high-ranking males – like Gongen in 1957, and Zao who left in 1959 – disappeared from their group. Chikusha was the first male known to desert his natal group as a young sub-adult and later enter another group elsewhere.

Since this observation, we accumulated many anecdotal observations of solitary males and all-male parties. Some of these males were found more than once and at present we know that for Japanese macaques, natal group desertion is a normal life process for males. However, even now we do not have sufficient knowledge of male life history, particularly regarding their lifetime reproductive success, and whether it is more or less effective to remain in one group for their entire life. We do not know yet why most males desert their group, and whether they would be more successful by staying in their group.

3.5   Culture

At the original feeding site, in the small flat area above the present main feeding ground – which used to be the observatory and resting place for people before reaching the feeding area – there was a small pond about 5m^2 and 20 cm in depth. Many monkeys, particularly females, would go there and usually rub provisioned sweet potatoes with both hands, perhaps to remove dried mud. Such behaviour is often observed in provisioned monkeys when they obtain food.

In May 1958, I observed an adult female named Cooper who began to wash sweet potatoes by dipping them in pond water. This behaviour was copied by her children, a few more juveniles and their mothers, but not many (Figure 3.3). Potato washing by Japanese monkeys had been known to occur at Koshima since 1954, and I understood that this was an extension of their habit to exhibit

Figure 3.3. Potato washing at Arashiyama, May 1958. Juvenile females are washing sweet potatoes (centre) in a small pond on the artificial feeding ground while others are eating their washed potatoes (left) (photo by Y. Sugiyama).

food-rubbing behaviour, effective for removing dirt from food items and making them easier to eat. Because food rubbing was common in many artificially fed Japanese monkey groups, I thought this behaviour may easily emerge simultaneously in many monkey groups, if similar conditions were present. The same occurs today with stone-handling behaviour (see chapter 13).

This observation at Arashiyama was used many years later by Lyall Watson to discuss the '100th monkey theory' in his book *Lifetide* (Watson, 1979). Sweet-potato washing behaviour was innovated by a particular individual and copied by other monkeys. This behaviour could be considered 'cultural' if it was transmitted to many group members and across generations, but the situation was based on unintentionally created artificial circumstances. I thought at the time that it was important to differentiate between natural and artificially induced situations and behaviours, even if they are not intended as experiments. Though provisioning created a good field laboratory situation, we must be careful about the phenomena created by artificial feeding and how they differ from natural situations.

3.6   Artificial feeding of wild boars

There was a tendency to provision the monkeys with more food than they could consume by themselves. Often, some food was left over on the ground in the evenings, attracting wild boar to the site, after the monkeys had returned to the forest. Being crepuscular, most of their activity occurs after dark. However, these boars gradually became habituated and would come to the feeding ground before sunset.

I tried to observe them with a night-vision scope, a new technology developed several years after the Korean War, but it was neither very cheap nor very effective. Wild boars were cryptic but I could observe them if I stayed still, although individual identification was not possible after dark. They came in groups of two to five. Perhaps the largest individual was the mother and the others were her offspring. Sometimes more than one group was seen simultaneously. Because wild boar hunting was allowed, continuous observations of the same individuals were difficult. Moreover, observation was restricted to the feeding grounds as wild boars were too shy to be observed in the forest. Due to my shift to Takasakiyama in February 1959, I had to abandon this study and even after I returned to Kyoto, I had no opportunity to start up again.

As my study at Arashiyama was suddenly cut short in the process of data collection, and I was busy with my research on other monkey groups elsewhere, I could not publish any scientific results from my observations at Arashiyama. My main interest was more on population dynamics with special reference to social organisation, than on individual behaviour. Therefore, I stepped away from 'personality' studies. However, the level of awareness of the park manager and staff at Takasakiyama of the influence of artificial feeding on the monkeys was much the same as that at Arashiyama. I could not ignore any longer the problem of unnatural population growth and its effect on the social organisation and social relationships among group members. Ever since then, I have been working towards reducing the effect of artificial feeding on monkeys at Takasakiyama.

References

Altmann, J. (1974). Observational study of behavior: Sampling methods. *Behaviour*, **49**, 227–267.

Delson, E., Terranova, C. J., Jungers, W. L. *et al.* (2000). Body mass in Cercopithecidae (Primates, Mammalia): Estimation and scaling in extinct and extant taxa. *Anthropological Papers of the American Museum of Natural History*, **83**, 1–159.

Hazama, N. (1964). Weighing wild Japanese monkeys in Arashiyama. *Primates*, **3–4**, 81–104.

Huffman, M. A. (1991). History of Arashiyama Japanese Macaques in Kyoto, Japan. In *The Monkeys of Arashiyama: Thirty-five Years of Research in Japan and the West*, eds. L. M. Fedigan and P. J. Asquith. Albany, NY: State University of New York Press, pp. 21–53.

Itani, J. (1954). *Takasakiyama no Saru* (Monkeys of Takasakiyama). Kobunsha. (in Japanese)

Izawa, K. (2009). *Studies on Wild Japanese Monkeys*. Dobutsu-sha. (in Japanese)

Sugiyama, Y. (1996). *Encyclopedia of Monkeys*. DataHouse. (in Japanese)

Sugiyama, Y. and Ohsawa, H. (1982). Population dynamics of Japanese monkeys with special reference to the effect of artificial feeding. *Folia Primatologica*, **39**, 238–263.

Watson, L. (1979). *Lifetide: A Biology of the Unconscious*. New York, NY: Hodder and Stoughton.

4   *Touches of humanity in monkey society*

NAOKI KOYAMA

(Translation of the original and footnotes by Michael A. Huffman)

Zola (male) at Arashiyama, January 1967 (photo by N. Koyama).

4.1   My first encounter

The first time I went to Arashiyama was in February 1964. It was from this time on that I began my friendship with the Japanese macaques. I was able to move about within the troop and identify the monkeys individually. The juvenile male group of the periphery gave me the most problems in the beginning. This elusive group, whether in the forest, the feeding station or on a steep mountain trail, would at the least expected moment jump out from nowhere

The Monkeys of Stormy Mountain: 60 Years of Primatological Research on the Japanese Macaques of Arashiyama, eds. Jean-Baptiste Leca, Michael A. Huffman and Paul L. Vasey. Published by Cambridge University Press. © Cambridge University Press 2012.

and threaten me. One particular monkey did this more than any of the others. With time, it became clear to me that he was the lowest ranking member of the male group. Whenever he did this, I would face him, and he would counter threat to intimidate me. In these encounters, if he found himself in a disadvantageous situation, he would seek help or encouragement from his companions. These confrontations between the monkeys and me continued for a long time, always occurring in the same pattern, the victor never being decided.

Then one day it happened. Ignoring the advancing bluffs of the group, I approached what appeared to be the highest-ranking individual. Catching him off guard, I got down on my hands and knees and kicked him in the jaw with my boot. He flew head over heals in a somersault. From that point on, all attacks from this group stopped. The monkey that I kicked is now the alpha male Matsu-59♂ of the Arashiyama B troop. I think it was because of this incident that my relationship with the troop proceeded so smoothly.

I finished my most intensive study of the Arashiyama-Kyoto monkeys in March 1967, but I have continued to observe the troop off and on since then. As an observer during this time, there have been several impressive monkeys. I would like to relate these incidents not as a scholarly endeavour, but from a personal perspective. One is apt to personify or anthropomorphise such impressions, making them hard to validate in a scientific fashion. Because of this, I have not had many chances to share my experiences publicly.

When I first began my study in February 1964, the alpha male then was Y♂. He was known by the attendants of the feeding station affectionately as 'Yaji-san'. Yaji-san and his one-rank-junior, X♂ (Kita♂-san), seemed to co-lead the troop as a pair, always supporting each other and frequently staying together.[1] In September of that year, Kita♂-san left the troop. After about a week, he returned. I noticed that he was behaving differently. He cautiously approached the alpha female Matsu♀ and sat down beside her. She gently touched his shoulder as if he had never been away and they separated. It was as if he was asking her for forgiveness for having been away. It conjured up images in my mind of a queen forgiving her knight for some indiscretion. A few days later, Kita♂-san left the troop a second time, never to return. After that Yaji♂-san showed little interest in leading the troop. He dropped down four rank positions. One of the males who rose above him was Ao♂ (nicknamed 'Medama'

---

[1] At the time, one of the assistants, Mr. Kunizawa was reading the Edo Period (1603–1867) classical novel 'Tokaido-chu hizakurige' (Travelling the East Sea Road by Shank's Mare' (on foot)) written by Ikku Jippensha (1765–1831). The comical antics of two inseparable travellers, Yajiroube and Kitahachi, along the Tokai-road are vividly described in the book and have become icons of Japanese culture. The names Yaji-san and Kita-san were given to this pair of males by the assistants from these two characters. Researchers adapted these nicknames to Y and X for official use.

meaning eyes). Ao♂ gradually entered the central part of the troop acting as a sub-leader. Because his entrance was not well received by all, he had a tough time. The next male in line was Zola♂, son of the previous alpha female, Tokiwa♀. Zola♂ was very sturdily built, strong and had a fine coat. He was the kind of monkey that would push others away to feed on provisioned food, but as a leader he was weak and lazy. The third individual was the son of Matsu♀, and my earlier sparring partner Matsu-59♂, nicknamed Hanatate (after the fact that he had a vertical cut on his nose).

In March 1965, there were 137 members in the troop. At this time, the 6-year-old Matsu-59♂, with the support of his high-ranking mother and older sister, was observed threatening 8-year-old Zola♂ and 9-year-old Ao♂. Yaji♂-san did not attempt to challenge or subordinate Matsu-59♂ for his actions. In September, exactly one year after his companion Kita-san♂ left the troop, Yaji♂-san disappeared. After the disappearance of Yaji♂-san, Zola♂ and Ao♂ rose in rank, but Zola♂ was rarely effective as the alpha male. Irrespective of actual rank status, beta male Ao♂ could effectively be called the leader of the troop. His behaviour decided the troop's movements and he seemed to be in general control.

4.2   Family lineages and troop fission

From the beginning of the mating season in October, to its height of intensity in December, Matsu-59♂ with the support of his mother and other family members was observed threatening Ao♂. However, Matsu-59♂'s mother came into oestrous and began mating with Ao♂. After this, Ao♂ clearly became dominant over Matsu-59♂. As Ao♂'s leadership consolidated, Matsu-59♂ and other family members began to support him and direct aggression towards the alpha male Zola♂. Zola♂ did not stand down and a change in status did not occur.

Ao♂ (son of Betta, seventh ranking family) had social bonds with many high-ranking central troop females, while Zola♂ only associated with lower to mid-ranking females, avoiding all contact with the higher ranking Matsu kin group. Especially in the beginning of 1966, he was consorting frequently with mid-ranking adult female Mino♀. This was the wedge that led to the splitting up of the Arashiyama-Kyoto troop into two groups. With Ao♂ and Matsu♀ as the alpha male and female, a new group was established including the top to lower high-ranking kin groups. Accordingly, Zola♂ and Mino♀ became the alpha male and female of the subordinate group.

By June 1966, these two groups had completely separated and formed two independently moving troops. Zola♂'s group (Arashiyama B troop) became dominant over Ao♂'s group (Arashiyama A troop). Behind the troop fission, there was a problem with family lineages. There were 16 kin-groups in the original Arashiyama-Kyoto troop. After the fission, Ao♂'s group consisted of lineages between the 1st-ranking Matsu family and 7th-ranking Betta family. Zola♂'s group consisted of lineages between the 8th-ranking Kojiwa family and the 16th-ranking Shirayuki family. The dividing line for this fission was formed between the 7th- and 8th-ranking lineages because of Zola♂'s close relationship with Mino♀, a member of the Kojiwa family. This resulted in the middle- to lower-ranking groups (8th–16th) rising above the higher-ranking groups. In effect, the Kojiwa family became 1st, the 9th-ranking family became 2nd, the 10th became 3rd, and so on, and the 16th became the 9th-ranking family. It is possible that the troop split in half because the 1st-ranking Matsu family fell down to the 10th-ranking position.

So, how did the adult males split up in accordance with this breakdown of the troop? Observing the adult males of both A and B troops just after the fission, there were three of them in B troop and five in A troop. Three months later in December, there were seven adult males in B troop and eight in A troop. In September 1967, nine months after the fission, there were 10 adult males in B troop and 13 adult males in A troop. In this way, the numbers increased. There were more males in the subordinate A troop. Undecided individuals temporarily formed all-male groups. In due course, males from higher-ranking lineages joined B troop while males from lower ranking lineages joined A troop. In short, A troop consisted of dominant lineage females and their dependent offspring plus subordinate lineage males. Conversely, B troop consisted of subordinate lineage females and their dependent offspring plus·dominant lineage males. Now then, let's see what happened to the friction present between Matsu-59♂ and Zola♂ after the troop fissioned. Matsu-59♂ went back and forth between A troop where his mother was, and B troop where Zola♂ was. Finally after about a month, in mid-July, Matsu-59♂ entered B troop and became the beta male under Zola♂.

4.3   The Biriken family moves west

In September 1967 Ao♂, the alpha male of A troop, left. Soon afterwards, W♂ (nicknamed 'Biriken' meaning a figure or statue of fortune, might be derived from Billiken) took his place. I would like to relate an interesting episode about W♂. In 1964, a monkey of unknown origin, W♂, was the 5th-ranking male under Yaji♂, Kita♂, Zola♂ and Ao♂. In August of the same year, a female named Shiro♀ died, leaving behind a 3-month-old female infant Shiro-64♀. W♂ was the first to hold and take care of her. Even though he was a male, W♂ took care of her by keeping her close to him all the time, just like a real mother would. I observed on many occasions that W♂'s nipples were red and swollen from the infant trying to nurse. From this event, I learned that three months is probably the very earliest age that an infant can continue to develop without its mother's milk.

The infant never received milk from any other females, and continued to be carried around by W♂. She matured without complications. I am not sure why this motherless infant was taken such good care of by W♂. After the troop's fission, W♂ moved with other males into Ao♂'s group. Shiro-64♀ had turned 2 years old and remained in Zola♂'s group with her grandmother and older sisters. However, one month after the troop split, Shiro-64♀ entered A troop staying close by to W♂. This and another instance where a 4-year-old female Meme-62♀ transferred with her older brother from A to B troop, were the only two cases in which a female transferred from one troop to another. Shiro-64♀ grew up with W♂, becoming a high-ranking female when he rose to become alpha male of A troop. A point of great interest is that there was no mating observed between them. In February 1972, W♂'s A troop was moved to a location outside of Laredo, Texas. According to American researchers, even after their arrival, the relationship between W♂ and Shiro-64♀ remained very close. As before, the two were never observed to mate.

I would like now to give an account of A troop's relocation to Texas. On 23 February 1972, 150 of the 158 A troop members were released into a compound near LaMoça, a town 30 miles north of Laredo, Texas. W♂'s loss of status after their arrival is of particular interest. Shortly after being put into the compound, the 2nd-ranking male Dai♂ challenged W♂ and became dominant over him. I have observed a similar situation with bonnet macaques. This happened in 1963. I was involved in a project capturing a troop of bonnet macaques in India. They were transported to Japan. When the troop arrived in Japan, there was a rank reversal between the beta and alpha males. The former alpha male appeared to have lost confidence after being captured. This might possibly be due to a change in environment or perhaps he was held responsible by the others for being captured and succumbed to the group's pressure. Of course we could never really know for sure. Perhaps what happened with W♂ when they moved to Texas is a universal phenomenon.

Once in Texas, Arashiyama A troop was renamed 'Arashiyama West'. Their new home was a 108-acre plot surrounded by a 2.5-metre high electric fence.[2] In the beginning there were major problems for the American researchers. One problem was poisonous vegetation and the other was a sickness caused by an imbalance of calcium and alkalinity. Both problems were eliminated. There

---

[2] The troop was later relocated to the Dilley ranch near San Antonio where they remain today (see Fedigan, 1991, for a detailed account of the Arashiyama West troop's history).

were also problems with parasites and bobcat predation. At times, the fence was damaged and not repaired. Even in these cases when the fence did function properly, the troop did not attempt to escape. In fact, only eight monkeys escaped. Two were recaptured, one came back on its own, three died and two have not been seen since. At any rate, by the end of 1978, Arashiyama West consisted of 252 individuals.

4.4   Various happenings at Arashiyama (B troop)

In July 1970, Zola♂ disappeared and Matsu-59♂ took his place as the alpha male of B troop. Presently, he still maintains this position.[3] During that period there was a change in status among the lower-ranking males. This change was influenced by the sickness and consequent capture of Matsu-59♂'s right-hand man, beta male Rheus-59♂, and the accidental death of the 3rd-ranking male, Nose-59♂. In January 1977, the then 5th-, 6th- and 7th-ranking adult males threatened and rose in rank above Kusha-65♂. Kusha-65♂, nicknamed 'Kusha Ototo' (little brother of Kusha), got along comparatively well with Rheus-59♂. It is thought that because Rheus-59♂ was suffering from a foot injury at the time, this left Kusha-65♂ with little strategic support. He was challenged by the older males and dropped in rank. Why was Rheus-59♂ not challenged too? This was probably due to a strong solidarity with Matsu-59♂ and Nose-59♂.

Shortly after this change in status occurred, Rheus-59♂ began limping on both hands. He was captured and put in a large holding cage to prevent him from falling prey to local feral dogs that roamed the mountains. Two months later on 5 March Rheus-59♂ died in the cage. His kidneys were swollen and he showed symptoms of urine poisoning. On the same day that Rheus-59♂ was removed from the group, Nose-59♂ was found dead in a wild boar snare. In this way, all of Matsu-59♂'s companions were lost. It looked as if Matsu-59♂ had lost his strong support, but this was not the case because he continued to be in control of the troop when the subordinate males, Kusha-63♂, Ran-63♂, Deko-64♂ and Kusha-65♂ rose in rank to occupy the 2nd, 3rd, 4th and 5th positions, respectively.

There is an interesting incident concerning the 4th-ranking Deko-64♂. It was a snowy day in February 1971. I discovered a 2-year-old monkey with a piece of his scalp torn off the top of his head.[4] This little male, unable to bend

---

[3] Eventually, Matsu-59♂ too disappeared from the troop in June 1981.
[4] The juvenile male was Cooper-65-73♂ and lived into adulthood becoming a central troop male. He moved into F troop when the troop fissioned and left with the group from Iwatayama around 1998.

its head over, sat with its chin rested tightly upon his tightly clenched fist. It appeared to have just happened. There was slimy blood and quivering white brain-like matter on the wound. Astonished, I approached with my camera in hand. When I would move away from the injured juvenile, Deko-64♂ would quiet down. If I moved closer, he would threaten me. Several times this happened. I was greatly impressed by the protective behaviour toward this unrelated juvenile that Deko-64♂ displayed. Earlier that day, I observed a large hawk circling around in the sky. An adult monkey was in the top of a large pine tree, threatening and grabbing out at this bird as it soared in close. This led me to believe that the injured juvenile had been attached by a hawk or other large bird of prey. That evening, when the troop returned to the mountain, the juvenile's mother and older sister quickly left together with the rest of the troop. This juvenile, unable to walk on all fours, crawled on its forearms, all the time keeping its head erect. I followed behind him and tried to take a picture, but as before, Deko-64♂ would not let me get close. Deko-64♂ stayed by the juvenile until he got into the forest. The weakened juvenile, that I thought would surely die, came back with the troop day after day. Deko-64♂ stayed by his side constantly. Shortly afterwards, a white membranous tissue formed over the wound, leaving a large bald spot. After about a month the wound was completely healed.

4.5   The birth of C troop: a second fission

With the change in the hierarchy and the deaths of two adult males mentioned earlier, Kusha-63♂ became beta male of B troop and Kusha-65♂ dropped down to 5th-ranking male. Of the five leader-class males in the troop, the Kusha brothers rose the highest in status among any of the males, compared with where they started from. The following big social change probably played a large role in accomplishing this. The Kusha brothers became quite active in the centre of the troop when their mother Kusha♀ and the other Kusha family females began frequently moving out into the troop's periphery. Eventually this kin group left the troop completely. The group became known as C troop. This all occurred between February and March 1978. Beginning from around the early autumn of the previous year, the Kusha family developed relations with Kojiwa-62–69♂, a solitary male who occasionally appeared in B troop's periphery. It is thought that possibly this solitary male joined the Kusha family when they moved out into the periphery forming a new troop. Another possible interpretation however is that the females were drawn into the periphery by this male. It is possible that either of these factors was involved. Later, C troop was captured and on 30 January 1979, the group was sent to a zoo in Korea.

4.6   Mothers and daughters

This section is about changes in status observed between mothers and daughters. In May 1966, before the Arashiyama-Kyoto troop underwent its first recorded fission, the adult female Mino♀, who was associating with Zola♂, attacked and bit her higher-ranking mother, Kojiwa♀. Taking this opportunity, she rose in status above her mother. When the troop fissioned, Mino♀ became the alpha female of B troop. In December 1975, Mino♀ was outranked by her daughter Mino-63♀. In January 1977, Mino-63♀ was outranked by her daughter, Mino-63–69♀. Mino-63–69♀ is presently the alpha female. In this instance, their young offspring were fighting with each other. Both mothers jumped in to protect their offspring. The two adults' ranks reversed subsequently to the conflict. All these changes in status occurred within one single family and occurred when the mother was approaching old age.

Next I would like to talk about the relationship between an extremely aged female Momo♀ and her daughters. Momo♀ was a senile female who died three years ago in 1976. During the last four or five years of her life she was the oldest monkey in the troop. Momo♀ had three adult daughters: Momo-67♀, Momo-61♀ and Momo-59♀. Momo♀ was always with her youngest daughter Momo-67♀. The aging Momo♀ became lower ranking than this youngest daughter, but because of Momo-67♀'s protection she was still dominant over her other daughters. Because Momo♀ was so old, it was not unusual that others would supplant her from the food she was eating. She lived off the leftovers of her daughter. Momo-67♀ protected and looked after her mother until her death. This is a rare example from my experience. Between mothers and daughters, the dominance hierarchy proceeds from mother to youngest daughter, continuing in order of birth with the firstborn being the lowest-ranking individual.[5] When Momo♀ grew very old her status did not drop below that of all her daughters. In short, one might think that this old female, by keeping close relations with her youngest daughter, prevented herself from sliding down to the bottom of her lineage.

When watching monkeys, one encounters many different situations and I have personally experienced all the things mentioned above. Even among these various instances that cannot always be systematically understood, there is much to be impressed by.

The original essay was published as:

---

[5] A phenomenon referred to as the 'youngest ascendancy principle' first described by Kawamura (1958) in the Minoo-B troop. It is common in provisioned troops especially among higher kin groups, but thus far not in wild troops of macaques (e.g. Hill and Okayasu, 1995; Kutsukake, 2000).

Koyama, N. (1980). Saru shakai ni okeru ningenmi no aru kannkei. *Hyojyunka to Hinshitsukanri*, **33**, 39–44.

References to footnotes

Fedigan, L. M. (1991). History of the Arashiyama West Japanese macaques in Texas. In *The Monkeys of Arashiyama: Thirty-five Years of Research in Japan and the West*, eds. L. M. Fedigan, and P. J. Asquith. Albany, NY: State University of New York Press, pp. 54–73.

Hill, D. A. and Okayasu, N. (1995). Absence of 'youngest ascendancy' in the dominance relations of sisters in wild Japanese macaques *(Macaca fuscata yakui)*. *Behaviour*, **132**, 367–379.

Kawamura, S. (1958). The matriarchal social order in the Minoo-B group: a study on the rank system of Japanese macaques. *Primates*, **1**, 149–156. (in Japanese with English abstract)

Kutsukake, N. (2000). Matrilineal rank inheritance varies with absolute rank in Japanese macaques. *Primates*, **41**, 321–335.

5   Fifty years of female Japanese macaque demography at Arashiyama, with special reference to long-lived females (> 25 years)

ALISA CHALMERS, MICHAEL A. HUFFMAN,
NAOKI KOYAMA AND YUKIO TAKAHATA

Rakushi-59–79–92 and her infant female Rakushi-59–79–92–09 at Arashiyama, May 2009 (photo by N. Gunst).

The Monkeys of Stormy Mountain: 60 Years of Primatological Research on the Japanese Macaques of Arashiyama, eds. Jean-Baptiste Leca, Michael A. Huffman and Paul L. Vasey. Published by Cambridge University Press. © Cambridge University Press 2012.

5.1   Introduction

For more than 50 years, generations of researchers and staff of the Iwatayama Monkey Park (IMP), Arashiyama, have been collaborating to produce an ever-growing collection of scientific knowledge and data for a population of Japanese macaques living in the mountains to the western side of Kyoto City in an area known as Arashiyama. Not only does this longitudinal database augment our understanding of this field site and its occupants, it has also provided a unique opportunity to add to our comprehension of primates in a more general way by paving the way for more extensive long-term studies of the life history of the species, spanning successive generations. Accumulation of numerous individual life histories is a difficult task to undertake, especially for such long-lived organisms as primates living under free-ranging conditions (Gage, 1998).

Records for this provisioned population have been built up since the 1950s providing invaluable data on individual- and troop-level activities over an unprecedented duration of time. The first 30 years of this dataset was previously presented by Koyama *et al.* (1992), but it covered less than the maximum lifetime known for this species, and in particular for some of the longer lived Arashiyama monkeys. The present chapter enhances the previous dataset by reporting on the life-history traits of members of the troop collected over a period of 53 years, from 1954 to those individuals born in the autumn of 2007. In particular, this chapter intends to report on age-specific reproductive parameters, infant survival and mortality, with special reference to long-lived individuals.

The first 30 years of the general history of the troop currently occupying this area (including fissioned troops removed from the area) has been described in detail elsewhere (Fedigan, 1991; Huffman, 1991). Since 1986, no fissions have occurred, with approximately 150 members known as the Arashiyama E troop remaining at IMP.

5.2   Materials and methods

The demographic data used in the present analysis come from a previously published genealogy (Asaba and Suzuki, 1984), records provided by the current management of the park and data provided by other researchers. The data were inclusive of individual ages at parturition, death, contraception and infant mortality (if an infant survived less than one year following birth). Where available, this dataset was supplemented with additional observations such as cause of death or health conditions preceding a female's disappearance from the troop. Captured individuals were included in the data up until the point of capture.

This chapter complements Koyama *et al.*'s (1992) report, which analysed life-history traits for this population from October 1954 to December 1983 for all troops feeding at this site. The current analysis concentrates only on the individuals of the Arashiyama E troop, whose members and ancestors have remained at IMP since 1954. Historically the troop was classified by recognised matrilines named after the progenitor females first identified when researchers arrived onsite in the early 1950s (see chapters 1, 2, 3 and 4). Of these original matrilines, 14 with descendants that can be directly traced back to the original progenitor females remain at IMP today. The following matrilines were included in this analysis: Kojiwa, Mino, Cooper, Yun, Chonpe, Kusha, Rakushi, Shiro, Momo, Blanche and Meme. Other matrilines present in the Arashiyama E troop were not included in the analyses due to a shortage of data. Of the females included in the analyses, some individual life histories were incomplete and portions could not be included due to either administration of birth control or a simple lack of adequate documentation.

Apart from the progenitors of these families, whose ages are not always known, this dataset encompasses the complete or partial reproductive histories of all 200 females whose birth years are known and qualified for analysis. An individual was included in the analysis if she gave birth at least once, therefore contributing at least one data point to the life-history data (i.e. age at primiparity). This parameter automatically excludes the use of the life histories of females who have never given birth, including female infants, juveniles and non-reproductive adult females.

Another factor in determining which individuals were included was whether or not they had received birth control. Starting in the mating season of autumn 1992, oral birth control was given to some females of all ages who had given birth in the previous spring of the same year. High-dose synthetic progesterone was administered in a small food item at the commencement of mating activity (around the first week of September) and this continued weekly for the duration of the season (until the beginning of January). Not all females who gave birth received birth control and higher-ranking females (such as the females of the Mino matriline) were more likely to receive birth control due to their already large family size. The majority of females given birth control did not give birth the following spring and birth control was not continued for these individuals until a subsequent birth, typically some 1–4 years after administration. Although effective in most cases, there were some instances where a female was able to conceive despite having been given contraception.

The data for individuals affected by birth control, such as interbirth interval and age at last birth, were not included in the analysis. However, unaffected traits such as age at death and age at first birth were included since they would not be affected by contraception (as mentioned above, a female must first have given birth to become eligible for birth control). For example, a female macaque who received birth control following her second offspring was included in the data for traits such as age at first parturition, interbirth interval following a first birth, and age at death (if known), but values for interbirth interval following administration of birth control (i.e. following the birth of her second offspring) were not counted, nor were these females used for the analysis of lifetime reproductive success. The data were selected in this way, to maximise sample sizes without allowing the more recent addition of birth control to affect the life-history data of individuals already well documented before the practice of birth control was begun.

Other than the administration of birth control, this population receives no medical intervention and is generally thought to be in good health. A random capture of 22 monkeys in the beginning of 1964 yielded no tuberculosis, dysentery bacillus, *Salmonella* bacillus, or B virus. Besides common intestinal parasites in 17 of the 22 individuals and one case of pyorrhoea, the captured individuals were in good health (Tanaka, 1964).

Death was also treated in a careful way so as to maximise the amount of data available for the analysis of this trait. Among Japanese macaque troops in the wild, females are seldom known to leave their natal troop except in cases of troop fission where often entire matrilines are prone to desertion (Koyama, 1970). For this study, a death was counted for each female of known age if the body was found or a single female disappeared from a troop unrelated to troop fission. This female may have exhibited a sickly demeanour or advanced signs of age-related decline in the days leading up to her disappearance, but this observation was not necessary for her departure to be counted as a death since female separation can be assumed to be a pre-emption to death based on the dynamics of female bonded macaque groups (Itani, 1985). Again, it is also of note that this value does not take into account females who died in infancy or during the juvenile period and only takes into account females who have given birth at least once before death.

Since not all individuals, matrilines, births or deaths were included in the analyses due to the above-mentioned constraints, this analysis does not encompass 100% of all female life histories that have occurred at this site and concentrates only on those matrilines and individuals for whom data are most complete. Additionally, since historical rank data are not available for this troop, rank was not included in the individual analyses even though rank is sometimes thought to have an effect on life-history traits, particularly fecundity (Fedigan, 1983).

The data were entered into Excel spreadsheets and arranged into matrilineal trees extending from 1954 to autumn of 2007. The trees reflect the individual's year of birth, years of birth for all offspring, cases of infant mortality, years of contraception (if applicable), year of death (if available), causes of death (if known) and matrilineal genealogy. Statistical analyses of the life-history parameters were conducted with Graphpad Prism, version 4.0 for MacIntosh. Outliers were removed from pooled data using standard Grubbs' test parameters for a significance level of 0.05.

5.3   Results and discussion

5.3.1   Age-specific fecundity and age at first birth

For identifiable females of known age, a total of 816 births were recorded for the study period and included births for females that had previously received birth control. Of these births, 200 first births were recorded accounting for about 25% of the total recorded births observed at IMP (n = 816). This is no surprise as sample sizes were largest for younger ages with sample sizes decreasing with age due to female death or simple lack of documentation after a certain age for some individuals (see Table 5.1 for specific number of females available for each age). Cases of infant mortality were also included in these birth rates.

Although females of provisioned Japanese macaques are thought to begin experiencing oestrus at 3.5 years of age based on behavioural mounting displays (Wolfe, 1978), age at first birth for this species is typically noted as being between 5 and 7 years based on nutritional conditions and body weight (Mori, 1979). Of the 200 first births recorded, the vast majority occurred at the ages of 5 and 6 accounting for 40% (80/200) and 31.5% (63/200) of the total first births, respectively (Table 5.2). The birth rate for 4-year-old females was very low (2%) involving only four of the 200 first births. The rest of the first births occurred between the ages of 7 and 10 (25.5%, 51/200) with a single birth occurring each for a female aged 11 and another female aged 12.

For these first few years of adult life, female productivity was fairly high and birth rates among all females ranged from 40% of 5-year-olds reproducing to more than half (51%) of all females giving birth at 10 years of age. Following this peak, birth rates for ages 11–14 remained high in the 40–45% range before dropping below 40% at the age of 15. Birth rates dropped to 23% for females

Table 5.1. *Age-specific fecundity and age-specific infant mortality (IM)*

Mother's age	Number of females	Number of births	Birth rate (%)	Number of IM	IM rate (%)
4	200	4	2%	0	0%
5	199	81	40%	7	8.6%
6	193	76	39%	6	7.8%
7	175	84	48%	14	16.6%
8	169	75	44%	7	9.3%
9	154	72	47%	6	8.3%
10	136	69	51%	8	11.5%
11	129	58	45%	1	1.7%
12	115	46	40%	5	10.9%
13	112	48	43%	2	4.2%
14	105	44	42%	2	4.5%
15	95	37	39%	1	2.7%
16	87	29	33%	1	3.4%
17	73	20	27%	2	10.0%
18	63	19	30%	2	10.5%
19	58	17	29%	2	11.8%
20	48	11	23%	1	9.1%
21	41	13	32%	0	0%
22	37	9	24%	0	0%
23	28	3	11%	1	33.3%
24	23	0	0%	–	–
25	18	1	6%	0	0%
Total	n = 200	n = 816		n = 68	

Table 5.2. *Age at first birth*

Age at first birth	Number of first births	Percentage of first births
4	4	2.0%
5	80	40.0%
6	63	31.5%
7	26	13.0%
8	14	7.0%
9	8	4.0%
10	3	1.5%
11	1	0.5%
12	1	0.5%
Total	n = 200	

aged 20 and continued to decline after that. Not a single female aged 24 was recorded to give birth among the sample while one individual female (named Mino) gave birth to her final infant at the age of 25. Of the 18 females who survived to this age, Mino was the only one that managed to give birth at this advanced age, and with the exception of a single birth at 26 years of age by a female in Katsuyama (Itoigawa *et al.*, 1992), age 25 is generally thought to be the age of reproductive termination in this species (cf. Pavelka and Fedigan, 1999).

Average age at first birth was 6.07 years (n = 200, s.d. = 0.85; Table 5.2) for this group and was comparable to Koyama *et al.*'s (1992) previously reported value of 5.39 years for the study period spanning 1954 to 1983. It should be noted that the range of age for first births in this study (age range: 4–12) is larger compared with Koyama *et al.* (1992) in which first births ranged from age 4 to 8. One explanation for this shift in ranges and differences in average values is a potential for energetic investments in reproduction and longevity later in life in response to stable resources in order to offset the costs of early maturation, as seen in other animal models (Stearns *et al.*, 1998). This trade-off would manifest similarly in Japanese macaques as increased adult longevity in females when reproductive maturation occurred at a later age, and has been shown to have a weak correlation in provisioned free-ranging rhesus macaques (Blomquist, 2009). This trade-off among traits is a common theory of life-history evolution and is thought to increase lifetime fitness as younger females have more risks associated with birth due to smaller uterine size, abnormal gestation and delivery which increases the chances of infant loss or even female death (Zhang *et al.*, 2007).

However, our dataset shows that no statistically significant correlation existed between age at first birth and age at death (Spearman rank-order correlation test, n = 34, $r = -0.23$, $p > 0.05$) suggesting no important trade-off in this population between first reproduction and survival. In the absence of a clear hypothesis for this difference in ranges and average values compared with previous analyses, one may speculate that this value may have stabilised in the recent data following the introduction of provisioning in the 1950s and unpredictable population booms and fissions which occurred starting in the 1960s and ending in the mid-1980s.

In fact, birth rates in general showed a marked difference in the current data compared with previously reported values from Koyama *et al.* (1992). These data reported birth rates about 10 percentage points higher than the birth rates reported here across all ages and there are a few explanations for the discrepancies in values. The most obvious is a difference in datasets used for analysis. As mentioned above, the current data omitted a large number of individuals that belong to lineages no longer in the current troop and even excluded matrilines considered to be not well documented. Still, this trimming of the data would not necessarily skew the current results in favour of lower birth rates. A more likely explanation is the impact of contraception on the reproductive rates of these monkeys which has decreased the birth rate since 1992. One can assume that under natural fertility conditions, birth rates for this population will more closely resemble the earlier dataset.

Data for age at last birth were also gathered for this population and were defined as the last birth occurring before a documented death or age 26, which is the oldest age recorded for birth in this species (Itoigawa *et al.*, 1992). Individuals who were given birth control were not included in this analysis. Average age at last birth was 15.73 years (n = 22, s.d. = 5.2) for this group with last birth ranging from age 6 to 25. The majority of last births occurred at the age of 14 (18.2%) with one individual each for a last birth at age 6 (4.5%) and age 25 (4.5%). The second highest percentage for last births occurred for females at age 16 (13.6%) and 18 (13.6%). Indeed, females are known to give birth at much later ages in this group but unfortunately many individuals could not be included due to the effects of contraception.

5.3.2   Interbirth interval

Of the 816 births recorded in this dataset, 481 could be used to calculate inter-birth interval (IBI). Although the contraceptive effects of birth control typic-ally wore off soon after the cessation of administration, individuals who ever received birth control were only included in this analysis up until the point that they were first given birth control and not included after that point.

Factors already known to affect birth spacing in this species include lacta-tional amenorrhoea (reduced fertility during lactation), female parity, infant loss and nutritional conditions (Fooden and Aimi, 2005). However, since pro-visioning has pretty much been consistent and abundant at this feeding site throughout its existence, the assumption is that IBI should not show seasonal or year-to-year variation based on food availability as is reported for natural troops (Suzuki *et al.*, 1998).

Since it normally takes an infant 1.5 years to become independent (Hasegawa and Hiraiwa, 1980), one would expect a standard IBI of between 1 and 2 years for this group. Indeed, we found that when looking at the data for age-specific IBI (minus cases of infant loss), birth intervals for most ages tended to remain within the expected range of 1 to 2 years throughout their entire reproductive period (Table 5.3).

The average IBI across all ages (not including cases of infant mortality) was 1.71 years for this dataset (n = 476, s.d. = 0.99), being slightly smaller than the

Table 5.3. *Age-specific interbirth interval (IBI) after live birth, averaged*

Mother's age	Number of females with subsequent recorded birth	Average IBI (in years)	s.d. value
4	2	1.50	0.71
5	47	1.89	0.48
6	41	2.15	1.06
7	45	1.60	0.58
8	46	1.54	0.66
9	43	1.67	1.67
10	43	1.47	0.74
11	38	1.42	0.55
12	36	1.61	0.77
13	31	1.45	0.57
14	24	1.38	0.58
15	27	1.48	0.58
16	16	2.31	1.62
17	13	1.54	0.66
18	10	1.10	0.32
19	9	1.56	0.73
20	3	1.53	1.15
21	1	2	–
22	1	3	–
Total	n = 476		

1.98 years reported for the females of the Arashiyama West group (cf. Fedigan *et al.*, 1986). Young females aged 5 and 6 exhibited an IBI slightly above the average (1.89 years and 2.15 years, respectively), a typical feature among adolescent mothers. This agrees with data for primiparous mothers which revealed an average IBI of 1.94 years (not including cases of infant death; n = 96, s.d. = 0.79) following a first birth. Multiparous females are quicker to recover reproductive ability after birth with an IBI of 1.7 years (not including cases of infant death; n = 91, s.d. = 0.54), very slightly below the average IBI for this dataset. Infant mortality caused an even greater reduction in IBI with average intervals falling to 1.15 years following an infant death within 1 year of birth (n = 46, s.d. = 0.36).

Except for an unusually long IBI seen at age 16 (2.31 years), age-specific IBI after age 6 did not rise above the mean value again until age 21 (IBI = 2 years) suggesting that older females experience longer IBIs. Although that is the expectation for females of advanced age, the sample size in this study was too small to draw any real conclusions. Although many females lived to old age, many either did not give birth again after age 21 or were administered birth control, therefore excluding them from this analysis. However, it is of note that in Koyama *et al.* (1992), 16 reproductively active females were recorded to live to age 20 but only 6 managed to give birth in subsequent years. Females in this age range were also noted to have longer IBIs (Koyama *et al.*, 1992).

5.3.3   Infant mortality

Of the 816 documented births for this dataset, 68 infants were known to have died within 1 year of birth, producing an infant mortality (IM) rate of 8.3%. This was slightly lower than the 10.3% reported by Koyama *et al.* (1992) and much lower than the 20.9% reported for the Arashiyama West group by Fedigan *et al.* (1986). In our dataset, primiparous mothers experienced a higher infant mortality rate of 10% (20/200) or roughly one in every 10 births. Multiparous mothers had the lowest rates of infant mortality with a rate of 7.7% (48/616) or roughly one in every 13 births. These rates were low in general compared with other provisioned groups of this species, and less than half of that reported for wild groups (cf. Fooden and Aimi, 2005).

Interestingly, a high IM rate was seen in females aged 7 (17%; see Table 5.1) with spikes again seen at ages 10 and 12. Following a rate of 10.9% at age 12 (2.2 percentage points above the average), rates dropped to 4.2% at the age of 13 and remained low until age 16. IM rates steadily rose again from the age of 17 to a peak of 33.3% at age 23, a rate that was 24.6 percentage points above the average. These data suggest that infant mortality is highest for older mothers and young mothers but adolescent mothers (age 4–6) seem to be unaffected by the high rates typical of primiparous mothers.

5.3.4   Total number of offspring

Females used in this analysis have never received birth control and have a completed life history ending in either documented death or having reached the age of 26, past which births were not recorded (whichever comes first). Here, births included the total number of live birth offspring regardless of whether they survived to the age of 1 or not.

Average total number of offspring for this group was 7.67 infants (n = 21, s.d. = 3.34) with a range of 1 to 13 infants total. This was slightly larger than the 6.56 infants reported for the females of the Arashiyama West group (cf. Fedigan *et al.*, 1986). The females of our dataset typically stopped reproducing in their late teens (age 17–19) before disappearing 1 to 2 years later. There were a number of births to mothers in their early 20s but unfortunately many of these mothers had been administered birth control at some point in their lives pointing to the conclusion that females are indeed capable of reproduction later in life and therefore producing more offspring under natural conditions. In fact, the two females who produced 13 infants each within a lifetime stopped reproducing at the ages of 22 and 23, and Mino gave birth to her 12th and final infant at the age of 25.

5.3.5   Age at death

Age at death was only counted for females of known age with a documented death or disappearance, and did not take into account females who died before giving birth. Therefore, this trait can be considered to be the average age at death for Japanese macaque mothers at this site.

Average age at death was 18.27 years (n = 36, s.d. = 8.64; Table 5.4) and included mothers who were administered birth control. This trait ranged in value from age 5 to age 33 with the higher death rates occurring after the age of 21. The survivorship data showed that the highest death rates occurred at age 30 (28.6%) and 33 (50%), and that almost half of the population will have died before reaching their 20s. With average age at death set at 18.27 years and survivorship reaching the halfway point at between ages 18 and 19, the data showed that only 46% of individuals made it to age 20 with only 19% of females surviving until the age of 26, the oldest known age for birth in this species (cf. Itoigawa *et al.*, 1992). Out of the 8.2% of individuals who managed to reach age 30, death rates fluctuated between 20–100% and so far, only four females have managed to make it to age 33. As of the beginning of 2010, no females have survived to age 34 (Mino-63–75 was the closest, dying just one month prior to reaching 34).

Age at death was positively correlated with age at last birth which was calculated for those individuals of known age at last birth, age at death, and who had never received contraception (Spearman rank-order correlation test, n = 17, $r = 0.91$, $p < 0.001$). Although most individuals typically died within 0–3 years following the last birth (10/17, i.e. 58.8%), five females continued to live well into their late 20s and early 30s, and there were two instances of females ceasing reproduction in their early teens, dying 9 years later. The period between last birth and death calculated for the total sample ranged from 0–10 years with an average of 4.53 years. However, we do think this mean is skewed to represent a longer value since birth control is unlikely to be administered to an older or historically infertile female, of which these individuals make up seven of the total data points with values ranging from 7–10 years. Without the addition of these latter points and with the addition of more data from individuals experiencing natural fertility (i.e. no birth control), one would expect that there would typically be a 0–3 year period between last birth and death.

Table 5.4. *Age-specific mortality and survivorship*

Mother's age	Number of females	Deaths	Death rate (%)	Survivorship (%)
4	200	0	0%	100%
5	199	1	0.5%	97.3%
6	193	1	0.5%	94.6%
7	175	2	1.1%	89.2%
8	169	4	2.4%	78.4%
9	154	1	0.6%	75.7%
10	136	1	0.7%	73.0%
11	129	1	0.8%	70.3%
12	115	2	1.7%	64.9%
13	112	0	0%	64.9%
14	105	1	1.0%	62.2%
15	95	0	0%	62.2%
16	87	2	2.3%	56.8%
17	73	2	2.7%	51.4%
18	63	0	0%	51.4%
19	58	1	1.7%	48.7%
20	48	1	2.1%	46.0%
21	41	0	0%	46.0%
22	37	4	10.8%	35.2%
23	28	3	10.7%	27.1%
24	23	1	4.3%	24.4%
25	18	1	5.6%	21.7%
26	16	1	6.3%	19.0%
27	14	1	7.1%	16.3%
28	10	1	10.0%	13.6%
29	7	0	0%	13.6%
30	7	2	28.6%	8.2%
31	5	1	20.0%	5.5%
32	4	0	0%	5.5%
33	4	4	100%	0%
Total	n = 200	n = 36		

5.3.6   *Characteristics of long-lived females (> 25 years)*

Of the 200 females who lived to give birth at least once, 154 females had records that were incomplete or were captured by the age of 25. Of the remaining 46 individuals, 29 died by the age of 25 and 16 were known to have survived to at least the age of 26. Of these 16 females, nine were alive as of the end of 2007 and seven had recorded deaths. Of the deceased individuals, the post-reproductive lifespan following age 26 ranged from 1–7 years, giving an average of 3.28 years of post-reproductive lifespan when using the age of 26 as a cut-off point for fertility (age 26 being the oldest recorded age for birth in this species; Itoigawa *et al.*, 1992). If last birth was used as an endpoint for fertility, five of the seven females had post-reproductive periods ranging from 7–10 years, while two were unfit for analysis since they were given birth control following their final births.

Looking at the lives of extraordinarily long-lived individuals, four females have been to known to reach the age of 33 as of the beginning of 2010, and this was the maximum lifespan ever recorded in this population. Interestingly, three of the individuals were from the Mino family, a matriline that became the highest-ranking at the time of fissioning between A and B troop in 1966. This is now the largest kin-group in E troop, which fissioned from B troop in 1986. Mino and Mino-63–69–74 were alpha females in their respective troops and survived to old ages. Mino survived to 33 years and 8 months, and Mino-63–69–74 survived to be 33 years and 10 months old. This suggests a possible correlation between longevity and rank, but this analysis is difficult to carry out since individual and matrilineal rank fluctuated during a female's lifetime due to fissioning or the capture of other individuals in the troop (Koyama, 1970; Huffman, 1991). In contrast to these high-ranking and long-lived females, the last recorded female to die at 33 years was descended from the Blanche lineage (Blanche–59–64–75 who died 33 years and 2 months) which has experienced rank shifts in the past but now belongs to a relatively low-ranking matriline. These extraordinarily long-lived individuals also exhibited other similar characteristics such as last births typically in the early 20s with a total number of offspring in the range of 9–12. Onset of parturition for these individuals was identical to the norm for the troop (age 5–6).

To summarise our data, reproduction slowed during the early 20s and was known to cease in this troop at the age of 25. For the most part, birth during the second decade did not seem to affect infant survival, although there seemed to be an increase in infant mortality for mothers after the age of 22, but sample sizes were too small to draw solid conclusions. Interbirth interval showed a marked increase beginning at the age of 21. Death rate showed a marked increase beginning at the age of 22 and experienced ups and downs until all individuals died by the age of 33. However, most females never reached their early or late 20s since survivorship was 46% at age 20 and dropped to 21.7% at age 25.

Given the current data for older females, there was still no evidence that this troop was undergoing any sort of reproductive cessation except in the case of a few very long-lived individuals who were not considered to be the norm for this population. For the majority of females, last birth occurred on average by 15.73 years of age with death following soon after at the age of 18.27 years, a difference not dissimilar from a long IBI. However, hormonal data taken in successive breeding seasons for all females will be needed to determine if any true reproductive senescence unrelated to overall bodily decline is occurring across the population, especially in light of the fact that these few cases can experience up to 7 years of non-reproduction after the age of 26. It is of note that studies on captive Japanese macaques are still inconclusive when it comes to follicular activities in the aged ovary (Nozaki *et al.*, 1995). Furthermore, at the Kyoto University Primate Research Institute in the summer of 2007, a single caged female aged 33 (Wakasa #278) was observed to menstruate during the non-breeding season (Chalmers, personal observation). This topic certainly needs more solid data before any conclusions can be drawn.

5.3.7   A brief note on rank

So far, no consistent relationship has been found between female rank and reproductive success in primates (Fedigan, 1983; but also see Harcourt, 1987). In some primate species, a positive association has been reported between these parameters (e.g. long-tailed macaques: van Noordwijk and van Schaik, 1999; yellow baboons: Cheney *et al.*, 2004), whereas there was no such association in other species (Barbary macaques: Kümmerli and Martin, 2005).

Although social variables are hard to apply to long-term studies such as this one, largely because dominance can change throughout an individual's lifetime, there are some interesting trends related to rank in the data that are worthy of mention (see also Wolfe, 1984; Fedigan *et al.*, 1986). The Mino family (the current highest-ranking matriline at IMP) has the longest-lived individuals, the largest number of descendants still living at the feeding site, and some of the most reproductively successful females (with the exception of Kojiwa–62–72 who birthed 14 infants before being given birth control at the age of 23). Unfortunately, the Mino family members are among the most common recipients of birth control since their females are so numerous. It would be difficult to say for certain that Mino is the most reproductively successful matriline and has been since the beginning, but this possibility is not out of the question and requires more study correlated with historical rank changes and careful trimming of the birth-control data.

However, what is most interesting is the propensity for long-lived females in this family, since three of the four females who reached 33 years of age came from this lineage. Perhaps access to high-quality foods or historically good relationships with high-ranking males due to familial dominance have allowed this family to flourish, but more data are needed relating historical social dominance to matriline longevity among all families at this site.

5.3.8   Future studies

Although the amount of life-history data presented in this chapter is useful in itself, these numbers would be more effective if more causative factors could be analysed to determine which environmental, nutritional or social pressures may shape these individuals' life histories. For example, there at least appears to be a heritable aspect or even a social pressure that exists to allow the Mino family to have increased longevity and fertility compared with other matrilines. Maybe there is even a genetic component that has allowed the family to maintain social dominance over the years. Whatever it is, accumulated social data for all families over the past 50 years combined with detailed life-history analysis corresponding to rank in any given period would be useful to see if there is a social component involved in lifespan and whether or not this actually raises reproductive success.

Also of interest is whether or not the potential for entering menopause exists in the Arashiyama macaque population. As of now, there is no doubt that there is a cut-off for fertility at around the age of 25 although this has not been accompanied by long-term hormonal studies (but see Shimizu: chapter 18). However, this age limit is generally unsupported on a species-wide scale if one turns to captive studies of this phenomenon or even other provisioned groups (Pavelka and Fedigan, 1999; Fedigan and Pavelka, 2001). As more long-lived individuals become available for study (and perhaps a more common occurrence in this population), opportunities for study of reproductively terminated females will be easier to carry out with adequate sample sizes.

In conclusion, probably the most important message of the foregoing is that long-term studies at this field site must absolutely continue to bolster sample sizes and determine how this population is changing over time. Fifty years of data is a good amount for any primate population, but since these monkeys are known to live to some 30+ years, more data collection is needed to more fully understand the various life-history traits of these very old individuals. What exactly is the reproductive potential of Japanese macaques? Can heritability of longevity be absolutely proven? Does longevity have a relationship with social rank? It goes without saying that this dataset is only a starting point for more interesting studies of this troop, especially in relation to potential longevity and lifetime reproductive success in macaques.

Acknowledgements

We wish to give our sincerest gratitude to IMP Director, Mr Shinsuke Asaba for allowing us access to park records and to the staff for helping with the identification of all 150 of its current members. We would also like to thank the researchers who kept meticulous records of this population and published their raw data for use, especially the contributors to the Iwatayama Shizenshi Kenkyujo Hokoku. Among the many individuals who have contributed to the building of this dataset, in particular we are indebted to K. Norikoshi, Baldev Singh Grewal, Hisayo Suzuki, Nobuo Asaba and Ryosuke Asaba. Thanks also to those who acted as knowledgeable guides in the pursuit of data, including Dr Juichi Yamagiwa, Eiji Inoue and Michio Nakamura. AC would like to thank members of the Laboratory of Human Evolution Studies at Kyoto University who contributed advice based on their own experiences at Arashiyama and provided helpful critiques throughout this study. Finally, we would like to thank Dr Jean-Baptiste Leca for his advice, encouragement and unwavering patience during the editing of this manuscript. Thank you.

References

Asaba, N. and Suzuki, H. [eds]. (1984). Birth lists and family trees for the Arashiyama monkeys. In *Bulletin of Iwatayama Institute of Natural History Iwatayama Shizenshi Kenkyujo Hokoku. Vol. 3.* Kyoto, Japan: Arashiyama Shizen Kenkyujo, pp. 83–114. (in Japanese) [note: there is no primary author(s) for this article]

Blomquist, G. E. (2009). Trade-off between age at first reproduction survival in a female primate. *Biology Letters*, **5**, 339–342.

Cheney, D. L., Seyfarth, R. M., Fischer, J. *et al.* (2004). Factors affecting reproduction and mortality among baboons in the Okavango Delta, Botswana. *International Journal of Primatology*, **25**, 401–428.

Fedigan, L. M. (1983). Dominance and reproductive success in primates. *Yearbook of Physical Anthropology*, **26**, 85–123.

(1991). History of the Arashiyama West Japanese macaques in Texas. In *The Monkeys of Arashiyama: 35 Years of Research in Japan and the West*, eds. L. M. Fedigan and P. J. Asquith. Albany, NY: State University of New York Press, pp. 54–73.

Fedigan, L. M. and Pavelka, M. S. M. (2001). Is there adaptive value to reproductive termination in Japanese macaques? A test of maternal investment hypotheses. *International Journal of Primatology*, **22**, 109–125.

Fedigan, L. M., Fedigan, L., Gouzoules, S., Gouzoules, H. and Koyama, N. (1986). Lifetime reproductive success in female Japanese macaques. *Folia Primatologica*, **47**, 143–157.

Fooden, J. and Aimi, M. (2005). Systematic review of Japanese macaques, *Macaca fuscata. Fieldiana Zoology*, **104**, 1–200.

Gage, T. B. (1998). The comparative demography of primates: with some comments on the evolution of life histories. *Annual Review of Anthropology*, **27**, 197–221.

Harcourt, A. H. (1987). Dominance and fertility among female primates. *Journal of Zoology, London*, **213**, 471–487.

Hasegawa, T. and Hiraiwa, M. (1980). Social interactions of orphans observed in a free-ranging troop of Japanese monkeys. *Folia Primatologica*, **33**, 129–158.

Huffman, M. A. (1991). History of the Arashiyama Japanese macaques in Kyoto, Japan. In *The Monkeys of Arashiyama: 35 Years of Research in Japan and the West*, eds. L. M. Fedigan and P. J. Asquith. Albany, NY: State University of New York Press, pp. 21–53.

Itani, J. (1985). The evolution of primate social structures. *Man*, **20**, 593–611.

Itoigawa, N., Tanaka, T., Ukai, N. *et al.* (1992). Demography and reproductive parameters of a free-ranging group of Japanese macaques (*Macaca fuscata*) at Katsuyama. *Primates*, **33**, 49–68.

Koyama, N. (1970). Changes in dominance rank and division of wild Japanese monkey troop in Arashiyama. *Primates*, **11**, 335–390.

Koyama, N., Takahata, Y., Huffman, M. A., Norikoshi, K. and Suzuki, H. (1992). Reproductive parameters of female Japanese macaques: Thirty years data from the Arashiyama troops, Japan. *Primates*, **33**, 33–47.

Kümmerli, R. and Martin, R. D. (2005). Male and female reproductive success in *Macaca sylvanus* in Gibraltar: No evidence for rank dependence. *International Journal of Primatology*, **26**, 1229–1249.

Mori, A. (1979). Analysis of population changes by measurement of body weight in the Koshima troop of Japanese monkeys. *Primates*, **20**, 371–397.

Nozaki, J., Mitsunaga, F. and Shimizu, K. (1995). Reproductive senescence in female Japanese monkeys (*Macaca fuscata*): age and season-related changes in hypothalamic-pituitary-ovarian functions and fecundity rates. *Biology of Reproduction*, **52**, 1250–1257.

Pavelka, M. S. M. and Fedigan, L. M. (1999). Reproductive termination in female Japanese monkeys: a comparative life history perspective. *American Journal of Physical Anthropology*, **109**, 455–464.

Stearns, S., Ackermann, M. and Doebli, M. (1998). The experimental evolution of aging in fruitflies. *Experimental Gerontology*, **33**, 785–792.

Suzuki, S., Noma, N. and Izawa, K. (1998). Inter-annual variation of reproductive parameters and fruit availability in two populations of Japanese macaques. *Primates*, **39**, 313–324.

Tanaka, T. (1964). Report on the general survey of the troop in Iwatayama Monkey Park. *Yaen*, **18**, 28–30. (in Japanese)

van Noordwijk, M. A. and van Schaik, C. P. (1999). The effects of dominance rank and group size on female lifetime reproductive success in wild long-tailed macaques, *Macaca fascicularis*. *Primates*, **40**, 105–130.

Wolfe, L. (1978). Age and sexual behavior of Japanese macaques. *Archives of Sexual Behavior*, **7**, 55–68.

  (1984). Female rank and reproductive success among Arashiyama B Japanese macaques (*Macaca fuscata*). *International Journal of Primatology*, **5**, 133–143.

Zhang, X., Cnattinguis, S., Platt, R., Joseph, K. and Kramer, M. (2007). Are babies born to short, primiparous, or thin mothers 'normally' or 'abnormally' small? *Journal of Pediatrics*, **150**, 603–607.

Part II

Sexual behaviour

6 Long-term trends in the mating relationships of Japanese macaques at Arashiyama, Japan

MICHAEL A. HUFFMAN AND

YUKIO TAKAHATA

Heterosexual consortship between Oppress-60–65–85 (male) and Mino-63–75–93–98 (female) at Arashiyama, October 2010 (photo by J.-B. Leca).

6.1 Introduction

The pioneering work of Zuckerman (1932) and Carpenter (1942) on mating relations in primates continues to be a topic of wide interest in primatology (e.g. *Pan troglodytes*: Nishida, 1997; *Macaca fuscata*: Stephenson, 1975; Sprague, 1991; Sprague *et al.*, 1998; Soltis *et al.*, 2001; Hayakawa, 2007; *M. mulatta*:

The Monkeys of Stormy Mountain: 60 Years of Primatological Research on the Japanese Macaques of Arashiyama, eds. Jean-Baptiste Leca, Michael A. Huffman and Paul L. Vasey. Published by Cambridge University Press. © Cambridge University Press 2012.

Altmann, 1962; Manson, 1992; Berard, 1999; *M. sylvanus*: Paul *et al.*, 1993; *Papio cynocephalus*: Hausfater, 1975; *P. hamadryas hamadryas*: Swedell, 2006; *Lemur catta*: Gould, 1991). Most studies, however, tend to analyse only a single to several years' data for any given troop. There are still many issues left to be analysed from the viewpoint of an individual's life-history perspective. For example, do males and females change their mating preferences over their reproductive lifetimes? Are there age-related differences in mating partners? Do they persist or change?

In wild groups of female bonded matrilineal cercopithecines, adult males typically stay in one group for a few years (e.g. Wrangham, 1980; Itani, 1985; Sprague, 1992, 1998; Sprague *et al.*, 1998). This poses logistical problems for long-term studies of male mating relations in wild populations, simply because males are harder to follow for long periods, making it nearly impossible to follow them through the completion of their life histories. Research on provisioned groups of macaques provides valuable data to address questions on long-term trends, because males tend to stay longer in one group or do not move too far away.

In the Arashiyama West troop of Japanese macaques, the same pairs tended to mate for at least three successive mating seasons (Fedigan and Gouzoules, 1978). This has been interpreted to indicate that Japanese macaques prefer to mate with the same 'good' partners to gain higher reproductive success (see Small, 1989 for a review of female choice). In the parent Arashiyama troop of Kyoto, Japan, Huffman (1991a) found that females tended to be responsible for the discontinuation of mating relations with male partners after 2–3 years. Interestingly, paternity analyses carried out in two captive groups of Japanese macaques showed that the same male was rarely, if ever, the father of multiple infants of the same female (Inoue, 1995). It seems clear that in this way, females are able to somehow determine who the fathers of their offspring are more likely to be, by changing partners from one season to the next (Inoue, 1995). Berard (1999) found that correlations between male dominance rank order, reproductive success and mating activity were not stable over a 4-year period, based on genetic analysis and behavioural observations of rhesus macaques on Cayo Santiago. Mating success declined after a few years in a troop, and Bercovitch (1997) suggested that female mate choice for novel males was a key factor in male dispersal. Similar results have been reported in several field sites of Japanese macaques (e.g. Huffman, 1991a; Hayakawa, 2008; Inoue and Takenaka, 2008) and Barbary macaques (Paul, 1997; Modolo and Martin, 2008).

In this chapter, we combine and analyse the published long-term data on males of the Arashiyama B troop, spanning seven mating seasons between 1968 and 1984 (Stephenson, 1975; Takahata, 1982a; Wolfe, 1984; Huffman, 1991b).

The Arashiyama-Kyoto troop has been provisioned by the Iwatayama Monkey Park, Arashiyama since 1954, and all the monkeys have been individually identified from that time on (Huffman, 1991b). In 1966, the group split into the Arashiyama A and B troops (Koyama, 1970). These two troops co-existed at the park until 1972, at which time A troop was captured and moved to Texas, where it became known as the Arashiyama West troop (Fedigan, 1991). During this period (1966–1972), many males left their natal troop around the age of sexual maturity (3.5–5.5 years) and immigrated into the neighbouring troop (Norikoshi and Koyama, 1975). Those males that stayed in one of the troops remained under observation by researchers and were found to attain high-ranking status in their new troop. Here, we investigate the mating preferences of individuals in each mating season, and examine how long mating pairs persist and how often they change over the long term.

6.2   Background and methods

We analysed the mating activity of Japanese macaques in the Arashiyama B troop (for troop history, see Huffman, 1991a; Huffman: chapter 1) over seven mating seasons between 1968 and 1984 (Stephenson, 1975; Takahata, 1982a; Wolfe, 1984; Huffman, 1991a, 1992). In total, 2823 copulations were recorded for 78 males (more than 3 years old) and 110 females (more than 2 years old).

At Arashiyama, males begin to ejaculate at the age of 4 years. Such males are divided into the following age classes: adults (more than 7 years old) and subadults (between 4 and 7 years old). Adult males were also divided into their relative social rank classes, i.e. alpha, high-ranking (2nd–4th rank), mid-ranking (5th–7th), and low-ranking (8th or more). In addition to the Arashiyama-born males, solitary males were recorded to visit and occasionally immigrate into the Arashiyama B troop. Such males were few, as the Arashiyama-Kyoto population is largely isolated from other local populations. Females were divided into three age classes according to their fecundity, namely sub-adult (between 3 and 6 years old), mid-aged (between 7 and 19 years), and old-aged (more than 19 years) (Takahata, 1980). They were also divided into three kin-group social rank classes: high-ranking (Ko, Co, Ch and Gl kin-groups), mid-ranking (Ku, Ra and Mo), and low-ranking (Ai, Sh, Bl and Me).

A copulation or mounting sequence was defined as one or more mounts terminated by a cessation of mounting for at least 5 minutes, or ejaculation (i.e. synonymous with 'mount event': Hanby and Brown, 1974). In each mating season, we calculated the relative mating success (RMS) for each male (i.e. the proportion of each male's copulations to the total number of observed copulations in the troop). We also calculated two rates of copulation as follows.

Re-copulation rate:
[number of pairs who mated in two consecutive mating seasons] ÷ [number of pairs who had mated in the previous mating season only] × 100

New copulation rate:
[number of mating pairs in a mating season, who did not mate in the previous season] ÷ [number of pairs who did not mate in the previous mating season] × 100

The data were analysed with non-parametric statistics using Excel (Microsoft, 2001) and Statistica (StatSoft Inc., 1999). The level of significance was set at $p < 0.05$.

6.3   Results

6.3.1   *Male age, rank and mating partners*

In the Arashiyama B group, males could attain the highest rank around the age of 10 years. Figure 6.1 shows the correlation between rank and age of our study subject males. Once they occupied the highest ranking position, they usually kept that for several years (see Figure 3 in Huffman, 1991b). Based on

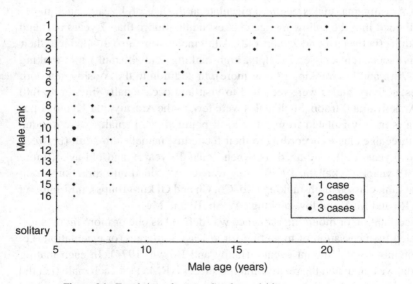

Figure 6.1. Trends in male age and rank acquisition.

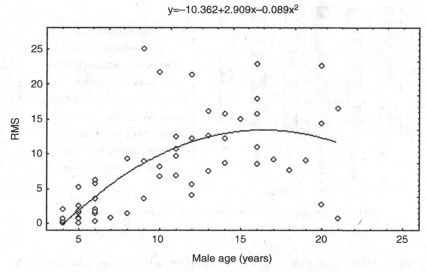

Figure 6.2. Trends in male age and relative mating success (RMS).

the combined data of seven mating seasons, the plot of relative mating success (RMS) showed an inverted U-shaped curve across male age. It approximated a second degree curve ($y = -10.36 + 2.91x - 0.089x^2$) and the correlation was statistically significant ($r^2 = 0.505$, $F = 29.533$, $p < 0.001$) (Figure 6.2). Males' RMS tended to peak at around the age of 9 or more.

Although the data of old males aged 20 years and above were insufficient, based on reported trends, it was expected that their RMS gradually decreased. For example, K-63♂ constantly attained high values of RMS (about 12–15) between the ages of 12 and 15 years, but some time after reaching alpha status, he began to show very low RMS values of 2.7–0.60 at the age of 20 and 21 years, respectively. It was at this time that refusals to mate by the high-ranking females began to increase (see Huffman, 1991a, 1991b).

A statistically significant positive relationship characterised mating patterns according to rank. The alpha males and other high-ranking males tended to mate with high-ranking, mid-aged females (Kruskal–Wallis test, $H = 208.57$, df = 5, n = 2818, $p = 0.0001$) (Figure 6.3). Copulations with such females accounted for 56.5% of the alpha males' mating activity and 58.0% of that of other high-ranking males. In particular, high-ranking males consistently copulated with high-ranking and mid-aged females throughout the seven mating seasons (49.0–71.6%).

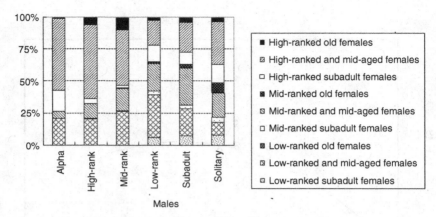

Figure 6.3. Age-rank class distribution of males' mating partners.

On the other hand, there was a large variation in the mating partners of the alpha males between mating seasons (Figure 6.3). In 1976, 1977 and 1983, their copulations with high-ranking and mid-aged females occupied 70.7%, 89.1% and 66.7% of their mating activity, respectively. Besides, in 1975, the alpha male (M-59) intensively mated with two high-ranking subadult females. Including mating with such females, high-ranking females occupied 81.3% of M-59's copulations. In contrast, copulations with high-ranking and mid-aged females occupied lower percentages of the alpha males' copula-tions in 1968 (Zola: 28.6%), in 1979 (M-59: 25.0%) and in 1984 (K-63: 0%). In particular, the two highest-ranking males (K-63 and Ran-63) were observed to perform frequent courtships toward oestrous females and were consistently rejected by them (Huffman, 1991a). It is noteworthy that such alpha males left the Arashiyama B troop within a few years after female mating rejection began. This was also the trend for two other previous alpha males, Zola who disappeared in July 1970, and M-59 who disappeared in June 1981 (Huffman, 1991b). K-63 intermittently left the troop between 1981 and 1986, gradually losing support from the troop's adult females. In September 1986, the Arashiyama B troop fissioned into the Arashiyama E and F troops. K-63 occupied the alpha male status of E troop, but he finally disappeared in December 1987.

As Norikoshi (1976) first pointed out and as was later confirmed by Huffman (1991a), a decline in the number of mating partners resulting from female rejection is often closely linked to the emigration of a troop's highest-ranking males. A similar pattern was noted for high-ranking males in the Takasakiyama troops, Beppu, Kyushu (Huffman, 1991a).

6.3.2   *Female age and mating partners*

It is noteworthy that high-ranking females did not always concentrate their mating activity with only high-ranking males. Out of the 1117 copulations recorded for high-ranking females, 56.6% were with mid- and low-ranking, sub-adult, and solitary males. Clearly, female choice did not always accord with male choice.

Figure 6.4 shows that the relation between female age and rank classes and their mating partner selection criteria seemed to change throughout their life history. Usually, sub-adult females aged 3–5 years tended to mate more with low-ranking or sub-adult males, while high-ranking females tended to mate more with the alpha and other high-ranking males. As they grew older, however, the number of their mid- or low-ranking male partners gradually increased, and females of all rank categories tended to avoid mating with the alpha male after the age of 14 years.

In general, there was considerable variation in females of different social rank classes in their transitions of mating partners throughout their life histories (Figure 6.4). High-ranking females tended to mate with alpha/high-ranking males at earlier ages than mid- and low-ranking females (Figure 6.4a).

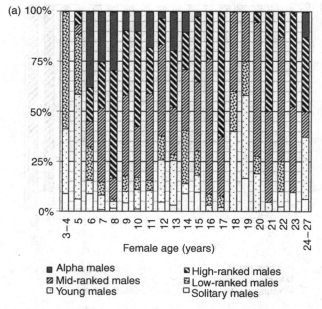

Figure 6.4. Age-rank class distribution of females' mating partners: (a) high-ranking females, (b) mid-ranking females and (c) low-ranking females.

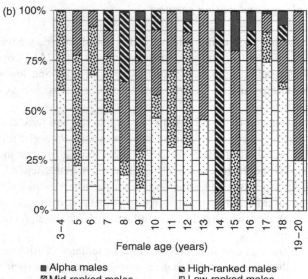

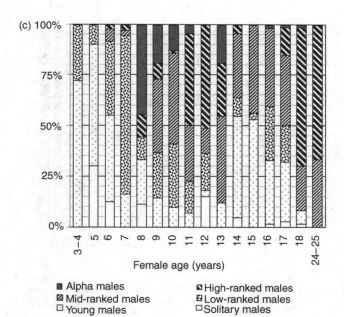

Figure 6.4. (*cont.*)

Copulations with mid-ranking males gradually increased after the ages of 11–12 years. Thus, they mated with a diverse range of males throughout their life history.

On the other hand, mid-ranking females infrequently mated with alpha/high-ranking males throughout their life histories (Figure 6.4b). Consequently, the diversity in the social ranks of their mating partners seemed to be lower than for high- or low-ranking females. Figure 6.4c shows that low-ranking females frequently mated with young and low-ranking males between the ages of 3 and 7 years. Then, they tended to mate with alpha males between the ages of 8 and 13 years, but not to the exclusion of males in other age or rank classes.

6.3.3   Lineage-specific mating

Does lineage-specific mating exist? Five adult males (M-59, K-63, Ran-63, De-64 and Sol-64) were observed in five or more mating seasons. Figure 6.5 shows the proportions of copulations with each female lineage recorded for Ran-63 in each mating season. There was a significant difference among seasons ($\chi^2 = 325.3$, df = 42, $p = 0.0001$), and there seemed to be no consistent tendency to mate with females from one particular lineage.

Similar tendencies were observed for the other four males (M-59: $\chi^2 = 234.9$, df = 32, $p < 0.0001$; K-63: $\chi^2 = 161.8$, df = 40, $p < 0.00001$; De-64: $\chi^2 = 255.2$, df = 40, $p < 0.0001$). This is consistent with the findings mentioned above that males changed mating partner age and rank preferences throughout their lives.

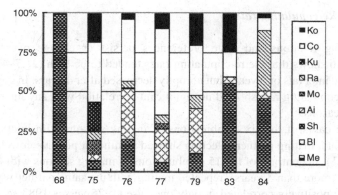

Figure 6.5. Distribution of mating partners by matriline over time for one adult male Ran-63. Full names of lineages are: Kojiwa (Ko), Cooper (Co), Kusha (Ku), Rakushi (Ra), Momo (Mo), Ai (Ai), Shiro (Sh), Blanche (Bl) and Meme (Me).

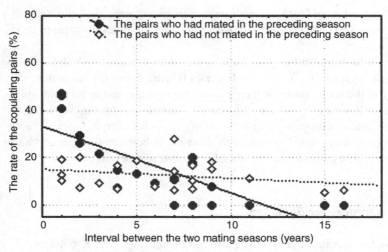

Figure 6.6. Distribution of the interval in years for re-copulation between males and females.

In the years when males were young and low-ranking, they tended to mate with sub-adult or lower-ranking females. With increasing age and rank, they began to mate with higher-ranking females. However, if they remained in the troop until reaching high-ranking status, mating frequency with high-ranking females decreased. Positive evidence for lineage-specific mating was not found from the long-term data.

6.3.4   Re-copulating rates

Do mating relations and mate preferences persist or change over time? Figure 6.6 shows that the re-copulating rates tended to be high (21.7–47.2%) within the intervals of 3 years, but abruptly decreased after 4 years. In contrast, the new copulating rate showed no significant correlation with any interval of mating seasons.

In five cases of the six combinations of mating seasons with intervals of 1 to 3 years, mating partner selection showed significant positive correlations (Table 6.1). In contrast, of the 15 combinations of mating seasons with longer intervals (more than 3 years), repeated mating with the same partner was significantly positively correlated in only one case (1975 versus 1983 seasons). There were significant correlations in 12 cases, and a significantly negative correlation was found in two cases.

Table 6.1. *Likelihood of pairs copulating again in the next 1 to 3 years*

	1968	1975	1976	1977	1979	1983	1984
1968		$\chi^2 = 4.52^*$ 7	$\chi^2 = 2.32$ 8	$\chi^2 = 2.02$ 9	$\chi^2 = 1.04$ 11	$\chi^2 = 0.06$ 15	$\chi^2 = 7.07^{**}$ 16
1975			$\chi^2 = 30.1^{**}$ 1	$\chi^2 = 2.94$ 2	$\chi^2 = 0.13$ 4	$\chi^2 = 4.58^*$ 8	$\chi^2 = 0.52$ 9
1976				$\chi^2 = 68.1^{**}$ 1	$\chi^2 = 12.3^{**}$ 3	$\chi^2 = 1.0$ 7	$\chi^2 = 1.13$ 8
1977					$\chi^2 = 32.6^{**}$ 2	$\chi^2 = 0.12$ 6	$\chi^2 = 0.26$ 7
1979						$\chi^2 = 0.02$ 4	$\chi^2 = 0.42$ 5
1983							$\chi^2 = 39.7^{**}$ 1
1984							

*: significant difference at $p < 0.05$, **: significant difference at $p < 0.01$.

6.4   Discussion

6.4.1   *Mating preferences of Japanese macaques*

The present data suggest that both male and female Japanese macaques change their mating partners throughout life as a consequence of their life-history trajectories and changes in their social status (Huffman, 1992; Inoue *et al.*, 1992). These two factors are likely inter-related. Such changes are assumed to reflect mating preferences. Young males mate primarily with younger and/or low-ranking females. As they attain higher status, males tend to prefer to mate with females of higher rank and older females, which is consistent with the conclusion by Inoue *et al.* (1992), stating that male choice in Japanese macaques can be divided into several patterns. A similar phenomenon has been reported for Barbary macaques (Taub, 1980) and yellow baboons (Smuts, 1985): (1) high-ranking males sired one or two offspring following frequent copulations with the mothers, the so-called 'secure' or 'best female' mating tactic; and (2) mid- and low-ranking males copulated with as many females as possible, or the so-called 'chance' or 'many female' mating tactic.

Why do higher-ranking males tend to choose to mate with high-ranking and mid-aged females? Ultimately, this mating tactic may result in higher mating/reproductive success. Indeed, mid-aged females show higher fecundity than younger and old-aged females in both wild and provisioned populations of Japanese macaques (Takahata, 1980; Koyama *et al.*, 1992; Takahata *et al.*, 1998). Although there was no significant difference in fecundity among female rank classes in Arashiyama (Koyama *et al.*, 1992), significant differences have been reported among Japanese macaques in Kinkazan (Saito, 1996) and Katsuyama (Itoigawa *et al.*, 1992). On the other hand, mating partners of the alpha or high-ranking males are fewer than those of the mid- or low-ranking males (Takahata, 1982a; Huffman, 1991a, 1992). Thus, high-ranking males seem to adopt the 'best female' strategy rather than the 'many female' strategy.

In spite of such persistent courtships toward better females, the alpha and high-ranking males are faced with a female counter strategy (Huffman, 1992). Even high-ranking females tended to mate with lower-ranking males after the age of about 14 years (Huffman, 1991a). There should be a difference in mating strategies between males and females (Small, 1989). Females seemed to opt for the 'many male' strategy, not the 'best male' strategy (Huffman, 1992).

The present analysis did not provide positive evidence of lineage-specific mating. While males tended to mate with females of the same lineages across several years, in a long time perspective, they changed the lineages of their mating partners over time, suggesting perhaps that mating partner rank was more important for males as their rank too increased. Inoue *et al.* (1992) carried out DNA analysis in a captive group of Japanese macaques and found that females tended not to conceive offspring with the same male over multiple years, indicating that there are few stable mating pairs. They also found no evidence of lineage-specific copulation (Figure 3 in Inoue *et al.*, 1992). On the contrary, all the evidence discussed above shows a strong tendency for assorted lineage mating, which will result in greater genetic diversity.

6.4.2   How long do mating relations or partner preferences persist?

In the Arashiyama B troop, we found that the same pairs tended to mate in one to two successive mating seasons, as was also pointed out by Fedigan and Gouzoules (1978) for the Arashiyama West troop. However, after 3 or more years, the re-copulation rate abruptly decreased, just as reported by Berard (1999) for the rhesus macaques of the Cayo Santiago colony. What could be the reason for the noted decrease in the re-copulation rate? Firstly, as males grow older, they attain higher rank and attain a 'priority of access' (Altmann, 1962) to high-ranking and mid-aged females. Consequently, their mating relations change with their social status. As Inoue *et al.* (1992) pointed out based on genetic analyses, Japanese macaque females continually change the fathers of their offspring throughout their life. They frequently reject the courtship of the highest-ranking males, who have stayed in the troop for a long period, although the reasons are still uncertain (Huffman, 1991a). Furthermore, males and females sometimes form peculiar proximate relations (PPR) through particularly strong mating relations, leading such pairs to begin to avoid mating (Takahata, 1982b), which is suggested to be an effect of changes in a female's mate preference (Huffman, 1991a, 1992).

That is to say, past mating relations negatively affect future mate choice for the same partners, and the mating success of higher-ranking males decreases. Indeed, there were few significant correlations between adult male rank order and the observed number of copulations in the Arashiyama B troop (Takahata, 1982a; Huffman, 1991a), which is in agreement with the Bercovitch–McMillan hypothesis (McMillan, 1989; Bercovitch, 1992). In such a situation, it is better for the alpha males to transfer to another group, increasing the likelihood of mating chances and ultimately reproductive success, at the cost of losing their high status. Irrespective of the ultimate causes and effects, these tendencies probably promote male transfer (Small, 1989; Huffman, 1992). Hence, male troop transfer appears regulated not by forced expulsion or male competition, but by female mate choice for novel mating partners and the male's reproductive strategy.

Both males and females change their mating partner selection tendencies throughout their life. This suggests some kind of principle driving mating partner selection in both sexes where females tend to seek novelty while males seek familiarity and perhaps in turn social stability within the troop by association with adult females. It stands to reason that while mating with high-ranking males provides stability or protection for females, once they become PPR partners with high-ranking males (cf. Takahata, 1982b), females have less of a social motivation to mate with them. On the other hand, it can be argued that males continue to need female support to maintain their stability within the troop, and sexual relations with high-ranking females would be the means by which this is attained. Such a system may, in the long run, work against a lineage-specific mating system. It may even be suggested to work against such a system from developing individuals. This makes intuitive sense from both a psychological and biological perspective to avoid mating with familiar individuals and with potential blood relations.

In conclusion, our longitudinal data buttress previous studies revealing that mating relationships do not persist over the long term, but in the short term, high-ranking and mid-aged males tend to have higher mating success; that declines with increased tenure. Further long-term studies of mating relations in other groups and in other species are clearly needed. Future studies are expected to elucidate the life-history transitions related to changes in mating strategies and resultant changes in reproductive success. A better grasp of the longitudinal aspects is key for understanding the relative roles of males and females in the maintenance and ultimate evolution of mating systems.

Acknowledgements

We thank N. Koyama and L. Wolfe, for cooperation in field work; N. Asaba, S. Asaba and R. Asaba, for their kind permission to carry out our study and maintenance of genealogical records. Last, we give our deep thanks to the monkeys of Arashiyama for the opportunity to learn about their lives.

References

Altmann, S. A. (1962). A field study of the sociobiology of rhesus monkeys, *Macaca mulatta*. *Annals of the New York Academy of Sciences*, **102**, 338–435.

Berard, J. (1999). A four-year study of the association between male dominance rank, residency, status, and reproductive activity in rhesus macaques (*Macaca mulatta*). *Primates*, **40**, 159–175.

Bercovitch, F. B. (1992). Re-examining the relationship between rank and reproduction in male primates. *Animal Behaviour*, **44**, 1168–1170.

  (1997). Reproductive strategies of rhesus macaques. *Primates*, **38**, 247–263.

Carpenter, C. R. (1942). Sexual behavior of free-ranging rhesus monkeys (*Macaca mulatta*). *Journal of Comparative Physiology*, **33**, 113–142.

Fedigan, L. M. (1991). History of the Arashiyama West Japanese macaques in Texas. In *The Monkeys of Arashiyama: Thirty-five Years of Research in Japan and the West*, eds. L. M. Fedigan and P. J. Asquith. Albany, NY: State University of New York Press, pp. 54–73.

Fedigan, L.M. and Gouzoules, H. (1978). The consort relationship in a troop of Japanese monkeys. In *Recent Advances in Primatology, Volume 1: Behaviour*, eds. D. J. Chivers and J. Herbert. London: Academic Press, pp. 493–495.

Gould, L. (1991). Reproductive behavior of free-ranging *Lemur catta* at Beza Mahafaly Special Reserve, Madagascar. *American Journal of Primatology*, **84**, 463–477.

Hanby, J. and Brown, C. E. (1974). The development of sociosexual behaviors in Japanese macaques (*Macaca fuscata*). *Behaviour*, **49**, 151–196.

Hausfater, G. (1975). Dominance and reproduction in baboons (*Papio cynocephalus*): A quantitative analysis. *Contributions to Primatology, Volume 7*. Basel: S. Karger.

Hayakawa, S. (2007). Female defensibility in small troops of Japanese macaques vis-a-vis nontroop males and copulation on the periphery of the troop. *International Journal of Primatology*, **28**, 73–96.

  (2008). Male-female mating tactics and paternity of wild Japanese macaques (*Macaca fuscata yakui*). *American Journal of Primatology*, **70**, 986–989.

Huffman, M. A. (1991a). Mate selection and partner preferences in female Japanese macaques. In *The Monkeys of Arashiyama: Thirty-five Years of Research in Japan and the West*, eds. L. M. Fedigan and P. J. Asquith. Albany, NY: State University of New York Press, pp. 101–122.

  (1991b). History of Arashiyama Japanese Macaques in Kyoto, Japan. In *The Monkeys of Arashiyama: Thirty-five Years of Research in Japan and the West*, eds. L. M. Fedigan and P. J. Asquith. Albany, NY: State University of New York Press, pp. 21–53.

(1992). Influences of female partner preferences on potential reproductive outcome in Japanese macaques. *Folia Primatologica*, **59**, 77–89.

Inoue, E. and Takenaka, O. (2008). The effect of male tenure and female mate choice on paternity in free-ranging Japanese macaques. *American Journal of Primatology*, **70**, 62–68.

Inoue, M. (1995). Application of paternity discrimination by DNA polymorphism to the analysis of the social behavior of primates. *Human Evolution*, **10**, 53–62.

Inoue, M., Mitsunaga, F., Ohasawa, H. *et al.* (1992). Paternity testing in captive Japanese macaques (*Macaca fuscata*) using DNA fingerprinting. In *Paternity in Primates: Genetic Tests and Theories*, eds. R. D. Martin, A. F. Dixson and E. J. Wickings. Basel: S. Karger, pp. 131–140.

Itani, J. (1985). The evolution of primate social structures. *Man*, **20**, 593–611.

Itoigawa, N., Tanaka, T., Ukai, N. *et al.* (1992). Demography and reproductive parameters of a free-ranging group of Japanese macaques (*Macaca fuscata*) at Katsuyama. *Primates*, **33**, 49–68.

Koyama, N. (1970). Changes in dominance rank and division of a wild Japanese monkey troop in Arashiyama. *Primates*, **11**, 335–390.

Koyama, N., Takahata, Y., Huffman, M. A., Norikoshi, K. and Suzuki, H. (1992). Reproductive parameters of female Japanese macaques: Thirty years data from the Arashiyama troops, Japan. *Primates*, **33**, 33–47.

Manson, J. H. (1992). Measuring female mate choice in Cayo Santiago rhesus macaques. *Animal Behaviour*, **44**, 405–416.

McMillan, C. A. (1989). Male age, dominance and mating success among rhesus macaques. *American Journal of Physical Anthropology*, **80**, 83–89.

Modolo, L. and Martin, R. D. (2008). Reproductive success in relation to dominance rank in the absence of prime-age males in Barbary macaques. *American Journal of Primatology*, **70**, 26–34.

Nishida, T. (1997). Sexual behavior of adult male chimpanzees of Mahale Mountains National Park, Tanzania. *Primates*, **38**, 379–398.

Norikoshi, K. (1976). Japanese monkeys. In *Hominizeishon Kenkyuukai. Scientific American: Special Volume on Animal Sociology*. Tokyo: Nihon Keizai Shinbunnsha, pp. 52–61. (in Japanese)

Norikoshi, K. and Koyama, N. (1975). Group shifting and social organization among Japanese monkeys. In *Symposium of the 5th Congress of the International Primatological Society*, eds. S. Kondo, M. Kawai, A. Ehara and S. Kawamura. Tokyo: Japan Science Press, pp. 43–61.

Paul, A. (1997). Breeding seasonality affects the association between dominance and reproductive success in non-human male primates. *Folia Primatologica*, **68**, 344–349.

Paul, A., Kuester, J., Timme, A. and Arnemann, J. (1993). The association between rank, mating effort and reproductive success in male barbary macaques (*Macaca sylvanus*). *Primates*, **34**, 491–502.

Saito, C. (1996). Dominance and feeding success in female Japanese macaques, *Macaca fuscata*: effects of food patch size and inter-patch distance. *Animal Behaviour*, **51**, 967–980.

Small, M. F. (1989). Female choice in nonhuman primates. *Yearbook of Physical Anthropology*, **32**, 103–127.

Smuts, B. B. (1985). *Sex and Friendship in Baboons*. New York: Aldine.

Soltis, J., Thomsen, R. and Takenaka, O. (2001). The interaction of male and female reproductive strategies and paternity in wild Japanese macaques, *Macaca fuscata*. *Animal Behaviour*, **62**, 485–494.

Sprague, D. S. (1991). Mating by nontroop males among the Japanese macaques of Yakushima Island. *Folia Primatologica*, **57**, 156–158.

    (1992). Life history and male intertroop mobility among Japanese macaques *(Macaca fuscata)*. *International Journal of Primatology*, **13**, 437–454.

    (1998). Age, dominance rank, natal status, and tenure among male macaques. *American Journal of Physical Anthropology*, **105**, 511–521.

Sprague, D. S., Suzuki, S., Takahashi, H. and Sato, S. (1998). Male life history in natural populations of Japanese macaques: Migration, dominance rank, and troop participation of males in two habitats. *Primates*, **39**, 351–363.

Stephenson, C. R. (1975). Social structure of mating activity in Japanese macaques. In *Symposium of the 5th Congress of the International Primatological Society*, eds. S. Kondo, M. Kawai, A. Ehara and S. Kawamura. Tokyo: Japan Science Press, pp. 65–115.

Swedell, L. (2006). *Strategies of Sex and Survival in Hamadryas Baboons: Through a Female Lens*. Upper Saddle River, NJ: Prentice Hall.

Takahata, Y. (1980). The reproductive biology of a free ranging troop of Japanese monkeys. *Primates*, **21**, 303–329.

    (1982a). The socio-sexual behavior of Japanese monkeys. *Ethology*, **59**, 89–108.

    (1982b). Social relations between adult males and females of Japanese monkeys in the Arashiyama B troop. *Primates*, **23**, 1–23.

Takahata, Y., Suzuki, S., Agetsuma, N. *et al.* (1998). Reproduction of wild Japanese macaque females of Yakushima and Kinkazan Islands: a preliminary report. *Primates*, **39**, 339–349.

Taub, D. M. M. (1980). Female choice and mating strategies among wild barbary macaques *(Macaca sylvanus)*. In *The Macaques: Studies in Ecology, Behavior and Evolution*, ed. D. G. Lindburg. New York: Van Nostrand Reinhold, pp. 287–344.

Wolfe, L. (1984). Female rank and reproductive success among Arashiyama B Japanese macaques *(Macaca fuscata)*. *International Journal of Primatology*, **5**, 133–143.

Wrangham, R. W. (1980). An ecological model of the evolution of female-bonded groups of primates. *Behaviour*, **75**, 262–300.

Zuckerman, S. (1932). The menstrual cycle of the primates, part 6: Further observations on the breeding of primates. *Proceedings of the Zoological Society of London*, **4**, 1059–1075.

7 Correlates between ovarian cycle phase and mating season behaviour in female Japanese macaques (Macaca fuscata)

ANN O'NEILL

Meme-62–80–91 (female) mounted by Oppress-60–65–85 (male) at Arashiyama, October 2010 (photo by J.-B. Leca).

The Monkeys of Stormy Mountain: 60 Years of Primatological Research on the Japanese Macaques of Arashiyama, eds. Jean-Baptiste Leca, Michael A. Huffman and Paul L. Vasey. Published by Cambridge University Press. © Cambridge University Press 2012.

7.1   Introduction

Sexual behaviours are conceptualised as those activities (e.g. mounting, holding or hip-touching) that are directly part of copulation (Beach, 1976). Three distinct characteristics of sexual behaviour have been proposed; attractivity, proceptivity and receptivity (Beach, 1976). Attractivity of a female can be measured by the number of solicitations received from males and is considered to be primarily regulated by oestrogens. The proceptivity of a female can be measured by the number of times a female initiates sexual interaction with a male. Receptivity is defined in terms of female responses necessary and sufficient for the male's success in achieving intravaginal ejaculation.

The sexual behaviour of non-primate mammals is thought to be under relatively strict hormonal control (Beach, 1976; Baum *et al.*, 1977). In contrast, hormones influence reproductive behaviour more indirectly in primates via complex interactions with the social environment. *Socioendocrinology* is the conceptual framework that examines the effects of social environment on the relationship between hormones and behaviour (Ziegler and Bercovitch, 1990).

Attractivity, proceptivity and receptivity have been shown to correspond to levels of circulating hormones. For example, research shows that in many Old World monkey species, attractivity increases during the oestrogen peak of the periovulatory phase of the menstrual cycle (e.g. Dixson, 1998). Additionally, the attractivity of ovariectomised monkeys increased significantly when injected with oestrogen (Dixson, 1998). Proceptive behaviours also tend to be exhibited by females when stimulated by oestrogen during the periovulatory phase (Dixson, 1998). Conversely, increased levels of progesterone during the luteal phase function to decrease attractivity and proceptivity in the female mammals, including macaques (Beach, 1976; Dixson, 1998).

Sexual receptivity appears to be dependent on oestrogenic stimulation, but the degree of dependence varies considerably in different species (Beach, 1976). It has also been suggested that testosterone, primarily of adrenal origin in the female, is the main libidinal hormone responsible for receptivity in primates (Trimble and Herbert, 1968; Dixson *et al.*, 1973; Johnson and Phoenix, 1976). In some non-human primates, such as the chacma baboon, receptivity in the female appears to be limited to the periovulatory period (Saayman, 1970), whereas rhesus macaque females may permit copulation at any stage of the cycle (Johnson and Phoenix, 1978). Consequently, the expression and functional significance of receptive behaviour vary from species to species (Dixson, 1998).

Sexual behaviour in many primate species occurs at higher rates than required to ensure conception and often occurs after the female has already conceived. Post-conceptive sexual behaviour appears to be influenced by fluctuations in progesterone and estrogen levels during pregnancy (Dixson, 1998; Soltis *et al.*, 1999). For example, captive rhesus macaques have been observed to copulate during pregnancy, exhibiting a secondary peak in mating between the 6th and 10th week of gestation, when progesterone levels temporarily drop (Bielert *et al.*, 1976). Wilson *et al.* (1982) reported a drop in progesterone concentrations in those rhesus monkeys who engaged in post-conceptive mating activity, whereas females who did not exhibit post-conceptive mating had comparatively higher concentrations of progesterone. Maestripieri (1999) observed a post-conceptive peak in mating activity in female pig-tailed macaques between weeks 4 and 10. However, a decrease in progesterone levels was not reported until week 16, when post-conceptive mating had ceased.

Same-sex mating behaviours have been observed in numerous primate species (Vasey, 1995) and have been particularly well studied in Japanese macaques (Fedigan and Gouzoules, 1978; Takahata, 1982; Wolfe, 1984; Vasey, 1996). Female–female mating behaviour in Japanese macaques is considered sexual because: (1) such behaviour is confined to the mating season and occurs when the females are showing other signs of proceptivity; (2) contact between the two females is maintained by courtship signals, mutual following and allogrooming, in a manner virtually indistinguishable from that of heterosexual pairs (Vasey *et al.*, 2008a, 2008b); (3) females avoid mounting interactions with close matrilineal female relatives, just as they avoid male relatives (Wolfe, 1984; Chapais and Mignault, 1991; Chapais *et al.*, 1997); and (4) females engage in genital stimulation during most same-sex mounts (Vasey and Duckworth, 2006). Females that engage in such behaviour are not exclusively homosexual and typically engage in heterosexual activity as well. A relationship between proceptivity and homosexual activity in female Japanese macaques has been suggested (Fedigan and Gouzoules, 1978; Wolfe, 1979), though there have been no direct studies of the relationship between hormonal status and same-sex mating published.

Although sexual activity is often the best indicator of the attractive and proceptive state of the female, other non-sexual behaviours, both affliative and aggressive, can also serve to indicate these states. This is because over the course of a mating season female primates demonstrate fluctuations in their patterns of affiliative and aggressive behaviours (Enomoto, 1981; D'Amato *et al.*, 1982). For example, the attractivity of the female can often be ascertained by the frequency with which male troop members approach her. Similarly, grooming among consorts is frequently observed throughout the mating season and the amount of time that a male spends grooming a female may be an indicator of the female's attractiveness. In some species male–female grooming may only occur in the context of mating and may decrease or disappear entirely after the courtship has ended. Grooming has been observed to increase among male and female primates during oestrus (e.g. *Cebus apella*: Linn *et al.*, 1995; *Macaca fuscata*: Enomoto *et al.*, 1979), particularly during the periovulatory phase of the cycle.

The frequency with which females display aggressive behaviours also varies over the course of a mating season. For example, there is extensive evidence that female–female competition for male mates occurs across a range of primate taxa and that such competition can influence a female's reproductive success (see Silk, 1992 for a review). Elevated levels of female–female mate competition and female mate choice during the mating season lead to increased aggression among female conspecifics. The precise relationship between female aggression and reproductive status remains unclear. Whereas most research on male aggressive behaviour has established testosterone as a fundamental component, comparable research on female aggression suggests no such role for testosterone or other androgens (Batty *et al.*, 1986; Martensz *et al.*, 1987; von Engelhardt·*et al.*, 2000; Beehner *et al.*, 2005). Instead, studies on female aggressive behaviour indicate that oestradiol may play a role in facilitating aggression (Michael and Zumpe, 1993; Beehner *et al.*, 2005). There is some evidence that oestrogen increases aggressiveness in female rhesus macaques (Michael and Zumpe, 1970), which supports captive studies suggesting increased aggression by females near mid-cycle.

7.2   Study objectives

Faecal steroid and observational analyses were conducted to assess the relationship between behavioural and endocrine changes during the ovarian cycles of female Japanese macaques in the semi-free ranging Arashiyama West (Texas) group. Observations of female pre-conceptive, post-conceptive and same-sex mating behaviour were conducted. Additionally, the rates of pericopulatory affiliative (approaching, grooming and proximity) and aggressive behaviours were analysed for hormonal variations in relation to the female reproductive cycle.

The primary research questions in this study were: (1) Does the frequency of specific sexual behaviours fluctuate over the course of the cycle in Japanese macaque females? (2) Do Japanese macaque females exhibit cycle-related variations in the frequency and duration of affiliative and aggressive behaviours? (3) How does non-conceptive mating behaviour (i.e. homosexual mating and mating during pregnancy) relate to ovarian hormone levels in female Japanese macaques? In addition, this research sought to assess the reliability of faecal analysis and enzyme immunoassay techniques for evaluating the hormonal status in this species.

7.3   Materials and methods

7.3.1   *Study site and animals*

Observations and sample collection were conducted on the semi-free ranging Arashiyama West group of Japanese macaques in south Texas during the 1997/1998 mating season (September–March). The monkeys reside in a 65-acre enclosure, surrounded by a 2.4 metre electrified fence just 15 km south of the town of Dilley, Texas, USA. The enclosure houses approximately 350 Japanese macaques in an environment of arid brush land, dense spiny shrubs and prickly pear cactus, combined with larger mesquite and acacia trees.

The subjects for this study consisted of eight non-lactating Japanese macaque females. Each female was between 10–11 years of age during the study period, and all were of similar rank. Seven females were multiparous, and one was primiparous. None of the eight females had offspring in the prior birth season (April–July, 1997).

7.3.2   *Data and faecal sample collection*

Faecal samples were collected 2–3 times per week from the eight target females between 8:00 a.m. and 8:00 p.m. Faecal collection was carried out during focal animal follows, as well as opportunistically. Samples were collected within 10 minutes of defecation, placed in 30-ml sterile plastic vials and frozen for preservation. Frozen samples were transported to the Wisconsin Regional Primate Research Center (WRPRC) in a plastic cooler containing dry ice, and stored in a freezer until analysis. Behavioural data were collected using focal animal sampling (Altmann, 1974). A total of 352 hours of focal animal data were collected (ranging from 41–48 hours per subject).

7.3.3   *Sample extraction and analysis*

Extraction of ovarian steroids was performed according to the procedure described by Strier and Ziegler (1997). Aliquots of 0.1 g wet faeces were extracted with 5 ml of distilled water and ethanol (50:50). The level of conjugation of the steroids was examined. It was determined that oestrone-glucuronide (E1) and pregnanediol-glucuronide (PdG) steroids were excreted in freeform. Therefore, ether extraction was used to prepare samples for EIA (enzyme immunoassay). High-pressure liquid chromatography (HPLC) was used to help identify which oestrogens and progestins were excreted in higher amounts in Japanese macaque faeces. The technique was described in Strier *et al.* (1999). The sample was separated on HPLC and collected in fractions to be run in the EIA. Ninety per cent of the cross-reactivity for oestrogen was found at the retention time of 22.6 min, where E1 elutes. Fractions from HPLC showed that PdG contributed to the majority of activity in the assay.

Tritiated steroids of known concentrations were added to a few representative faecal samples before extractions to determine procedural losses for extraction techniques. Tritiated 20 alpha hydroxyprogesterone (for pregnanediol) and tritiated oestrone (18 000 cpm) were added to 0.1 g samples of faeces. Pooled faecal samples were used to validate the PdG and E1 assays. PdG and E1 concentrations were determined by EIA methods, running assays with volumes from 5–100 μl from ether-extracted pools. The accuracy with which the assay measures steroids in each sample was assessed by adding small volumes of the faecal pool to the standard curve points. Parallelism was determined through serial dilutions of the faecal pool along the standard curve. In addition to running a standard curve, faecal pool samples were run in duplicate on each assay in order to establish a mean intra- and inter-assay coefficient of variation (CV) for E1 and PdG. Mean steroid recoveries were high (PdG = 90.00%; E1 = 91.80%). Accuracy assessment was also found to be within acceptable parameters (above 90%). A mean per cent accuracy was determined for PdG (102.69 ± 3.38%) and E1 (91.76 ± 1.44%) assays. Pooled samples were parallel to the standard curve for both PdG and E1 (PdG: $T = -0.50$, df = 28, $p < 0.05$; E1: $T = -1.70$, df = 28, $p < 0.05$). In addition, mean intra- and inter-assay coefficients of variation (CV) values were within acceptable limits (no difference in slope; PdG: intra = 3.92%, inter = 11.41%; E1: intra= 7.62%, inter = 21.40%). Oestrone was measured in faeces, employing an E1 EIA previously reported (Ziegler *et al.*, 1997). Pregnanediol was measured in faeces by a PdG EIA also described previously (Carlson *et al.*, 1996).

7.3.4   Interpretation of hormonal data

Due to the delayed excretion of E1 (approximately 24 hours in macaques; Dixson, 1998; Fujita *et al.*, 2001) and the infrequency of faecal collection, the onset of PdG increase was considered to be the best estimate of the day of ovulation (Strier and Ziegler, 1997). Progesterone synthesis from the ovary

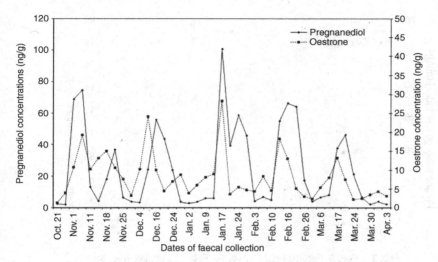

Figure 7.1. Representative hormone profile for one female Japanese macaque. Base levels of pregnanediol indicate the follicular phase of the cycle, the day of the Oestrone surge (or the first day pregnanediol levels begin to rise) ±3 days represents the periovulatory phase and a sustained increase in pregnanediol represents the luteal phase.

actually begins to increase prior to ovulation; within hours of the gonadotropin surge (Espey and Lipner, 1994). Consequently, the sample preceding the first day of the PdG increase was considered to represent the most likely day of ovulation. The periovulatory phase was therefore conservatively defined as the estimated day of ovulation ± 3 days (as per Strier and Ziegler, 1997).

The earliest possible day of conception was considered to be the estimated day of ovulation if steroid levels failed to return to their lowest baseline levels. The length of gestation was calculated as the interval between estimated ovulation during the cycle in which conception occurred and the date of parturition. Cycle lengths were calculated as the intervals between successive PdG surges. There were 1–5 cycle lengths calculated for each female (Figure 7.1). Figure 7.2 represents one complete ovarian cycle.

7.3.5 Behavioural analysis

The results of the hormone analysis were used to create hormone profiles for each female. Hormone profiles were used to calculate ovulatory cycle and gestation lengths. The cycles were divided into the three phases: (1) follicular, when pregnanediol (PdG) has dropped to baseline levels; (2) luteal, when PdG levels stay well above baseline for an extended duration and oestrone levels

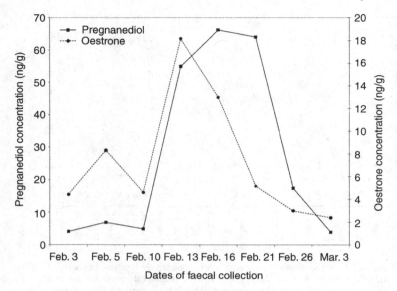

Figure 7.2. One representative ovarian cycle for a Japanese macaque female.

return to baseline; and (3) periovulatory, which was conservatively estimated as the day of ovulation ±3 days. In all, 26 complete cycles from eight females were identified and used in behavioural analyses for this study.

Analyses were carried out on the behavioural variables listed in Table 7.1. Those activities indicating the female's attractivity included sexual and non-sexual behaviours affiliated with consortships *received* by the target female from either male or female conspecifics, such as mounts, holds, approaches, grooms, as well as 'other' sexual behaviours (e.g. lip-smacking). Proceptive activities included both sexual and non-sexual behaviours *directed* by the target female toward males or females, such as mounts, holds, oestrus calls, approaches, grooms, as well as 'other' sexual behaviours. Sexual and non-sexual behaviours were differentiated by intent; approaches and grooming activity that were part of a consortship, or led to sexual activity, were considered sexual in nature, while behaviours that were not affiliated with a consort were considered non-sexual. Data were not normally distributed and as a result non-parametric statistics were used for this analysis. The mean scores obtained for each cycle phase were compared for cycle phase variation using Friedman analysis of variance by ranks (Siegel and Castellan, 1988). Behavioural variables showing statistical significance were further analysed with Wilcoxon signed ranked tests (Siegel and Castellan, 1988) to determine where the variation existed.

Spatial proximity between the female subject and male troop members was analysed using three categories (contact, less than one metre, between one

Table 7.1. *Behavioural variables used for analyses*

Behaviour*	Description
Sexual behaviours	
Mount	One individual climbs either ventrodorsally or ventoventrally upon a standing partner, with or without pelvic thrusts. Partners may be male to female, female to male or female to female
Hold	One individual sits in body contact with a partner, with arms around waist area. Usually observed in conjunction with a mount series
Other	Other sexual behaviours include presenting, lip smacking, or hip touching and can be directed or received
Non-sexual behaviours†	
Groom	One individual inspects and cleans the fur of another. May be observed in conjunction with sexual consorts
Approach	One individual advances toward another to within one metre
Proximity	Two individuals maintain proximity to each other. Three criteria for proximity were used in the analysis; contact, less than one metre and between one and four metres
Chase	One individual pursues another with accompanying aggressive signals
Threat	Visual or vocal signal consisting of a stare, raised brows or growl
Lunge	One individual feints a sudden charge toward another, stopping abruptly before reaching the other
Supplant	One individual moves toward another who immediately moves out of the former's way. The displacer frequently sits or stands in the exact spot vacated by the displaced monkey
Grab/Bite	One individual either pinches, grabs or bites another

* Behaviours may be directed or received by female subjects. Sexual behaviours received by the female represent attractivity of the female and sexual behaviours directed represent proceptivity.
† Non-sexual behaviours may also be distinguished by the attractive and proceptive state of the female.

and four metres). A mean frequency per hour score for each of the behavioural variables was calculated from each focal animal's interaction with male troop members. Data were analysed to determine whether females were more likely to be found in proximity to males and females during certain phases of the ovarian cycle. This was accomplished by tallying the total time the target female was in proximity to a male or a female, respectively, during focal samples and calculating a mean percentage of time for each female for each phase of the cycle. Friedman and Wilcoxon tests were performed on spatial proximity data in order to detect variability between the phases.

The frequency of non-sexual approaches, those not associated with sexual objectives, *directed* toward males and females by the target female and *received* by the target female from male troop members was analysed. Similarly, agonistic signals both *directed* toward others and *received* by target females from others were analysed. In addition, the non-sexual grooming behaviour of the target females in relation to male troop members was analysed for cyclic patterns. This was achieved by calculating the mean percentage of time each female spent *directing* or *receiving* non-sexual grooming from male troop members and separating the data into the three phases of the cycle. Friedman and Wilcoxon tests were performed on approach, agonistic and grooming data in order to detect variability between the phases.

Level of significance was set at $p < 0.05$ and all tests were two-tailed. All statistical analyses were performed using an SPSS 8.0 statistical software package.

7.4   Results

7.4.1   Hormone profiles

The profiles of immunoreactive E1 and PdG showed clear cyclic patterns in which the follicular and luteal components of the cycles could be clearly distinguished. One to six ovarian cycles were calculated for each of the target females for a total of 26 complete cycles (1–6 cycles were calculated for each female; mean = 3.4) (Table 7.2). The mean ovarian cycle length (± s.d.) was 27.6 ± 4.2 days (range 19–35 days). Cycle lengths did not significantly differ among the eight females ($\chi^2 = 5.31$, df = 7, $p = 0.623$). The mean follicular and luteal cycle lengths (± s.d.) were 8.4 ± 3.4 and 12.3 ± 3.8 days (± s.d.), respectively and did not differ significantly among females (follicular: $\chi^2 = 5.694$, df = 7, $p = 0.576$; luteal: $\chi^2 = 6.092$, df = 7, $p = 0.529$).

7.4.2   Sexual behaviour

Target females were observed to engage in sexual behaviour during all three phases of the ovarian cycle (with the exception of one female who was not observed participating in sexual activity but did conceive during this mating season). Mounts *received* by females varied throughout the ovarian cycle, with a significant increase during the follicular and periovulatory phases compared with the luteal phase of the cycle. Although there was no significant difference in the rate of mounts *received* by females in the follicular phase compared to the periovulatory ($p = 0.53$; Table 7.3), the mean frequency of mounts per hour showed a 56% increase in the periovulatory compared to the follicular phase.

Holding behaviour in relation to consorts varied significantly throughout the cycle, with considerably more holds *received* by males during the periovulatory and follicular phases compared to the luteal (Table 7.3). Sexual solicitations (e.g. lip-smacking) *received* by females from males occurred very infrequently. As a consequence, these behaviours were not found to show significant differences

Table 7.2. *Ovarian cycle and gestation lengths for Japanese macaque females*

Female	Cycle 1	Cycle 2	Cycle 3	Cycle 4	Cycle 5	Cycle 6	Gestation
Trisha	30	-	-	-	-	-	192
Saskia	35	22	25	-	-	-	170
Zoe	29	27	-	-	-	-	176
Tyler	33	-	-	-	-	-	163
Shy Mug	22	19	28	34	28	27	-
Tantalis	32	26	35	28	-	-	-
Lisa*	27	-	29	29	21	-	-
Bo Nose	23	24	27	31	28	28	-

Full names of individuals are Wania–65–80–87 (Trisha), Petitemon–64–72–87 (Saskia), Matsu–58–63–69–77–82 (Zoe), Pelka–60–78–86 (Tyler), Wania–65–87 (Shy Mug), Matsu–58–68–81–86 (Tantalis), Matsu–64–74–87 (Lisa) and Nose–62–79–87 (Bo Nose).

* Sample collection for Lisa was very inconsistent for cycle 2 and was consequently removed from analysis.

among the three phases of the cycle. Nevertheless, the few times these behaviours were displayed occurred during the follicular and periovulatory phases.

Proceptive behaviours (i.e. approaches, grooming and sexual solicitations *directed* toward males and females) also varied significantly throughout the three phases in the cycle. Females displayed increased sexual activity toward males and females during the follicular and periovulatory phases compared to the luteal. Although there was a slight increase in sexual behaviour *directed* by focal females toward others during the periovulatory compared with the follicular phase, the difference was not significant.

7.4.3   *Post-conceptive sexual behaviour*

During the study period, four of the eight females conceived after one to four cycles. The mean (± s.d.) gestation length for the four females that carried to term was calculated to be 175 ± 12.8 days (median = 173 days).

Mean PdG levels during early pregnancy were 50.01 ng/g in the first week of gestation, increasing to 100.70 ng/g by the 5th week, followed by a temporary drop in weeks 6 through 8 (averaging 56.03 ng/g), and then steadily increasing to 131.18 ng/g by the 18th week. E1 levels were highly variable for the initial weeks of gestation, but then showed a steady increase by the 14th week. A positive correlation was found between PdG and E1 levels (Spearman: $r_s = 0.51$, $p < 0.05$), showing a parallel rise in both hormones as the pregnancy nears term.

As previously reported (O'Neill *et al.*, 2006), the majority of post-conceptive mating occurred during weeks 6–11 of gestation. The rate of post-conceptive

Table 7.3. *Summary statistics for behavioural variables*

Sexual behaviours	Differences across three cycle phases	Follicular vs periovulatory	Follicular and periovulatory vs luteal
	Friedman analyses	Wilcoxon analyses	Wilcoxon analyses
Behaviours indicating attractivity:			
Mount received	$\chi^2 = 8.00$, df = 2, $p = 0.02*$	$Z = -0.63$, df = 2, $p = 0.53$	$Z = -2.20$, $p = 0.03*$
Hold received	$\chi^2 = 8.27$, df = 2, $p = 0.02*$	$Z = -1.78$, df = 2, $p = 0.08$	$Z = -2.80$, $p = 0.005*$
Other received	$\chi^2 = 2.00$, df = 2, $p = 0.37$	N/A	N/A
Behaviours indicating proceptivity:			
Mount directed	$\chi^2 = 6.71$, df = 2, $p = 0.03*$	$Z = -0.11$, df = 2, $p = 0.91$	$Z = -2.02$, $p = 0.04*$
Hold directed	$\chi^2 = 7.18$, df = 2, $p = 0.02*$	$Z = -1.36$, df = 2, $p = 0.17$	$Z = -2.80$, $p = 0.005*$
Other directed	$\chi^2 = 6.33$, df = 2, $p = 0.04*$	$Z = -0.83$, df = 2, $p = 0.41$	$Z = -1.89$, $p = 0.06$
Oestrus call	$\chi^2 = 9.33$, df = 2, $p = 0.01*$	$Z = -0.82$, df = 2, $p = 0.41$	$Z = -2.27$, $p = 0.02*$
Non-sexual behaviours	Differences across 3 cycle phases	Follicular vs periovulatory	Follicular and periovulatory vs luteal
Approach received	$\chi^2 = 1.23$, df = 2, $p = 0.54$	N/A	N/A
Approach directed	$\chi^2 = 6.07$, df = 2, $p = 0.05*$	$Z = -0.07$, $p = 0.94$	$Z = -2.08$, $p = 0.04*$
Groom received	$\chi^2 = 11.08$, df = 2, $p = 0.001*$	$Z = -1.69$, $p = 0.09$	$Z = -2.41$, $p = 0.02*$
Groom directed	$\chi^2 = 8.86$, df = 2, $p = 0.01*$	$Z = -0.85$, $p = 0.40$	$Z = -2.43$, $p = 0.02*$
Proximity (contact)	$\chi^2 = 10.90$, df = 2, $p = 0.001*$	$Z = -0.30$, $p = 0.76$	$Z = -2.43$, $p = 0.02*$
Proximity (<1 m)	$\chi^2 = 10.40$, df = 2, $p = 0.01*$	$Z = -0.10$, $p = 0.27$	$Z = -2.25$, $p = 0.02*$
Proximity (1–4 m)	$\chi^2 = 6.65$, df = 2, $p = 0.04*$	$Z = -0.65$, $p = 0.52$	$Z = -2.20$, $p = 0.03*$

$* = $ significant difference at $p < 0.05$.

mounts was similar for both heterosexual and same-sex pairs (mounts *received* by female target: 13.17/hr for same-sex and 16.86/hr for heterosexual; mounts performed by female target: 5.13/hour for same-sex and 5.67/hr for heterosexual). It was found that weeks with higher levels of mounting were significantly associated with decreases in PdG levels relative to the previous week (Spearman: $r_s = 0.59$, $p < 0.05$). E1 profiles were too variable to make definitive statements concerning their causal influences on post-conception sexual behaviour.

7.4.4   Same-sex sexual behaviour

Five of the eight subject females engaged in same-sex sexual behaviour during the mating season. Of these five, three were involved in same-sex activity during cycling (n = 16 cycles) and the other two exhibited such behaviour during pregnancy. The following comparison was focused on the three individuals that engaged in same-sex mating during cycling. As with heterosexual mounting behaviour, same-sex mounting was not distributed across the three cycle phases proportionately to time observed; again, mounts during the luteal phase occurred less frequently than expected (mounts *directed* to other females: $\chi^2 = 109.22$, df = 2, $p < 0.001$; mounts *received* from other females: $\chi^2 = 40.26$, df = 2, $p < 0.001$). It was observed that same-sex mounting behaviour showed the same general pattern as heterosexual mounting and occurred somewhat during the follicular phase, increased modestly in the periovulatory phase and dropped to zero occurrences in the luteal phase.

The comparative rates of heterosexual vs. same-sex mounting were examined within each phase of the cycle. During the follicular phase, the three target females *directed* a mean of 4.42 and 7.92 mounts/hr to males and females, respectively, and they *received* 1.54 and 1.92 mounts/hr from males and other females, respectively. During the periovulatory phase, the target females *directed* 5.45 and 8.86 mounts/hr to males and other females, respectively, and they *received* 8.18 and 4.65 mounts/hr from males and other females, respectively. Thus, during the two phases of the cycle in which mounting occurred, the rates of same-sex mounts were very similar to the rate of heterosexual mounts. Although not statistically significant, the only time the rate of heterosexual mounting was higher than that of same-sex mounting was during the periovulatory phase, when females *received* more mounts from males than from other females.

Two of the eight subjects engaged in same-sex mounting after they became pregnant, rather than during cycling. The rate of mounts *directed* by pregnant females was similar for same-sex and heterosexual mating pairs (5.13/hr and 5.67/hr, respectively) and the rate of mounts *received* by pregnant females was also similar for same-sex and heterosexual mating pairs (13.17/hr and 16.86/hr).

Same-sex mounting, like heterosexual mounting, took place between the 6th and 10th weeks of gestation and coincided with a marked decrease in PdG levels at that time (Gouzoules and Goy, 1983; O'Neill, 2000).

7.4.5   Non-sexual behaviour

The proportion of time in which females spent in close proximity (less than one metre) to males and females varied considerably throughout the cycle (female/male $\chi^2 = 10.90$, $p = 0.001$; female/female $\chi^2 = 6.26$, $p = 0.04$; Table 7.3). Females were observed to spend significantly more time in close proximity to males during the follicular and periovulatory compared with the luteal phase ($Z = -2.43$, $p = 0.02$), while they spent significantly more time in proximity to other females (primarily kin) during the luteal phase of the cycle ($Z = -2.38$, $p = 0.02$). Females were also observed to approach males significantly more during the follicular and periovulatory periods compared with the luteal ($Z = -2.08$, $p = 0.04$), while they approached other females (primarily kin) significantly more during the luteal phase of the cycle (luteal versus follicular: $Z = -1.96$, $p = 0.05$; luteal versus periovulatory: $Z = -2.52$, $p = 0.01$. Although there was no statistical significance in frequency of approaches *received* by males ($\chi^2 = 1.23$, $p = 0.54$), trends showed a 71% increase in mean frequency of approaches *received* during the follicular compared with the luteal and a 69% increase in the periovulatory compared with the luteal. Females *received* significantly more approaches by other females during the luteal phase of the cycle compared with the follicular phase ($Z = -2.52$, $p = 0.01$), but no significant differences were found between the luteal and periovulatory phases ($Z = -1.54$, $p = 0.12$).

The proportion of time which females spent being groomed by male troop members varied significantly across the three cycle phases ($\chi^2 = 11.08$, df = 2, $p = 0.001$). Specifically, females received grooming from males substantially more during the follicular and periovulatory phases compared with the luteal phase of the cycle ($Z = -2.41$, $p = 0.02$; Table 7.3). Grooming *directed* by females towards males also showed significant variation across the three phases ($\chi^2 = 8.86$, df = 2, $p = 0.01$), with substantial increases once again in the follicular and periovulatory phases compared with the luteal phase ($Z = -2.43$, $p = 0.02$).

7.4.6   Aggressive behaviour

Inter-sexual aggression

Since the target females for this study ranked in the lower part of the hierarchy, occurrences of aggression *directed* toward male conspecifics were virtually non-existent. Females were not observed to supplant males during focal periods and were rarely observed *directing* other agonistic signals toward males. Consequently, *directed* aggression by target females toward male conspecifics was not found to significantly differ across the three phases of the cycle (chases: $\chi^2 = 5.43$, df = 7, $p = 0.07$; threats: $\chi^2 = 2.30$, df = 7, $p = 0.32$; grab/bite: $\chi^2 = 5.20$, df = 7, $p = 0.07$; lunge: $\chi^2 = 2.36$, df = 7, $p = 0.31$).

The predominant form of aggression between male and female Japanese macaques involved males chasing females. Approximately 15% of chases included contact, whereby the male would grab the female and proceed to bite and pinch her. Males chased females frequently throughout the mating season and the occurrence of these chases varied significantly among the three phases of the cycle ($\chi^2 = 9.75$, df = 7, $p = 0.01$). Females were chased by male troop members significantly more often during the follicular and periovulatory phases compared with the luteal phase of the cycle ($Z = -2.59$, df = 7, $p = 0.01$). The frequency of chases, threats, grabs, bites and lunges *received* during the follicular and periovulatory phases did not significantly differ ($Z = -0.91$, df = 7, $p = 0.37$; $\chi^2 = 1.20$, df = 7, $p = 0.55$; $\chi^2 = 1.08$, df = 7, $p = 0.58$; $\chi^2 = 5.20$, df = 7, $p = 0.07$ respectively). However, mean values for chases *received* increased 35% from the follicular to the periovulatory phase.

Intra-sexual aggression

Aggression between female Japanese macaques often involves subtle facial threats, or supplants, but during the mating season aggression between females may escalate into chases and occasional contact fights. The frequency of agonistic and submissive signals such as grabbing, biting, lunging and fear grimaces occurred too infrequently between target females and other female conspecifics for statistical analysis. Therefore, statistical analyses were limited to chases, threats, supplants and avoids.

Chases between female Japanese macaques occurred frequently throughout the mating season and rarely resulted in physical contact. The frequency of chases *directed* by target females toward other females differed significantly among the three phases of the cycle ($\chi^2 = 6.26$, df = 7, $p = 0.04$). Target females were observed to chase other females more often during the periovulatory phase of the cycle (periovulatory versus luteal: $Z = -2.10$, df = 7, $p = 0.04$; periovulatory versus follicular: $Z = -2.03$, df = 7, $p = 0.04$; Table 7.3).

The frequency of chases *received* by target females varied significantly across the three phases of the female's cycle ($\chi^2 = 12.25$, df = 7, $p < 0.01$). In contrast to *directed* chases, females were observed to be chased significantly more often by other females during the luteal phase of the cycle (luteal versus periovulatory: $Z = -2.25$, df = 7, $p = 0.01$; luteal versus follicular: $Z = -2.25$, df = 7, $p = 0.01$).

The frequency of threats *directed* also differed significantly among the three phases of the cycle (χ^2 = 5.85, df = 7, p = 0.05). The target females were observed to threaten (both facially and vocally) other females significantly more during the follicular phase of the cycle (Z = −2.46, df = 7, p = 0.01; Table 7.3). There was a trend toward increased threats *received* by target females during the follicular and periovulatory phases compared with the luteal, however this trend did not reach statistical significance (χ^2 = 5.25, df = 7, p = 0.07).

Due to the low rank of the target females, the occurrence of *directed* supplants toward other females was infrequent and consequently did not significantly differ among the three phases (χ^2 = 1.73, df = 7, p = 0.42). Similarly, females were not supplanted significantly more by female conspecifics during any of the three phases (χ^2 = 1.93, df = 7, p = 0.38).

7.5   Discussion

7.5.1   *Hormone profiles*

Results of this study indicate these Arashiyama West Japanese macaque females experience cycle and gestation lengths similar to those previously reported for this species (Nigi, 1976; Fujita *et al.*, 2001). The employment of a simple extraction technique, coupled with established enzyme immunoassay methods, proved very successful for measuring faecal steroids in Japanese macaques. This method yielded high total steroid recoveries (> 90%). The sustained increase in estrogens, which typically occurs in other macaques in a 1–2-day period at time of ovulation (Dixson, 1998; Fujita *et al.*, 2001), was not always present in the sample collection. Therefore, given the irregular rate of faecal sample collection (i.e. not daily), PdG levels proved the most reliable means of determining cyclic patterns in the ovarian cycle. Average cycle lengths were calculated at 27.6 ± 4.2 days, which is comparable to those reported in an earlier study (Nigi, 1975; 26.3 ± 5.4 days) that employed vaginal swabs and sexual swellings. Although not statistically significant, ovarian cycle lengths varied both within and between female subjects.

Gestation lengths varied considerably among the four females who conceived. Mean gestation length for females in this study, as determined by hormonal profiles, was 175.25 ± 12.8 days (± s.d.; median = 173 days; O'Neill *et al.*, 2006), which was also comparable to that reported in an earlier study (173 ± 6.9 days; Nigi, 1976) that employed vaginal swabs and sexual swellings. The mean gestation length found in this study was also comparable to previously reported hormonal estimates (170–189 days; Fujita *et al.*, 2001). PdG and E1 levels showed fluctuations in the initial 100 days of gestation.

PdG concentrations rose steadily from the time of conception to approximately the 5th week of gestation, followed by a drop in concentration and low levels between weeks 6 and 14, after which PdG levels continued to increase steadily toward term. Although E1 levels followed a similar pattern as that of PdG, the changes in concentration were not as marked during the temporary decline in the 6th to 12th week. This pattern of hormone fluctuations throughout pregnancy is similar to findings on other macaque species (*Macaca mulatta*: Wilson *et al.*, 1982; *Macaca fascicularis*: Stabenfelt and Hendrikx, 1973; *Macaca radiata*: Stabenfelt and Hendrikx, 1972).

7.5.2   Sexual behaviour

It has been previously reported that attractivity in primates is regulated by circulating oestrogen levels (Johnson and Phoenix, 1976; Enomoto *et al.*, 1979; Michael and Zumpe, 1993). In Japanese macaques, oestrogens begin to increase late in the follicular phase and peak toward mid-cycle (Enomoto *et al.*, 1979). Consistent with past research, this study demonstrated increased oestrone levels toward the periovulatory phase of the cycle. Females exhibited a significant increase in frequency of mounting and holding *received* from males in the follicular and periovulatory phases, followed by an absence of this behaviour in the luteal phase of the cycle. This pattern indicates increased attractivity in the follicular and periovulatory phases of the cycle.

The role of progesterone in the expression of sexual behaviour is not entirely clear. This study found that pregnanediol (PdG) increases during the luteal phase, while maintaining low levels in the follicular and periovulatory phases of the cycle. Earlier findings with rhesus macaques showed that progesterone had an inhibitory effect on attractivity (Baum *et al.*, 1977; Michael and Zumpe, 1993). In this study, mounts were primarily restricted to the follicular and periovulatory phases, when PdG concentrations are low, followed by a significant decrease in the behaviour during the luteal phase, when PdG levels reach their peak. Holding behaviour generally occurred during a mount series. Consequently the rate of holding behaviour followed a similar pattern to that observed in 'mounts *received*' by the female. 'Other' sexual behaviours, such as hip touch, lip-smacking and presenting, were rarely observed in target females. Thus, no significant patterns were detected in the rate of these sexual behaviours *received* by target females.

Although proceptivity in primates has not been linked to any single hormone (Baum, 1983), it is suggested that females are not only more attractive during the periovulatory phase, when oestrogen levels are increased, but that they also show increases in proceptive behaviour during this phase (Beach, 1976;

Wallen *et al.*, 1984; Chambers and Phoenix, 1987; Wallen and Tannenbaum, 1997). This study revealed significant increases in proceptive behaviours during the follicular and periovulatory phases of the cycle. Females demonstrated an increased rate of mounting behaviour toward males in the follicular and periovulatory phases, during the decrease in PdG and towards the E1 peak. Females also *directed* holds and 'other' sexual behaviours toward males during the follicular and periovulatory phases and showed a complete absence of these behaviours during the luteal phase.

7.5.3   Post-conceptive sexual behaviour

In addition to cyclic patterns in reproductive sexual behaviour, females also showed indications of hormonal influence on non-conceptive sexual behaviour during this breeding season. A relationship was found between dropping PdG levels during gestation and the rate of non-conceptive mounting. Decreases in PdG levels at certain times during gestation may allow for the expression of proceptivity in females, similar to the effects of decreased PdG levels during the follicular and periovulatory phases of the cycle (O'Neill *et al.*, 2004).

The majority of post-conceptive mating occurred during weeks 6–11 of gestation, when PdG levels are relatively low. Similar results have been reported previously for rhesus monkeys (Bielert *et al.*, 1976) and Hanuman langurs (Sommer, 1993). It is worth noting that, while most heterosexual mating activity occurred during post-conception weeks 6 and 7, same-sex mating took place later in post-conception, during weeks 8 to 11. The female's pregnant state and closer proximity to females at this time may influence her attractivity to males, causing the female to gravitate less toward non-conceptive heterosexual behaviour and more towards homosexual behaviour to satisfy any sexual impulses. The ovaries are essential for the production of progestins and oestrogens during the first few weeks of gestation, after which the placenta assumes the production of these hormones. The findings of this study which indicate an increase in PdG and E1 between the 3rd and 5th week of gestation may represent a temporary dual production of these hormones by both the ovaries and the placenta. The subsequent drop in hormones between weeks 6 and 11 may represent a loss of the ovaries' contribution, while the placenta attempts to take its place in the production of progestins and oestrogens. Both PdG and E1 levels then begin to recover and continue to rise (O'Neill *et al.*, 2006).

In terms of a proximate explanation for the widely reported pattern of post-conceptive mating in Japanese macaques, results indicate that the temporary drop in PdG levels during the 6th and 11th weeks of gestation may facilitate a 'false oestrus' or period of proceptivity and attractivity in the pregnant female.

Furthermore, from an ultimate perspective, if post-conceptive mating does encourage males to behave more tolerantly toward pregnant females and their subsequent infants, then a hormonally induced period of proceptivity/attractivity during gestation may also be selectively advantageous. More data are required to determine whether a peak in mating activity between weeks 6 and 11 is prevalent in Japanese macaques and other Old World monkey species.

7.5.4   Same-sex sexual behaviour

More than half of the females in this sample engaged in same-sex mounting behaviour either before or after conception. The females that engaged in same-sex mounting during cycling exhibited it in the follicular and periovulatory phases, with a complete absence of such activity during the luteal phase. In this regard, the cyclical pattern of same-sex mounts mirrored heterosexual mounts. Similarly, rates of same-sex mounts were not perceptibly different from rates of heterosexual mounting. This pattern supports earlier suggestions that homosexual activity is linked to periods of sexual proceptivity (Fedigan and Gouzoules, 1978; Wolfe, 1984). Furthermore, the same-sex mating observed in this study included most of the behavioural elements that are typical of heterosexual mating in Japanese macaques (e.g. courtship signals, series mounts, pelvic thrusts, formulation of consort bonds, etc.). Taken together, these findings suggest a link between hormonal patterns and sexual proceptivity, regardless of partner preferences.

Several reasons have been postulated to explain why females choose same-sex partners: because there is a higher ratio of preferred female to preferred male partners (Vasey and Gauthier, 2000), for novelty (Wolfe, 1984), for sexual gratification (Vasey and Duckworth, 2006), to minimise interruptions from unwelcome male soliciting (O'Neill, unpublished data), or for all of these reasons. By demonstrating a relationship between same-sex mounting and endocrine condition, as well as similar cyclic patterns in homosexual and heterosexual mounting, this study supports the argument that same-sex mounting in Japanese macaque females is primarily sexual in motivation. The occurrence of such behaviour in other non-human primates remains insufficiently documented and inadequately understood. Further research is needed to better understand the hormonal correlates of same-sex mating, as well as the underlying factors that influence partner choice.

7.5.5   Non-sexual behaviour

A female's attractivity can often be ascertained by the frequency with which males approach her. Additionally, grooming among consorts is frequently observed throughout the mating season and the amount of time a male spends grooming a female may be an indicator of the female's attractivity.

The frequency with which males approached females did not vary significantly throughout the cycle. Although grooming patterns were not always consistent and not all females *received* grooming from male partners during consortships, the frequency with which females were groomed by males increased significantly during the follicular and periovulatory phases compared with the luteal phase. In this study, females approached males more often in the follicular and periovulatory phases than in the luteal phase, suggesting increased proceptivity when PdG levels are low and E1 levels are increasing. In addition, females groomed males significantly more in the follicular and periovulatory phases, with a notable, but not significant, increase in the periovulatory phase.

A further analysis was conducted to examine spatial proximity between males and females during the three phases of the cycle. Although the maintenance of spatial proximity cannot be attributed to either individual, it does require the cooperation of both individuals to sustain proximity. The results of this analysis clearly demonstrate that females spend significantly more time in proximity (contact, < 1 m and 1–4 m) with adult males during the follicular and periovulatory phases, when E1 levels peak.

7.5.6   Aggressive behaviour

The aggressive behaviours of the target females showed cyclic changes throughout the cycle and mating season. The target females for this study all ranked in the lower part of the hierarchy and, as such, they rarely *directed* aggression toward male troop members. General trends in the data show an increase in aggression by females toward male conspecifics in the follicular and periovulatory phases of the cycle, coinciding with increased proximity to males at this time. These differences were not statistically significant, however.

The most prevalent form of aggression *received* by the females from male troop members was chases. Females became a target for male aggression during the follicular and, most notably, the periovulatory phase of the cycle. Of the six more serious wounds which target females incurred during the study, four of them occurred during the late follicular phases, and the other two during the periovulatory phase. During the late follicular and periovulatory phases of the cycle females begin to increase their proximity to males, which in effect increases their vulnerability to attacks from male troop members. Moreover, if females are not receptive to male advances during these peak periods of female attractivity, then the male may use aggression to sexually coerce the female into compliance (Enomoto, 1981; Soltis, 1999). There is no evidence to support a direct influence of the female's hormonal levels on aggression. Rather, the increased proximity to males during the copulatory period increases the occurrence of aggression between male and female macaques.

Females were observed to chase other females more often during the periovulatory phase of the cycle. Although, not analysed statistically, anecdotal observations suggest that females *direct* aggression toward other females more often than when they were in a heterosexual consortship. The increased level of aggression during the periovulatory phase and within the context of heterosexual consortships may indicate active competition between cycling females for proximity to male mates and may function to deter other females from mating partners (Smuts, 1987).

In contrast to chases *directed* at other females, females *receive* chases from other females more often in the luteal phase of the cycle. This suggests consorting partners may protect their female partners from direct agonistic encounters with other females. Close qualitative observations of the females in this study suggested a shift in the female's rank when affiliated with higher-ranking males, eliciting a more aggressive disposition by the female. Similarly, Vasey (1996) reported that, during homosexual consortships, subordinate female partners often increase in rank temporarily owing to the support they obtain from their more dominant female consort partner and this, in turn, results in the subordinate consort partner *directing* more aggression to third-party individuals, while *receiving* less aggression from such individuals. The increase in chases *received* by females in the luteal phase may also reflect an increase in social interactions with female conspecifics during this period. Therefore, the increased aggression at this time may simply be a function of increased exposure opportunities for conflict that are associated with social activity.

Females threatened (i.e. facial, vocal threats) other female troop members more often during the follicular phase of the cycle. Although females *receive* fewer overt aggressive signals (i.e. chases) from other females during the follicular and periovulatory phases, there was a trend (although not statistically significant) indicating that they *received* more visual threats from other females.

7.6   Conclusions

Results of this study are consistent with previous research on primate sexual behaviour, which has shown that rates of copulatory and pericopulatory behaviour are consistently higher during the follicular and periovulatory phases and lower during the luteal phase of the ovarian cycle (Dixson, 1998). Likewise, results of this study are consistent with previous research showing that rates of aggressive and affiliative behaviour show distinct fluctuations across the ovulatory cycle.

Taken together, data from this and other studies demonstrate that it is clearly the luteal phase that is behaviourally distinct from the other two phases of the cycle. If differences in the expression of sexual behaviour exist between the follicular and periovulatory phases, they were not detected during this study. Some increases in sexual behaviours during the periovulatory phase as compared with the follicular phase were found, but these differences did not reach statistical significance.

Much of the sexual activity that was exhibited during the follicular phase of the cycle occurred near the end of that phase and close to the onset of the periovulatory phase, suggesting that the boundary of the two phases is not behaviourally discrete. Functionally, it makes sense on an adaptive level to attract males just prior to, as well as during, ovulation. As Dixson (1998) suggested, attractivity and proceptivity may begin to increase during the follicular phase in order to locate and attract males who will then be in place and mating when ovulation does occur.

References

Altmann, J. (1974). Observational study of social behavior: sampling methods. *Behaviour*, **49**, 227–265.

Batty, K. A., Herbert, J., Keverne, E. B. and Vellucci, S. V. (1986). Differences in blood levels of androgens in female talapoin monkeys related to their social status. *Neuroendocrinology*, **44**, 347–354.

Baum, M. J. (1983). Hormonal modulation of sexual behavior in female primates. *Bioscience*, **33**, 578–582.

Baum, M. J., Keverne, E. B., Everitt, B. J., Herbert, J. and DeGreef, W. J. (1977). Effects of progesterone and estradiol on sexual attractvity of female rhesus monkeys. *Physiology of Behavior*, **18**, 659–670.

Beach, F. A. (1976). Sexual attractivity, proceptivity and receptivity in female mammals. *Hormones and Behavior*, **7**, 105–138.

Beehner, J. C., Phillips-Conroy, J. E. and Whitten, P. L. (2005). Female testosterone, dominance rank and aggression in an Ethiopian population of hybrid baboons. *American Journal of Primatology*, **67**, 101–119.

Bielert, C., Czaja, J. A., Eisele, S., Scheffler, G., Robinson, J. A. and Goy, R. W. (1976). Mating in the rhesus monkey (*Macaca mulatta*) after conception and its relationship to oestradiol and progesterone levels throughout pregnancy. *Journal of Reproduction and Fertility*, **46**, 179–187.

Carlson, A. A., Ginther, A. J., Scheffer, G. R. and Snowdon, C. T. (1996). The effects of infant births on the sociosexual behavior and hormonal patterns of a cooperatively breeding primate. *American Journal of Primatology*, **40**, 23–39.

Chambers, K. C. and Phoenix, C. H. (1987). Differences among ovariectomized female rhesus macaques in the display of sexual behavior without and with estradiol treatment. *Behavioral Neuroscience*, **101**, 303–308.

Chapais, B. and Mignault, C. (1991). Homosexual incest avoidance among females in captive Japanese macaques. *American Journal of Primatology*, **23**, 177–183.

Chapais, B., Gauthier, C., Prud'homme, J. and Vasey, P. (1997). Relatedness threshold for nepotism in Japanese macaques. *Animal Behaviour*, **53**, 1089–1101.

D'Amato, F. R., Troisi, A., Scucchi, S. and Fuccillo, R. (1982). Mating season influence on allogrooming in a confined group of Japanese macaques: A quantitative analysis. *Primates*, **23**, 220–232.

Dixson, A. F. (1998). *Primate Sexuality: Comparative Studies of the Prosimians, Monkeys, Apes, and Human Beings*. Oxford: Oxford University Press.

Dixson, A. F., Everitt, B. J., Herbert, J., Rugman, S. M. and Scruton, D. M. (1973). Hormonal and other determinants of sexual attractiveness and receptivity in rhesus and talapoin monkeys. In *Symposia of the Fourth International Congress of Primatology, Volume 2: Primate Reproductive Behavior*, ed. C. H. Phoenix. Basel: S. Karger, pp. 36–63.

Enomoto, T. (1981). Male aggression and the sexual behavior of Japanese monkeys. *Primates*, **22**, 15–23.

Enomoto, T., Seiki, K. and Haruki, Y. (1979). On the correlation between sexual behavior and ovarian hormone level during the menstrual cycle in captive Japanese monkeys. *Primates*, **20**, 563–570.

Espey, L. L. & Lipner, H. (1994). Ovulation. In *The Physiology of Reproduction*, eds. E. Knobil and J. D. Neill. New York: Raven, pp. 725–780.

Fedigan, L. M. and Gouzoules, H. (1978). The consort relationship in a troop of Japanese monkeys. In *Recent Advances in Primatology, Volume 1: Behaviour*, eds. D. J. Chivers and J. Herbert. London: Academic Press, pp. 493–495.

Fujita, S., Mitsunaga, F., Sugiura, H. and Shimizu, K. (2001). Measurement of urinary and fecal steroid metabolites during the ovarian cycle in captive and wild Japanese macaques, *Macaca fuscata. American Journal of Primatology*, **53**, 167–176.

Gouzoules, H. and Goy, R. W. (1983). Physiological and social influences on mounting behavior of troop-living female monkeys (*Macaca fuscata*). *American Journal of Primatology*, **5**, 39–49.

Johnson, D. F. and Phoenix, C. H. (1976). Hormonal control of female sexual attractiveness, proceptivity, and receptivity in rhesus monkeys. *Journal of Comparative Physiology and Psychology*, **90**, 473–483.

(1978). Sexual behavior and hormone levels during the menstrual cycles of rhesus monkeys. *Hormones and Behavior*, **11**, 160–174.

Linn, G. S., Mase, D., Lafrancois, D., O'Keeffe, R. T. and Lifshitz, K. (1995). Social and menstrual cycle phase influences on the behavior of group-housed *Cebus apella. American Journal of Primatology*, **35**, 41–57.

Maestripieri, D. (1999). Changes in social behavior and their hormonal correlates during pregnancy in pig-tailed macaques. *International Journal of Primatology*, **20**, 707–718.

Martensz, N. D., Vellucci, S. V., Fuller, L. M. *et al.* (1987). Relation between aggressive behaviour and circadian rhythms in cortisol and testosterone in social groups of talapoin monkeys. *Journal of Endocrinology*, **115**, 107–120.

Michael, R. P. and Zumpe, D. (1970). Aggression and gonadal hormones in captive rhesus monkeys (*Macaca mulatta*). *Animal Behaviour*, **18**, 1–10.

(1993). A review of hormonal factors influencing the sexual and aggressive behavior of macaques. *American Journal of Primatology*, **30**, 213–341.

Nigi, H. (1975). Menstrual cycle and some other related aspects of Japanese monkeys (*Macaca fuscata*). *Primates*, **16**, 207–216.

(1976). Some aspects related to conception of the Japanese monkey (*Macaca fuscata*). *Primates*, **17**, 81–87.

O'Neill, A. C. (2000). The relationship between ovarian hormones and the behavior of Japanese macaque females (*Macaca fuscata*) during the mating season. Master's thesis, University of Alberta. *Master's Abstracts*, 2001, **39**(6), 1494.

O'Neill, A. C., Fedigan, L. M. and Ziegler, T. E. (2004). Relationship between ovarian cycle phase and sexual behavior in female Japanese macaques (*Macaca fuscata*). *American Journal of Physical Anthropology*, **125**, 352–362.

(2006). Hormonal correlates of post-conceptive mating in female Japanese macaques. *Laboratory Primate Newsletter*, **45**, 1–4.

Saayman, G. S. (1970). The menstrual cycles and sexual behavior in a troop of free-ranging chacma baboons (*Papio ursinus*). *Folia Primatologica*, **12**, 81–110.

Siegel, S. and Castellan, Jr. N. J. (1988). *Nonparametric Statistics for the Behavioral Sciences*. 2nd edition. New York, NY: McGraw-Hill.

Silk, J. B. (1992). The evolution of social conflict among female primates. In *Primate Social Conflict*, eds. W. A. Mason and S. P. Mendoza. New York, NY: State University of New York Press, pp. 49–83.

Smuts, B. B. (1987). Sexual competition and mate choice. In *Primate Societies*, eds. B. B. Smuts, D. L. Cheney, R. M. Seyfarth, R. W. Wrangham and T. T. Struhsaker. Chicago: The University of Chicago Press, pp. 385–399.

Soltis, J. (1999). Measuring male-female relationships during the mating season in wild Japanese macaques (*Macaca fuscata yakui*). *Primates*, **40**, 453–467.

Soltis, J., Mitsunaga, F., Shimizu, K., Yanagihara, Y. and Nozaki, M. (1999). Female mating strategy in an enclosed group of Japanese macaques. *American Journal of Primatology*, **47**, 263–278.

Sommer, V. (1993). Infanticide among the langurs of Jodhpur: testing the sexual selection hypothesis with a long-term record. In *Infanticide and Parental Care*, eds. S. Parmigiani and F. vom Saal. London: Harwood Academic Publishing, pp. 155–198.

Stabenfelt, G. H. and Hendrikx, A. G. (1972). Progesterone levels in the bonnet monkey (*Macaca radiata*) during the menstrual cycle and pregnancy. *Endocrinology*, **91**, 614–619.

(1973). Progesterone studies in *Macaca fascicularis*. *Endocrinology*, **92**, 1296–1300.

Strier, K. B. and Ziegler, T. E. (1997). Behavioral and endocrine characteristics of the reproductive cycle in wild muriquis monkeys, *Brachyteles arachnoides*. *American Journal of Primatology*, **42**, 299–310.

Strier, K. B., Ziegler, T. E. and Wittwer, D. J. (1999). Seasonal and social correlates of fecal testosterone and cortisol levels in wild male muriquis (*Brachyteles arachnoides*). *Hormones and Behavior*, **35**, 125–134.

Takahata, Y. (1982). The socio-sexual behavior of Japanese monkeys. *Zeitschrift für Tierpsychologie*, **59**, 89–108.

Trimble, M. R. and Herbert, J. (1968). The effects of testosterone or oestradiol upon the sexual and associated behaviour of the adult rhesus monkey. *Journal of Endocrinology*, **42**, 171–185.

Vasey, P. L. (1995). Homosexual behavior in primates: A review of evidence and theory. *International Journal of Primatology*, **16**, 173–204.

(1996). Interventions and alliance formation between female Japanese macaques, *Macaca fuscata*, during homosexual consortships. *Animal Behaviour*, **52**, 539–551.

Vasey, P. L. and Duckworth, N. (2006). Sexual reward via vulvar, perineal and anal stimulation: a proximate mechanism for female homosexual mounting in Japanese macaques. *Archives of Sexual Behavior*, **35**, 523–532.

Vasey, P. L. and Gauthier, C. (2000). Skewed sex ratios and female homosexual activity in Japanese macaques: An experimental analysis. *Primates*, **41**, 17–25.

Vasey, P. L., Rains, D., VanderLaan, D. P., Duckworth, N. and Kovacovsky, S. D. (2008a). Courtship behavior during heterosexual and homosexual consortships in Japanese macaques. *Behavioural Processes*, **78**, 401–407.

Vasey, P. L., VanderLaan, D. P., Rains, D., Duckworth, N. and Kovacovsky, S. D. (2008b). Inter-mount social interactions during heterosexual and homosexual consortships in Japanese macaques. *Ethology*, **114**, 564–574.

von Engelhardt, N., Kappeler, P. M. and Heistermann, M. (2000). Androgen levels and female social dominance in *Lemur catta*. *Proceedings of the Royal Society of London* B, **267**, 1533–1539.

Wallen, K. and Tannenbaum, P. L. (1997). Hormonal modulation of sexual behavior and affiliation in rhesus monkeys. *Annuals of the New York Academy of Sciences*, **807**, 185–202.

Wallen, K., Winston, L., Gaventa, L., Davis-DaSilva, M. and Collins, D. C. (1984). Periovulatory changes in female sexual behavior and patterns of steroid secretion in group-living rhesus monkeys. *Hormones and Behavior*, **18**, 431–450.

Wilson, M. E., Gordon, T. P. and Collins, D. C. (1982). Variations in ovarian steroids associated with the annual mating period in rhesus macaques (*Macaca mulatta*). *Biology of Reproduction*, **27**, 530–539.

Wolfe, L. (1979). Behavioral patterns of estrous females of the Arashiyama West troop of Japanese macaques. *Primates*, **20**, 525–534.

(1984). Japanese macaque female sexual behavior: A comparison of Arashiyama east and west. In *Female Primates: Studies by Women Primatologists*, ed. M. E. Small. New York, NY: Alan R. Liss, pp. 141–157.

Ziegler, T. E. and Bercovitch, F. B. (1990). *Socioendocrinology of Primate Reproduction*. New York, NY: Wiley-Liss, Inc.

Ziegler, T. E., Santos, C. V., Pissinatti, A. and Strier, K. B. (1997). Steroid excretion during the ovarian cycle in captive and wild muriquis, *Brachyteles arachnoides*. *American Journal of Primatology*, **42**, 311–321.

8   *Factors influencing mating frequency of male Japanese macaques* (Macaca fuscata) *at Arashiyama West*

KATHARINE M. JACK

Several members of the Arashiyama West troop (photo by K. Jack).

8.1   Introduction

The main problem encountered by a male primate in regards to mating is locating willing partners. Males can employ a variety of behaviours when attempting to attract a mate; however, the bulk of data available on male mating strategies is directly related to male–male competition and dominance relations. In populations of provisioned or captive Japanese and rhesus macaques, it has been

The Monkeys of Stormy Mountain: 60 Years of Primatological Research on the Japanese Macaques of Arashiyama, eds. Jean-Baptiste Leca, Michael A. Huffman and Paul L. Vasey. Published by Cambridge University Press. © Cambridge University Press 2012.

demonstrated that there is no consistent relationship between high dominance rank and male mating success (Japanese macaques, *Macaca fuscata*: reviewed by Takahata *et al.*, 1999; see also Fedigan *et al.*, 1986; Huffman, 1987, 1991a; rhesus macaques, *Macaca mulatta*: Berard *et al.*, 1993; Smith, 1994; Huffman and Takahata: chapter 6).

Previous research on the Arashiyama West and other provisioned/captive groups of Japanese macaques has found that the majority of sexually mature troop members, high ranking or low ranking, are active participants during the mating season (e.g. Jack and Pavelka, 1997; Inoue and Takenaka, 2008; Garcia *et al.*, 2009). If dominance is not a consistently significant variable in determining which males will gain mating access to oestrous females, male macaques must be employing alternative strategies to secure mates. Possible behaviours males may be utilising include: consort formation (Wolfe, 1991; Pavelka, 1993), consort monopolisation (Huffman, 1987, 1992), consort intrusion and takeover (Huffman, 1987), aggressive intimidation or sexual coercion (Smuts and Smuts, 1993; Soltis *et al.*, 1997a, 1997b; Muller *et al.*, 2009), display behaviours (Wolfe, 1981; Sprague, 1991), and/or formation of affiliative bonds prior to mating (Smuts, 1985). None of these behaviours appears to singularly affect a male's mating success indicating that other tactics, or a combination of several tactics, are being employed by males to influence female mate choice (e.g. Soltis *et al.*, 1997b).

The primary objective of this study is to investigate and describe the mating strategies and behaviours of male Japanese macaques residing in the Arashiyama West colony and determine how these influence their access to mates. I examined the influence of male rank, age, spatial distribution (peripheral/heterosexual troop), affiliative and aggressive interactions with females, and display performance (long-distance and courtship) on male mating success.

8.2   Methods

8.2.1   *Study population*

The study was conducted at the site of the Arashiyama West colony of Japanese macaques over a 6-month period between September 1994 and January 1995. This troop of Japanese macaques was relocated from Kyoto, Japan in 1972 (see Fedigan, 1976; Pavelka, 1993; Huffman *et al.*: chapter 1, for detailed histories of the troop). At the time of the study, the Arashiyama West colony was located in Dilley, Texas, which is approximately 125 km southwest of San Antonio, Texas. The monkeys had fully adapted to the desert-like environment of south Texas and at the time of this study the population had grown from an original 150 animals to over 300. The colony was divided into two distinct social groups: the Pelka troop (n ~ 50) and the Main troop (n ~ 300). At the time of the study, the troops were not contained within any physical barriers and they ranged over an area of approximately 100 hectares of ranch land. The troops were able to forage extensively on native vegetation; however, this was supplemented daily with prepackaged monkey chow pellets and grain. Fresh produce was also provided to the monkeys once per week.

Genealogical information along maternal lines was recorded for all animals in the Arashiyama West population through 1991. Up until that time, the Arashiyama West naming system followed that of the Arashiyama East (i.e. Arashiyama-Kyoto) colony: all animals were assigned a name which reflected the maternal kin group they belonged to, their birth year, the birth year of their mother, grandmother, etc. All animals were also tattooed with a number on their inner thigh and corresponding number markings on their faces for easy identification of individuals and as a permanent reference to their genealogical name (see Pavelka, 1993 for more detail).

A total of 25 males were chosen as focal animals for this study. All males were selected from the Main troop and its peripheral male subgroup. In September 1994 there were ~ 67 breeding aged males (≥ 4 years) including Main troop (n = 24), peripheral (n = 37), and solitary (n = 6) males. Focal males were selected to represent a variety of ages, ranks and family lines. Fifteen Main troop males and ten peripheral males were chosen as subjects. Solitaries were not selected as focal animals due to the inherent difficulty in both locating and following these animals.

8.2.2   Data collection

Japanese macaques are seasonal breeders with a distinctive mating season beginning in late September or early October, and lasting through late January or early February (Napier and Napier, 1967). This pattern of fall mating accompanied by a spring birth season is maintained by the Japanese macaques at Arashiyama West. Observations of the group began prior to the start of the mating season in August 1994 with data being collected from 19 September to 13 January 1995. A total of twenty 15-min focal animal samples were conducted for each of the study subjects, yielding 125 hours of focal animal data (Altmann, 1974). An additional 36.75 hours of scan data (49 scans each of 45 minutes in duration) were also collected. The 45-min scans were undertaken at the start of every data collection day, during which time I tried to locate each focal male. When a focal animal was encountered, his geographic location was mapped and any consorts or mating behaviours were noted. As these monkeys were free-ranging at the time of the study, not all subjects could be located during each daily scan, in which case they were recorded as 'not observed'.

8.2.3 Research variables and behavioural categories

Male rank

Among Japanese macaques, males occupy a position, or dominance rank, within their troop. Subjects could not be ranked in relation to one another in a linear fashion (i.e. 1–25) as peripheral males seldom, if ever, interacted with members of the Main troop. In fact, the relative ranks of males in this large free-ranging troop are only obvious for the first four or five males and become increasingly difficult to discern as one moves down the hierarchy (Pavelka, 1993). Huffman (1991a) encountered similar problems in determining male ranks in his study at the Arashiyama B colony in Kyoto, Japan. Consequently, he divided males into two groups, either high or low ranking, according to their socio-spatial locations in the non-mating season. Males found in the central troop area were categorised as high ranking and those in the peripheral areas as low ranking. This methodology was followed in the present study, with all peripheral focal males (n = 10) categorised as low ranking. Main troop males, however, were categorised as either medium (n = 9) or high ranking (n = 6) according to the direction of submissive behaviours exchanged among them (Koyama, 1967). Peripheral males were categorised as such during pre-mating season observations.

Male age

Ages and genealogies of all focal males were gathered from the Arashiyama West genealogical data base. For analyses, absolute ages of focal males were used (log transformed due a positively skewed distribution; Zar, 2009). Focal males ranged in age from 4.5–26.5 years (mean ± s.d. = 9.78 ± 4.198).

Male spatial distribution

At sexual maturity (~ 4 years) wild male Japanese macaques emigrate from their natal troop and either directly join another heterosexual troop, an all-male peripheral group, or become solitary for a period of time. There are two distinct troops at Arashiyama West, making all of the above options available for dispersing sexually mature males (≥ 4 years). However, many males in this colony do remain in their natal troop, or transfer back into it after a period of living

Figure 8.1. Peripheral male subgroup prior to the onset of the mating season. This subgroup is comprised of both juvenile and fully adult males (photo by K. Jack).

as a solitary or peripheral male. A peripheral male subgroup is a spatially distinct subgroup comprised of sub-adult males (4.5–9.5 years), adult males (10+ years), and occasionally 'floating females' (i.e. low-ranking females found away from the Main troop area; Fedigan, 1976) (Figure 8.1). Peripheral males are both spatially and socially separated from the Main troop and they rarely have interactions with them. Main troop males, on the other hand, do not normally leave the core area of the group (i.e. the area where females and their immature offspring spend their time).

Affiliative behaviour

The total amount of time each male spent in proximity to ($\leq$ 1 metre), or in non-aggressive contact with, unrelated females was calculated over all focal sessions. Mating between close kin is rare among Japanese macaques and due to the focus of this study (male mating strategies and behaviours) their interactions with related individuals were not of primary interest. Male/female dyads with a coefficient of kinship less than 1/8 (e.g. aunt:nephew and half-siblings are considered 0.125) were considered unrelated, as evidence suggests that kin recognition does not extend past this level of relatedness (see Chapais *et al.*, 1997). Affiliative behaviours include: grooming, resting in contact/proximity and foraging in contact/proximity.

Aggressive behaviour

Aggressive behaviours directed towards females were scored as events and included: gape, growl, lid, stare, head bob, chase, lunge, bluff charge, pinch, slap, grab, push, pull and bite. An aggression score was calculated as the sum of all aggressive acts each focal male directed against any unrelated female during focal animal sampling. Several aggressive acts carried out in succession were scored as a single aggressive interaction. That is, if a male were to lunge, chase, and grab a female this would be scored as one aggressive act, rather than three.

Long-distance displays

These displays involve the climbing of a flexible structure and forcefully shaking the structure with the entire weight of the body, often accompanied by loud guttural vocalisations. These loud, vigorous displays are most commonly carried out in tall trees and are observable, and often audible, from great distances. Shaking displays accompanied by vocalisations are unique to the mating season (Modahl and Eaton, 1977; Wolfe, 1981).

Courtship displays

Courtship displays are performed at close range to the target female and are only observed during the mating season (Pavelka, 1993). Courtship displays vary in behavioural sequences from male to male, but the most common sequence of behaviours exhibited during the displays are as follows: strut/approach, lip quiver/stare, strut/leave. After such a behaviour sequence has been performed, a male will usually stand (approximately one metre away from the female) for several seconds in what is termed a 'bird-dog' pose. This pose involves a frozen stance with the tail up, as the displaying male stares into the distance. Following the performance of a courtship display, a male will attempt to sit in contact with, or in close proximity to, the female to which his display was directed.

Male mating frequency

Using data collected during both scan and focal sessions, a mating frequency score (MFS) was calculated for each focal male. Japanese macaques engage in series-mounting, which requires several non-ejaculatory mounts with intromission before ejaculation can occur. In tabulating MFS, any sexual mount of a female by a male was considered a mating. MFS was calculated as the percentage of samples (focal and scan) for which each focal male was observed mating with each and any individual female (Inoue *et al.*, 1993), with an individual female being scored as a mating partner only once per day. For example, focal male, Betta–58–68–86, was observed mating during 28 of the 59 sample sessions (20 focal and 39 scans) making his MFS 47.5%. On several occasions a male was observed mating with more than one female during the 45-minute scan. Because both focal and scan samples were used to calculate MFS, it is possible for a male to have a score that exceeds 100%. MFS was calculated as a percentage because not all focal males were sampled equally.

Note that male reproductive success was not determined in the study. Inoue *et al.* (1991) demonstrated, through paternity discrimination, that mating and reproductive success are not synonymous and it is not implied in the present study.

8.3   Results

A total of 575 matings were recorded for the 25 focal males during the 1993/1994 mating season at Arashiyama West, and all but two of the focal males were observed mating. The MFS of focal males ranged from 0–104.9% (mean ± s.d. = 39.4 ± 28.3). Using multiple regression analysis male age, affiliative and aggressive behaviour were all found to be significant in predicting male mating success. Male age was negatively correlated with MFS, while both male affiliation and aggression scores were positively correlated (Table 8.1). Male affiliation scores ranged from 0–12412 s or ~3.5 hours (mean ± s.d. = 4849.9 ± 2866.7) and male aggression scores ranged from 0–27 aggressive acts (mean ± s.d. = 14.2 ± 7.1). Male rank (negative correlation), spatial distribution (positive correlation), the frequency of long-distance displays (positive correlation; range 0–11; mean ± s.d. = 3.9 ± 3.3) and the frequency of courtship displays (negative correlation; range 0–12; mean ± s.d. = 5.0 ± 3.6), all had a non-significant effect on male MFS. Together the seven independent variables tested explained 70.7% of the observed variability in male mating success ($R^2 = 0.707$; $F = 5.869$; $p = 0.001$).

Further analysis of these variables was run (step-wise regression analysis) in order to eliminate any confounding correlation between variables, and to obtain a better fit for the regression model. The resulting model included only those variables significantly correlated with male mating success: male age, affiliation and aggression (Table 8.2). Together these three variables accounted for 64% of the variation observed in male mating success ($R^2 = 0.64$). Although this is a lower R^2 than that achieved by the multiple regression including all seven independent variables, this model, with an

Table 8.1. *Independent variables predicting male mating frequency scores (MFS) in a multiple regression analysis (7 independents)**

Independent variables	Regression coefficient	SE	p
Rank	−1.280	1.003	0.219
Age	−6.220	2.889	0.046
Spatial distribution	2.339	1.639	0.172
Affiliation	5.072E-04	0.000	0.024
Aggression	0.201	0.061	0.004
Long-distance displays	9.370E-02	0.130	0.480
Courtship displays	−6.140E-02	0.133	0.650

* Model for multiple regression is $R^2 = 0.707$, s.e. $= 1.7612$, $F = 5.869$, $p = 0.001$.

Table 8.2. *Independent variables predicting male mating frequency scores (MFS) in a step-wise regression analysis (3 independents)**

Independent variables[+]	Regression coefficient	SE	p
Aggression	0.200	0.055	0.001
Affiliation	4.546E-04	0.000	0.006
Age	−5.661	2.647	0.044

* Model for step-wise regression is $R^2 = 0.64$, s.e. $= 1.7568$, $F = 12.457$, $p < 0.001$.
+ Variables are listed in the order in which they were added to the model.

increased F-value ($F = 12.457$) and a higher significance level ($p < 0.001$) was a better fit.

8.4   Discussion

In most primate species where multiple males reside together in groups with multiple females, not all males experience equal access to mating opportunities. A male's ability to successfully gain mating access is largely determined by his ability to influence female mate choice, which in turn may be determined by the mating strategies he employs (e.g. Dunbar, 2001; Soltis *et al.*, 2001; Lynch-Alfaro, 2005). These strategies may vary according to male life-history stage, dominance rank and even personality. In the present study, the variability in mating success for male Japanese macaques residing in the Arashiyama West colony was very high; the number of matings observed per individual male during the 5-month mating season ranged from 0 to 64 (mating frequency scores ranged from 0–104.9%). This is a large range of variation and invites the question of why some males are more successful at attracting mates than others. Are there particular characteristics of a male that attract a female (e.g. age, familiarity and rank)? Do males engage in particular behaviours in order to gain access to or attract the attention of breeding females (e.g. affiliation, aggression and displays)?

Of the mating strategies available to primate males, the influence of rank on individual mating and/or reproductive success has received the most attention from primatologists. Generally, attaining a high dominance rank confers priority of access to desired resources including food, mates and space (Altmann, 1962) and in many species increased access to mates results in greater mating and reproductive success (reviewed by Ellis, 1995; Di Fiore, 2003). Male dominance rank in the present study, however, was a non-significant factor influencing male mating frequency. In fact, during the present study the alpha male at Arashiyama West was only observed mating with a single female throughout the entire mating season (most of his mating attempts were ignored or avoided).

While the relationship between male rank and mating success in wild populations of Japanese macaques appears to be more clearcut (e.g. Soltis et al., 2001) the lack of correlation between male dominance rank and mating frequency in the present study is similar to earlier reports on captive and provisioned populations of Japanese macaques (Inoue et al., 1993; Soltis et al., 1997b; but see Garcia et al., 2009). For example, in their study of a captive group of Japanese macaques, Soltis et al. (1997b) found that females showed a preference for subordinate males. The authors suggest, as does Rubenstein (1980), that the alternative mating strategies of low-ranking males may dilute the effects of male–male competition thus leading to a non-significant, and often negative, correlation between mating success and rank.

However, in the present study subordinate males did not engage in different mating strategies than higher-ranking males (see Jack and Pavelka, 1997). In fact, many of the matings by medium- and low-ranking males occurred in full view of the alpha and beta males. Also, instances of male–male aggression were extremely rare in the study population: a total of 33 aggressive acts were observed between males (mean ± s.d. = 1.3 ± 1.6) and only six of these aggressive interactions involved contact between the males (e.g. slap, pinch, push, bite) (Jack, unpublished data). This low rate of male–male aggression demonstrates that mating competition within the group was minimal. This finding is curious given that the sex ratio of 1:2.1 males to females at Arashiyama West is considerably lower than the reported ratio of 1:3.4 for wild groups (Kurland, 1977). With a lower sex ratio one would expect a higher occurrence of male mating competition in the Arashiyama West colony and perhaps, as a consequence, a positive linear relationship between male rank and mating success. The finding of a negative correlation (non-significant) between these two variables may indicate that, as Soltis *et al.* (1997b) concluded, female mate choice is a more powerful determinant of male mating success than male–male competition, at least among captive and provisioned groups of Japanese macaques.

Indeed male age was negatively associated with male mating frequency in the present study indicating female mating preference for younger, more novel, males. At the time of this study, the alpha male was 26 years old. This old male did not attain his high rank by engaging in high rates of male–male competition as is the case in wild populations (Sprague, 1992). Rather, this male attained his alpha male status when the four males that ranked above him perished in a mountain lion attack at the site in 1993. After the attack, he was the highest ranked male and the entire group deferred to him and he received no challenges to his rank. It is common in captive or provisioned groups of macaques for male rank to be positively correlated to their age (Berard, 1999) rather than physical condition (Takahashi, 2002) or 'alpha male attitude' as is the case in a wild population (Sprague, 1992).

It is also likely that, in the present study, any effects of rank on male mating frequency are stymied because of the long tenure of group males (Inoue and Takenaka, 2008). At Arashiyama West there are two distinct troops: Main troop and Pelka troop. However, in this population after a sexually mature male has lived as a peripheral or solitary for some period of time and subsequently joined one of the two troops, there is surprisingly little movement of males between them. Also, at the time of the study, the Pelka troop had moved out of range (> 15 km away), further limiting mate choice for the females of the Main troop (matings between members of the two groups had been observed in previous years: Griffin, pers. comm.).

In several male-dispersed primate species, male mating success has been shown to wane with increasing tenure length (reviewed in Jack, 2003). Female sexual interest in unfamiliar males appears to be a widespread phenomenon throughout the primate order, may be a means of inbreeding avoidance, and among female-bonded species, this preference has been hypothesised as the driving force behind male mobility (Berard, 1991; Huffman, 1992; Zumpe and Michael, 1996). In wild populations, male Japanese macaques are reported to change groups several times throughout their lives, approximately every 3 or 4 years, and they almost never return to their natal troops (Sprague, 1992; Sprague *et al.*, 1998). Resident males have been noted as experiencing decreased mating success over time (Huffman, 1991a, 1991b) and female avoidance of males with whom they were sexually active in the preceding 4 or 5 years has also been well documented in this species (Furuichi, 1985; Huffman, 1991a, 1992;

Perloe, 1992). Given this evidence for free-ranging groups, it seems natural that when dispersal is restricted, as it is in the Arashiyama West population, males will experience a decline in mating success as they age regardless of their dominance rank (see Takahata *et al.*, 1999 for similar conclusions).

Access to mates has traditionally been assumed to be the incentive for male sociality in primates and troop membership was once considered a requisite for male mating success (see Wrangham, 1980). However, accumulating reports of females mating with males from outside of their groups are beginning to challenge this assumption (e.g. Japanese macaques: Huffman, 1991a; Jack and Pavelka, 1997; Sprague *et al.*, 1998; Soltis *et al.*, 2001; Hayakawa, 2007, 2008; rhesus macaques: Berard, 1993; white-faced capuchins, *Cebus capucinus*: Rose and Fedigan, 1995; patas monkeys, *Erythrocebus patas*: Olson, 1985; red colobus monkeys, *Procolobus rufomitratus*: Mbora and McGrew, 2002; Verreaux's sifaka, *Propithecus verreauxi verreauxi*: Lawler, 2007). Peripheral male Japanese macaques have been reported as successful in mating but they are generally assumed to be less successful than troop-living males (Sugiyama, 1976; Pavelka, 1993). In the present study, peripheral males (n = 10) accounted for 38% of all observed matings by focal males (weighted percentage to account for the lower number of peripheral focal males) and spatial distribution was not a significant factor in explaining variance in male mating frequency.

The finding of non-significant differences between the mating frequency of peripheral and Main troop males may be a direct result of peripheral males increasing their proximity to the Main troop during the first few weeks of the mating season. Prior to the onset of the 1994–1995 mating season, peripheral males were never observed within the Main troop core area, never interacted with members of the Main troop, and always maintained approximately 0.5 km between themselves and any members of the Main troop. However, 3 weeks into the mating season, an observable change in the daily locations of peripheral males had taken place. Examination of daily scan maps of focal male locations revealed that most peripheral males had decreased the distance between themselves and the Main troop. This increase in spatial proximity was the first noticeable attempt by peripheral males to gain access to, and the attention of, oestrous females (see Jack and Pavelka, 1997 for details on the movement of peripheral males during the mating season).

The formation of affiliative bonds prior to mating has been recognised as a mating strategy in many primates (Smuts, 1985; Strum, 1987). Previous studies of female mate choice in Japanese macaques (Soltis *et al.*, 1997a) and rhesus macaques (Manson, 1992) demonstrate that males with whom females preferentially maintained proximity have higher ejaculatory/copulatory rates. The results of this study support these previous findings as the amount of time a male spent affiliating with unrelated females was a positive significant factor

Figure 8.2. An extended consortship between a male and female; the pair is taking a break between mount series (photo by K. Jack).

in predicting male mating success (Figure 8.2). It may be argued that because Japanese macaques are series mounters, and often form consorts which may last for several hours or days, affiliation is a prerequisite of mating success. Extended consortships are not, however, necessary for ejaculation to occur and opportunistic matings are common (Wolfe, 1991; Jack, pers. obs.). At Arashiyama West, male affiliation with unrelated females during the mating season may serve as a tactic through which some males are able to secure mating partners. However, analyses in the present study did not differentiate between female and male proximity maintenance (see Soltis *et al.*, 1997a) and the affiliation score reflects overall rates of male affiliation with all unrelated females rather than affiliation that occurred solely within male–female mating dyads. The results, therefore, indicate that the overall amount of time a male spends affiliating with unrelated adult females is a good predictor of his mating success and may be a factor influencing female mate choice.

Figure 8.3. An adult male engaging in stalking behaviour during the mating season (photo by K. Jack).

During the 1994/1995 mating season most males, both peripheral and Main troop, underwent a drastic alteration in their behaviour. Normally docile males were observed stalking (an aggressive mode of travelling employed only during the mating season: Pavelka, 1993; Figure 8.3) and engaging in aggressive behaviours (e.g. chasing, biting). In particular, male aggression directed towards females increased dramatically during the mating season and female wounding began to appear on a daily basis, a rare occurrence outside of the mating season (Griffin and Pavelka, pers. comm.). Smuts and Smuts (1993) have suggested that evidence of increased male aggression directed at females during mating seasons or times of oestrus supports the hypothesis that aggression directed towards females may function to increase a male's access to mates. They have labelled such aggression as 'sexual coercion'. Male aggression may keep females from leaving males once mounting interactions have begun (Soltis, 1999). Furthermore, in those dyads in which male-to-female aggression is observed, males appear to experience increased ejaculation rates (Enomoto, 1981).

Of the five variables examined in this study, male aggression directed towards unrelated adult females was the most significant predictor of male mating frequency. The significant positive linear relationship between male aggression and mating frequency may provide support for Smuts and Smuts' (1993) hypothesis: however, caution is warranted in this interpretation. Not all male aggression towards females functions as a form of sexual coercion and it is still unclear how increased aggression is related to male courtship behaviours or mating success. Soltis *et al.* (1997a), in their study of captive Japanese macaques, concluded that male aggression was a by-product of increased proximity during mating and did not function as sexual coercion. The same may hold for some of the aggression observed in the present study, as overall rates of both affiliation and aggression were significant factors in predicting variance in male mating success. However, not all of the male aggression towards females observed during this study was the result of proximity. For example, during the mating season most males are observed stalking and engaging in very aggressive behaviours.

Stalking is a mode of travelling during which a male keeps his body low to the ground with his head protruding out in front of him as he constantly scans his surroundings (Figure 8.3). A male engaged in stalking moves quickly and appears to be searching for something or someone. Although the target of stalking is rarely evident such behaviour often terminates in aggression directed towards females. When stalking does end in aggressive interactions, the male does not usually aggress against a single female. Instead, he generally runs around chasing every female in sight, often slapping and biting them. Male stalking behaviour is unique to the mating season (Pavelka, 1993) and in the present study, it was never observed to terminate in aggression towards other males.

Neither courtship nor long-distance displays were significant in predicting male mating success, and both displays were performed at relatively low frequencies (means = 5 and 3.9 respectively). The function of both of these types of displays is still unclear. Courtship displays were often performed by males after an oestrous female had already begun to approach and court them. These displays may, therefore, serve as a signal of reassurance to approaching females. Interpreting the function of long-distance displays is even more problematic. Although these displays are believed to be a means of gaining the attention of oestrous females (Modahl and Eaton, 1977; Wolfe, 1981; Sprague, 1991; Pavelka, 1993), it is unclear whether they are directed at particular females or if they are a means of male advertising. The focal male in the study group who performed the greatest number of long-distance displays was a peripheral male named Rotte–63–85. Although Rotte–63–85 did increase his proximity to the Main troop during the first month of the mating season, he was never observed closer than 25 m to the Main troop. This male performed the highest number of long-distance displays (n = 11) perhaps in the attempt to attract the attention of oestrous females to his location. Unfortunately for Rotte–63–85, his displays appeared to go unnoticed as he was never observed interacting with a female, let alone mating.

8.5   Conclusions

1. Male age was a significant variable in predicting male mating success in the Arashiyama West population. The negative correlation between male age and mating success may reflect female sexual aversion to mates they have been sexually active with in preceding years and may be due to the limitations on male immigration at the site.
2. Male affiliation with and aggression towards unrelated females were also significant variables in predicting male mating success. More successful males spent more time affiliating with unrelated group females (not necessarily mates) and they were also the most aggressive males. This aggression may be a form of sexual coercion and/or it may be the by-product of increased proximity between males and females during the mating season.
3. Male rank, spatial distribution (peripheral or Main troop member), and the frequency with which long-distance and courtship displays were performed, were found to be non-significant variables in predicting male mating success in the study population.

Acknowledgements

I thank Lou Griffin and Tracy Wyman, who served as Director and Assistant Director of the South Texas Primate Observatory/Arashiyama West when this study was conducted, for their assistance and support during all stages of this research. Thanks are also extended to Mary Pavelka, Lisa Hansen, Craig Lamarsh, Tracy Wyman and three anonymous reviewers for their comments on earlier versions of this manuscript. I also thank the editors (J.-B. Leca, M. A. Huffman and P. L. Vasey) for inviting me to contribute to this volume. Financial support for this project was provided by the Faculty of Graduate Studies, Research Services, and the Department of Anthropology at the University of

Calgary. Preparation of this manuscript was made possible with the support of the Department of Anthropology at Tulane University.

References

Altmann, J. (1974). Observational study of social behavior: sampling methods. *Behaviour*, **49**, 227–265.

Altmann, S. A. (1962). A field study of the sociobiology of rhesus monkeys, *Macaca mulatta*. *Annals of the New York Academy of Sciences*, **102**, 338–435.

Berard, J. D. (1991). The influence of mating success on male dispersal in free-ranging rhesus macaques. *American Journal of Primatology*, **24**, 89–90.

(1993). Male rank, mating success and dispersal: a four-year study of mating patterns in free-ranging rhesus macaques. *American Journal of Primatology*, **30**, 298.

(1999). A four-year study of the association between male dominance rank, residency status, and reproductive activity in rhesus macaques (*Macaca mulatta*). *Primates*, **40**, 159–175.

Berard, J. D., Nürnberg, P., Epplen, J. T. and Schmidtke, J. (1993). Male rank, reproductive behavior, and reproductive success in free-ranging rhesus macaques. *Primates*, **34**, 481–489.

Chapais, B., Gauthier, C., Prud'homme, J. and Vasey, P. L. (1997). Relatedness threshold for nepotism in Japanese macaques. *Animal Behaviour*, **53**, 1089–1101.

Di Fiore, A. (2003). Molecular genetic approaches to the study of primate behavior, social organization, and reproduction. *Yearbook of Physical Anthropology*, **46**, 62–99.

Dunbar, R. I. M. (2001). The economics of male mating strategies. In *Economics in Nature: Social Dilemmas, Mate Choice and Biological Markets*, eds. R. Noe, J. A. R. A. M. van Hooff and P. Hammerstein. Cambridge: Cambridge University Press, pp. 245–269.

Ellis, L. (1995). Dominance and reproductive success among nonhuman animals: a cross-species comparison. *Ethology and Sociobiology*, **16**, 257–333.

Enomoto, T. (1981). Male aggression and the sexual behavior of Japanese monkeys. *Primates*, **22**, 15–23.

Fedigan, L. M. (1976). A study of roles in the Arashiyama West troop of Japanese monkeys (*Macaca fuscata*). A Monograph in the Series: *Contributions to Primatology*, Volume 9. Basel: S. Karger.

Fedigan, L. M., Gouzoules, S., Gouzoules, H. and Koyama, N. (1986). Life time reproductive success in female Japanese macaques. *Folia Primatologica*, **47**, 143–157.

Furuichi, T. (1985). Inter-male associations in a wild Japanese macaque troop on Yakushima Island, Japan. *Primates*, **26**, 219–237.

Garcia, C., Shimuzu, K. and Huffman, M. A. (2009). Relationship between sexual interactions and the timing of the fertile phase in captive female Japanese macaques (*Macaca fuscata*). *American Journal of Primatology*, **71**, 868–879.

Hayakawa, S. (2007). Female defensibility in a small troop of Japanese macaques vis-à-vis nontroop males and copulation on the periphery of the troop. *International Journal of Primatology*, **28**, 73–96.

(2008). Male-female mating tactics and paternity of wild Japanese macaques (*Macaca fuscata yakui*). *American Journal of Primatology*, **70**, 986–989.

Huffman, M. A. (1987). Consort intrusion and female mate choice in Japanese macaques (*Macaca fuscata*). *Ethology*, **75**, 221–234.

(1991a). Mate selection and partner preferences in female Japanese macaques. In *The Monkeys of Arashiyama: Thirty-five Years of Research in Japan and the West*, eds. L. M. Fedigan and P. J. Asquith. Albany, NY: State University of New York Press, pp. 101–122.

(1991b). Consort relationship duration, conception, and social relationships in female Japanese macaques. In *Primatology Today*, eds. A. Ehara, T. Kimura, O. Takenaka and M. Iwamoto. Amsterdam: Elsevier Science Publishers, pp. 199–202.

(1992). Influences of female partner preference on potential reproductive outcome in Japanese macaques. *Folia Primatologica*, **59**, 77–88.

Inoue, E. and Takenaka, O. (2008). The effect of male tenure and female mate choice on paternity in free-ranging Japanese macaques. *American Journal of Primatology*, **70**, 62–68.

Inoue, M., Mitsunaga, F., Ohsawa, H. *et al.* (1991). Male mating behavior and paternity discrimination by DNA fingerprinting in a Japanese macaque group. *Folia Primatologica*, **56**, 202–210.

Inoue, M., Mitsunaga, F., Nozaki, M. *et al.* (1993). Male dominance rank and reproductive success in an enclosed group of Japanese macaques: with special reference to post-conception mating. *Primates*, **34**, 503–511.

Jack, K. (2003). Males on the move: Evolutionary significance of secondary dispersal in male nonhuman primates. *Primate Report*, **67**, 61–83.

Jack, K. M. and Pavelka, M. S. M. (1997). The behavior of peripheral males during the mating season in *Macaca fuscata*. *Primates*, **38**, 369–377.

Koyama, N. (1967). On dominance rank and kinship of a wild troop in Arashiyama. *Primates*, **8**, 189–216.

Kurland, J. A. (1977). Kin selection in the Japanese monkey. *Contributions to Primatology*, **12**, 1–145.

Lawler, R. R. (2007). Fitness and extra-group reproduction in male Verreaux's sifaka: an analysis of reproductive success from 1989–1999. *American Journal of Physical Anthropology*, **132**, 267–277.

Lynch-Alfaro, J. W. (2005). Male mating strategies and reproductive constraints in a group of wild tufted capuchin monkeys (*Cebus apella nigritus*). *American Journal of Primatology*, **67**, 313–328.

Manson, J. H. (1992). Measuring female mate choice in Cayo Santiago rhesus macaques. *Animal Behaviour*, **44**, 405–416.

Mbora, D. N. M. and McGrew, W. C. (2002). Extragroup sexual consortship in the Tana River red colobus (*Procolobus rufomitratus*)? *Folia Primatologica*, **73**, 210–213.

Modahl, K. B. and Eaton, G. G. (1977). Display behavior in a confined troop of Japanese macaques (*Macaca fuscata*). *Animal Behaviour*, **25**, 525–535.

Muller, M. N., Kahlenberg, S. M. and Wrangham, R. W. (2009). Male aggression and sexual coercion of females in primates. In *Sexual Coercion in Primates and Humans: An*

Evolutionary Perspective on Male Aggression Against Females, eds. M. N. Muller and R. W. Wrangham. Cambridge, MA: Harvard University Press, pp. 3–22.

Napier, J. R. and Napier, P. H. (1967). *Handbook of Living Primates*. London: Academic Press.

Olson, D. (1985). The importance of female choice in the mating system of wild patas monkeys. *American Journal of Physical Anthropology*, **66**, 211.

Pavelka, M. S. M. (1993). *Monkeys of the Mesquite: The Social Life of the South Texas Snow Monkeys*. Iowa: Kendall/Hunt Publishing.

Perloe, S. I. (1992). Male mating competition, female choice and dominance in a free ranging group of Japanese monkeys. *Primates*, **33**, 289–304.

Rose, L. M. and Fedigan, L. M. (1995). Vigilance in white-faced capuchins, *Cebus capucinus*, in Costa Rica. *Animal Behaviour*, **49**, 63–70.

Rubenstein, D. (1980). On the evolution of alternative mating strategies. In *Limits to Action: The Allocation of Individual Behavior*, ed. J. E. R. Staddon. New York, NY: Academic Press, pp. 65–100.

Smith, D. G. (1994). Male dominance and reproductive success in a captive group of rhesus macaques (*Macaca mulatta*). *Behaviour*, **129**, 225–242.

Smuts, B. B. (1985). *Sex and Friendship in Baboons*. New York: Aldine.

Smuts, B. B. and Smuts, R. W. (1993). Male aggression and sexual coercion of females in nonhuman primates and other mammals: evidence and theoretical implications. *Advances in the Study of Behavior*, **22**, 1–63.

Soltis, J. (1999). Measuring male–female relationships during the mating season in wild Japanese macaques (*Macaca fuscata yakui*). *Primates*, **40**, 453–467.

Soltis, J., Mitsunaga, F., Shimizu, K., Yanagihara, Y. and Nozaki, M. (1997a). Sexual selection in Japanese macaques I: female mate choice or male sexual coercion? *Animal Behaviour*, **54**, 725–736.

Soltis, J., Mitsunaga, F., Shimizu, K. *et al.* (1997b). Sexual selection in Japanese macaques II: female mate choice and male-male competition. *Animal Behaviour*, **54**, 737–746.

Soltis, J., Thomsen, R. and Takenaka, O. (2001). The interaction of male and female reproductive strategies and paternity in wild Japanese macaques, *Macaca fuscata*. *Animal Behaviour*, **62**, 485–494.

Sprague, D. S. (1991). Mating by nontroop males among the Japanese macaques of Yakushima Island. *Folia Primatologica*, **57**, 156–158.

(1992). Life history and male inter-troop mobility among Japanese macaques (*Macaca fuscata*). *International Journal of Primatology*, **13**, 437–451.

Sprague D. S., Suzuki S., Takahashi H. and Sato, S. (1998). Male life history in natural populations of Japanese macaques: migration, dominance rank, and troop participation of males in two habitats. *Primates*, **39**, 351–363.

Strum, S. C. (1987). *Almost Human: A Journey into the World of Baboons*. New York: W. W. Norton and Co.

Sugiyama, Y. (1976). Life history of male Japanese monkeys. In *Advances in the Study of Behaviour, Volume 7*, eds. J. S. Rosenblatt, R. A. Hinde, E. Shaw and C. Beer. New York, NY: Academic Press, pp. 255–284.

Takahashi, H. (2002). Changes of dominance rank, age, and tenure of wild Japanese macaque males in the Kinkazan A troop during seven years. *Primates*, **43**, 133–138.

Takahata, Y., Huffman, M. A., Suzuki, S., Koyama, N. and Yamagiwa, J. (1999). Why dominants do not consistently attain high mating and reproductive success: a review of longitudinal Japanese macaque studies. *Primates*, **40**, 143–158.

Wolfe, L. D. (1981). Display behavior of three troops of Japanese monkeys (*Macaca fuscata*). *Primates*, **22**, 24–32.

    (1991). Human evolution and the sexual behavior of female primates. In *Understanding Behavior: What Primate Studies Tell us About Human Behavior*, eds. J. D. Loy and C. B. Peters. New York, NY: Oxford University Press, pp. 121–151.

Wrangham, R. (1980). An ecological model of female-bonded primate groups. *Behaviour*, **75**, 262–299.

Zar, J. H. (2009). *Biostatistical Analysis*, 5th edn. Upper Saddle River, NJ: Prentice-Hall.

Zumpe, D. and Michael, R. (1996). Social factors modulate the effects of hormones on the sexual and aggressive behavior of macaques. *American Journal of Primatology*, **38**, 233–261.

9 Costs and benefits of old age reproduction in the Arashiyama West female Japanese macaques

MARY S. M. PAVELKA AND
LINDA M. FEDIGAN

Lady Di (Betta–59–66–78), alpha female of the Arashiyama West troop (photo by the late Karen Dickey, courtesy of Linda Fedigan and Tracy Wyman).

The Monkeys of Stormy Mountain: 60 Years of Primatological Research on the Japanese Macaques of Arashiyama, eds. Jean-Baptiste Leca, Michael A. Huffman and Paul L. Vasey. Published by Cambridge University Press. © Cambridge University Press 2012.

9.1   Introduction

Do non-human primate females experience menopause? Has natural selection favoured the cessation of reproduction in grandmother monkeys and apes because the benefits of caring for grandchildren outweigh the benefits of continuing to produce more babies? Various attempts to answer these questions have produced a variety of conflicting results based on different interpretations of the significance of a small number of old, mostly captive individual monkeys and apes that appear to have stopped reproducing (Pavelka and Fedigan, 1991; Caro *et al.*, 1995; Fedigan and Pavelka, 2011). The kinds of data that are needed to address these questions are rare: a large sample of individuals of known age at death and reproductive history, living under free-ranging conditions. Fortunately, the maintenance of genealogical records on all members of the Arashiyama monkey group since 1954 has provided us with just such a database and thus the opportunity to shed light on ageing and reproduction in non-human primates and to highlight some of the unique characteristics of human ageing. In this paper, we review the results of a series of studies of ageing and reproduction, and present new data on the costs of old age reproduction utilising the Arashiyama West monkey database. Many of the individual monkeys in this database were ones that we each knew personally due to years of behavioural study on the colony.

Early descriptions of ageing monkeys and apes were surprisingly familiar to those living in conditions of modernisation, at least in the west. Old monkeys were described as socially withdrawn and peripheral, less active and socially disengaged (Waser, 1978; Hrdy, 1981; Hauser and Tyrell, 1984; Huffman, 1990). Indeed these reports of isolated monkeys paralleled what had long been proposed (and hotly debated) in social gerontology: that successful ageing in humans would be characterised by social disengagement, a mutual withdrawal of the individual and society in preparation for the ultimate withdrawal of the individual through death (Cumming and Henry, 1961; Cumming, 1975). However the early reports of old monkeys were based on descriptions of small numbers of individuals whose exact ages were not known. Identifying individuals as old because they look and act the way one expects an old monkey (or human) to act, and then using them to describe what old animals look and act like leaves open the strong possibility for circularity, subjectivity and human bias in descriptions of ageing in non-human primates.

A year long study of a large sample (n = 20) of old female Japanese monkeys of known age, compared with the same number of younger adults, was conducted on the Arashiyama West colony in the mid 1980s. This study (Pavelka, 1991) showed that old females in general were not less socially integrated than younger adult females. And they did not fall in rank. In retrospect, why would we expect them to? Japanese monkey society, especially in large provisioned groups, is organised into matrilines. Kinship and the number of available close female relatives, regardless of age, will determine the size of the social network, and likely the time spent in social contact (Pavelka, 1994). Simply getting older did not trigger changes in social relationships, as one might argue it can in humans. When comparing old female Japanese monkeys to old female humans, a striking contrast emerged: most of the old monkeys were still producing babies. There did not appear to be a class of females that were post-reproductive, as all elderly human females are.

9.2   Post-reproductive female Japanese monkeys compared with menopausal women

This observation led to the investigation of the reproductive performance of old female Japanese monkeys in the Arashiyama West colony, using the many decades of genealogical and life-history data that were available. Behavioural observations at the time of the ageing study suggested that most old females continued to produce infants, however the rich database allowed us to investigate the actual relationship between age and reproduction using a large sample (n = 95) of completed lives with complete reproductive histories. The first challenge was to find a way to distinguish females who were post-reproductive (presumably no longer able to reproduce, like menopausal women) from those who still possessed these capabilities. In the absence of endocrinological and histological data, which are typically only available for small numbers of captive animals (Atsalis *et al.*, 2008), reproductive termination must be identified from reproductive records. Some studies have assumed that the entire time lag between the birth of the last infant and the death of the mother is post-reproductive; however this is clearly not the case. By this measure, every female would be considered post-reproductive (or menopausal), regardless of the age at which she died. Death within a normal interbirth interval is not evidence of the termination of reproductive abilities, and the time lag between last parturition and death is not necessarily post-reproductive.

Therefore, we used a technique developed by Caro *et al.* (1995) in which individual females are classified as reproductively terminated when the time lag between last parturition and death of the mother exceeds that of the female's own average interbirth interval by more than two standard deviations. This method requires that an individual female live significantly longer than her own lifetime interbirth interval, and thus reduces the likelihood of categorising females as reproductively terminated simply because they died before having another baby. It also allowed us to explore questions about variation in the age at which reproductive termination can occur.

Using the Arashiyama West genealogical records from 1954–1996, we extracted the data on 95 females whose reproductive histories, along with age at birth and death were known (Pavelka and Fedigan, 1999). The youngest female in the sample died at the age of 5 years, and the oldest female in the sample lived to 32.6 years. In order to use the criteria described above for identifying reproductive termination, a female had to have given birth to at least three infants, which 70 of the 95 had done. Because the 1-year behavioural study had given us the impression that old females continue to reproduce, we were surprised when this analysis revealed that 20 of the 70 females had died after living significantly longer than their own lifetime interbirth interval. In other words, they appeared to have stopped reproducing. Who were these females, and where were they during the 1-year behavioural study?

We investigated the age distribution of these reproductively terminated, or post-reproductive females, and discovered that they were mainly a small group of very old individuals who lived exceptionally long lives. Post-reproductive females ranged in age from 14.5 to 32.7 years, although the vast majority of females under age 25 were still reproducing (Figure 9.1). After age 25, reproductive termination became population-wide, as menopause is for women over age 50. Age 25 appears to be the biological endpoint for parturition in

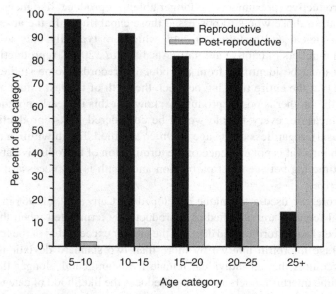

Figure 9.1. Percentage of females in the Arashiyama West population that are reproductive versus post-reproductive in each age category (years).

free-ranging Japanese macaques, as has been shown for the closely related rhesus monkey as well (Walker, 1995; Johnson and Kapsalis, 1998, 2008; Walker and Herndon, 2008). Interestingly, 24 is also the age at which the oldest known male rhesus (Bercovitch *et al.*, 2003; Bercovitch and Harvey, 2004) and Barbary (Kuester *et al.*, 1995) macaque have sired offspring.

However, age 25 for a Japanese monkey is much older than is age 50 for a human female, and in fact only a small proportion of the population, less than 3%, ever reach this age (Pavelka and Fedigan, 1999). Most females continue to reproduce until death even if they die in old age. Female Japanese monkeys aged 20 are considered to be old, as they are in the 3rd trimester of the life course, yet most of them were still producing infants, as was observed in the behavioural study. At any one time there might be only one or two of the very old and post-reproductive females in the group, and these females spend less than 10% of their lives in a post-reproductive state, compared with up to 50% of life spent post-menopausal for women. Bercovitch and Harvey (2004) also conclude that very few female non-human primates live long enough to become post-reproductive.

Thus the conclusion from this first set of analyses was that, as a life-history characteristic of the species, in which universality and phase of the life course are important, reproductive termination in Japanese monkeys bears few similarities to human female menopause. Menopause is universal and occurs only half way through the maximum lifespan of humans. It is a feature of healthy middle age (Peccei, 2001; Fedigan and Pavelka, 2011). Reproductive termination in Japanese monkeys is idiosyncratic, characteristic of a very small number of very old individuals, and not distinct from the overall late biological decline of the organism.

9.3 The evolution of menopause in humans and reproductive termination in Japanese monkeys

Menopause in humans is a challenge to explain. How can natural selection have favoured phenotypes that are not reproducing (see van Noordwijk and de Jong, 1986)? One general approach is to view menopause as a by-product of something else, such as the evolution of a long lifespan in humans, rather than as directly adaptive. Data on brain to body ratios in fossil hominids suggest that the maximum human lifespan increased from about 50 years in early hominids to approximately 120 years in *Homo sapiens* (Bogin and Smith, 1996; Hammer and Foley, 1996). However, the age at which human females cease to reproduce appears to have remained stationary at approximately 50 years. Female *H. sapiens* stop reproducing at about the same age as great apes (e.g.

Nishida *et al.*, 2003; Wich *et al.*, 2004; Robbins *et al.*, 2006; Thompson *et al.*, 2007), however 50 years of age is also near the maximum lifespan of great apes. In contrast, women are capable of living for up to another 70 years after the age of 50 in a post-reproductive state! It has been suggested that there may be inherent constraints on the mammalian reproductive system that made reproductive function unable to keep pace with the increase in human longevity (Weiss, 1981; Pavelka and Fedigan, 1991).

The other approach is to view menopause as something that is directly adaptive. Commonly (and collectively) known as the 'grandmother hypothesis', this line of argument is basically that there are greater inclusive fitness benefits to be had by ceasing to produce new infants and instead investing energy in enhancing the survival and reproductive performance of children and grandchildren (e.g. Lancaster and King, 1985; Hawkes *et al.*, 1989, 1997; Alvarez, 2000; Hawkes, 2003). Hawkes *et al.* (1989) provided evidence that post-menopausal Hadza (Tanzanian foragers) grandmothers supply sufficient surplus calories and babysitting services to allow their daughters to successfully raise more offspring. Blurton-Jones *et al.* (1999, 2005) concluded that grandmothering facilitated the evolution of earlier weaning in hominids, and Mace (2000) showed that babies in the Gambia are more likely to survive if their maternal grandmother is alive. Lahdenpera *et al.* (2004) documented that pre-modern Finnish and Franco-Canadian women with a prolonged post-reproductive lifespan had more grandchildren. There is scattered evidence from the ethnographic literature on a variety of societies that older women are a substantial help to their progeny, but whether or not this has a significant effect on inclusive fitness is still a matter of debate (e.g. compare Kaplan *et al.*, 2000 to Hawkes, 2003, 2004). There is a growing body of literature that addresses the costs and benefits of grandmothers in human populations (e.g. see chapters in volumes edited by Voland *et al.*, 2005; Hawkes and Paine, 2006).

While the grandmother hypothesis is still debated as an explanation for human menopause, reports of post-reproductive monkeys and apes (even if they are individual cases) have led to the suggestion that termination of reproduction may have been favoured by natural selection in other primates as well (Hrdy, 1981; Sommer *et al.*, 1992; Paul *et al.*, 1993; Paul, 2005). Here again the Arashiyama West database proved to be invaluable in addressing these questions. Even though the pattern of reproductive termination that we found in this population was significantly different (from a life-history perspective) from human female menopause, we did have lifetime reproductive records on 20 females who appear to have stopped reproducing at some point before death (they did not just die in an interbirth interval) and 50 others who continued to reproduce until death. Female Japanese macaques

Table 9.1. *Offspring survivorship in post-reproductive versus reproductive females*

Generation	Females	n	% survival to age 1	% survival to age 5	W	df	p
Offspring	Post-reproductive	191	0.83	0.71	0.806	1	0.396
	Reproductive	379	0.85	0.79			
Final offspring	Post-reproductive	20	0.85	0.80	1.408	1	0.235
	Reproductive	50	0.72	0.67			
Daughters' offspring	Post-reproductive	338	0.86	0.80	1.246	1	0.264
	Reproductive	555	0.83	0.77			

are good candidates for evaluating the grandmother hypothesis because they live in matrilineal societies and engage extensively in kin-directed affiliative behaviours.

We compared the survivorship of descendants of those females who ceased to reproduce (post-reproductive females or PR) with those who continued to reproduce until death (R), looking at three measures: mean survival of all offspring; mean survival of matrilineal grandchildren (daughters' offspring); and survival of final offspring, using SPSS-SURVIVAL (Fedigan and Pavelka, 2001). We found no significant difference in offspring survivorship between PR and R females. In fact infant survival rates are remarkably similar for PR and R females: 85 versus 83% to age 1, and 71 versus 79% for survival to age 5 (Table 9.1).

Although not statistically significant, the data suggest that survival of the final infant of PR females is actually 13% better than it is for the younger, still reproductive females. However by definition, PR females live significantly longer than their own average lifetime IBI after the birth of their last infant than do R females, which means that PR females live well past the age at which their last offspring is weaned. (This will be discussed further below.) We ran a regression on the length of maternal care (defined as the number of months the mother lived after the birth of her final infant) versus the lifespan of the final offspring and found a significant positive relationship between length of maternal care and infant survivorship ($F = 12.485$, df = 68, $p = 0.001$, $R = 0.394$). Since the longer the mother lives, the more care she will provide, the survivorship of final offspring is strongly related to the length of time the mothers survive after giving birth. This same pattern is probably reflected in the association found between longer interbirth intervals and greater infant survival in mountain baboons (Lycett *et al.*, 1998).

Overall, cessation of reproduction did not result in significantly improved survivorship of offspring or maternal grandchildren. Thus we explored a variety of different variables in trying to determine what was different about PR versus R females. For example, heavier female macaques might continue to cycle and reproduce, since per cent of body fat has been linked to later menopause in human females (Brambilla and McKinley, 1989). Or, if there were a genetic proclivity to become post-reproductive (Caro *et al.*, 1995), perhaps it would occur more often in one matriline than another.

Because the Arashiyama West dataset is so rich, we were able to test our two groups of females for differences in a wide variety of social and life-history variables, including dominance rank, matrilineal affiliation, cause of death, body weight, sex ratio of offspring, age at first birth and fecundity rates. We found almost no significant differences in any of these variables. What we did find was that PR females lived significantly longer (24.6 years as opposed to 17.4 years for R females; $t = -5.475$, df = 68, $p < 0.001$) and gave birth to significantly more offspring (9.7 compared to 7.7 for R females; $t = -2.157$, df = 68, $p = 0.035$).

From these data it certainly seems that natural selection is favouring PR females, however the real benefit is due to longevity. If a female Japanese monkey lives to be over 25 she will have had that many more years to produce infants. Because age at first birth varies little (age 5 or 6) a longer lifespan is the same as a longer reproductive lifespan. So old post-reproductive females live longer and have more offspring; however, neither these offspring nor the maternal grandchildren have better survivorship that can be tied to their post-reproductive status. Likewise the non-significant trend for the last infants of PR females to survive better than the last infants of R females is again due to them living longer after the last baby is born and thus not orphaning it when it is too young to survive.

The fact that we did not find any way to distinguish between these two categories of females other than variables that follow logically from differential longevity suggests that reproductive termination in this species is associated with enhanced longevity and its consequences rather than with enhanced investment in care of offspring or any adaptive or genetic package that distinguishes post-reproductive females. Packer *et al.* (1998), who provide the only comparable test of the grandmother hypothesis in non-humans, concluded that post-reproductive olive baboon and African lion females did not enhance the fitness of grandoffspring or older offspring, and that reproductive cessation in these species results from the general ageing process, rather than from direct selection favouring post-reproductive grandmothers.

Why might post-reproductive human females be able to improve survivorship of descendants but monkeys (and other non-human primates) are not?

This difference may be tied to another distinction between human and non-human primate ageing that was highlighted in the behavioural study of ageing in the Arashiyama West female Japanese macaques: the absence of food sharing (Pavelka, 1991). Japanese macaque females extend affiliative and potentially beneficial behaviours to their younger matrilineal kin, but they do not provision juvenile kin, which are capable of feeding themselves from weaning age. Macaques do not share food with or provide food for juvenile offspring and grandoffspring, the primary caregiving behaviour that is described in the human literature on the grandmother hypothesis (Lancaster and Lancaster, 1983; Lancaster and King, 1985; Hawkes *et al.*, 1989, 1997, 1998; Hill and Hurtado, 1991, 1996; Kaplan, 1997; O'Connell *et al.*, 1999).

9.4 How common are post-reproductive grandmother Japanese macaques?

Anthropological research specifically targets post-reproductive grandmothers and in humans, there is a strong correlation between being post-reproductive (menopausal) and being a grandmother. However, in our analyses described above, we did not distinguish the grandmothers from the post-reproductive females. Not all post-reproductive females were necessarily grandmothers, and not all grandmothers were necessarily post-reproductive. Thus we next used the rich Arashiyama West database to target the theoretically important post-reproductive mothers and grandmothers and to find out how common they are, and how great is their potential to have an impact on the reproductive output of their descendants (Pavelka *et al.*, 2002).

Eight of the 70 females who could be categorised as reproductive or post-reproductive did not produce any daughters (and hence grandchildren that were known to us), so these next analyses are based on 62 grandmothers and their 175 daughters and 905 grandchildren. For tests in which we needed the daughters' death dates, we were able to use only 88 of the 175 daughters, and only 74 of these produced infants. Of the 905 grandchildren born, we have complete information on 886 individuals who were included in the survival analyses. Of these, 504 survived to age 5. The probability of survival to age 5 was calculated based on the survivorship of all of the 886 individuals in the sample.

Post-reproductive grandmothers turned out to be very rare (Figure 9.2). Of the 175 daughters in our sample, 37% had no mother (their mother was dead), 60% had a reproductive mother (present but still dealing with infants of her own), and only 3% (five females) had a post-reproductive mother available to help them when they reached reproductive age. We also looked at how much time they would have their mothers available to them and again, it was not

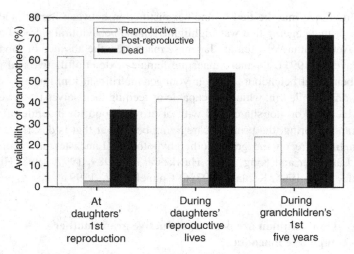

Figure 9.2. The availability (alive versus dead) and reproductive state of grandmothers for daughters at the time of their first reproduction, during their entire reproductive lives and at the time of their daughters' first reproduction.

much. Only 0.41/9.71 years or 4.2% of the reproductive lifespan of the daughters were spent with a post-reproductive mother available to help them. This represents less than 5 months for the average female in the population. This pattern was further reflected in the next generation, in the time available for a post-reproductive grandmother to have an impact on her grandchildren's survival – the essence of the grandmother hypothesis. Most of the grandchildren's first 5 years of life was spent without a living grandmother (3.6 years or 72%), reproductive or otherwise. The grandchildren who lived to age 5 actually had a grandmother available to them for, on average, less than a year and a half. For most of this time, a grandmother had her own baby to care for. The amount of time that these grandchildren had with a post-reproductive grandmother was less than 2 months.

9.5   The value of mothers

This analysis did show some interesting benefits to the 63% of females who had a mother available to them when they reached sexual maturity. Females who had a live mother (regardless of her reproductive status) when they reached reproductive age were significantly more likely to have their first baby at age 5 instead of age 6 (Chi-square: $\chi^2 = 8.942$, $p = 0.030$). How does the presence of a mother result in a significantly earlier age at first birth for her daughters?

Table 9.2. *Effect of mother's status on daughter's age at first birth and interbirth intervals (IBI)*

Daughter's reproduction	Mother's status			
	Dead	Alive	Alive and reproductive	Alive and post-reproductive
Age at first birth (years)[1]				
$\bar{x} \pm$ s.d.	6.0 ± 1.0	5.6 ± 0.9	5.6 ± 0.9	6.1 ± 0.7
n	64	111	106	5
Average interbirth interval (months)[2]				
$\bar{x} \pm$ s.d.	19.2 ± 8.5	18.1 ± 7.3	18.5 ± 7.4	16.0 ± 5.7
No. of IBIs	451	273	239	34

[1] *t*-test: Dead vs. Alive, $t_{173} = 2.565$, $p = 0.01$; ANOVA: Dead/Alive-R/Alive-PR, $F_{2,172} = 3.952$, $p = 0.02$.

[2] *t*-test: Dead vs. Alive, $t_{722} = 1.702$, $p = 0.09$; ANOVA: Dead/Alive-R/Alive-PR, $F_{2,271} = 2.858$, $p = 0.06$.

It may be that the presence of the mother improves a young female's chance of having a 'successful' first oestrus (one that results in a conception) at age 4.5. The first oestrus period for female Japanese monkeys requires that these young females venture out of their tight female kinship units for the first time in order to establish contact with unrelated adult males. Most young females would have had little need or opportunity to interact with unrelated adult males previously. The inexperienced behaviour of these young Japanese monkey females increases the likelihood that adult males will target them for aggression (McDonald, 1985).

In vervet monkeys, adult daughters whose mothers were still living in the group received less aggression and were defended more often than were young adult females whose mother had died (Fairbanks, 1988). Thus, the mother of the young female, through the agonistic support she provides to her daughter, may help to increase confidence on the part of the daughter and/or reduce the frequency and intensity of serious aggression from adult males, thus increasing the likelihood of the daughter forming a successful consort. As Bercovitch and Harvey (2004) pointed out, it is important to note that although life-history theory proposed that lifetime reproductive success is improved with earlier age at first birth (Stearns, 1992), this has not been shown to be the case in female primates (cf. Fedigan *et al.*, 1986 for Arashiyama West Japanese macaques).

The presence of a live mother also appeared to be of some benefit to females in terms of shortening their interbirth intervals (Table 9.2), although this trend was not quite statistically significant. Females had, on average, shorter inter-birth intervals when their mothers were alive (18.1 months) compared with when they were dead (19.2 months). This trend became even greater when the mother became post-reproductive, and the average interbirth intervals of her daughter dropped from 19 to 16 months.

9.6   The adaptive value of grandmothers

The essence of the grandmother hypothesis is that grandmothers – specific-ally post-reproductive grandmothers – are able to enhance the survivorship of their grandchildren. Thus we compared the probability of survivorship to age 1 and to age 5 by status of grandmother during first year of the grandchild's life (Dead, Alive R, Alive PR). Here we found a somewhat surprising significant difference in survivorship to age 1 depending on the status of the grandmother (Wilcoxon–Gehan test, $W2 = 6.29$, $p = 0.043$). Specifically, grandchildren with a living post-reproductive grandmother were significantly more likely to live to age 1 than were grandchildren with either a dead grandmother or a live one who was herself still reproducing ($W1 = 3.99$, $p = 0.046$ and $W1 = 6.47$, $p = 0.011$, respectively).

Nakamichi *et al.* (2010) described two instances of grandmother Japanese monkeys, without dependent young, providing essential care to their two granddaughters (one aged 2 months, the other aged 14 months). Whatever the PR grandmother was doing to help her grandchild reach age 1 at Arashiyama West did not extend to helping it reach age 5: there were no differences in the survivorship of grandchildren to age 5 among the three groups of grandmoth-ers ($W2 = 1.272$, $p = 0.53$). But recall that on average, most grandchildren had a post-reproductive grandmother for only a couple of months, so clearly what-ever she was able to do was centred on the first year, in the pre-weaning stage of the grandchild's life.

9.7   Costs of old age reproduction

All of the analysis and discussion so far has focused on the possible benefits that might accrue to females who ceased reproducing. However, in order for natural selection to favour cessation of reproduction over continued reproduc-tion, there would not only have to be benefits to the former, but costs to the latter. All other things being equal, continued reproduction, which we know to be the norm for Japanese macaque females until well into old age, is what we would expect, unless continuing to produce babies were to become too costly.

In humans the latter may well be the case, given the difficulties in labour and delivery that are side-effects of the evolution of bipedalism and enceph-alisation. Giving birth to a large-brained infant through the narrow pelvis of a biped puts human females of all ages at considerable risk, and perhaps these risks increase as women get older. We certainly know that children born to older human mothers carry a significant risk of genetic defects, although these are likely a by-product of the same constraints on the mammalian system mentioned earlier: a fixed number of eggs are available at birth and these are suspended in anaphase. The older the woman, the longer the eggs have been suspended and the greater the chance of chromosomal mutation.

Are there costs associated with continued reproduction in older Japanese monkeys that might tip the balance in favour of caring for descendants instead? It is generally recognised that reproduction, particularly lactation, is energetically costly for primate females (e.g. Altmann, 1983; Lee, 1996a; Dunbar *et al.*, 2002; Miller *et al.*, 2006), and these costs would be greater for females in poorer overall body condition (Lee, 1996b) as ageing and/or post-reproductive females might be expected to be. Thus first, we looked at weight loss. Lactation in females is associated with loss of body mass in chacma baboons (Barrett *et al.*, 2006) and this is an effect which might be more pronounced in older individuals.

In the Arashiyama West females, we compared the body weights of reproductive and post-reproductive females, the latter being significantly older than the former, and found no difference. Both weighed close to 8 kg. PR females ($n = 19$) who were weighed at least once in adulthood weighed on average 7.96 kg (s.d. $= 1.81$), and R females ($n = 32$) weighed on average 8.15 kg (s.d. $= 3.7$; $t = -0.392$; df $= 49$; $p = 0.697$).

Secondly, the most obvious way to diminish future reproduction would be for the female herself to die. For example, there may be a cost to a fast pace of reproduction (Bercovitch and Berard, 1993). Females who produce infants in close succession might be expected to pay a price in terms of their own life-span. However, this was not the case with the Arashiyama West females. We calculated the average interbirth interval for all females in our sample and ran a regression of interbirth interval on age at death, and found no relationship ($F_{1,72} = 0.776$, $p = 0.381$). There appeared to be no trade-off for females who had infants more closely spaced than those who did not. This result may be tied to the fact that the Arashiyama West monkeys were provisioned and the artificially food-enhanced environment may have permitted enhanced reproduction as well. Asquith (1989) argued that an increase in population size occurs in provisioned populations due to reduced infant mortality and shortened interbirth intervals.

Do females who continue to reproduce when they are old pay a cost in terms of their own survival? We addressed this question by first running a logistic regression to determine whether maternal survival for 12 months subsequent to parturition declines with age. We found that if a female gave birth, her age did not affect her probability of surviving for 12 months after parturition. Older females were no more likely to die after giving birth than were younger females (*Wald* = 0.409, df = 1, p = 0.520). However, for females who failed to give birth, their probability of surviving for 12 months subsequent to a given age (e.g. 6 years, 7 years, 8 years) did decline as they got older (*Wald* = 66.1, df = 1, p < 0.001). Furthermore, we used a logistic covariance analysis to compare females who gave birth versus those who did not give birth at ages 4–25 and found that at every age (except age 5), the probability of survival was greater for those females that gave birth than those that did not give birth. This result suggests that it is not costly to give birth, rather it is beneficial.

However, another explanation could be that healthy females are the ones giving birth, and thus also the ones likely to survive. This pattern too might be related to the specific conditions at Arashiyama West where the animals were provisioned and predation was minimal. In wild baboons, females with small infants were more vigilant and spent less time feeding (Barrett *et al.*, 2006), and again, this effect might be expected to be more pronounced in older females who might be more vulnerable to predation and weight loss.

Thirdly, is there a cost to old age reproduction in terms of infant survival? Do older mothers provide less or lower-quality infant care? For the most part, the literature suggests not. Quite a few studies have shown that older females experience higher offspring survival, especially in birds (e.g. Crawford, 1977; De Steven, 1977; Blus and Keahey, 1978; Curio, 1982; Pugesek and Diem, 1983) and also in some mammals (e.g. Ozoga and Verme, 1986; Festa-Bianchet, 1988; Green, 1990; Hastings and Testa, 1998; but see Ericsson *et al.*, 2001).

Reproductive costs are generally assumed to be highest in primiparous females due to inexperience, small body size and infant suckling patterns (Bercovitch *et al.*, 1998). Not surprisingly, when we ran survival analyses, we found no significant differences in infant survival to 1 year comparing mothers grouped into four age classes (n = 474 infants born to mothers 5–10 years; 257 infants to mothers 11–15 years; 132 infants to mothers 16–20 years; 30 infants to mothers 20–25 years, $W2$ = 1.670; df = 3, p = 0.644). Likewise there was no effect of maternal age on survival of offspring to age 5. The survival rates were very high in all cases: 85% of infants survive to age 1, regardless of their mother's age, and 77–83% survive to age 5 regardless of their mother's age.

Overall infant survival may be unaffected by mother's age, or indeed be enhanced by it because older mothers have more experience. But no amount of maternal experience is going to help the last infant if its mother dies. A fourth obvious cost of old age reproduction would be the greater likelihood that a female would die and leave behind an unweaned orphan. One might suggest that the death of a final infant due to being orphaned is a waste, rather than a cost, however it is a cost if those females could have had a significant impact on the survivorship of a grandchild, had they not had that last baby. We know that the presence of a post-reproductive grandmother does significantly improve the survival to age 1 of their grandchildren.

In the data presented above comparing the survivorship of the last infant of post-reproductive (older) and reproductive (younger) females, we in fact found the opposite trend: post-reproductive females had a significantly higher survival of their last offspring than did reproductive females. However, keep in mind that post-reproductive females were identified by the fact that they lived a long time after the birth of their last baby, so by definition these females were not dying and leaving unweaned orphans. Only reproductive females, by definition, could do that.

We looked more closely at all cases of infants who died in their first year of life. Weaning is typically completed when the infant is 1 year of age. We divided the data into individuals who died before reaching age 1 (n = 196) and those who did not (n = 438) and found that the average age of the mothers of those who died unweaned was younger (10.8 years) than those whose offspring lived longer than 1 year (18 years). Again, losing infants appeared more characteristic of younger adult females than of older ones. We ran a logistic regression to test the effect of maternal age on the likelihood of an offspring dying young (less than 1 year of age versus more than 1 year of age), and neither the first (*Wald* = 135; df = 1, *p* = 0.245) nor the second (*Wald* = 2.25; df = 1; *p* = 0.135) order term was significant. In other words, there was no effect of maternal age. Infant mortality was 13% and it was unrelated to the age of the mother. Indeed whatever trend there might be, the data suggest that infants of young mothers are more likely to die than offspring of older females (Figure 9.3).

Finally, arguments have been made that the sex of offspring may confer different costs on the mother. For example Bercovitch *et al.* (2000) and Mueller (2001) have suggested that male infants are more costly to their mothers, especially in dimorphic species, and production of more sons than daughters has been shown to affect the longevity of women in premodern human populations (Helle *et al.*, 2002a, 2002b; but see Beise and Voland, 2002). Takahata *et al.* (1995) and van Schaik *et al.* (1989) however have argued that macaque daughters are more costly.

Thus we investigated whether mothers who produced more sons than daughters during their reproductive lives experience shorter lifespans. We ran a survival analysis on females categorised into those that had more sons than daughters during their lifetimes (n = 47, $\overline{x}_{lifespan}$ = 270 months), those that produced more daughters than sons (n = 67, $\overline{x}_{lifespan}$ = 245 months) and those

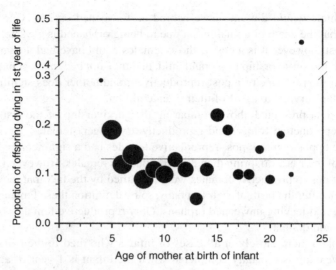

Figure 9.3. The proportion of infants born to mothers of different ages that died at younger than 1 year of age.

that produced equal sex ratios of offspring (n = 23, $\overline{x}_{lifespan}$ = 293 months). We found no significant differences in longevity among these three categories (*W2* = 3.225, df = 2, *p* = 0.1194).

9.8   Closing comments

The maintenance of long-term reproductive records on the large sample of Arashiyama West females resulted in a rich database that has allowed us to gain considerable insight into ageing in old females and the possible evolution of post-reproductive lifespans in monkeys and humans. In the Arashiyama West population, as is the case for most published accounts of free- and semifree-ranging primates (Bercovitch and Harvey, 2004; Fedigan and Pavelka, 2011) most females continue to give birth until the end of their lifespans and few will live to reach a post-reproductive stage of life.

The age at which termination of reproduction occurs in Japanese monkeys is 25 years, although this is very close to the maximum lifespan for the species, with few individuals living past it. Reproductive termination was characteristic of only a small cohort of very old females and post-reproductive females who were significantly older at time of death than were reproductive ones. When we investigated post-reproductive grandmothers specifically, we found them to be extremely rare. However, in the rare cases in which a female found herself post-reproductive and with a grandchild, she did appear to be able to increase its chance of surviving to age 1.

Nonetheless, the reproductive termination that we have documented in female Japanese macaques of the Arashiyama West population probably occurs too late in life, with too few females reaching and remaining in this stage for any substantial proportion of their descendant's lives, to have sufficient inclusive fitness effects to compensate for the loss of the grandmother's direct reproductive output, as theorised by the grandmother hypothesis. Reproductive termination in this population appears to be a by-product of selection favouring longevity rather than as a result of direct selection for reproductive termination (Fedigan and Pavelka, 2011). This conclusion is further supported by the new data presented here in which there do not appear to be reproductive costs that increase with age. Continuing to reproduce until death is beneficial, not costly, to the fitness of Arashiyama West female Japanese macaques.

Acknowledgements

We are grateful to the many scientists and caretakers who have studied the Arashiyama monkeys in both Japan and Texas over the past half century and who have made invaluable contributions to the Arashiyama West life-history database. We would also like to thank Alison Behie, Greg Bridgett, Tracy Wyman and John Addicott for assistance with the analyses and preparation of this manuscript. Our research has been supported over many years by Discovery Grants from the Natural Sciences and Engineering Research Council of Canada (NSERC).

References

Altmann, J. (1983). Costs of reproduction in baboons (*Papio cynocephalus*). In *Behavioral Energetics: The Cost of Survival in Vertebrates*. 7th Biosciences Colloquium. Colombus, OH: Ohio State University Press, pp. 67–88.

Alvarez, H. (2000). Grandmother hypothesis and primate life histories. *American Journal of Physical Anthropology*, **113**, 435–450.

Asquith, P. (1989). Provisioning and the study of free-ranging primates: history, effects and prospects. *Yearbook of Physical Anthropology*, **32**, 129–158.

Atsalis, S., Margulis, S. and Hof, P. R. (2008). *Primate Reproductive Aging: Cross-taxon Perspectives*. Basel: S. Karger.

Barrett, L., Halliday, J. and Henzi, S. P. (2006). The ecology of motherhood: the structuring of lactation costs by chacma baboons. *Journal of Animal Ecology*, **75**, 875–886.

Beise, J. and Voland, E. (2002). A multilevel event history analysis of the effects of grandmothers on child mortality in a historical German population (Krummhorn, Ostfriesland, 1720–1874). *Demographic Research*, **7**, 469–498.

Bercovitch, F. and Berard, J. (1993). Life history costs and consequences of rapid reproductive maturation in female rhesus macaques. *Behavioral Ecology and Sociobiology*, **32**, 103–109.

Bercovitch, F. and Harvey, N. (2004). Reproductive life history. In *Macaque Societies: A Model for the Study of Social Organization*, eds. B. Thierry, M. Singh and W. Kaumans. New York, NY: Cambridge University Press, pp. 61–80.

Bercovitch, F., Lebron, M., Martinez, H. S. and Kessler, M. J. (1998). Primigravidity, body weight, and costs of rearing first offspring in rhesus macaques. *American Journal of Primatology*, **46**, 135–144.

Bercovitch, F., Widdig, A. and Neuberg, P. (2000). Maternal investment in rhesus macaques (*Macaca mulatta*): reproductive costs and consequences of raising sons. *Behavioral Ecology and Sociobiology*, **48**, 1–11.

Bercovitch, F. B., Widdig, A., Trefilov, A. *et al.* (2003). A longitudinal study of age-specific reproductive output and body condition among male rhesus macaques, *Macaca mulatta. Naturwissenschaften*, **90**, 309–312.

Blurton Jones, N. G., Hawkes K. and O'Connell, J. F. (1999). Some ideas about evolution of the human life history. In *Comparative Primate Socioecology*, ed. P. C. Lee. Cambridge: Cambridge University Press, pp. 140–166.

    (2005). Hadza fathers and grandmothers as helpers: residence data. In *Hunter-Gatherer Childhoods: Evolutionary, Developmental and Cultural Perspectives*, eds. B. S. Hewlett and M. E. Lamb. Brunswick: Transaction, pp. 214–236.

Blus, L. and Keahey, J. (1978). Variation in the reproductivity with age in the brown pelican. *The Auk*, **95**, 128–134.

Bogin, B. and Smith, B. (1996). Evolution of the human life cycle. *American Journal of Human Biology*, **8**, 703–716.

Brambilla, D. and McKinley, S. (1989). A prospective study of factors affecting age at menopause. *Journal of Clinical Epidemiology*, **42**, 1031–1039.

Caro, T., Sellen, D. W., Parish, A. *et al.* (1995). Termination of reproduction in non-human and human female primates. *International Journal of Primatology*, **12**, 205–220.

Crawford, R. (1977). Breeding biology of year-old and older female red-winged and yellow-headed blackbirds. *Wilson Bulletin*, **89**, 73–80.

Cumming, E. (1975). Engagement with an old theory. *International Journal of Aging and Human Development*, **6**, 187–191.

Cumming, E. and Henry, W. (1961). *Growing Old: The Process of Disengagement*. New York, NY: Basic Books.

Curio, E. (1982). Why do young birds reproduce less well? *Ibis*, **125**, 400–404.

De Steven, D. (1977). The influence of age on the breeding biology of the tree swallow (*Iridoprocne bicolor*). *Ibis*, **120**, 516–523.

Dunbar, R., Hannah-Stewart, L. and Dunbar, P. (2002). Forage quality and the costs of lactation for female gelada baboons. *Animal Behaviour*, **64**, 801–805.

Ericsson, G., Kjell, W., Ball, J. P. and Broberg, M. (2001). Age-related reproductive effort and senescence in free-ranging moose, *Alces alces*. *Ecology*, **82**, 1613–1620.

Fairbanks, L. (1988). Vervet monkey grandmothers: effects on mother–infant relationships. *Behaviour*, **104**, 176–188.

Fedigan, L. and Pavelka, M. (2001). Is there adaptive value to reproductive termination in Japanese macaques? A test of maternal investment hypotheses. *International Journal of Primatology*, **22**, 109–125.

(2011). Menopause: interspecific comparisons of reproductive termination in female primates. In *Primates in Perspective*, eds. C. J. Campbell, A. Fuentes, K. C. MacKinnon, S. K. Bearder and R. Stumpf. New York, NY: Oxford University Press, pp. 488–499.

Fedigan, L. M., Fedigan, L., Gouzoules, H., Gouzoules, S. and Koyama, N. (1986). Lifetime reproductive success in female Japanese macaques. *Folia Primatologica*, **47**, 143–157.

Festa-Bianchet, M. (1988). Age-specific reproduction of bighorn ewes in Alberta, Canada. *Journal of Mammalogy*, **69**, 157–160.

Green, W. (1990). Reproductive effort and associated costs in bison (*Bison bison*): do older mothers try harder? *Behavioral Ecology*, **1**, 148–160.

Hammer, M. and Foley, R. (1996). Longevity and life history in human evolution. *Human Evolution*, **11**, 61–66.

Hastings, K. and Testa, J. (1998). Maternal and birth colony effects on survival of Weddell seal offspring from McMurdo Sound, Antarctica. *Journal of Animal Ecology*, **67**, 722–740.

Hauser, M. and Tyrell, G. (1984). Old age and its behavioral manifestations: a study on two species of macaque. *Folia Primatologica*, **43**, 24–35.

Hawkes, K. (2003). Grandmothers and the evolution of human longevity. *American Journal of Human Biology*, **15**, 380–400.

(2004). The grandmother effect. *Nature*, **428**, 128–129.

Hawkes, K. and Paine, R. (2006). *The Evolution of Human Life History*. Santa Fe, NM: School of American Research Press.

Hawkes, K., O'Connell, J. F. and Blurton-Jones, N. G. (1989). Hardworking Hadza grandmothers. In *Comparative Socioecology of Mammals and Man*, eds. V. Standen and R. Foley. London: Blackwell, pp. 341–366.

(1997). Hadza women's time allocation, offspring provisioning and the evolution of post-menopausal life spans. *Current Anthropology*, **18**, 551–577.

Hawkes, K., O'Connell, J., Blurton Jones, N. G., Alvarez, H. P. and Charnov, E. L. (1998). Grandmothering, menopause, and the evolution of human life histories. *Proceedings of the National Academy of Sciences USA*, **95**, 1336–1339.

Helle, S., Kaar, P. and Jokela, J. (2002a). Human longevity and early reproduction in pre-industrial Sami populations. *Journal of Evolutionary Biology*, **15**, 803–807.

Helle, S., Lummaa, V. and Jokela, J. (2002b). Sons reduced maternal longevity in pre-industrial humans. *Science*, **296**, 1085.

Hill, K. and Hurtado, A. (1991). The evolution of premature reproductive senescence and menopause in human females: an evaluation of the 'grandmother hypothesis'. *Human Nature*, **2**, 313–350.

(1996). *Ache Life History. The Ecology and Demography of a Foraging People*. New York, NY: Aldine de Gruyter.

Hrdy, S. (1981). 'Nepotists' and 'Altruists': The behavior of old female monkeys among macaque and langur monkeys. In *Other Ways of Growing Old: Anthropological Perspectives*, eds. P. Amoss and S. Harrell. Stanford, CT: Stanford University Press, pp. 59–96.

Huffman, M. A. (1990). Some socio-behavioral manifestations of old age. In *The Chimpanzees of the Mahale Mountains, Sexual and Life History Strategies*, ed. T. Nishida. Tokyo: University of Tokyo Press, pp. 235–255

Johnson, R. and Kapsalis, E. (1998). Menopause in free-ranging rhesus monkeys: estimated incidence, relation to body condition and adaptive significance. *International Journal of Primatology*, **19**, 751–765.

(2008). Heterogeneity of reproductive aging in free-ranging female rhesus monkeys. In *Primate Reproductive Aging: Cross-taxon Perspective*, eds. S. Atsalis, S. W. Margulis and P. R. Hof. Basel: S. Karger, pp. 62–79.

Kaplan, H. (1997). The evolution of the human life course. In *Between Zeus and the Salmon – The Biodemography of Longevity*, eds. K. W. Wachter and C. E. Finch. Washington, DC: National Academy Press, pp. 175–211.

Kaplan, H., Hill, K., Lancaster J. and Hurtado, A. M. (2000). A theory of human life history evolution: diet, intelligence and longevity. *Evolutionary Anthropology*, **9**, 156–185.

Kuester, J., Paul, A. and Arnemann, J. (1995). Age-related and individual differences of reproductive success in male and female Barbary macaques. *Primates*, **36**, 461–476.

Lahdenpera, M., Lummaa, V., Helle, S., Tremblay, M. and Russell, A. F. (2004). Fitness benefits of prolonged post-reproductive lifespan in women. *Nature*, **428**, 178–181.

Lancaster, J. and King, B. (1985). An evolutionary perspective on menopause. In *A New View of Middle-aged Women*, eds. J. K Brown and V. Kerns. Hadley: Bergin and Garvey Press, pp. 13–20.

Lancaster, J. and Lancaster, C. (1983). The Hominid adaptation. In *How Humans Adapt: A Biocultural Odyssey, Proceedings of the Seventh International Smithsonian Symposium*. Washington, DC: Smithsonian Institute, pp. 33–66.

Lee, P. (1996a). Lactation, condition and sociality: constraints on fertility of non-human mammals. In *Variability in Human Fertility*, eds. L. Rosetta and C. G. N. Mascie-Taylor. Cambridge: Cambridge University Press, pp. 25–45.

(1996b). The meanings of weaning: growth, lactation and life history. *Evolutionary Anthropology*, **5**, 87–96.

Lycett, J. S., Henzi, P. and Barrett, L. (1998). Maternal investment in mountain baboons and the hypothesis of reduced care. *Behavioral Ecology and Sociobiology*, **42**, 49–56.

Mace, R. (2000). Evolutionary ecology of human life history. *Animal Behaviour*, **59**, 1–10.

McDonald, M. (1985). Courtship behavior of female Japanese monkeys. *Canadian Review of Physical Anthropology*, **4**, 67–75.

Miller, K., Bales, K. Ramos, J. H. and Dietz, J. M. (2006). Energy intake, energy expenditure and reproductive costs of female wild golden lion tamarins (*Leontopithecus rosalia*). *American Journal of Primatology*, **68**, 1037–1053.

Mueller, T. (2001). Reproductive costs in captive female baboons (*Papio hamadryas anubis*). *American Journal of Primatology*, **54**, 35–36.

Nakamichi, M., Onishi, K. and Yamada, K. (2010). Old grandmothers provide essential care to their young granddaughters in a free-ranging group of Japanese monkeys (*Macaca fuscata*). *Primates*, **51**, 171–174.

Nishida, T., Corp, N., Hamai, M. *et al.* (2003). Demography, female life history, and reproductive profiles among the chimpanzees of Mahale. *American Journal of Primatology*, **59**, 99–121.

O'Connell, J., Hawkes, K. and Blurton Jones, N. G. (1999). Grandmothering and the evolution of *Homo erectus*. *Journal of Human Evolution*, **36**, 461–485.

Ozoga, J. and Verme, L. (1986). Relation of maternal age to fawn-rearing success in white-tailed deer. *Journal of Wildlife Management*, **50**, 480–486.

Packer, C., Tatar, M. and Collins, A. (1998). Reproductive cessation in female mammals. *Nature*, **392**, 807–811.

Paul, A. (2005). Primate predispositions for human grandmaternal behavior. *Grandmotherhood: The Evolutionary Significance of the Second Half of Female Life*, eds. E. Voland, A. Chasiotis and W. Schiefenhovel. New Brunswick: Rutgers University Press, pp. 21–37.

Paul, A., Kuester, J. and Podsuweit, D. (1993). Reproductive senescence and terminal investment in female Barbary macaques (*Macaca sylvanus*) at Salum. *International Journal of Primatology*, **14**, 105–124.

Pavelka, M. S. M. (1991). Sociability in old female Japanese monkeys: human vs. nonhuman primate aging. *American Anthropologist*, **93**, 588–598.

  (1994). The nonhuman primate perspective: old age, kinship and social partners in a monkey society. *Journal of Cross-cultural Gerontology*, **9**, 219–229.

Pavelka, M. S. M. and Fedigan, L. M. (1991). Menopause: a comparative life history perspective. *Yearbook of Physical Anthropology*, **34**, 13–34.

  (1999). Reproductive termination in female Japanese monkeys: a comparative life history perspective. *American Journal of Physical Anthropology*, **109**, 455–464.

Pavelka, M. S. M., Fedigan, L. M. and Zohar, S. (2002). Availability and adaptive value of reproductive and post-reproductive Japanese macaque mothers and grandmothers. *Animal Behaviour*, **67**, 407–414.

Peccei, J. (2001). Menopause: Adaptation or epiphenomenon? *Evolutionary Anthropology*, **10**, 43–57.

Pugesek, B. and Diem, K. (1983). A multivariate study of the relationship of parental age to reproductive success in California Gulls. *Ecology*, **64**, 829–839.

Robbins, A. M., Robbins, M. M. and Gerald-Steklis, N. (2006). Age-related patterns of reproductive success among female mountain gorillas. *American Journal of Physical Anthropology*, **131**, 511–521.

Sommer, V., Srivastava, A. and Borries, C. (1992). Cycles, sexuality and conception in free-ranging langurs (*Presbytis entellus*). *American Journal of Primatology*, **28**, 1–27.

Stearns, S. (1992). *Life History Evolution*. New Haven, CT: Yale University Press.

Takahata, Y., Koyama, N., Huffman, M., Norikoshi, K. and Suzuki, H. (1995) Are daughters more costly to produce for Japanese macaque mothers? Sex of the offspring and subsequent interbirth intervals. *Primates*, **35**, 571–574.

Thompson, M. E., Jones, J. H., Pusey, A. E. *et al.* (2007). Aging and fertility patterns in wild chimpanzees provide insights into the evolution of menopause. *Current Biology*, **17**, 2150–2156.

van Noordwijk, A. and de Jong, G. (1986). Acquisition and allocation of resources: their influence on variation in life history tactics. *American Naturalist*, **128**, 137–142.

van Schaik, C., Netto, W., van Amerongen, A. and Westland, H. (1989). Social rank and sex ratio of captive long-tailed macaque females (*Macaca fascicularis*). *American Journal of Primatology*, **19**, 147–161.

Voland, E., Chasiotis, A. and Schiefenhovel, W. (2005). *Grandmotherhood: The Evolutionary Significance of the Second Half of the Female Life*. Piscataway, NJ: Rutgers University Press.

Walker, M. (1995). Menopause in female rhesus monkeys. *American Journal of Primatology*, **35**, 59–71.

Walker, M. and Herndon, J. (2008). Menopause in nonhuman primates. *Biology of Reproduction*, **79**, 398–406.

Waser, P. (1978). Post-reproductive survival and behavior in a free-ranging female mangabey. *Folia Primatologica*, **29**, 142–160.

Weiss, K. (1981). Evolutionary perspectives on human aging. In *Other Ways of Growing Old: Anthropological Perspectives*, eds. P. Amoss and S. Harrell. Stanford, CA: Stanford University Press, pp. 25–58.

Wich, S. A., Utami-Atmoko, S. S., Mitra Setia, T. *et al.* (2004). Life history of wild Sumatran orangutans (*Pongo abelii*). *Journal of Human Evolution*, **47**, 385–398.

10 Is female homosexual behaviour in Japanese macaques truly sexual?

PAUL L. VASEY AND
DOUG P. VANDERLAAN

Female–female mounting between Kusha-59–71–76–87–01 and Momo-59–78–92–99 at Arashiyama, November 2010 (photo by J.-B. Leca).

10.1   Introduction

A large body of observational evidence indicates that interactions involving same-sex courtship, mounting and pair-bonding are phylogenetically widespread (e.g. Vasey, 1995; Dixson, 1998; Bagemihl, 1999; Sommer and Vasey, 2006; MacFarland *et al.*, 2010; Poiani, 2010). The frequency with which such

The Monkeys of Stormy Mountain: 60 Years of Primatological Research on the Japanese Macaques of Arashiyama, eds. Jean-Baptiste Leca, Michael A. Huffman and Paul L. Vasey. Published by Cambridge University Press. © Cambridge University Press 2012.

behaviours are manifested varies greatly both inter-specifically and intra-specifically, but the existence of such interactions seems to be relatively uncontested by the scientific community. Although it is recognised that manipulation of steroid hormones (e.g. Adkins-Regan, 1988, 1998; Adkins-Regan *et al.*, 1997), lesions to the medial preoptic–anterior hypothalamus (Paredes and Baum, 1995; Kindon *et al.*, 1996; Paredes *et al.*, 1998) and rearing in uni-sexual social groups can promote the expression of same-sex courtship, mounting and pair-bonding in certain species (Zenchak *et al.*, 1981; Price and Smith, 1984/1985; Price *et al.*, 1988; Adkins-Regan and Krakauer, 2000), it is commonly acknowledged that such behaviours also occur in hormonally and neurologically unmanipulated animals raised in species-typical social environments.

In most species, same-sex mounting, courtship and pair-bonding are not expressed frequently (Vasey and Sommer, 2006; Vasey, 2006b). In this respect, Japanese monkeys (*Macaca fuscata*) are quite unusual because, in addition to engaging in heterosexual behaviour, hormonally and neurologically unmanipulated females, in certain captive and free-ranging populations (Vasey and Jiskoot, 2010), *routinely* engage in homosexual behaviour, which takes the form of same-sex mounting (with pelvic thrusting) and courtship within the context of temporary, but exclusive, sexual relationships called 'consortships' (Vasey, 2006a). Consortships can be brief, lasting less than one hour, or they can continue for over a week (Wolfe, 1984, 1991). During same-sex consortships, females solicit each other to mount, and to be mounted, using vocalisations and a variety of postural and facial gestures. Sexual solicitations include pushing, hitting, grabbing, slapping the ground, head bobbing, screaming, lip quivering, body spasms and intense gazing (Vasey *et al.*, 2008a). Single mounts between females rarely occur. Instead, female consort partners engage in 'series mounting' during which they mount each other, in succession, tens or even hundreds of times. During these interactions, females mount each other in a bi-directional manner and routinely employ up to eight different types of mounting postures (Vasey *et al.*, 1998). During mounts, mounters often thrust against mountees who, in turn, commonly reach back to grasp the mounter and gaze intently into her eyes. As such, female mountees do not try to push or shake off the mounter. Instead, they typically brace their bodies to facilitate mounts-in-progress.

Mounting between consort partners and the accompanying sexual solicitations can continue uninterrupted for well over an hour. Thus, sexual behaviour is clearly the defining feature of same-sex consortships. It would be erroneous, however, to characterise these relationships solely in terms of sex. Between bouts of sexual activity, partners exhibit highly synchronised activity, huddling and sleeping together, as well as grooming, following and defending each other for prolonged periods of time (Vasey, 1996; Vasey *et al.*, 2008b). Partners will also forage side-by-side during same-sex consortships, although this form of co-feeding is not a prominent feature of these relationships.

In some populations of Japanese macaques, female homosexual behaviour has never been reported and appears to be non-existent (Vasey and Jiskoot, 2010). In other populations, some females engage in same-sex sexual behaviour, while others do not, and the proportion of females engaging in the behaviour varies from year to year. In the Arashiyama West (Texas) population for example, the proportion of females engaging in same-sex consortship activity has been reported to vary from 61% (Fedigan and Gouzoules, 1978), to 51% (Gouzoules and Goy, 1983), to 78% (Wolfe, 1984). Similarly, in the Arashiyama-Kyoto troop of Japanese macaques, the proportion of females engaging in same-sex consortship activity has been reported to vary from 47% (Takahata, 1982) to 23% (Wolfe, 1984). In other populations, *every* sexually mature female engages in same-sex consortship activity and they do so frequently. For example, Vasey (2002a) reported that during a typical mating season (October 1993–February 1994) every sexually mature female in the University of Montréal colony (n = 14 sexually active females; n = 5 sexually active males) consorted with other females. During these consortships, females mounted each other, on average, 31 times per hour of observation (n = 129 observation hours). Vasey *et al.* (2006a) reported even higher rates of female–female mounting for the Arashiyama-Kyoto group of Japanese macaques (40 times per hour of observation; n = 37 females).

The reason why animals engage in same-sex mounting, courtship and pair-bonding is a subject of debate (Vasey, 1995; Dixson, 1998; Bagemihl, 1999; Sommer and Vasey, 2006; MacFarland *et al.*, 2010; Poiani, 2010). Although some researchers have interpreted these interactions as sexual (Chevalier-Skolnikoff, 1974; Goldfoot *et al.*, 1980; Wolfe, 1984, 1991), a common theme throughout the animal behaviour literature has been to characterise these interactions as 'anything but sex' (reviewed in Bagemihl, 1999, p. 106). This stance is somewhat paradoxical in light of the evidence that has been presented throughout the animal behaviour literature for male and female sexual response during many same-sex mounting interactions (Hrdy, 1981; Dixson, 1998; Bagemihl, 1999). For example, female stumptail macaques (*Macaca arctoides*) exhibit copulatory facial expressions and undergo orgasmic uterine contractions when mounting other females (Goldfoot *et al.*, 1980).

How do researchers reconcile an 'anything but sex' perspective with the evidence for sexual response during same-sex mounting interactions? Traditionally, they have done so by characterising these exchanges as 'sociosexual'. In ethological parlance, sociosexual behaviours are those that are sexual in terms of their external form, but are enacted to mediate some sort of adaptive social goal (e.g. dominance demonstration; Wickler, 1967) or breeding strategies (e.g. proceptive behaviour in the presence of potential male mates; Beach, 1968; Parker and Pearson, 1976). Sexual motivation is rarely ascribed to these interactions because their sociosexual functions are often seen as their primary purpose, thus diminishing or even negating any sexual component that such activity might have. As such, these behaviours are typically characterised as desexualised mediums by which adaptive social goals are achieved. For example, in a discussion of female sexual response during same-sex mounting, Dixson (1998) conceded that such interactions might be pleasurable, but then went on to state that there was no evidence that such responses were anything more than epiphenomenal aspects of an activity that was, first and foremost, sociosexual in nature. In contrast, male–female mounting, courtship and pair-bonding are typically characterised as primarily sexual, with the possibility of some secondary sociosexual functions.

In many animal species same-sex mounting, courtship and pair-bonding do indeed appear to facilitate adaptive, sociosexual functions (e.g. Vasey, 1995; Dixson, 1998; Bagemihl, 1999; Sommer and Vasey, 2006; MacFarland *et al.*, 2010; Poiani, 2010). Nevertheless, what we know about female homosexual behaviour in Japanese macaques does not fit easily into the sociosexual interpretive framework. Despite over 40 years of continuous research on some populations of Japanese macaques (e.g. Arashiyama-Kyoto: Huffman *et al.*: chapter 1), there are no studies indicating that female–female mounting in this species is a sociosexual adaptation. Vasey (2006a) reviewed the available evidence, which demonstrates that female Japanese macaques do not use same-sex mounting or courtship to attract male mates (Gouzoules and Goy, 1983; Vasey, 1995), impede reproduction by same-sex competitors (Vasey, 1995), form alliances inside or outside of the mating season (Vasey, 1996), communicate about dominance relationships (Vasey *et al.*, 1998), obtain alloparental care (Vasey, 1998), reduce social tension associated with incipient aggression (Vasey *et al.*, 1998; Vasey, 2006a), practise for heterosexual activity (i.e. female mounting of males) and reconcile conflicts (Vasey, 2004). Taken together, this body of research suggests that female homosexual behaviour in Japanese macaques was not designed by natural selection as a sociosexual adaptation for alliance formation.

Given that female–female mounting, courtship behaviour and consortships do not appear to serve any sociosexual function, does this mean we can conclude that they are *truly* sexual behaviours? Several lines of evidence indicate that the mounting and courtship that occur during female–female consortships are, indeed, sexual behaviours. First, these interactions mirror male–female sexual behaviour in many aspects of their behavioural expression. With this caveat in mind, it is interesting to note that like heterosexual behaviour:

(1) female homosexual behaviour is never observed outside of the species' autumn–winter mating season (Wolfe, 1984, 1991);

(2) females that engage in homosexual behaviour exhibit a reddening of the face and perineum that is indicative of increased sexual receptivity (Wolfe, 1984, 1991);

(3) female homosexual behaviour occurs within the context of temporary, but exclusive, consortships (Wolfe, 1984, 1991);

(4) female mountees do not try to push or shake off the mounter; instead, they typically brace their bodies to support the mounters. Often the mountees will facilitate mounts-in-progress by reaching back and clasping the mounters while gazing into the mounters' eyes (Wolfe, 1984, 1991);

(5) prior to conception, female–female mounting peaks during the periovulatory period (i.e. day of ovulation ±3 days) when pregnanediol levels begin to rise slightly from baseline levels and oestrone levels reach their peak. Following conception, there is a second peak in mounting (i.e. female–female, male–female and female–male) between the sixth and tenth weeks of gestation, coinciding with a marked decrease in pregnanediol levels (Gouzoules and Goy, 1983; O'Neill *et al.*, 2004; O'Neill: chapter 7).

Second, when considering the issue of sexual motivation, it is also important to note that patterns of affiliation during homosexual consortships do not simply mimic patterns of social affiliation that occur outside of these associations. Rather, patterns of affiliation between females represent a radical departure from normal patterns of affiliation that are widely recognised as social. This suggests that homosexual consortships are not social relationships, but rather are sexual ones. For example,

(1) female Japanese macaques exercise incest avoidance with close female kin ($r \leq 0.25$; Chapais *et al.*, 1997). Consequently, mounting, sexual solicitations and consortships are never observed between sisters, mothers and daughters, or grandmothers and granddaughters (Wolfe, 1984; Chapais and Mignault, 1991; Chapais *et al.*, 1997). In contrast, these close categories of female kin commonly engage in other types of affiliative behaviour with each other, such as grooming, huddling and interventions (Kurland, 1977; Baxter and Fedigan, 1979; Koyama, 1991; Chapais, 1992);

(2) female consort partners intervene[1] in conflicts to support each other at consistently high rates during homosexual consortships. Outside the

---

[1] Interventions refer to the intrusion of an individual (supporter) into an ongoing conflict between two opponents, during which the intruding animals supported one of the opponents (recipient of support) against the other (target; Chapais, 1992).

consortship period, these same females rarely support each other, if at all (Vasey, 1996);

(3)   during homosexual consortships, a dramatic increase occurs in the number of bridging interventions[2] performed by dominant consort partners for their subordinate counterparts (Vasey, 1996). This pattern is highly unusual because outside of homosexual consortships, non-kin female Japanese macaques almost never perform bridging interventions for each other. Instead, they perform conservative interventions[3] (Chapais, 1992). Yet, when opportunities for support are considered, dominant consort partners are as likely to perform bridging interventions (Vasey, 1996);

(4)   normal dominance relationships are temporarily destabilised during homosexual consortships because subordinate consort partners receive support from their dominant partners against targets that normally rank above them, or with whom they share ambiguous dominance relationships (Vasey, 1996);

(5)   dominant consort partners groom their subordinate partners significantly more often than the reverse (Vasey, 1996). Outside of homosexual consortships the opposite pattern holds true: female Japanese macaques overwhelmingly direct grooming up the hierarchy not down (Koyama, 1991; Chapais *et al.*, 1995);

(6)   During homosexual consortships, females are almost exclusively in proximity with their non-kin consort partners (Vasey *et al.*, 2008b). Outside of the consortship period, however, female Japanese macaques spend most of their time in proximity with close kin (Kurland, 1977; Singh *et al.*, 1992).

In what follows, we review some evidence from our research on the sexual behaviour of the Arashiyama-Kyoto macaques, which further corroborates the conclusion that female–female mounting, courtship and pair-bonding can be objectively and accurately described as 'sexual' in character.

10.2   Courtship behaviour

Although heterosexual copulation should not be the yardstick by which all other types of sexual behaviour are measured, heterosexual copulation is considered, a priori, to be sexual (e.g. Dixson, 1998; for discussion see

---

[2] Bridging interventions occurred when the target ranked between the supporter and the recipient of support (Chapais, 1992).

[3] Conservative interventions occurred when the target ranked below the supporter and the recipient of support (Chapais, 1992).

Bagemihl, 1999). Consequently, demonstrating that homosexual behaviour parallels heterosexual behaviour in many aspects of its expression represents powerful evidence for most researchers that the former is, indeed, a sexual behaviour.

With this in mind, Vasey *et al.* (2008a) undertook a detailed, quantitative comparison of the courtship behaviour that is exchanged between partners in both heterosexual and homosexual consortships. We reasoned that if females were exhibiting similar patterns of courtship behaviour in both heterosexual and homosexual consortships, then this would furnish further evidence that homosexual consortships are, indeed, sexual. Courtship behaviour occurred during inter-mount intervals and functioned as sexual solicitations to prompt mounting activity. It is not unusual for various courtship behaviours to occur simultaneously or sequentially. The courtship behaviours analysed by Vasey *et al.* (2008a) included: (1) lip quivering; (2) head bobbing; (3) pushing; (4) grabbing; (5) ground smacking; (6) body spasms; (7) intense gazing; (8) inclined-back presentations[4]; (9) hindquarter presentations[5]; (10) hands-on-hindquarters solicitations[6]; and (11) sexual vocalisations. In most instances, it could not be discerned if a particular courtship signal functioned as a request to mount or to be mounted. Exceptions included *hindquarter presentations* and *inclined-back presentations*, which functioned specifically as requests to be mounted. Conversely, *hands-on-hindquarters solicitations* functioned specifically as requests to mount.

Sex differences existed for heterosexually consorting males and females for only two of the 11 courtship behaviours (i.e. inclined-back presentations and sexual vocalisations). Heterosexual consorting males performed inclined-back presentations during significantly fewer inter-mount intervals compared with heterosexually and homosexually consorting females. Inclined-back presentations, like hindquarter presentations, function as requests to be mounted. Some might argue that a sex-difference in inclined-back presentations is hardly surprising given that, in most species, males mount females and not vice versa (reviewed in Baum, 1979). As such, one would not expect males to request to be mounted by directing inclined-back presentations toward females. In reality, however, female mounting of males is quite common in the Arashiyama

---

[4] *Inclined-back presentations* were performed by potential mountees that sat with their forearms slightly bent, and their backs inclined and oriented towards the potential mounters (Vasey *et al.*, 2008a).

[5] *Hindquarter presentations* were performed by potential mountees, which stood quadrupedally with their arms and legs flexed and their perineums oriented towards the potential mounters (Vasey *et al.*, 2008a).

[6] *Hands-on-hindquarters solicitations* were performed by potential mounters that grasped or placed both hands on the hindquarters of the potential mountees (Vasey *et al.*, 2008a).

populations of Japanese macaques (Gouzoules and Goy, 1983; Vasey and Duckworth, 2008). For example, during heterosexual consortships, Vasey and Duckworth (2008) observed approximately eight female–male mounts per observation hour. Sometimes males actively solicit their female consort partners to mount them as can be seen from the fact that heterosexually consorting males and females did not differ in terms of the proportion of intermount intervals during which they executed hindquarter presentations. As such, there is no a priori reason why sex difference in inclined-back presentations should exist. Homosexually consorting females did not differ from heterosexually consorting females in terms of the proportion of inter-mount intervals in which they executed inclined-back presentations. As such, homosexually consorting females exhibited a female-typical pattern of inclined-back presentation sexual solicitations.

Heterosexually consorting males also performed sexual vocalisations during significantly fewer inter-mount intervals than heterosexually and homosexually consorting females. Homosexually consorting females did not differ from heterosexually consorting females in terms of the frequency with which they performed sexual vocalisations. As such, homosexually consorting females exhibited a female-typical pattern of sexual vocalisations.

It is striking how few differences we found in courtship behaviour between heterosexual and homosexual consortships. In those few instances when differences were exhibited, the courtship behaviour manifested within the context of female–female consortships appeared to be identical to that exhibited by females in the context of male–female consortships. Given that the courtship behaviour exchanged between heterosexual partners is characterised, first and foremost, as sexual, the similarities in courtship behaviour across heterosexual and homosexual lend further support to the conclusion that female homosexual activity in Japanese macaques is also sexual in character.

10.3   Genital stimulation

From an observational point of view, genital stimulation is, arguably, the *sine qua non* of sexual behaviour. If it could be demonstrated that female Japanese macaques engage in genital stimulation during same-sex mounting, then this would provide what many researchers would consider the strongest and most conclusive evidence that these interactions are sexual in character. The vulvar region of Japanese macaques consists of the clitoris, the clitoral prepuce and the labia minora. The labia majora are absent in Japanese macaques and the glans clitoris projects approximately 4.5 mm beyond the rima of the labia minora (Hill, 1974). As in humans, the vulvar, perineal and anal (VPA) region of female macaques is richly innervated with nerve endings that are special-ised for receiving and transmitting erotic sensations of touch, tension, pres-sure and friction (Winkelmann, 1959; Hill, 1974; Campbell, 1976; Cold and Tarara, 1997; Cold and McGrath, 1999). On the basis of his comparative work, Winkelmann (1959) argued that specialised patterns of VPA innervation in pri-mates render members of that group better able to appreciate erotic sensations than non-primates.

In addition to this work, the role of the VPA in mediating sexual reward in female macaques has been well documented (Burton, 1970). Experimental work demonstrates that stimulation of the female macaque's VPA region clearly induces three of Masters and Johnson's (1966) four copulatory phases: excitement, plateau and resolution (Burton, 1970). During the excitement and the plateau phases the labia minora and the clitoris engorge and the perineal region deepens in colour (Burton, 1970). Some experimental subjects also exhibited behavioural indicators of orgasm including spasmatic body move-ments and contractions of the perineal muscles (Burton, 1970). These types of anatomical changes and behavioural patterns have also been observed in free-ranging female Japanese macaques during sexual interactions (Wolfe, 1984, 1991).

Vasey and Duckworth (2006) documented patterns of VPA stimulation dur-ing same-sex mounts. During the majority of female–female mounts analysed, female mounters engaged in repetitive VPA stimulation. Females stimulated their VPA regions approximately 2–3 times per same-sex mount and they did so during episodes of series mounting during which they mounted each other tens or even hundreds of times in succession. From a human perspec-tive, the amount of stimulation obtained during each mount may appear lim-ited. However, it is important to keep in mind that the pattern and duration of stimulation necessary for female (and male) sexual arousal vary tremendously across species (Hrdy, 1981; Wolfe, 1991; Dixson, 1998, 2009). As such, we should not use human sexual response as a yardstick for measuring what does, and does not, constitute sufficient sexual arousal in other species.

Two forms of VPA stimulation by female Japanese macaque mounters were observed, and these occurred alone, simultaneously or sequentially during a single mount. First, female mounters rubbed their VPA regions against their partners' backs during sitting mounts[7] with pelvic thrusting. This particular mount posture and motor pattern was specific to female Japanese macaque

---

[7] *Sitting mounts* occurred when the mounter sat on the mountee's back in a jockey-like position while grasping the mountee's upper back with his or her hands and the mountee's lower back with his or her feet (Vasey *et al.*, 2006).

mounters because male mounters were never observed to engage in sitting mounts, with or without thrusting. Second, female mounters stroked their VPA region with their tails during same-sex mounts. During episodes of VPA manipulation, females' clitorises and labia minora were sometimes visibly engorged and contractions of the muscles in the area of the VPA were sometimes observed (Vasey and Duckworth, 2006).

In order to examine alternative explanations as to why female Japanese macaque mounters were moving their tails during same-sex mounts, we posed a series of questions aimed at refuting the conclusion that such tail movement was indicative of female sexual motivation. First, we asked whether all Japanese macaque mounters moved their tails regardless of their sex, and whether they did so in the same way. Results of this study indicated that both male and female mounters moved their tails during mounts, but they did so in different ways (Vasey and Duckworth, 2006). Both sexes moved their tails upward and downward in the sagittal plane, but female mounters routinely hooked their tails downward and forward in between their legs, whereas male mounters almost never did so. For their part, male mounters routinely arch their tails upward and forward over their backs, while female mounters were never observed to do so. In addition, female mounters often swept their tails across their perineums in a circular manner, executing sideward movement to the left and to the right within the vertical plane, while male mounters rarely moved their tails in this manner. These results indicate that tail movement by female mounters was sex-specific and not part of a more generalised pattern of behaviour that was indicative of all Japanese macaque mounters, regardless of their sex.

Next, we asked whether tail movement by female mounters was simply a biomechanical by-product of pelvic thrusting. Most tail movement by female mounters was associated with pelvic thrusting. However, almost all female mounters executed tail movement without pelvic thrusting, and thrusting without tail movement. These results demonstrate that tail movement and pelvic thrusting were dissociable, indicating that tail movement by female mounters was a voluntary motor pattern and not merely an involuntary reflex associated with pelvic thrusting. The markedly precise manner in which female mounters moved their tails, particularly in terms of tail movement in the vertical plane (e.g. sideward left and sideward right movement), further suggested that these motor patterns were voluntary.

We also asked whether tail movement by female mounters during same-sex mounts was simply part of a generalised pattern of behaviour that occurred in all sorts of non-sexual situations. It is striking that tail movement, such as that observed by female mounters, was *never* performed by consorting females in non-sexual, inter-mount contexts involving movement alone (e.g. walking) or movement with inter-individual contact (e.g. grooming). These results indicate that tail movement by female mounters during same-sex mounts was specific to a sexual context (i.e. mounting) and not part of a more generalised pattern of non-sexual behaviour. Furthermore, Japanese macaques possess small, stubby tails and, as such, it seems unlikely that tail movement played any significant role in aiding mounters, of either sex, maintain their balance during mounting interactions.

If, as we argue here, female Japanese macaques mount other females because they derive immediate sexual reward from such interactions, why was it that approximately one fifth of female–female mounts involved no observable VPA manipulation on the part of the mounter? Female mounters may have obtained VPA stimulation during some of these mounts, but this may have occurred in such an inconspicuous manner as to go unrecognised by the observer. For example, during sitting mounts without thrusting, female mounters may have pressed their VPA region against their partners' backs, thus stimulating these erogenous zones. During other types of mounts in which the VPA region of the mounter was not in direct contact with the mountee's body, the mounter may have experienced stimulation of these erogenous zones via contraction of the muscles in these areas.

Although the focus of our investigation pertained to why female Japanese macaques mount same-sex partners, an equally interesting question is: Why would female Japanese macaques allow themselves to *be* mounted by a same-sex partner? During female–female mounts, we observed that mountees moved their tails against their VPA regions, but systematic data collection on this phenomenon was hindered because the mounter's body often blocked the observer's view of the mountee's tail region. In addition, during double foot-clasp mounts[8] and standing mounts[9] female mounters routinely thrusted against the mountee's VPA region, which may have afforded direct stimulation of these erogenous zones. These observations lead us to believe that female mountees also derive sexual reward via VPA stimulation during some same-sex mounting episodes. Moreover, like mounters, mountees may have experienced VPA stimulation during some same-sex mounts via contractions of the muscles in these regions.

---

[8] *Double-foot clasp mounts* involved the mounter grasping with his or her feet between the mountee's ankles and hips, and with his or her hands on the mountee's back. As such the mounter's groin region was in contact with the mountee's perineum. Double-foot clasp mounts could be performed with or without pelvic thrusting (Vasey *et al.*, 2006).

[9] *Standing mounts* involved the mounter standing bipedally with his or her feet on the ground and his or her knees slightly bent, while grasping the mountee's lower back with his or her hands. As such the mounter's groin region was in contact with the mountee's perineum. Standing mounts could be performed with or without pelvic thrusting (Vasey *et al.*, 2006).

In light of the VPA region's primary role in mediating female sexual arousal in human and non-human primates (Winkelmann, 1959; Masters and Johnson, 1966; Hill, 1974; Campbell, 1976; Rogers, 1992; Shafik, 1993; Cold and Tarara, 1997; Dixson, 1998; Cold and McGrath, 1999), the repetitive and precise patterns of vulvar, perineal and anal stimulation observed during this study indicate that immediate sexual reward provides a robust proximate explanation as to why female Japanese macaques engage in same-sex mounting.

10.4   Mount postures and pelvic movement

Given differences in the genital architecture of males and females, there is no a priori reason to expect that the sexes would stimulate their genitals in the same manner. Vasey *et al.* (2006) furnished evidence bearing on this possibility by undertaking a Laban Movement Analysis (LMA) of mounting behaviour by male and female Japanese macaques. Laban Movement Analysis is a universal language for movement that describes quantitative features of movement, such as changes in the relation of the body segments, as well as qualitative features, such as the style of movements (Laban, 1971). This analysis indicated that females engaged in the same mount posture that males almost invariably perform (i.e. double-foot clasp mounts with thrusting) and they executed the entire range of male-typical pelvic movement, but the converse was not true. Given that males were able to perform pelvic movements in all directions, it seems unlikely that they were biomechanically incapable of executing the entire range of diverse mount postures and the more complex pelvic movement patterns observed among females. Nevertheless, they did not do so.

What factors might account for these sex differences? We believe that both males and females sought sexual reward via genital stimulation during mounts. However, the precise motor requirements imposed by penile–vaginal intromission ultimately constrained male pelvic movement during male–female mounts. Unfettered by the constraints imposed by penile–vaginal penetration, female mounters were able to employ a far greater range and complexity of postures and movements aimed at stimulating their genital regions. Not surprisingly, one of the mount postures that females performed most frequently, sitting mounts with thrusting, afforded them the opportunity to rub their genital region against the back of the mountee (Vasey and Duckworth, 2006; Vasey *et al.*, 2006). It is noteworthy, however, that female mounters did not need to make direct contact with the mountee's body to obtain genital stimulation. This is because, during many mounts, female mounters stimulated their clitorises by masturbating with their tails (Vasey and Duckworth, 2006). In effect, female mounters who sought sexual reward during mounts were freed from the constraint of obtaining direct genital contact with their sexual partners. This ability provided females with the means to execute a wide diversity of postures and pelvic movements in their quest for sexual reward during mounts. For their part, male Japanese macaques are incapable of this sort of behaviour because their tails are located in a position that is too dorsal in orientation from their more ventrally located genitalia to allow for tail–penile or tail–scrotal stimulation.

10.5   Discussion

Taken together, the information provided here furnishes further support for the conclusion that female–female mounting in Japanese macaques is sexual. First, research on the Arashiyama-Kyoto macaques demonstrates that homosexually consorting females routinely exchange courtship behaviours that are virtually identical to those that are exchanged within heterosexual consortships. Second, homosexually consorting females in this population routinely perform double-foot clasp mounts with thrusting, a pattern of mounting that males almost always perform during heterosexual consortships. Third, during the vast majority of female–female mounts, the mounter engages in vulvar, perineal and anal (VPA) stimulation either by rubbing this body region on the back of the mountee, or by stimulating this body region with their tail. Fourth, during these female–female consortships, mounters employ a much wider variety of mount postures and pelvic movements than do male mounters. These mount postures and pelvic movements appear to be aimed at achieving stimulation of their VPA region.

In the vast majority of species, individuals prefer opposite-sex sexual partners over same-sex ones (Adkins-Regan, 1998; Vasey, 2002b). However, female Japanese macaques seem particularly unusual in that they do not merely mount, court and form consortships with same-sex partners. Rather, they sometimes *prefer* female sexual partners even when given the simultaneous choice of an acceptable and motivated, opposite-sex alternative (Vasey, 1998; Vasey and Gauthier, 2000). Females' facultative preference for same-sex sexual partners, over opposite-sex mates, creates a unique mating problem for male Japanese macaques in the Arashiyama populations. Namely, males not only have to compete intra-sexually for female mates, but they must do so *inter*-sexually as well. Indeed, based on research conducted on the Montréal colony of Japanese macaques, Vasey (1998) demonstrated that inter-sexual competition for female sexual partners was a salient interaction throughout the mating season that occurred approximately once per observation hour.

Ad libitum observations indicate that these interactions occur in the Arashiyama-Kyoto macaques as well (Vasey, pers. obs. 2000–2010).

Inter-sexual competition for female sexual partners occurred when a sexually motivated male and female (competitors) simultaneously sought access to the same female sexual partner (focus of competition). Competitors were considered to be sexually motivated vis-à-vis the focus of competition if they solicited her for sex while competing inter-sexually for her, or if they solicited her for sex at some point during the same day. Inter-sexual competition for female sexual partners was initiated when a sexually motivated male intruded on a same-sex consortship, targeting one of the females as the focus of competition and the other as his competitor. Intrusions took the form of approaches and sexual solicitations directed at the focus of competition, as well as displacements and aggression directed at the female competitor. They functioned to increase the male's access to a female sexual partner (i.e. the focus of competition), while decreasing or preventing her same-sex consort partner from doing the same.

Intrusions on same-sex consortships did not reflect a generalised motivation among males to disrupt any type of affiliation between females. Males intruded on same-sex consortships for female sexual partners significantly more often than they interfered with non-kin female grooming dyads observed outside the mating season. The tendency to disrupt females engaged in sexual affiliation, but not those engaged in social affiliation, strongly suggests that intrusions are correctly identified as sexual, not social, competition.

Following 62% of intrusions, inter-sexual competition for female sexual partners terminated. In such cases, the consorting females would: (1) ignore the male who would eventually leave, (2) move away from the male altogether, (3) separate and then rejoin to continue their consortship or (4) terminate their consortship. Inter-sexual competition for female sexual partners could, however, escalate with the female competitor posing a counter-challenge against the intruding male. During counter-challenges, the female competitor attempted to maintain exclusive access to her consort partner, the focus of competition, while preventing the male competitor from doing the same. Counter-challenges took the form of approaches and sexual solicitations directed at the focus of competition, as well as displacements and aggression directed at the intruding male competitor. During some counter-challenges, female competitors would also interpose themselves between the male competitor and the focus of competition. Overall, 38% of intrusions by male competitors provoked counter-challenges by female competitors.

Proceptive females did not perform counter-challenges simply because they were in a heightened state of arousal and thus, more assertive. They retaliated, in kind, against males that aggressed or displaced them significantly more often when they were engaged in same-sex consortships, compared with when they were not. Furthermore, females were never seen to retaliate against males that disrupted their grooming interactions with non-kin females outside the mating season. The females' tendency to challenge males that disrupted their sexual, but not their social, affiliation strongly suggests that counter-challenges are correctly identified as sexual competition and are not part of a more generalised pattern of social competition.

Following episodes of inter-sexual competition, females that were the foci of competition could choose between two sexually motivated competitors, one male and the other female, who were simultaneously available. In only one instance of inter-sexual competition for female sexual partners (n = 120) was a female focus of competition sexually coerced by a female competitor. Consequently, these interactions represented an excellent 'natural experiment' for testing whether females preferred certain female sexual partners relative to certain male alternatives when simultaneously given a choice. By examining sexual partner choice in anovulatory females, Vasey (1998) was able to ensure that subjects were choosing between two sexual partners, one male and one female. In contrast, had sexual partner preference been examined in females that were ovulating, then the subjects would have had to choose between a female sexual partner and a male reproductive partner. Obviously, this would have introduced the confounding effect of reproductive preference. In the overwhelming majority of cases (92.5%), the foci of competition chose to continue mounting with their female sexual partners significantly more often than they chose to begin mounting with the male competitor, or to cease mounting altogether. Taken together, this research suggests that in some populations, female Japanese macaques sometimes prefer same-sex sexual partners even when they are simultaneously presented with sexually motivated, opposite-sex sexual partners.

The substantial body of evidence bearing on the sexual (not sociosexual) nature of female–female mounting, courtship and consortship activity in the Arashiyama macaques provides support for the conclusion that the term 'homosexual behaviour' is an accurate and appropriate label for these interactions. Some researchers have disputed this terminology (Wallen and Parsons, 1997; Dixson, 1998) arguing that it is misleading because homosexuality is a uniquely human phenomenon. This argument is predicated, in part, on the assumption that homosexual behaviour is synonymous with homosexuality when, in reality, these two terms are not conceptually equivalent. Homosexual behaviour refers to discrete acts or interactions, whereas homosexuality refers to a specific type of sexual orientation (i.e. an individual's pattern of sexual attraction/arousal). Care should be taken not to make conceptual leaps from terms that denote behaviour, to those that denote sexual orientation (i.e. an overall pattern of sexual attraction and arousal), sexual orientation identity (i.e. the sexual orientation that an individual perceives themselves to have), categories of sexual beings (i.e. homosexual, heterosexual, lesbians, gays, etc.), sexual preference or any sort of temporally stable sexual pattern. We remind the reader that this study examines female homosexual *behaviour* in Japanese macaques. As such, our research was *not* a study of lesbian macaques, homosexual macaques or macaque homosexuality.

In addition, we stress that use of the term 'homosexual behaviour' in reference to various species, including humans, does not imply evolutionary homology. Researchers routinely utilise identical terminology to describe evolutionarily *analogous* characteristics. For example, entomologists talk about the evolution of flight in insects without presupposing it is homologous to flight in birds or bats. As such, we remind readers that female homosexual behaviour in Japanese macaques should not, a priori, be considered the evolutionary homologue of human homosexuality or human homosexual behaviour.

Acknowledgements

We are grateful to the following individuals, without whom this research would not have been possible: Shigeru Suzuki, Shinsuke Asaba, Michael Huffman, Juichi Yamagiwa, Syuhei Kobatake, Eiji Inoue, Eiji Enomoto and the other members of the Enomoto family of Arashiyama, and the Sakami family of Tokyo. This research was funded by grants to P.L.V. by the University of Lethbridge, the L.S.B. Leakey Foundation, the Natural Science and Engineering Research Council (NSERC) of Canada, by an Alberta Graduate Scholarship and a NSERC Canada Graduate Scholarship to D.P.V.

References

Adkins-Regan, E. (1988). Sex hormones and sexual orientation in animals. *Psychobiology*, **16**, 355–347.
    (1998). Hormonal mechanisms of mate choice. *American Zoologist*, **38**, 166–178.
Adkins-Regan, E. and Krakauer, A. (2000). Removal of adult males from the rearing environment increases preference for same-sex partners in the zebra finch. *Animal Behaviour*, **60**, 47–53.
Adkins-Regan, E., Mansukhani, V., Thompson, R. and Yang, S. (1997). Organizational actions of sex hormones on sexual partner preference. *Brain Research Bulletin*, **44**, 497–502.

Bagemihl, B. (1999). *Biological Exuberance: Animal Homosexual and Natural Diversity*. New York, NY: St. Martin's Press.

Baum, M. J. (1979). Differentiation of coital behavior in mammals. *Neuroscience and Biobehavioral Reviews*, **3**, 65–84.

Baxter, M. J. and Fedigan, L. M. (1979). Grooming and consort partner selection in a troop of Japanese monkeys (*Macaca fuscata*). *Archives of Sexual Behavior*, **8**, 445–458.

Beach, F. A. (1968). Factors involved in the control of mounting behavior by female mammals. In *Perspectives in Reproduction and Sexual Behavior*, ed. M. Diamond. Bloomington, IN: Indiana University Press, pp. 83–131.

Burton, F. D. (1970). Sexual climax in female *Macaca mulatta*. *Proceedings of the 3rd International Congress of Primatology*, **3**, 180–191.

Campbell, B. (1976). Neurophysiology of the clitoris. In *The Clitoris*, eds. T. P. Lowry and T. S. Lowry. St. Louis, MO: Warren H. Green, pp. 35–74.

Chapais, B. (1992). The role of alliances in the social inheritance of rank among female primates. In *Coalitions and Alliances in Humans and Other Animals*, eds. A. Harcourt and F. M. B. de Waal. Oxford: Oxford University Press, pp. 29–60.

Chapais, B. and Mignault, C. (1991). Homosexual incest avoidance among females in captive Japanese macaques. *American Journal of Primatology*, **23**, 171–183.

Chapais, B., Gauthier, C. and Prud'homme, J. (1995). Dominance competition through affiliation and support in Japanese macaques: an experimental study. *International Journal of Primatology*, **16**, 521–536.

Chapais, B., Gauthier, C., Prud'homme, J. and Vasey, P. L. (1997). Relatedness threshold for nepotism in Japanese macaques. *Animal Behaviour*, **53**, 533–548.

Chevalier-Skolnikoff, S. (1974). Male-female, female-female, and male-male sexual behavior in the stumptail monkey, with special attention to the female orgasm. *Archives of Sexual Behavior*, **3**, 95–116.

Cold, C. J. and McGrath, K. A. (1999). Anatomy and histology of the penile and clitoral prepuce in primates: an evolutionary perspective of the specialised sensory tissue of the external genitalia. In *Male and Female Circumcision*, eds. G. C. Denniston, F. M. Hodges and M. F. Milos. New York: Kluwer Academic, pp. 19–29.

Cold, C. and Tarara, R. (1997). Penile and clitoral prepuce mucocutaneous receptors in *Macaca mulatta*. *Veterinary Pathology*, **34**, 506.

Dixson, A. F. (1998). *Primate Sexuality*. Oxford: Oxford University Press.

(2009). *Sexual Selection and the Origin of Human Mating Systems*. New York, NY: Oxford University Press.

Fedigan, L. M. and Gouzoules, H. (1978). The consort relationship in a troop of Japanese monkeys. In *Recent Advances in Primatology, Volume 1: Behavior*, eds. D. J. Chivers and J. Herbert. New York, NY: Academic Press, pp. 493–495.

Goldfoot, D., Westerborg-van Loon, H., Groenevelde, W. and Slob, A. K. (1980). Behavioral and physiological evidence of sexual climax in female stump-tailed macaques (*Macaca arctoides*). *Science*, **208**, 1477–1479.

Gouzoules, H. and Goy, R. W. (1983). Physiological and social influences on mounting behavior of troop-living female monkeys (*Macaca fuscata*). *American Journal of Primatology*, **5**, 39–49.

Hill, W. C. O. (1974). *Primates, Comparative Anatomy and Taxonomy, Volume 7. Cynopithecinae, Cercocebus, Macaca, Cynopithecus.* Edinburgh: Edinburgh University Press.

Hrdy, S. B. (1981). *The Woman That Never Evolved.* Cambridge, MA: Harvard University Press.

Kindon, H. A., Baum, M. J. and Paredes, R. G. (1996). Medial preoptic/anterior hypothalamic lesions induce a female-typical profile of sexual partner preference in male ferrets. *Hormones and Behavior*, **30**, 514–527.

Koyama, N. (1991). Grooming relationships in the Arashiyama group of Japanese monkeys. In *The Monkeys of Arashiyama: Thirty-five Years of Research in Japan and the West*, eds. L. M. Fedigan and P. J. Asquith. Albany, NY: State University of New York Press, pp. 211–226.

Kurland, J. A. (1977). *Kin Selection in the Japanese Monkey.* Basel: S. Karger.

Laban, R. (1971). *The Mastery of Movement* (Rev. 3rd edn by Ullman, L.). Boston, MA: Plays.

MacFarland, G. R., Blomberg, S. P. and Vasey, P. L. (2010). Homosexual behavior in birds: frequency of expression is related to sex-specific relative parental care. *Animal Behaviour*, **80**, 375–390.

Masters, W. H. and Johnson, V. E. (1966). *Human Sexual Response.* Boston, MA: Little Brown & Co.

O'Neill, A. C., Fedigan, L. M. and Ziegler, T. E. (2004). Ovarian cycle phase and homosexual behavior in Japanese macaque females. *American Journal of Primatology*, **63**, 25–31.

Paredes, R. G. and Baum, M. J. (1995). Altered sexual partner preference in male ferrets given excitotoxic lesions of the preoptic area/anterior hypothalamus. *Journal of Neuroscience*, **15**, 6619–6630.

Paredes, R. G., Tzschentke, T. and Nakach, N. (1998). Lesions of the medial preoptic area/anterior hypothalamus (MPOA/AH) modify partner preference in male rats. *Brain Research*, **813**, 1–8.

Parker, G. A. and Pearson, R. G. (1976). A possible origin and adaptive significance of the mounting behavior shown by some female mammals in oestrous. *Journal of Natural History*, **10**, 241–245.

Poiani, A. (2010). *Animal Homosexuality: A Biosocial Perspective.* Cambridge: Cambridge University Press.

Price, E. O. and Smith, V. M. (1984/1985). The relationship of male-male mounting to mate choice and sexual performance in male dairy goats. *Applied Animal Behaviour Science*, **13**, 71–82.

Price, E. O., Katz, L. S., Wallach, S. J. R. and Zenchak, J. J. (1988). The relationship of male-male mounting to the sexual preferences of young rams. *Applied Animal Behaviour Science*, **21**, 347–355.

Rogers, J. (1992). Testing for and the role of anal and rectal sensation. *Baillière's Clinical Gastroenterology*, **6**, 179–191.

Shafik, A. (1993). Vaginocavernosus reflex: clinical significance and role in sexual act. *Gynecologic and Obstetric Investigation*, **35**, 114–117.

Singh, M., D'Souza, L. and Singh, M. R. (1992). Hierarchy, kinship and social interaction among Japanese monkeys, *Macaca fuscata. Journal of Biosciences*, **17**, 15–27.

Sommer, V. and Vasey, P. L., eds. (2006). *Homosexual Behaviour in Animals: An Evolutionary Perspective*. Cambridge: Cambridge University Press.

Takahata, Y. (1982). The sociosexual behavior of Japanese macaques. *Zeitschrift für Tierpsychologie*, **59**, 89–108.

Vasey, P. L. (1995). Homosexual behaviour in primates: a review of evidence and theory. *International Journal of Primatology*, **16**, 173–204.

(1996). Interventions and alliance formation between female Japanese macaques (*Macaca fuscata*) during homosexual consortships. *Animal Behaviour*, **52**, 539–551.

(1998). Female choice and inter-sexual competition for female sexual partners in Japanese macaques. *Behaviour*, **135**, 579–597.

(2002a). Sexual partner preference in female Japanese macaques. *Archives of Sexual Behavior*, **31**, 45–56.

(2002b). Same-sex sexual partner preference in hormonally and neurologically unmanipulated animals. *Annual Review of Sex Research*, **13**, 141–179.

(2004). Pre- and post-conflict interactions between female Japanese macaques during homosexual consortships. *International Journal of Comparative Psychology*, **17**, 351–359.

(2006a). The pursuit of pleasure: Homosexual behaviour, sexual reward and evolutionary history in Japanese macaques. In *Homosexual Behaviour in Animals: An Evolutionary Perspective*, eds. V. Sommer and P. L. Vasey. Cambridge: Cambridge University Press. pp. 191–219.

(2006b). Where do we go from here? Research on the evolution of homosexual behaviour in animals. In *Homosexual Behaviour in Animals: An Evolutionary Perspective*, eds. V. Sommer and P. L. Vasey. Cambridge: Cambridge University Press, pp. 349–364.

Vasey, P. L. and Duckworth, N. (2006). Sexual reward via vulvar, perineal and anal stimulation: A proximate mechanism for female homosexual mounting in Japanese macaques. *Archives of Sexual Behavior*, **35**, 523–532.

(2008). Female-male mounting in Japanese macaques: The proximate role of sexual reward. *Behavioural Processes*, **77**, 405–407.

Vasey, P. L. and Gauthier, C. (2000). Skewed sex ratios and female homosexual activity in Japanese macaques: An experimental analysis. *Primates*, **41**, 17–25.

Vasey, P. L. and Jiskoot, H. (2010). The biogeography and evolution of female homosexual behavior in Japanese macaques. *Archives of Sexual Behavior*, **39**, 1439–1441.

Vasey, P. L. and Sommer, V. (2006). Homosexual behaviour in animals: topics, hypotheses, and research trajectories. In *Homosexual Behaviour in Animals: An Evolutionary Perspective*, eds. V. Sommer and P. L. Vasey. Cambridge: Cambridge University Press, pp. 3–42.

172   *Y. Takenoshita*

Vasey, P. L., Chapais, B. and Gauthier, C. (1998). Mounting interactions between female Japanese macaques: testing the influence of dominance and aggression. *Ethology*, **104**, 387–398.

Vasey, P. L., Foroud, A., Duckworth, N. and Kovacovsky, S. D. (2006). Male-female and female-female mounting in Japanese macaques: a comparative analysis of posture and movement. *Archives of Sexual Behavior*, **35**, 116–128.

Vasey, P. L., Rains, D., VanderLaan, D. P., Duckworth, N. and Kovacovsky, S. D. (2008a). Courtship behavior during heterosexual and homosexual consortships in Japanese macaques. *Behavioural Processes*, **78**, 401–407.

Vasey, P. L., VanderLaan, D. P., Rains, D., Duckworth, N. and Kovacovsky, S. D. (2008b). Inter-mount social interactions during heterosexual and homosexual consortships in Japanese macaques. *Ethology*, **114**, 564–574.

Wallen, K. and Parsons, W. A. (1997). Sexual behavior in same-sexed nonhuman primates: Is it relevant to understanding homosexuality? *Annual Review of Sex Research*, **7**, 195–223.

Wickler, W. (1967). Socio-sexual signals and their intra-specific imitation among primates. In *Primate Ethology*, ed. D. Morris. London: Weidenfeld and Nicolson, pp. 69–147.

Winkelmann, R. K. (1959). The erogenous zones: their nerve supply and significance. *Mayo Clinic Proceedings*, **34**, 39–47.

Wolfe, L. D. (1984). Japanese macaque female sexual behavior: A comparison of Arashiyama East and West. In *Female Primates: Studies by Women Primatologists*, ed. M. F. Small. New York, NY: Alan R. Liss, pp. 141–157.

(1991). Human evolution and the sexual behavior of female primates. In *Understanding Behavior: What Primate Studies Tell us About Human Behavior*, eds. J. D. Loy and C. B. Peters. New York/Oxford: Oxford University Press, pp. 121–151.

Zenchak, J. J., Anderson, G. C. and Schein, M. W. (1981). Sexual partner preference of adult rams (*Ovis aries*) as affected by social experiences during rearing. *Applied Animal Ethology*, **7**, 157–167.

10 Box essay   *Male homosexual behaviour in Arashiyama macaques*

YUJI TAKENOSHITA

10 Box.1 Introduction

Homosexual activities between females are 'common' among Japanese macaques in Arashiyama (Vasey, 2006; Vasey and VanderLaan: chapter 10). Takahata (1980, 1982) and Wolfe (1984) reported high frequency of female homosexual behaviours during the late 1970s, when 23–46% of adult females engaged in homosexual behaviours at least once during a mating season. In the 1990s when I had been studying sexual behaviours in Arashiyama, female homosexual behaviours were still common. From 1992 to 1998, these behaviours were observed in every mating season without exception, although the number of females who engaged in these activities was smaller than in the 1970s.

In contrast, male homosexual behaviour in the Arashiyama macaques has only been reported once (Takenoshita, 1998). This report highlights that male homosexual behaviour exists in this population, but such behaviour appears to be much less frequent than homosexual behaviour between females. In this article, I describe the details of male homosexual behaviours that I observed in Arashiyama during the 1993–94 and 1994–95 mating seasons. Then I examine how these behaviours can be explained.

10 Box.2 Methods

Field research was conducted on the Arashiyama E-troop in the 1992–93, 1993–94, 1994–95 and 1996–97 mating seasons. Male homosexual activities were only observed during the 1993–94 and 1994–95 mating seasons. Observations of male homosexual behaviours were made ad libitum, i.e. whenever I suspected some males of engaging in homosexual activities, I followed them to observe their behaviours in detail.

Four males, two extra-troop adult males (ETMs) from the adjacent Arashiyama F-troop and two E-troop natal prepubescent males (NPMs), were observed to engage in homosexual mounting accompanied with ejaculation by one of the participants. In addition, two natal males, an adult peripheral male

173

Table 10 Box.1. *List of males who engaged in homosexual mounting series*

Name	Age (years)*	Class**
Momo-67–75–84	10.5	ETM
Mino-63–69–74–83–89	5.5	NPM
Mino-63–75–90	5.5	NPM
Glance-64–71–88	6.5	ETM
Oppress-75–81–87***	7.5	E troop peripheral male
Meme-62–80–88***	6.5	NPM

* Age when homosexual behaviour occurred. ** ETM: extra-troop male; NPM: natal prepubescent male. *** Confirmed by indirect evidence (see text).

and a natal prepubescent male, were believed to be involved in homosexual intercourse because semen was observed on their lower backs. Information pertaining to the name, age and class (e.g. ETM versus NPM) of these males is summarised in Table 10 Box.1.

As is the case with heterosexual intercourse (i.e. copulation) in Japanese macaques, a male homosexual intercourse involved series mountings. Here I applied the definition of 'mount event' used by Hanby (1974) and Hanby and Brown (1974) to define a homosexual series mounting in my observations. Homosexual series mounting took the form of an interaction between two males which involved one or more mountings. These interactions ceased when: (1) one of the participants ejaculated, (2) the participants were separated by at least 15 metres or (3) mounting ceased for at least 5 minutes. Duration of a homosexual series mounting was measured by the interval between the first and the last mounting, regardless of the mountee. Thus, the duration of a single mount event was 0 seconds.

10 Box.3 Results

10 Box.3.1 Overview of male homosexual series mounting

Eleven male homosexual series mounting events were observed in total (Figure 10 Box.1). During these 11 series mounting events, 152 mountings were observed in three dyads: Momo-67–75–84 and Mino-63–69–74–83–89 (Dyad A); Glance-64–71–88 and Mino-63–75–90 (Dyad B); Momo-67–75–84 and Mino-63–75–90 (Dyad C). All the dyads were combinations of an ETM and a NPM. The participants did not share maternal kinship in any dyad. The number of mountings in series mounting events ranged from one to 44 times. Mean

Table 10 Box.2. *Number of mountings by extra-troop adult males (ETMs) and natal prepubescent males (NPMs)*

Dyad	Series	Number of mounts		Duration (seconds)	Cessation
		by ETMs	by NPMs		
A	1	4	3	236	Absence of mounting > 5 min
	2	13	9	1129	Absence of mounting > 5 min
	3	15	27	1351	Ejaculation of ETMs
B	4	6	0	237	Separation > 15 metres
	5	4	0	229	Ejaculation of ETMs
C	6	0	1	0	Absence of mounting > 5 min
	7	2	5	428	Absence of mounting > 5 min
	8	1	0	0	Absence of mounting > 5 min
	9	6	7	793	Absence of mounting > 5 min
	10	0	4	162	Separation > 15 m
	11	18	26	3014	Ejaculation of ETMs

Figure 10 Box.1. Male homosexual mounting. A natal prepubescent male (Mino-63–75–90) mounts an extra-troop male (Momo-67–75–84) (photo by Y. Takenoshita).

duration of series mounting was 689 seconds, ranging from 0 to 3014 seconds. In two cases, I did not observe the events from the beginning.

In Dyad B, mountings were unidirectional, i.e. all the mountings are performed by the ETM. On the other hand, the series mounting were bidirectional in Dyads A and C (Table 10 Box.2). In Dyads A and C, there were no

Table 10 Box.3. *Intermount behaviour during homosexual mount series*

| Dyad | Grooming | Ventro-dorsal contact | | Ventro-ventral contact | Sit separately |
		ETM behind NPM	NPM behind ETM		
A	11	58	0	0	1
B	3	3	0	1	1
C	11	49	0	0	10

significant differences between the number of mountings by the extra-troop adult and the natal prepubescent (Binomial test, $p > 0.05$). In all three dyads, I confirmed that ETMs ejaculated in the lower back of their partners. Anal intromission was not observed in any of the mountings.

10 Box.3.2 Mount posture and inter-mount posture/behaviour

All the males mounted with double-foot-clasp posture. 'Sit on' or 'lie on' postures that are frequent in female homosexual mountings (Wolfe, 1979; Vasey *et al.*, 2006) were not observed during the male homosexual interactions.

The most frequent behaviour exhibited by the participants during inter-mount intervals was sitting in ventro-dorsal contact (Table 10 Box.3, Figure 10 Box.2). The ETM always sat behind the NPM. As such, the ETM sat in the position that males typically adopt during heterosexual intercourse (Vasey *et al.*, 2008). In other cases, they groomed each other or sat apart. Sitting in ventro-ventral contact, which is occasionally observed during heterosexual and female homosexual interactions (Vasey *et al.*, 2008), was seen only once.

10 Box.3.3 Responses by third-party individuals

During a mount series by the Dyad C, the second ranking E-troop male, Shiro-62–74–79 incidentally passed by the dyad at a distance of about 5 metres. However, Shiro-62–74–79 completely ignored the dyad. This incident was curious because Momo-67–75–84 was an ETM. Normally, high-ranking troop males react aggressively to the presence of ETMs within their troops and chase such males away.

In contrast, troop females were not tolerant of ETMs engaged in homosexual series mounting. During one mount series within the Dyad C, a high-ranking E-troop female, Kojiwa-62–72–77–85, came into sight of the dyad. She then

Figure 10 Box.2. Typical intermount behaviour during male homosexual series mounting. A natal prepubescent male (Mino-63–75–90) sits in front of an extra-troop male (Momo-67–75–84) with ventro-dorsal contact (photo by Y. Takenoshita).

screamed and approached to within 5 metres of the dyad. The dyad left the area together without any vocalisation or aggressive response toward Kojiwa-62–72–77–85. It is curious that Mino-63–75–90 did not strike back at Kojiwa-62–72–77–85, because he was a son of a female who is higher ranking than Kojiwa-62–72–77–85, and he was also dominant over Kojiwa-62–72–77–85.

On one occasion, a male (Momo-67–75–84) who had formerly engaged in a homosexual series mounting was observed to intrude on another pair of males (Glance-64–71–98 and Mino-63–75–90) engaged in series mounting. On the day when Glance-64–71–88 and Mino-63–75–90 (Dyad B) were in consort, Momo-67–75–84 chased the dyad. Although Momo-67–75–84 did not perform distinct aggressive behaviour, the dyad avoided Momo-67–75–84 and finally terminated the mount event. After several weeks, Momo-67–75–84 and Mino-63–75–90 (Dyad C) formed a consort relationship.

10 Box.3.4 Temporary 'consort' relations

On the days when the homosexual series mounting was observed, the partici-pants of the homosexual dyads were observed to be in 'consort relationship', i.e. they exhibited frequent synchronisation of feeding, resting and travelling (Huffman, 1991). It means that they stayed, moved and fed together and followed each other between the series mounts. During these homosexual 'consortships', none of the participants directed sexual solicitations towards troop females, nor did they receive sexual solicitations from females. It also seemed that NPMs avoided close contact with their kin individuals.

10 Box.3.5 Prepubescent males in 'oestrus'

It is noteworthy that NPMs observed to engage in homosexual activities showed several characteristics that are typical of oestrous females. Namely, they had bright red faces and emitted vocalisations similar to 'oestrous calls' (Figure 10 Box.3). Such 'oestrous' behaviours were observed when they

Figure 10 Box.3. A natal prepubescent male (Mino-63–75–90) in 'oestrus' (photo by Y. Takenoshita).

were separated from their homosexual partners during consortships. It is uncertain if such 'oestrous' behaviours occurred on days when they were not engaged in homosexual consortships. The younger males of homosexual pairs behaved like females and this is consistent with observations of male homosexual interactions among mountain gorillas (Yamagiwa, 2006).

10 Box.4 How can male homosexual behaviours be explained?

Several explanations have been proposed for male homosexual behaviours in animals. Here I examine if some of these explanations fit with my observations of male homosexual behaviours in the Arashiyama-Kyoto macaques. Note that this examination is not empirical, because male homosexual behaviours were observed only incidentally.

10 Box.4.1 Proximate causes

Biased sex ratio

It has been pointed out that if the adult sex ratio in a population is biased toward a particular sex, the probability and/or frequency of occurrence of homosexual activities by the individuals of the more numerous sex increases. In Japanese macaques, the frequency of female homosexual behaviours increases when adult sex ratio is biased toward females (Wolfe, 1984; Vasey and Gauthier, 2000). In group-living captive rhesus macaques, males engage in homosexual behaviour when their access to females is physically limited (Gordon and Bernstein, 1973).

In the present case, biased sex ratio does not account for the occurrence of male homosexual activities. Adult male-to-female sex ratios of Arashiyama E-troop in 1993–4 and 1994–5 mating seasons were 0.36 and 0.32, respectively. Thus, the adult sex ratio of the study group was not male-biased during my study period. In addition, during 1993–4 and 1994–5 mating seasons, female homosexual activities were also observed in Arashiyama E-troop at the same time as male homosexual behaviours.

It might be possible that ETMs and NPMs were not able to gain access to oestrous females, partly because the central adult males tend to attack peripheral males and/or males from outside the troop if they approach oestrous females. In addition, females do not tend to prefer immature males who, for the most part, were the ones participating in the observed homosexual consortships. Thus, it is possible that the males I observed engaging in homosexual activities were limited in their ability to access the oestrous females (Takenoshita, 1998).

Sexual frustration

Yamane (2006) argued that male homosexual mounting in feral cats could be explained, in part, by the 'sexual frustration hypothesis'. This hypothesis is based on the observations that a certain amount of male–male mounting occurs when courting males lose access to an oestrous female because the female flees the area. The courting male then becomes sexually frustrated and redirects his sexual behaviour toward an adjacent male.

However, this hypothesis cannot explain the cases I observed. Male homosexual series mounting occurred not after failed attempts to perform heterosexual intercourse but, rather, it occurred in the context of temporary consortships as described above. Thus, it seems that the participants in male homosexual activities purposely chose same-sex partners.

Mistaken sexual identity

Yamane (2006) also argued that the majority of male–male mounts in feral cat were caused by mistaken sexual identity. That is, large males sometimes mistake smaller males for females. With the present cases, accordingly, one might argue that ETMs mistook NPMs for females because the latter exhibited behavioural and visual features indicating the female-typical sexual state of 'oestrus'. However, Momo-67–75–84, an ETM, reached back and manipulated the penis of his partners when he was being mounted. Therefore, he apparently recognised his partners were indeed males.

Response to imbalanced physiological state

The fact that the NPMs exhibited not only behavioural but also visual features (i.e. reddening of their face) that resemble female oestrus, suggested that there was some physiological basis influencing their behaviour, at least in part. Imbalance of sexual hormones during adolescence is common among humans and non-human primates. Such imbalance of hormones may cause temporary development of secondary sexual characteristics of the opposite sex (e.g. gynaecomastia). 'Oestrus' and male-oriented sexual behaviour in the NPMs might be explained as a response to such an imbalanced physiological state.

Effect of early experiences during development

Exclusive social contacts between males sometimes lead to homosexual relationships in rhesus macaques. Erwin and Maple (1976) reported that a pair of male rhesus macaques who were reared together in a laboratory during early stages of their life formed a very intimate social bond and engaged in frequent series mounting.

In my observations, all the dyads were combinations of an E-troop NPM and an ETM from F-troop. It is therefore impossible that these males grew up showing exclusive close contact with each other during the early stages of their development. Moreover, homosexual consortships between males were only temporary, not exclusive over the long term. Momo-67–75–84 had different same-sex partners in the 1993–94 and 1994–95 mating seasons, respectively, and Mino-63–75–90 changed same-sex partners during the 1994–95 mating season.

10 Box.4.2 Ultimate factors

Vasey (2006) summarised functional explanations for animal homosexual behaviours and conducted empirical tests of these explanations for homosexual behaviours between female Japanese macaques. Here I follow his arguments to investigate my observations. Again, it should be noted that the following investigations are not empirical.

Alliance formation

The participants in homosexual dyads were not observed to co-attack other individuals, nor did I ever observe one of the participants support the other in a conflict against a third-party individual. The only exception took place when I approached Momo-67–75–84 and Mino-63–69–74–83–89 (Dyad A) too closely in order to confirm whether anal intromission had occurred, and they attacked me together. In other cases, rather than forming alliances against others, they seemed to actively avoid any aggressive interactions with others.

However, from the perspective of ETMs, their NPM partners might be 'passive allies' in the sense that NPMs did not join with other troop members to drive their ETM partners out of the troop, as is typically the case. As described previously, there were two cases in which homosexually consorting males encountered other E-troop individuals. In both cases, the NPM just left with his ETM partner. If they had not been in consort with the ETM, the NPM would have normally intervened with other group members against the ETM. Consequently, ETMs might have been able to stay in the range of E-troop for a longer time because their NPM partners did not actively drive them away.

Dominance demonstration

In the present cases, male–male homosexual mounting does not seem to be dominance demonstration. In Dyad A and C, mountings were bidirectional. In other words, both the dominant partners and the subordinate ones mounted each other. In Dyad B, only the ETM (Glance-64–71–88) mounted, but dominance relations between Glance-64–71–88 and Mino-63–75–90 outside the consort periods were uncertain. So it is unclear as to whether Glance-64–71–88 was dominant to Mino-63–75–90 or vice versa.

Acquisition of alloparental care

It does not seem that the NPMs solicited homosexual partners in order to derive alloparental care for the infants of their close kin. During consortships, the dyads tended to stay in peripheral regions of E-troop range and did not approach the maternal relatives of NPMs.

Acquisition of opposite-sex mates

ETMs might gain more chances to access oestrous females by forming consortship with NPMs. As mentioned above, Momo-67–75–84 was once ignored by higher-ranking males. This interaction was odd, given that higher-ranking group males typically drive ETMs away. Consequently he was able to stay within E-troop for a longer duration than when alone. On the other hand, for NPMs, homosexual activities might not have increased their chances of acquiring female sexual partners, because acting like oestrous females might not be 'sex appealing' to females.

Reconciliation, tension regulation and enhancement of social bonding

The observed homosexual series mounting was not limited to post-conflict periods, nor to post-feeding periods when many monkeys gather on the feeding ground and when most conflicts over food occur. So it does not seem that the observed series mounting was related to conflict or social tension.

All the dyads were combinations of an ETM and a NPM, i.e. combinations of males of different groups. So they only had to regulate tensions between themselves, and did not need to reconcile. In case of tension escalation or even conflicts, all they have to do is to separate. Moreover, they might not share stable social bonds that would need to be enhanced through homosexual relationships.

Inhibition of competitor's reproduction

There was no need for ETMs to impede reproduction by their NPM partners because the latter were prepubescent and still incapable of ejaculating. For NPMs, the case is ambiguous. One might infer that NPMs disturbed ETMs' attempts to solicit E-troop oestrous females by soliciting ETMs with female-like 'oestrous' behaviours. However, it would have been much easier for

NPMs to inhibit ETMs' reproduction simply by chasing ETMs away from the E-troop's range. Moreover, homosexual consortships with NPMs might have increased the time during which ETMs were able to stay within the range of E-troop and, as such, possibly provided them with opportunities to access E-troop oestrous females. Unfortunately, there were not enough data on the basis of the observed cases to confirm or disprove this hypothesis.

Practice for heterosexual activities

The interactions observed in this study included several behavioural elements which are rarely seen in heterosexual intercourse, i.e. mutual mounting, thrusting without intromission and emission of female-like oestrous calls by NPMs. If these behavioural elements were seen in heterosexual intercourse, the interaction would become less harmonious and would impede fertilisation. Thus, it seems unlikely that these behaviours were training for heterosexual intercourse.

10 Box.5 Is male homosexual behaviour in Japanese macaques so special or abnormal?

In summary, one might not be able to specify the adaptive function or the proximate mechanisms of the male homosexual activities that I observed in Arashiyama. The only plausible function, if any, is that the ETMs were able to increase the amount of time they spent within the range of E-troop, but that is nothing more than speculation at this point. The only possible proximate factor identified is a potential imbalance in the physiological state of NPMs. However, such imbalance might simply be, if any, a normal part of the course of puberty rather than a disorder, per se. After the maturation of NPMs, none of them exhibited 'oestrus-like' behaviours.

Hanby (1974) investigated male–male mounting interactions in captive Japanese macaques and argued that most male–male mounts should not be considered sexual behaviours. His arguments were that (1) the majority of male–male mounting interactions consisted of a single mount, (2) male–male mounting interactions occurred more frequently in non-breeding seasons than in breeding seasons and (3) the predominant context in which male–male mounting interactions occurred was different from that of male–female copulations.

At the same time, however, Hanby (1974) claimed he did not intend to judge that all male–male mounting interactions were asexual, and clearly stated that 'male–male mounting obviously had a multiplicity of causal factors'. During his study, Hanby (1974) observed 1374 male–male mounting interactions and of these, 769 occurred during the breeding seasons. Overall, 16 of these inter-actions consisted of three or more mounts (i.e. mount series). Ejaculation was seen in two cases in one pair. The pair consisted of the 'second highest ranking and one of the most vigorously heterosexual males in the troop' and 'a middle-ranking and moderately heterosexually active male'. During inter-mount inter-vals, these male partners were often in proximity or in contact. These facts suggest that Hanby observed at least two episodes of series mounting, and possibly more, during the breeding seasons and these 'homosexual' activities were similar to the ones I observed in Arashiyama.

Homosexual behaviour is far less frequent in male than in female Japanese macaques. More research is needed to draw definitive conclusions about homosexual activity in male Japanese macaques. As such, male homosexual behaviour in Japanese macaques is exhibited under natural circumstances and, while it occurs only infrequently, it does not appear to be 'abnormal'. Male homosexual behaviour in Japanese macaques, whether expressed in captiv-ity or in free-ranging conditions, may have been (and will be) overlooked or ignored by the researchers to a certain degree. In captivity, male homosexual behaviour may be considered to be a kind of stereotyped behaviour caused by inappropriate captive conditions, and the pair may be separated to break up the consort. In wild or provisioned populations, when a researcher incidentally observes male homosexual mount series, he/she may regard the interaction as asexual because of the preconception that the majority of male–male mounts are asexual. Researchers interested in sexual behaviour may then cease paying further attention to such interaction.

In order to further investigate male homosexual behaviour in Japanese macaques, whether it is considered an evolutionary by-product or an adap-tive behaviour, it might be important to raise researchers' awareness that such behaviour exists. Those who conduct behavioural studies of Japanese macaques, especially those who study sexual behaviours in Arashiyama, should keep in mind the following: 'When you observe male–male mounting, it might be a homosexual interaction.'

References

Erwin, J. and Maple, T. (1976). Ambisexual behavior with male-male anal penetration in male rhesus monkeys. *Archives of Sexual Behavior*, **5**, 9–14.

Gordon, T. P. and Bernstein, I. S. (1973). Seasonal variation in sexual behavior of all-male rhesus troops. *American Journal of Physical Anthropology*, **38**, 221–225.

Hanby, J. P. (1974). Male-male mounting in Japanese monkeys (*Macaca fuscata*). *Animal Behaviour*, **22**, 836–849.

Hanby, J. P. and Brown, C. E. (1974). The development of sociosexual behaviours in Japanese macaques *Macaca fuscata*. *Behaviour*, **49**, 152–196.

Huffman, M. A. (1991). Mate selection and partner preference in female Japanese macaques. In *The Monkeys of Arashiyama: Thirty-five Years of Research in Japan and the West*, eds. L. M. Fedigan and P. J. Asquith. Albany, NY: State University of New York Press, pp. 101–122.

Takahata, Y. (1980). The reproductive biology of a free-ranging troop of Japanese monkeys. *Primates*, **21**, 303–329.

(1982). The socio-sexual behavior of Japanese monkeys. *Zeitschrift für Tierpsychologie*, **59**, 89–108.

Takenoshita, Y. (1998). Male homosexual behaviour accompanied by ejaculation in a free-ranging troop of Japanese macaques (*Macaca fuscata*). *Folia Primatologica*, **69**, 364–367.

Vasey, P. L. (2006). The pursuit of pleasure: an evolutionary history of female homosexual behaviour in Japanese macaques. In *Homosexual Behaviour in Animals: An Evolutionary Perspective*, eds. V. Sommer and P. L. Vasey. Cambridge: Cambridge University Press, pp. 191–219.

Vasey, P. L. and Gauthier, C. (2000). Skewed sex ratios and female homosexual activity in Japanese macaques: an experimental analysis. *Primates*, **41**, 17–25.

Vasey, P. L., Foroud, A., Duckworth, N. and Kovacovsky, S. D. (2006). Male-female and female-female mounting in Japanese macaques: a comparative study of posture and movement. *Archives of Sexual Behavior*, **35**, 117–129.

Vasey, P. L., VanderLaan, D. P., Rains, D., Duckworth, N. and Kovacovsky, S. D. (2008). Inter-mount social interactions during heterosexual and homosexual consortships in Japanese macaques. *Ethology*, **114**, 564–574.

Wolfe, L. (1979). Sexual maturation among members of a transported troop of Japanese macaques (*Macaca fuscata*). *Primates*, **20**, 411–418.

(1984). Japanese macaque female sexual behavior: a comparison of Arashiyama East and West. In *Female Primates: Studies by Women Primatologists*, ed. M. Small. New York, NY: Alan R. Liss, pp. 141–157.

Yamagiwa, J. (2006). Playful encounters: the development of homosexual behaviour in male mountain gorillas. In *Homosexual Behaviour in Animals: An Evolutionary Perspective*, eds. V. Sommer, and P. L. Vasey. Cambridge: Cambridge University Press, pp. 273–293.

Yamane, A. (2006). Frustrated felines: male-male mounting in feral cats. In *Homosexual Behaviour in Animals: An Evolutionary Perspective*, eds. V. Sommer, and P. L. Vasey. Cambridge: Cambridge University Press, pp. 172–189.

11 A theoretical model of the development and evolution of non-conceptive mounting behaviour in Japanese macaques

DOUG P. VANDERLAAN, SERGIO M. PELLIS
AND PAUL L. VASEY

Two juvenile males (Ai-61–72–83–08 and Bl-59–64–75–91–08) engaged in play mounting (photo by N. Gunst).

11.1   Introduction

In mammals, sexual interactions between individuals overwhelmingly involve two opposite-sex adult conspecifics copulating. For those with even the most

The Monkeys of Stormy Mountain: 60 Years of Primatological Research on the Japanese Macaques of Arashiyama, eds. Jean-Baptiste Leca, Michael A. Huffman and Paul L. Vasey. Published by Cambridge University Press. © Cambridge University Press 2012.

basic understanding of Darwinian evolution, this widespread pattern of sexual interaction is hardly surprising and presents no great mystery. Such behaviour is necessary. It facilitates reproduction and species would simply fail to exist without it. The development of opposite-sex sexual behaviour between adults of the same species is, therefore, unlikely to be flagged as an evolutionary puzzle because it is entirely consistent with the basic principles of Darwinian evolution.

In stark contrast, the existence of same-sex sexual behaviour is simply perplexing. Because it cannot result in conception, and thus does not directly contribute to reproduction, same-sex sexual behaviour appears to challenge the basic principles of Darwinian evolution. Consequently, when faced with the existence of same-sex sexual behaviour, one must reconcile how such behaviour comes into existence despite the fact that it is at odds with the traditional emphasis on the evolution of behaviours that enhance reproduction. By taking up this challenge, one has the opportunity to expand theoretical understanding of how Darwinian evolution occurs and gain significant insight into the behaviours that it produces.

To take advantage of the potential insights provided by examining same-sex sexual behaviour, one requires an appropriate model species. While there are numerous examples of species that exhibit same-sex sexual behaviour (Vasey, 1995; Bagemihl, 1999), relatively few exist that do so often enough to render such behaviour amenable to quantitative study (for discussion see Vasey, 2006a). Japanese macaques at Arashiyama represent an ideal species and population in which to empirically examine questions concerning non-conceptive mounting behaviour. To begin with, non-conceptive mounting behaviour occurs in the Arashiyama macaques with great enough frequency that hypothesis testing is possible via quantitative methods. Same-sex mounting has been observed in sexually mature males (Takenoshita, 1998) and females (for review, see Vasey, 2006b). Non-conceptive *heterosexual* behaviour in the form of adult female–male mounting has also been documented (Vasey and Duckworth, 2008). Juvenile–adult mounting in the Arashiyama macaques has been described qualitatively, but to date, has not been the focus of quantitative analysis (e.g. Wolfe, 1978). Examining how evolution produces such behaviour (or the capacity for such behaviour) in most, if not all, members of a population, as opposed to select segments of populations, has the potential to yield even more substantial insight into developmental and evolutionary processes.

In the present chapter, we detail preliminary findings from our budding programme of research focusing on juvenile mounting in the Arashiyama-Kyoto macaques. In addition to providing basic information about such mounting and its behavioural correlates, we provide direction for future research by outlining possible motivational bases for such behaviour. Previous authors have noted that mounting behaviour during the juvenile period might serve as a developmental stage upon which later adulthood patterns of mounting behaviour critically depend (Hanby and Brown, 1974). Taking this suggestion as our lead, we also present a tenable theoretical model that offers a testable account of how juvenile mounting might be related to non-conceptive adult mounting patterns. This model may thus provide an explanation of how widespread patterns of non-conceptive mounting behaviour emerged in certain populations of Japanese macaques. Hence, the present chapter is intended as a forum for proposing ideas that may serve as a launching pad for the further study of non-conceptive mounting behaviour in Japanese macaques.

11.2   Mounting in juvenile Japanese macaques

Qualitative reports describe how infant or young juvenile male and female Japanese macaques often mount while in the process of grasping onto the back of their mothers or other close female kin (Figure 11.1a; Hanby and Brown,

Figure 11.1. Depictions of infantile and juvenile mounting at Arashiyama.
(a) An infant male exhibiting a 'boarding-style' mount; (b) 2-year-old juvenile males: double-foot-clasp mounting; (c) 2-year-old juvenile males: shoulder biting from a mounting position (photos by D. P. VanderLaan).

Figure 11.1. (*cont.*)

1974; Wolfe, 1978). Presumably, such infantile mounting facilitates 'boarding' behaviour and subsequently being carried on the mother's, or another close female kin's, back during locomotion (Hanby and Brown, 1974; Wolfe, 1978). Following this initial period of sexually undifferentiated 'boarding-style' mounting, sex differences in mounting behaviour begin to emerge as males and females enter the juvenile phase of development.

Quantitative research conducted at the Oregon Primate Research Institute by Hanby and Brown (1974) indicated that as male Japanese macaques progressed through infancy and into the juvenile period (i.e. beginning at approximately 6 months of age), they exhibited mounting patterns that were less strictly oriented toward female kin. Instead, mounting behaviour shifted and became increasingly focused toward other juveniles, and particularly other juvenile males who are the frequent recipients of mounts.

In contrast, as females enter the juvenile period of development, they then cease mounting (Eaton et al., 1985), only to begin mounting again, albeit at low frequencies, when they experience their first oestrous period (Eaton, 1978). Furthermore, compared with males, juvenile females are less likely to be the recipients of mounting, but then as they mature reproductively they experience a substantive increase in the mounts they receive (Hanby and Brown, 1974). These findings indicate that when mounting is expressed by juvenile Japanese macaques, it is males that are overwhelmingly the participants.

Same-sex mounting interactions in animals, such as the juvenile male–male mounting in Japanese macaques, have traditionally been interpreted as 'sociosexual behaviours'. Sociosexual behaviours are those that are sexual in terms of their superficial form, but that are enacted to achieve some sort of social goal (Wickler, 1967). Although sociosexual behaviours are regularly conceptualised as being devoid of any sexual motivation, there is no a priori reason why mounts could not be sexually motivated, in part, and also serve some sociosexual function (Wickler, 1967; Vasey and Sommer, 2006). What evidence, if any, is there that juvenile male–male mounting serves some sociosexual function?

The form that same-sex mounting takes between juvenile males suggests that it may serve as an effective vehicle for communication. Effective communication involves the use of signals that are reliable and unambiguous as well as of sufficient intensity to be detected by intended receivers (Wiley, 2006). Research on the Oregon troop of Japanese macaques indicates that double-foot-clasp mounts are the most commonly employed mount posture during male–male mounting. These mounts involve the mounter placing his hands on the hindquarters of the recipient and grasping the recipient's legs with his feet (Figure 11.1b; Hanby and Brown, 1974). The double-foot-clasp mount is unique relative to other mount postures (e.g. single-foot-clasp, standing, sitting or reclining mounts) in that it involves making physical contact using all four limbs, each with a separate point of contact. The structural features of double-foot-clasp mounts may render such interactions more optimal mediums for transmitting signals associated with attracting a receiver's attention. For example, because double-foot-clasp mounting involves contact between multiple body parts, it may provide a more intense signal than alternative forms of physical contact that employ a single point of physical contact (e.g. grabbing some part of another individual's body with a hand). In addition, mounters may be better able to control a recipient's movement from the mounting position than from a non-mounting position and this may, in turn, augment a mounter's ability to channel a recipient's attention, thereby rendering any signal they produce more reliable. From the recipient's perspective, restricted mobility and exposure of vulnerable body parts that are more difficult to defend may intensify the salience of any signals performed by the mounter.

It is also noteworthy that double-foot-clasp mounting between juvenile males in the Oregon troop is often accompanied by pelvic thrusting during which the mounter moves his genital region back and forth against the recipient's perineal region. In addition, the participants routinely exhibit erections (Hanby and Brown, 1974; personal observations, 2007). In macaques, as in humans, the genital and perineal regions are richly innervated with nerve endings (genital end bulbs) that are specialised for receiving and transmitting erotic sensations of touch, tension, pressure and friction (Winkelmann, 1959; Hill, 1974). On the basis of his comparative work, Winkelmann (1959) argued that specialised patterns of genital/perineal innervation render primates better able to appreciate erotic sensations than non-primates. The occurrence of pelvic thrusting and penile erection during mounts may further contribute to the overall intensity and salience of any signals performed by the mounter.

Given that the form juvenile male–male mounting takes would seem to serve as an effective vehicle for communicating a signal, the question arises as to what that signal might be. Research on the Oregon troop indicates that mounting between sexually mature males (i.e. ≥5 years old) declines sharply during the mating season, but in contrast, juvenile male–male mounting occurs at a relatively consistent rate across the year. Hence if juvenile male–male mounting in Japanese macaques is a sociosexual behaviour, then its function probably involves the attainment of social goals that are salient throughout the year. Social play is a male-typical behaviour beginning from early in life in a number of Old World primates (see Mitchell, 1979), including Japanese macaques (Eaton *et al.*, 1985). Furthermore, social play and mounting behaviours seem to be developmentally linked. Exposing female rhesus macaques to elevated levels of androgens *in utero* produces behavioural masculinisation with respect to both of these behaviours (Goy *et al.*, 1988). In line with this logic, in Japanese macaques, juvenile male–male mounting commonly occurs in the context of social play and typically involves the mounter approaching the recipient (Hanby, 1974; Hanby and Brown, 1974). Furthermore, mounting is often accompanied by the mounter biting the recipient's shoulder (Figure 11.1c; Hanby, 1974). In macaque species, biting of the shoulder is the primary goal or target of social play (Reinhart *et al.*, 2010). On the basis of this evidence, we hypothesise that mounting in juvenile male Japanese macaques might function as a signal to facilitate social play.

11.3   Juvenile male–male mounting and social play at Arashiyama

Our research was carried out on Japanese macaques residing within the provisioned, free-ranging Arashiyama-E troop, which ranges in the mountains on the northwest outskirts of Kyoto, Japan, near the suburb of Arashiyama (Huffman, 1991). During the study period, the troop consisted of 140 individuals. More specifically, there were 109 females, 23 of which were juveniles (i.e. 0.5 to 5 years of age), and 31 males, 21 of which were juveniles. Social interactions between juvenile males were filmed between 08:00 h and 18:00 h from 24 September to 7 December 2007.

Videotaped recordings were made using a Sony digital video camera recorder (DCR-TRV19) with a colour LCD monitor. Sampling was opportunistic in that recording began when the observer (DPV) witnessed a congregation of juvenile males (i.e. two or more within 2 m of one another). Once such a congregation was observed, social interactions between males were recorded until cessation. Also, recording was limited to those social interactions clear of visual obstructions. In total, 29 hours of videotaped data were collected. Digital cassettes were converted to VHS format and a time code (1/30th of a second) was added using a Horita TRG-50 time encoder (Horita, Mission Viejo, CA). To reduce bias in scoring the behaviour observed during juvenile male–male social interactions, all interactions were viewed at normal speed, in slow motion, and frame-by-frame by the same observer (DPV).

Only dyadic interactions that took place on a relatively flat surface and from which behaviour could be scored from the initiation to the termination of the interaction were included. A *mount* consisted of a double-foot-clasp mount in which the mounter placed both hands on the recipient's hindquarters and used his feet to grasp the recipient's legs. *Social play* consisted of a bi-directional exchange of at least one bite or grab, and the interaction had to be of a non-aggressive nature and could not merely be an instance of social grooming. *Aggression* was assayed via the presence of aggressive (i.e. grunt vocalisations, open mouth stare threats) and submissive (i.e. screaming, fear grimacing) gestures. *Social grooming* consisted of removing matter (e.g. dirt, parasites) from the skin or fur of another individual. We also documented whether the mounter bit the recipient's shoulder during the mount.

We began our research on juvenile male–male mounting behaviour in the Japanese macaques at Arashiyama by assessing whether patterns of such mounting in this population were similar to those reported for the captive Oregon troop. The first step in doing so involved assessment of whether juvenile male–male mounting at Arashiyama was associated with social play. Hence, we examined the social contexts in which such mounting was exhibited for the 120 dyadic interactions observed. Combined, aggression and social grooming comprised the social context for a statistically significant minority of interactions (4.17%; n = 5; $\chi^2[1]$ = 30.63, $p < 0.001$). Mounting was followed by no alternative form of social interaction at chance levels (i.e. only mounting and no additional form of social interaction occurred; 43.33%; n = 52; $\chi^2[1]$ = 3.6, $p = 0.06$). The remaining interactions involved social play, and, statistically speaking, social play was the most common social context following mounting (52.5%; n = 63; $\chi^2[1]$ = 13.23, $p < 0.001$).

Given that biting of the shoulder is the primary goal or target of social play in macaque species (Reinhart *et al.*, 2010), we reasoned that if juvenile male–male mounting is associated with play, it should co-occur relatively frequently with shoulder biting. To test this possibility, we compared shoulder biting during the 92 mount interactions beginning with the mounter approaching the recipient from the rear to a random sample of 111 non-mount interactions in which one juvenile male approached another from behind and performed an alternative form of physical contact (i.e. bite or grab). This comparison showed that shoulder biting was statistically significantly more likely to be associated with mounting (16.3%; n = 15) than alternative forms of physical contact (1.8%; n = 2; $\chi^2[1]$ = 13.80, $p < 0.001$). Furthermore, the rate of shoulder biting in the Arashiyama-Kyoto sample was similar to that reported for the Oregon troop (i.e. 10%; Hanby, 1974).

Having established that, like the Oregon troop of Japanese macaques, juvenile–juvenile mounting behaviour in the Arashiyama-Kyoto group typically occurred within the context of play interactions, we then aimed to determine whether these mounts were used as signals for acquiring attention from recipients. Effective signalling that efficiently acquires another's attention involves the use of the appropriate behavioural signal in the appropriate context (Horowitz, 2009). Consequently, in addition to the mount itself, other features of mounting interactions should be indicative of signalling behaviour. For example, if an individual is approaching a potential recipient of a mount from the front, the use of a mount to acquire attention might be regarded as gratuitous. Rather, a visually conspicuous display such as a play face, or some other alternative form of physical contact, would be a more economical and efficient attention-getting behaviour compared with walking around to the rear of an already attentive individual and then performing a mount.

Hence, we hypothesised that a mounting interaction should be relatively unlikely to begin with the mounter approaching the recipient from the front or side. Instead, if mounting is performed to acquire a recipient's attention, it should be observed more often when the mounter approaches the recipient from the rear. During our fieldwork at Arashiyama, 120 dyadic interactions involving juvenile male–male mounts were observed. Of these, the statistically significant majority (76.67%; n = 92) consisted of the mounter approaching the recipient from the rear ($\chi^2[1] = 17.07$, $p < 0.001$). This pattern of approach from the rear is consistent with principles of efficient signalling (Horowitz, 2009) and the notion that juvenile male mounting is enacted as a behavioural signal to acquire the attention of juvenile male recipients.

These initial data suggest that juvenile male–male mounting is similar across the Arashiyama-Kyoto and Oregon troops despite the fact that the former is free-ranging and the latter is captive. As detailed above, observations of the Oregon troop indicated that juvenile male–male mounting might often facilitate social play. In the Arashiyama-Kyoto macaques, juvenile male–male mounting was also most commonly followed by social play. Further, shoulder biting, the target of social play in macaques, was a relatively common occurrence during mounting, with a consistent incidence across the Arashiyama-Kyoto and Oregon troops. The fact that juvenile male mounters in Arashiyama tended to approach recipients from the rear provides some additional evidence to suggest that juvenile males might employ double-foot-clasp mounts as behavioural signals for acquiring attention, thereby facilitating subsequent play.

11.4   Developmental and motivational aspects of juvenile male–male mounting

If juvenile male–male mounting is related to the facilitation of social play, one might question the nature of the developmental and motivational bases that underlie the link between these two social behaviours. Given the available evidence, it seems reasonable to rule out the possibility that juvenile male–male mounting is simply a precocious manifestation of adult male sexuality. Juvenile male mounting patterns differ from normative, adult male sexual behaviour in a number of respects. First, compared with adult male–female mounting, juvenile male–male mounting involves thrusting less often (Hanby and Brown, 1974). Second, juvenile males show less seasonality in their male–male mounting patterns compared with adult males (Hanby and Brown, 1974).

Third, during male–male mounting, adolescent and adult males will sometimes exhibit series mounting (i.e. a number of mounts over a period of time
with a mounting partner), which is typical of male–female mounting; occasionally this male–male series mounting is also associated with ejaculation
(Takenoshita, 1998). In contrast, juvenile male–male mounting is most often
comprised of a single mount (Hanby, 1974; personal observation, 2007). In all,
there is little evidence to support the claim that juvenile male–male mounting
behaviour is related to the early emergence of adult male sexuality.

Another possible explanation is that juvenile male mounting behaviour is
somehow about asserting dominance. Dixson (2010) noted, however, that
although many same-sex sexual behaviours have been associated with social
dominance relationships in a number of Old World primate species (e.g.
mounting, touching or handling of the external genitalia), such is not always
the case. Indeed, for a number of reasons, it appears likely that dominance has
little bearing on juvenile male mounting behaviour in Japanese macaques. For
the majority of juvenile males, particularly those under 3 years of age, status
and rank are most likely to be affected by their mothers' ranks given the matri-
archic social structure of Japanese macaques (Kawai, 1958; Kawamura, 1958;
Nakamichi, 1996; Rizaldi and Watanabe, 2010). As such, for a juvenile male
there would be little social benefit to be gained via mounting of other males to
manipulate status.

For males older than 3 years in the Hanby (1974) study, the relationship
between rank and male–male mounting behaviour was considered. Overall,
there was no consistent relationship between rank and the frequency of mounting or between rank and number of mounting partners. Rather, it was common
for these males to have mounted other males both above and below them in
the dominance hierarchy. Only when interactions involving both mounting and
social play were excluded from analysis was there some slight effect of dominance on mounting, and dominant males seemed to be particularly likely to
mount subordinate males during interactions with an agonistic element. Yet,
data presented on the Oregon troop (Hanby, 1974; Hanby and Brown, 1974)
as well as data presented above on the Arashiyama-Kyoto troop indicated
that social play was the most common social context associated with juvenile male–male mounting. Moreover, agonistic interactions are highly unlikely
to be associated with juvenile male–male mounting. It appears, then, that in
juvenile male–male mounting the assertion of dominance is more likely to be
the exception as opposed to the rule.

A more likely motivational basis for juvenile male–male mounting is
tension reduction. The best evidence in support of this claim comes from
observations of the Oregon troop of Japanese macaques, which indicate that
male–male mounting increases in frequency immediately after a disturbance

(e.g. aeroplane flying overhead, people entering the compound in which the monkeys were housed; Hanby, 1974). Increases in male–male mounting also occur during periods of excitement such as before or following feeding (Hanby, 1974). In the Arashiyama-Kyoto population, anecdotal observations by one of the present authors (PLV) indicate that juvenile male–male mounting may also increase in the presence of adult males. During these interactions, the juvenile males appear nervous and they alternately mount and direct lip-trembling – a type of submissive display – towards an adult male who typically ignores them. The mounting appears to 'calm' the juvenile males who eventually refocus their attention away from the adult male and often initiate play. The stimulation of genital and perineal regions during mounting is likely experienced as rewarding, and this positive emotional valence may function to reduce tension by ameliorating negative emotional valence.

Given that tense situations and social play are both apparently related to juvenile male–male mounting, the question arises as to what, if anything, these two correlates of mounting behaviour might have to do with one another. If the phenomena of interest (i.e. mounting, playing, tension reduction) are related, then there must be some logical sequence by which they become associated during development. As highlighted above, nascent mounting behaviour by males most often consists of infant or young juvenile male Japanese macaques mounting while in the process of grasping onto the back of their mothers to facilitate 'boarding' behaviour (Hanby and Brown, 1974; Wolfe, 1978). Such behaviour might be proximally reinforced, at least in part, via the emotional reward associated with stimulation of male genitalia. If so, then at later developmental stages, males might use the emotionally rewarding properties of mounting to reduce tension in tension-inducing situations.

Research on the Oregon troop indicates that tension-inducing situations produce increases in mounting between juvenile males, but do not produce increases in mounting between juvenile males and females (Hanby, 1974). Why might this be the case? The manner in which juvenile males and females are spatially distributed might provide the answer. Compared with males, young females play in smaller social groups, play approximately half as often, and break contact with their mothers less frequently (Eaton *et al.*, 1985; Nakamichi, 1989). Hence, when tension-inducing disturbances arise and individuals congregate to investigate the source, or when congregations form for other reasons during times of elevated excitement such as during feeding, juvenile males would be relatively more available for social interaction. Thus, as a juvenile male, engaging in male–male mounting during such instances might be an available and effective means of ameliorating tension.

Because double-foot-clasp mounting involves distinctive posturing that constrains the movement of recipients, it is likely that mounting results in orienting the attention of the recipient toward the mounter. Hence, as a consequence of mounting, males might learn that mounting facilitates the acquisition of attention. Over time, then, males may come to employ mounts to solicit play, perhaps along with additional play-oriented contact such as shoulder biting. At this developmental time point, such mounting could be termed multi-functional.

A strength of this hypothesis is that it posits relatively simple mechanisms as being the critical components that link the different stages in the developmental sequence. These mechanisms include manipulating emotional valence via the physical stimulation associated with mounting, and learning the contingency between the enactment of behaviour and its outcome. Moreover, the efficacy of this developmental hypothesis can be tested because it leads to a number of specific predictions. For example, if correct, mounting should enhance acquisition of attention from recipients as well as facilitate play between mounters and recipients. Another testable prediction is that the proportion of mounts performed to facilitate play (i.e. not for reducing tension and occurring outside of tension-inducing situations) should increase with age as individuals increasingly learn the play-facilitating nature of mounting. Similarly, increases in age should also be associated with other behavioural events indicative of social play during mounting (e.g. shoulder biting). Regardless of whether future studies support, refute or refine this particular developmental hypothesis, it offers a viable point from which to direct the investigation of juvenile male–male mounting in Japanese macaques. The qualities held by this hypothesis (i.e. agreement with the available data, reliance on simple mechanisms and openness to falsification) should also be sought after by any alternative hypotheses that attempt to explain these behavioural phenomena.

11.5 Juvenile male–male mounting and population-wide, non-conceptive mounting patterns

Hanby and Brown (1974) studied sexual behaviour in infant, juvenile, adolescent and adult members of the captive Oregon troop to gain insight regarding how early sexual behaviour might provide the building blocks for adult sexual behaviour. Indeed, early mounting and associated social experiences do appear to be important for the skilled execution of adult male mounting behaviour in closely related primate species such as rhesus macaques (Goy and Wallen, 1979). However, motivational bases such as those highlighted by the developmental account presented above seem to offer a more complete account of why juvenile males exhibit such behaviour. That is, rather than simply serving as practice for adult mounting, juvenile male–male mounting serves its own, more temporally proximal functions. These functions, which we postulate are related to tension reduction and social play, may still inevitably aid in the development of skilled adult mounting behaviour, which would be beneficial to male reproductive success. Hence, one benefit of shifting focus toward the temporally proximal functions of juvenile male–male mounting is that those developmental mechanisms and properties of the organism that are actually being favoured by evolution begin to come into clearer focus.

Another benefit of highlighting the temporally proximal functions of juvenile male–male mounting is that doing so may shed light on other non-conceptive mounting patterns exhibited by certain populations of Japanese macaques, including the Arashiyama-Kyoto and Oregon troops. Specifically, if our developmental model is correct, juvenile male–male mounting might also relate to adult female–male and female–female mounting in certain Japanese macaque populations. Compared with juvenile, adolescent and adult male–male mounting, adult female–female mounting has been more thoroughly studied. Adult female–female mounting typically consists of bidirectional series mounting during which females mount one another in succession tens or even hundreds of times over a period of time that can range from minutes to days (Fedigan and Gouzoules, 1978; Gouzoules and Goy, 1983; Wolfe, 1978, 1984). These mounting relationships are referred to as *consortships*. Unlike male–male mounting, which might be sociosexual behaviour for reducing tension and facilitating play, adult female–female mounting does not appear to relate to any sociosexual function for facilitating social or reproductive goals (for review, see Vasey, 2006b). For example, females do not use these behaviours to attract male sexual partners, impede reproduction by same-sex competitors, form alliances, foster social relationships outside these consortships, communicate about dominance relationships, obtain alloparental care, reduce social tension, practise for heterosexual activity or reconcile conflict.

Four lines of evidence indicate that adult female–female mounting behaviour is sexually motivated. First, these interactions mirror adult heterosexual behaviour in that female–female consortships never occur outside of the breeding season. Second, variations in female–female mounting across the ovarian cycle parallel the pattern found for heterosexual mounting (O'Neill *et al.*, 2004). Third, females exercise incest avoidance in that they do not form consortships with close female kin such as daughters, sisters, mothers or grandmothers (Chapais and Mignault, 1991; Chapais *et al.*, 1997). This pattern of avoidance differs from their normative pattern of social affiliation during grooming, co-sleeping, huddling and interventions (Kurland, 1977; Koyama, 1991; Chapais, 1992). Fourth, during the majority of female–female mounts, female mounters engage in vulvar, perineal and anal (VPA) stimulation by rubbing their VPA regions against the backs of recipients or stroking their VPA regions with their tails while mounting (Vasey and Duckworth, 2006). This sort of VPA stimulation was not observed during other forms of social interaction (i.e. grooming), nor was it observed in other situations involving non-social movement (i.e. walking). Given the VPA region's primary role in mediating sexual response (Burton, 1970; Dixson, 1998; Cold and McGrath, 1999), these observations provide direct evidence bearing on the sexual nature of the majority of female–female mounts in Japanese macaques.

Some research has also documented the phenomenon of adult female–male mounting in Japanese macaques, both within the Arashiyama-Kyoto (Vasey and Duckworth, 2008) and captive Oregon troops (Hanby and Brown, 1974). The study by Vasey and Duckworth (2008) provides the most detailed analysis of adult female–male mounting behaviour to date. The authors found that, similar to adult female–female mounting, female mounters stimulate their VPA region while performing such mounts, but not in other social situations or during other types of movement such as walking. Interestingly, however, VPA stimulation during female–male mounting occurred less frequently compared with that observed by Vasey and Duckworth (2006) during female–female mounting (i.e. approximately 45% of female–male mounts versus 78% of female–female mounts).

Vasey and Duckworth (2008) proposed that VPA stimulation by female mounters differs depending on the sex of the mounting partner because female–male mounts may serve certain sociosexual functions, whereas female–female mounts do not. Vasey (2006b) hypothesised that when female–female competition for male mates is high, female consort partners might mount males to restrict their movement, focus their attention on the consortship and prompt them to mount. If this hypothesis is accurate, adult female–male mounting has certain parallels with our propositions concerning juvenile male–male mounting. In both cases, mounting is proximally reinforced via sexual reward and manipulates the attention of the recipient and directs it toward the mounter to facilitate interaction.

Indeed, a male's propensity to allocate attention toward a female mounter during a consortship in adulthood might be developmentally predicated on the propensity to allocate attention toward a juvenile male mounter during that earlier developmental stage. If so, the emergence of juvenile male–male mounting may have created the opportunity for females to manipulate male attention via mounting in adulthood. Once the ability to derive such reward from mounting males emerged, an overspill consequence might have been that females began to do the same through female–female mounting (Vasey and Duckworth, 2006). Further, they may do so despite the availability of sexually motivated male alternatives because such male alternatives are less preferred than the female alternatives (Vasey and Gauthier, 2000).

11.6   Conclusions

In the present chapter, we presented a theoretical model concerning proximate and evolutionary bases for the emergence of widespread patterns of non-conceptive mounting behaviour in certain populations of Japanese macaques. In contrast to the more common emphasis on adult mounting behaviour, our model relies heavily on the full range of developmental stages in order to explain the non-conceptive patterns of mounting documented. Specifically, our model focuses on the potential importance of more fully comprehending the nature of juvenile male–male mounting in relation to social play, and how those behaviours might relate to the phenomena of adult female–male and female–female mounting.

In addition to assessing the hypotheses and predictions outlined above, perhaps another key test of the model we propose here will be its ability to account for why such non-conceptive patterns of mounting only appear in certain Japanese macaque populations, including the population at Arashiyama and the captive Oregon troop. In a recent study examining adult female–female sexual behaviour across Japanese macaque populations, Vasey and Jiskoot (2010) highlighted that such behaviour appears to be specific to Japanese macaque groups residing in or captive troops derived from central to western Honshu. Our model posits that the emergence of adult female–female mounting can be traced back to adult female–male mounting and juvenile male–male mounting. As such, we expect that in populations where female–female mounting exists, we will also observe female–male mounting and juvenile male–male mounting.

In any case, our model provides a launching pad for further research into non-conceptive patterns of mounting behaviour in Japanese macaques. Moreover, it serves as an illustration of how perplexing phenomena such as non-conceptive behaviour not only emerge through development and evolve, but also of how seemingly disparate phenomena, like that of varying mounting patterns among different segments of the same population, might be inextricably linked.

Acknowledgements

All three authors were funded by the University of Lethbridge and the Natural Science and Engineering Research Council of Canada. They thank the editors and Alan Dixson for helpful comments and suggestions on an earlier draft of this chapter.

References

Bagemihl, B. (1999). *Biological Exuberance: Animal Homosexuality and Natural Diversity*. New York, NY: St. Martin's Press.

Burton, F. D. (1970). Sexual climax in female *Macaca mulatta*. *Proceedings of the 3rd International Congress of Primatology*, **3**, 180–191.

Chapais, B. (1992). The role of alliances in the social inheritance rank among female primates. In *Coalitions and Alliances in Humans and Other Animals*, eds. A. Harcourt and F. M. B. de Waal. Oxford: Oxford University Press, pp. 29–60.

Chapais, B. and Mignault, C. (1991). Homosexual incest avoidance among females in captive Japanese macaques. *American Journal of Primatology*, **23**, 171–183.

Chapais, B., Gauthier, C., Prud'homme, J. and Vasey, P. L. (1997). Relatedness threshold for nepotism in Japanese macaques. *Animal Behaviour*, **53**, 1089–1101.

Cold, C. J. and McGrath, K. A. (1999). Anatomy and histology of the penile and clitoral prepuce in primates. In *Male and Female Circumcision*, eds. G. C. Denniston, F. M. Hodges and M. F. Milos. New York, NY: Kluwer Academic, pp. 19–30.

Dixson, A. F. (1998). *Primate Sexuality: Comparative Studies of Prosimians, Monkeys, Apes and Human Beings*. Oxford: Oxford University Press.

Dixson, A. (2010). Homosexual behaviour in primates. In *Animal Homosexuality: A Biosocial Perspective*, ed. A. Poiani. Cambridge: Cambridge University Press, pp. 381–400.

Eaton, G. G. (1978). Longitudinal studies of sexual behaviour in the Oregon troop of Japanese macaques. In *Sex and Behaviour*, eds. T. E. McGill, D. A. Dewsbury and B. D. Sachs. New York, NY: Plenum, pp. 35–59.

Eaton, G. G., Johnson, D. F., Glick, B. B. and Worlein, J. M. (1985). Development in Japanese macaques (*Macaca fuscata*): sexually dimorphic behavior during the first year of life. *Primates*, **26**, 238–248.

Fedigan, L. M. and Gouzoules, H. (1978). The consort relationship in a troop of Japanese monkeys. In *Recent Advances in Primatology, Volume 1: Behaviour*, eds. D. J. Chivers and J. Herbert. New York, NY: Academic Press, pp. 493–495.

Gouzoules, H. and Goy, R. W. (1983). Physiological and social influences on mounting behaviour of troop living female monkeys (*Macaca fuscata*). *American Journal of Primatology*, **5**, 39–49.

Goy, R. W. and Wallen, K. (1979). Experiential variables influencing play, foot-clasp mounting and adult sexual competence in male rhesus monkeys. *Psychoneuroendocrinology*, **4**, 1–12.

Goy, R. W., Bercovitch, F. B. and McBrair, M. C. (1988). Behavioral masculinization is independent of genital masculinization in prenatally androgenized female rhesus macaques. *Hormones and Behavior*, **22**, 552–571.

Hanby, J. P. (1974). Male-male mounting in Japanese monkeys (*Macaca fuscata*). *Animal Behaviour*, **22**, 836–849.

Hanby, J. P. and Brown, C. E. (1974). The development of sociosexual behaviours in Japanese macaques (*Macaca fuscata*). *Behaviour*, **49**, 152–196.

Hill, W. C. O. (1974). *Primates, Comparative Anatomy and Taxonomy, Volume 7. Cynopithecinae, Cercocebus, Macaca, Cynopithecus.* Edinburgh: Edinburgh University Press.

Horowitz, A. (2009). Attention to attention in domestic dog dyadic play. *Animal Cognition*, **12**, 107–118.

Huffman, M. A. (1991). History of the Arashiyama Japanese macaques in Kyoto, Japan. In *The Monkeys of Arashiyama: Thirty-five Years of Research in Japan and the West*, eds. L. M. Fedigan and P. J. Asquith. Albany, NY: State University of New York Press, pp. 21–53.

Kawai, M. (1958). On the system of social ranks in a natural group of Japanese monkeys. *Primates*, **1**, 111–130.

Kawamura, S. (1958). Matriarchal social ranks in the Minoo-B troop: a study of the rank system of Japanese monkeys. *Primates*, **2**, 181–252.

Koyama, N. (1991). Grooming relationships in the Arashiyama group of Japanese monkeys. In *The Monkeys of Arashiyama: Thirty-five Years of Research in Japan and the West*, eds. L. M. Fedigan and P. J. Asquith. Albany, NY: State University of New York Press, pp. 211–226.

Kurland, J. A. (1977). *Kin Selection in the Japanese Monkey*. Basel: S. Karger.

Mitchell, G. (1979). *Behavioral Sex Differences in Nonhuman Primates*. New York, NY: Litton Educational Publishing.

Nakamichi, M. (1989). Sex differences in social development during the first 4 years in a free-ranging group of Japanese monkeys, *Macaca fuscata. Animal Behaviour*, **38**, 737–748.

  (1996). Proximity relationships within a birth cohort of immature Japanese monkeys (*Macaca fuscata*) in a free-ranging group during the first four years of life. *American Journal of Primatology*, **40**, 315–325.

O'Neill, A. C., Fedigan, L. M. and Ziegler, T. E. (2004). Ovarian cycle phase and same sex mating behaviour in Japanese macaque females. *American Journal of Primatology*, **63**, 25–31.

Reinhart, C. J., Pellis, V. C., Thierry, B. *et al.* (2010). Targets and tactics of play fighting: Competitive versus cooperative styles of play in Japanese and Tonkean macaques. *International Journal of Comparative Psychology*, **23**, 166–200.

Rizaldi and Watanabe, K. (2010). Early development of peer dominance relationships in a captive group of Japanese macaques *Macaca fuscata. Current Zoology*, **56**, 190–197.

Takenoshita, Y. (1998). Male homosexual behavior accompanied by ejaculation in a free-ranging troop of Japanese macaques (*Macaca fuscata*). *Folia Primatologica*, **69**, 364–367.

Vasey, P. L. (1995). Homosexual behaviour in primates: a review of evidence and theory. *International Journal of Primatology*, **16**, 173–203.

  (2006a). Where do we go from here? Research on the evolution of homosexual behaviour in animals. In *Homosexual Behaviour in Animals: An Evolutionary*

Perspective, eds. V. Sommer and P. L. Vasey. Cambridge: Cambridge University Press, pp. 349–364.

(2006b). The pursuit of pleasure: homosexual behaviour, sexual reward and evolutionary history in Japanese macaques. In *Homosexual Behaviour in Animals: An Evolutionary Perspective*, eds. V. Sommer and P. L. Vasey. Cambridge: Cambridge University Press, pp. 191–219.

Vasey, P. L. and Duckworth, N. (2006). Sexual reward via vulvar, perineal and anal stimulation: a proximate mechanism for female homosexual mounting in Japanese macaques. *Archives of Sexual Behavior*, **35**, 523–532.

(2008). Female-male mounting in Japanese macaques: the proximate role of sexual reward. *Behavioural Processes*, **77**, 405–407.

Vasey, P. L. and Gauthier, C. (2000). Skewed sex ratios and female homosexual activity in Japanese macaques: an experimental analysis. *Primates*, **41**, 17–25.

Vasey, P. L. and Jiskoot, H. (2010). The biogeography and evolution of female homosexual behavior in Japanese macaques. *Archives of Sexual Behavior*, **39**, 1439–1441.

Vasey, P. L. and Sommer, V. (2006). Homosexual behaviour in animals: topics, hypotheses, and research trajectories. In *Homosexual Behaviour in Animals: An Evolutionary Perspective*, eds. V. Sommer and P. L. Vasey. Cambridge: Cambridge University Press, pp. 3–42.

Wickler, W. (1967). Socio-sexual signals and their intra-specific imitation among primates. In *Primate Ethology*, ed. D. Morris. Chicago: Aldine, pp. 69–79.

Wiley, R. H. (2006). Signal detection and animal communication. *Advances in the Study of Behavior*, **36**, 217–247.

Winkelmann, R. K. (1959). The erogenous zones: their nerve supply and significance. *Proceedings of the Mayo Clinic*, **34**, 39–47.

Wolfe, L. D. (1978). Age and sexual behavior of Japanese macaques (*Macaca fuscata*). *Archives of Sexual Behavior*, **7**, 55–68.

(1984). Japanese macaque female sexual behavior: a comparison of Arashiyama East and West. In *Female Primates: Studies by Women Primatologists*, ed. M. F. Small. New York, NY: Alan R. Liss, pp. 141–157.

12 Male masturbation behaviour of Japanese macaques in the Arashiyama E troop

EIJI INOUE

Mino-63–69–74–83–90, beta male at Arashiyama, July 2003 (photo by E. Inoue).

The Monkeys of Stormy Mountain: 60 Years of Primatological Research on the Japanese Macaques of Arashiyama, eds. Jean-Baptiste Leca, Michael A. Huffman and Paul L. Vasey. Published by Cambridge University Press. © Cambridge University Press 2012.

12.1   Introduction

Male masturbation has been reported in various primate species such as chimpanzees, gibbons, red colobus, rhesus macaques, stump-tailed macaques and Japanese macaques (Carpenter, 1942; Kollar *et al.*, 1968; Hanby *et al.*, 1971; Enomoto, 1974; Linnankoski *et al.*, 1981; Bielert and van der Walt, 1982; Nieuwenhuijsen *et al.*, 1986, 1987; Mootnick and Baker, 1994; Strain, 2004; Thomsen and Soltis, 2004). A questionnaire survey conducted by Thomsen *et al.* (2003) showed male masturbation was observed in 34 out of 52 species of primates (65%) and in 22 out of 30 species studied in the wild (73%). The same survey found that masturbation with ejaculation was observed in 21 out of 52 species (40%). A significant association was found between the occurrence of masturbation and the multi-male–multifemale (MM–MF) breeding system. Masturbation was observed in 27 of 33 species (82%) forming a MM–MF breeding system and in only seven of 18 species (39%) forming other breeding systems. Noting that sperm competition is probably higher in MM–MF groups, Thomsen *et al.* (2003) concluded that masturbation is an adaptation to sperm competition whereby old ejaculate is discarded. In contrast, Dixson and Anderson (2004) have argued that sperm competition is unlikely to account for the existence of masturbation in human males because the relative weight of their testes is low within one-male groups. Dixson and Anderson (2004) suggested that male masturbation is more likely to occur in MM–MF groups because males in MM–MF groups possess neuroendocrine specialisations for greater sexual arousal and performance. Thus, the function of male masturbation is still unclear.

This chapter will mainly focus on the contexts in which male non-human primates masturbate. Research on captive primates showed that the sight of females triggered male masturbation. For example, two male gibbons were reported to masturbate in the presence of female gibbons (Mootnick and Baker, 1994). Another study found that the frequency of masturbation performed by caged male chacma baboons who had visual access to females varied according to changes in perineal swelling of oestrous females (Bielert and van der Walt, 1982). Kollar *et al.* (1968) described that the frequency of masturbation bouts by a male chimpanzee increased in the presence of an oestrous female. Hanby *et al.* (1971) reported that in 10 of 27 cases, male Japanese macaques masturbated while sitting less than one metre from a sexually mature female. Male stump-tailed macaques are known to engage in a special type of masturbation called 'directed' masturbation, i.e. while staring at a female and performing teeth-chattering, while holding a female's fur, or while watching a copulating dyad (Linnankoski *et al.*, 1981; Nieuwenhuijsen *et al.*, 1986, 1987). Linnankoski *et al.* (1993) described how masturbation in male stump-tailed macaques was triggered by the presence of a female outside the cage where males were kept, and the authors noted that eye contact with the female was an important factor in male sexual arousal.

Male masturbation of captive primates was also observed in other contexts. Mootnick and Baker (1994) described that two male gibbons, which were reared by humans and had no contact with other conspecifics until the age of 5, masturbated in the presence of humans with whom they maintained eye contact while masturbating and one masturbated in the presence of food. In captive stump-tailed macaques, masturbation was associated with particular events such as taking an infant from her mother, and occasional drilling and hammering in the room (Linnankoski *et al.*, 1981). Carpenter (1942) reported three cases of male masturbation in captive rhesus macaques, and in one case a male masturbated soon after a female (who appeared pregnant) groomed him.

Masturbation was rarely reported in wild primates. Two studies reported that male masturbation occurred in the presence of oestrous females. In wild Japanese macaques (Nina-A troop) on Yakushima Island, Thomsen and Soltis (2004) observed 108 cases of male masturbation with ejaculation and 73 cases (68%) appeared to be triggered by oestrous females. Forty cases (37%) and 19 cases (18%) occurred in the presence of an oestrous female and a mating pair, respectively. In 14 cases (13%), males started masturbating after an oestrous female had passed them. In wild Temmnick's red colobus, male masturbation with ejaculation was observed only five times during approximately 10 000 hours of observation and all cases happened during aggressive inter-troop encounters, in which oestrous females and rival males were present (Strain, 2004).

Many studies suggest that the presence of females triggers male masturbation. In a group containing a few oestrous females or in a confined group, males have less chance to copulate if they observe oestrous females with higher-ranking males. In contrast, in a group containing many oestrous females, males may search for other females for maximising their mating success when they observe oestrous females, and male masturbation probably occurs only when copulation is difficult. In Japanese macaques, the number of females simultaneously in oestrous varies across study sites. This number is considerably larger in Arashiyama E troop (16.3 on average: Inoue and Takenaka, 2008) than that in Nina-A troop in Yakushima (2.4 on average: Soltis *et al.*, 2001), where Thomsen and Soltis (2004) reported that many cases of masturbation appeared to be triggered by oestrous females. Males in Arashiyama may masturbate only when they have less chance to copulate, regardless of the presence of oestrous females.

In wild Japanese macaques, Thomsen and Soltis (2004) also showed that: (1) masturbation frequency was higher in mating seasons than in non-mating seasons, (2) masturbation with ejaculation occurred only during mating seasons, and (3) masturbation increased as male social rank and male mating success declined. Hanby *et al.* (1971) also reported that low-ranking males, whose copulation frequency was low, tended to show the highest masturbation frequencies in a group of captive Japanese macaques. However in those studies, it was difficult to determine whether some males masturbated more frequently because they were lower-ranking or because they showed lower copulation rates. In Arashiyama-Kyoto macaques, the rate of copulation was not correlated with male dominance rank (Takahata, 1982), which was probably due to a large number of oestrous females (Inoue and Takenaka, 2008). It should also be noted that male age generally affects male dominance rank in macaques (Sprague, 1998) and the sexual activity of old males may be low. Thus, age may affect masturbation frequency in males.

Overall, although various factors may trigger male masturbation in several primate species, basic information on masturbation is still lacking and few detailed studies have investigated the proximate and ultimate causation of this behaviour. For example, little is known about masturbation frequency in free-ranging primates, the relationship between male masturbation and male copulation, the effect of age or dominance rank on male masturbation, and the effect of male masturbation on male reproductive success.

In this chapter, I first provide a descriptive account of male masturbation behaviour in the Arashiyama E troop of Japanese macaques (i.e. frequency, duration and hand preference while masturbating). Second, I present detailed information about the contexts in which males masturbated with ejaculation. Third, I investigate some of the mechanisms underlying male masturbation (i.e. frequency and time interval between ejaculation via masturbation versus ejaculation via copulation, and effect of dominance rank and age), as well as a possible functional aspect of the behaviour, i.e. male reproductive success, by using the data on paternity analyses from the same group and the same seasons (cf. Inoue and Takenaka, 2008).

12.2   Materials and methods

12.2.1   Study group

The free-ranging Japanese macaques of the Arashiyama E troop in the Iwatayama Monkey Park, Arashiyama, Kyoto, Japan, have been continuously provisioned since 1954 (Huffman, 1991). All the troop members have been individually identified and their demography and maternal relatedness have been recorded by the park's staff. During the study periods (see below for detail), 93–100 females and 24–27 males over 4 years of age lived in the troop. The mating season ranged from late September to March and the birth season ranged from April to August.

Table 12.1. *Individual differences in masturbation behaviour*

	Rank			Masturbation frequency		No. masturbation**			
	Period I	Period II	Observation time (hour)	Total	With ejaculation	Right hand	Left hand	Both hands***	No data****
Glance-78*	1	-	48.3	0.02	0.02				1 (1)
Mino-94	2	1	112.3	0.09	0.01	3 (0)	6 (1)		3 (0)
Mino-90	3	2	114.0	0.21	0.15	7 (5)	12 (10)	3 (2)	2 (0)
Blanche-80	4	3	115.3	0.10	0.03	12 (4)			0 (0)
Shiro-83	5	4	108.6	0.05	0.01		3 (1)	3 (0)	0 (0)
Glance-81	6	5	112.9	0.33	0.12 ·	38 (12)			1 (1)
Oppress-75–83	7	6	104.8	0.12	0.03		9 (2)	2 (1)	1 (0)
Shirayuki-84	8	7	109.0	0.14	0.01	14 (1)			1 (0)
Oppress-83*	9	-	50.7	0.00	0.00				0 (0)
Oppress-85	10	8	111.2	0.12	0.02		11 (1)		4 (1)
Blanche-84	11	9	109.5	0.05	0.00	3 (0)	1 (0)	1 (0)	1 (0)
Glance-89*	12	10	57.7	0.12	0.07	3 (2)	4 (2)		0 (0)

Full names of individuals are: Glance–64–71–78 (Glance–78), Mino–63–69–74–94 (Mino–94), Mino–63–69–74–83–90 (Mino–90), Blanche–59–64–75–80 (Blanche–80), Shiro–60–68–77–83 (Shiro–83), Glance–71–81 (Glance–81), Shirayuki–60–65–73–84 (Shirayuki–84), Oppress–60–75–83 (Oppress–83), Oppress–60–65–85 (Oppress–85), Blanche–59–64–75–84 (Blanche–84) and Glance–64–76–89 (Glance–89)

*Glance-78 and Oppress-83 disappeared from Arashiyama E troop in December 2001 and February 2002, respectively. I observed Glance-89 only in period II.

**The numbers in parentheses indicate the cases of masturbation with ejaculation.

*** This does not mean that males used both hands simultaneously, but they changed hands during the same masturbation bout.

****I did not collect the data on manual laterality during the first few weeks.

12.2.2 Data collection

Using focal animal sampling (Altmann, 1974), I sampled 11 males from 22 September to 29 December 2001 (Period I) and ten males from 22 September 2002 to 6 March 2003 (Period II). The total time of focal samples was 1154 hours collected in 147 days (Table 12.1). All the target males were adults (over 7 years old). The last two numbers of a male's name indicate his birth year. On each observation day, a target male was selected for observation so that the distribution of observation periods across males was unbiased. Target males were then followed for as long as possible. All the behaviours of the target male and the identity of all individuals within 5 metres were recorded.

12.2.3 Data analysis

Japanese macaques are series mounters (Tokuda, 1961–62). A copulation bout referred to one or more series of mounts performed by a mature male towards an oestrous female. A copulation bout ended when the male ejaculated or changed mating partners, or when the intervals between mounts were longer than 5 minutes. In this study, only data pertaining to copulation ending with ejaculation were analysed. A masturbation bout was defined as a male rubbing his erect penis with his hands, and ended when the male ejaculated or stopped rubbing his penis for more than 5 minutes. Thus, in this study, the term 'masturbation' included bouts with and without ejaculation. In some analyses, I used only the data on masturbation with ejaculation. Male reproductive success referred to the number of sired infants. The paternity of infants born in 2002 and 2003 was determined using 11 microsatellite markers (cf. Inoue and Takenaka, 2008).

In order to account for a possible manual preference in masturbating behaviour, I recorded which hand (left or right) was used in each masturbation bout. Some males changed hands during the same bout. In such cases, the bout was labelled as 'both hands'. A male was categorised as significantly lateralised when he used one hand over the other significantly (Binomial test, $p < 0.05$). Kendall rank correlations were used to analyse correlations. Wilcoxon matched-pairs signed-ranks tests were used to compare two matched datasets. Chi-square tests were conducted to compare the frequency distribution. Mann–Whitney U tests were conducted to compare two independent datasets.

12.3   Results

12.3.1   *Masturbation frequency and duration*

All the sampled males were observed masturbating, except one (Table 12.1). In total, 149 masturbation bouts were recorded, and 47 bouts of these were masturbation ending with ejaculation. The average frequency of masturbation and masturbation with ejaculation was 0.12 and 0.04 bouts per hour, respectively. The maximum number of masturbation and masturbation with ejaculation bouts per day was eight and three times, respectively. Mi-90 and Gl-81 masturbated more frequently than other males.

The duration of masturbation with ejaculation was significantly shorter than that of copulation with ejaculation (median values: 69 and 524 seconds, respectively, Wilcoxon matched-pair signed-rank test, $Z = 2.8$, $p < 0.01$, n = 10 males). It took less than 3 minutes in 42 of 47 masturbation bouts with ejaculation (89%) and in 24 cases of 209 copulation bouts (11%).

12.3.2   *Hand preference*

Some males used only one hand when they masturbated (Table 12.1). I observed more than 10 masturbation bouts for each of the four males named Blanche-80,

Table 12.2. *Manual laterality in masturbation compared with stone-handling behaviour*

| Name | Masturbation | Stone handling behaviour** | |
		Two-handed patterns***	One-handed patterns
Blanche-80	AR*	–	SL*
Glance-81	AR*	–	AL*
Oppress-75–83	SL*	–	SL*
Oppress-85	AL*	SL*	SR*
Blanche-84	A*	A* ·	SL*

*A, ambi-preference (no significant hand use); SL, significant left preference; SR, significant right preference; AL, always left preference; and AR, always right preference.
**Data from Leca *et al*. (2010).
***Leca *et al*. (2010) selected the patterns in which both hands operated simultaneously but complementarily on the same stones with one hand (termed dominant hand) performing the finer, more skilful component, while the other hand (termed subordinate hand) had a supportive role.

Glance-81, Shirayuki-84 and Oppress-85. Of these four males, three (Blanche-80, Glance-81, Shirayuki-84) only used their right hands while masturbating, whereas the other male (Oppress-85) only used his left hand. Two males (Shiro-83, Oppress-75–83) used either their left hands or both hands. Of the ten males observed masturbating five times or more, five males (50%) were significantly lateralised. Three of them showed a right-hand preference, and the other two showed a left-hand preference. When considering the five males for which data on manual laterality in masturbation behaviour (this study) and stone-handling behaviour (cf. Leca *et al*., 2010) were available, it appeared that these individuals did not show consistent lateral biases across the two types of behaviour (Table 12.2).

12.3.3   *Contexts in which males masturbated with ejaculation*

Males masturbated to ejaculation while alone (i.e. without individuals in 5-metre proximity) in 28 of 47 cases (60%). In 12 cases (26%) males were observed masturbating with non-oestrous females within 5 metres. The presence of oestrous females might have triggered male masturbation in only seven of 47 cases (15%). In three of these cases, males ejaculated by masturbation with oestrous females within 5 metres. Among the other four cases, a male masturbated at the bottom of a tree in which an oestrous female was sitting (one case), a male started to masturbate a few minutes after he stopped following an oestrous female (two cases), and a male started to masturbate 6 minutes after he passed near a dominant male and an oestrous female (one case).

Out of these seven cases when oestrous females might have triggered male masturbation, three cases happened on 29 November 2001 (cf. Glance-81, CASE 1 in Table 12.3) and two cases happened on 28 October 2002 (cf. Blanche-80, CASE 2 in Table 12.3). Although the males were close to oestrous females for a long period of time during both days, they did not copulate with these females (probably due to the female rejection of the male copulation approach in CASE 1, and to the interruption by dominant males in CASE 2; cf. underlined sentences in Table 12.3).

Males started to masturbate when they were groomed in seven cases (15%) and in an eighth case, a male started to masturbate 2 minutes after being groomed. In only one of these eight cases, an oestrous female groomed a male. Among the other seven cases, females who were not in oestrus groomed males (six cases), and a 2-year-old female groomed a male (one case).

I compared the duration of masturbation with ejaculation in different contexts. When oestrous females might have triggered masturbation with ejaculation, this behaviour was not significantly shorter than in other situations (Mann–Whitney, $Z = 0.7$, $p = 0.47$). The duration of masturbation when males were groomed was significantly shorter than the duration in other situations (Mann–Whitney, $Z = 2.2$, $p < 0.05$).

12.3.4 Ejaculation via copulation and ejaculation via masturbation

The frequency of masturbation with ejaculation was lowest in November and December, i.e. when the frequency of copulation was highest (Figure 12.1). There was a tendency for a negative correlation between monthly average ejaculation frequency by masturbation versus by copulation, although this tendency was not statistically significant ($\tau = -0.45$, $p = 0.052$). Males who ejaculated by copulation less often masturbated to ejaculation more frequently ($\tau = -0.39$, $p = 0.02$). However, the frequency of male masturbation, including both with and without ejaculation, was not significantly correlated with that of ejaculation by copulation ($\tau = -0.27$, $p = 0.09$).

Males were more likely to ejaculate by masturbation on days when they did not ejaculate by copulation ($\chi^2 = 12.9$, $p < 0.001$, Table 12.4). During the study period, there were 14 days during which males ejaculated by masturbation twice or more, and for 13 of these days, males did not ejaculate by copulation. Males never masturbated to ejaculation on days when they copulated to ejaculation three times or more.

During the study period, there were 10 days during which males ejaculated by both masturbation and copulation. Ejaculation by masturbation

Table 12.3. *Two particular cases in which male masturbation was triggered by oestrous females. The significance of underlined sentences is explained in the text*

CASE 1. Glance-81 (6th ranking male), 29 December 2001

Time	Behaviour
7:31	Glance-81 (6th ranking male) was found.
7:40	Glance-81 went where Blanche-84 (11th ranking male) and Mino-75 (oestrous females) stayed together. After that, Glance-81 followed Mino-75 whenever Mino-75 moved.
9:47	Glance-81 started grooming to Mino-75.
9:58	<u>Glance-81 pushed the back of Mino-75, but she did not accept his mounting.</u>
9:59	<u>Glance-81 pushed the back of Mino-75, but she did not accept his mounting.</u>
10:02	<u>Glance-81 pushed the back of Mino-75, but she did not accept his mounting.</u> Mino-75 moved and Glance-81 followed her.
10:08	Glance-81 started masturbation with Mino-75 in approximately 10 cm proximate, and ejaculated by masturbation. After that, he followed Mino-75 whenever she moved. I twice observed that Glance-81 lost Mino-75 and then he found Mino-75 with Blanche-84. He followed Mino-75 and in the way he masturbated at 11:08 and 13:36. Glance-81 almost kept the proximity with Mino-75 until I stopped observation of Glance-81 at 15:31, but he did not copulate.

CASE 2. Blanche-80 (3rd ranking male), 28 October 2002

Time	Behaviour
7:17	Blanche-80 (3rd ranking male) was found with Chonpe-88 (oestrous female) Chonpe-88 followed Blanche-80. There was Mino-90 (2nd ranking male) around them.
7:57	Blanche-80 started grooming to Chonpe-88.
8:03	During grooming, <u>Mino-90 chased Chonpe-88.</u> Blanche-80 and Chonpe-88 again stayed together and sometimes groomed.
8:27–29	Blanche-80 masturbated with Chonpe-88 in proximity, but did not ejaculate. Blanche-80 and Chonpe-88 sometimes groomed.
8:58	<u>Mino-90 again chased Chonpe-88.</u>
9:22	Blanche-80 went near Chonpe-88, and then moved together.
12:28	<u>Mino-90 chased Chonpe-88 who stayed near Blanche-80.</u>
12:33	Blanche-80 and Chonpe-88 stayed together and then started grooming.
12:37	During grooming, <u>Mino-94 (first ranking male) chased Chonpe-88.</u>
12:40	Blanche-80 masturbated with ejaculation. After that, he moved where some oestrous females stay.
12:53	Blanche-80 started masturbation and ejaculated 3 minutes later.
13:04–13	Blanche-80 masturbated, but did not ejaculate.
13:15	Blanche-80 was groomed by Mino-83 (females not in oestrus) with Chonpe-88 in 5 m proximity.
13:25	During grooming, Blanche-80 masturbated with ejaculation.

Table 12.4. *Relationship between the number of ejaculations by copulation and by masturbation*

| | | No. ejaculations by masturbation | | | |
		0	1	2	3
No. ejaculations	0	34	5	8	5
by copulation	1	27	8	1	0
	2	24	1	0	0
	3–7	32	0	0	0

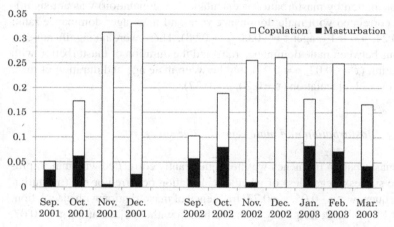

Figure 12.1. Monthly frequency of ejaculation by masturbation and by copulation.

preceded ejaculation by copulation on 8 of 10 days, including 1 day when a male ejaculated by masturbation once and then by copulation twice. On 2 other days, ejaculation by copulation preceded ejaculation via masturbation, including the day a male ejaculated by copulation once and by masturbation twice.

The time interval between ejaculations by masturbation was significantly longer than that between ejaculations by copulation (median values: 188.0 and 75.5, respectively; Mann–Whitney, $Z = 2.6$, $p < 0.01$). There were only two cases in which ejaculation by copulation preceded ejaculation by masturbation. The time interval between these two bouts (217 minutes) was longer than between two ejaculations by copulation, although statistical analysis would not be valid due to the small sample size. In only 17 of 120 cases (14%), time interval between ejaculations by copulation was longer than 217 minutes.

12.3.5   Effect of dominance rank and age on masturbation behaviour

Masturbation frequency was not significantly correlated with male dominance rank, regardless of the outcome, i.e. with ejaculation or not (masturbation frequency: $\tau = -0.14$, $p = 0.40$; frequency of masturbation with ejaculation: $\tau = -0.19$, $p = 0.25$; frequency of masturbation without ejaculation: $\tau = -0.12$, $p = 0.47$). Masturbation frequency was not significantly correlated with the age of males either (masturbation frequency: $\tau = -0.12$, $p = 0.45$; frequency of masturbation with ejaculation: $\tau = -0.06$, $p = 0.73$; frequency of masturbation without ejaculation: $\tau = -0.04$, $p = 0.81$). The total ejaculation frequency (i.e. ejaculation by masturbation + ejaculation by copulation) was not significantly correlated with male dominance rank and male age (dominance rank: $\tau = -0.24$, $p = 0.13$; age: $\tau = -0.00$, $p = 0.98$). There were no significant correlations between male dominance rank and the duration of masturbation with ejaculation ($\tau = -0.02$, $p = 0.93$), and between male age and duration of masturbation with ejaculation ($\tau = 0.20$, $p = 0.37$).

12.3.6   Masturbation and male reproductive success

Male masturbation frequency was not significantly correlated with male reproductive success, regardless of whether ejaculation occurred or not (masturbation frequency: $\tau = -0.05$, $p = 0.77$; frequency of masturbation with ejaculation: $\tau = -0.12$, $p = 0.52$; frequency of masturbation without ejaculation: $\tau = -0.07$, $p = 0.72$).

12.4   Discussion

12.4.1   Frequency and duration of masturbation behaviour

In the Japanese macaques of Arashiyama, male masturbation average frequency was 0.12 (times/hour). Considering the average daily observation time (7.85 hours), male masturbation was observed once a day on average. In Yakushima, males spent on average 4.2 minutes per hour masturbating, calculated by one-zero sampling within one minute-periods (Thomsen and Soltis, 2004). Assuming that the average duration of masturbation bouts in Yakushima is similar to that in Arashiyama (69 seconds), the rate of male masturbation at Yakushima was much higher than that at Arashiyama. This difference may be due to the high rates of masturbation by young males at Yakushima.

Thomsen and Soltis (2004) proposed that masturbation without ejaculation could function to increase sexual arousal of young males. I focused on adult males and did not collect data on young males in Arashiyama-Kyoto monkeys. This may explain why the masturbation frequency was lower in Arashiyama than in Yakushima. In captive stump-tailed macaques, male masturbation was observed a total of 1694 times in 1065 observation hours (Nieuwenhuijsen *et al.*, 1986). This study group contained 18–23 males. This means that the average masturbation frequency was 0.08 times per individual per hour, and this value was similar to that of male Japanese macaques in Arashiyama.

The duration of masturbation with ejaculation was shorter than that of copulation with ejaculation. In stump-tailed macaques, the duration of uninterrupted penis manipulation during masturbation was 20–60 seconds, which was similar to that of copulation with ejaculation (Linnnankoski *et al.*, 1981). Japanese macaques are multi-mount ejaculators, which results in longer copulation bouts. Multi-mounting macaque species show high degrees of male–male competition (Shively *et al.*, 1982). In rhesus macaques, higher-ranking males engaged in longer mounting series than lower-ranking males (Manson, 1996). These findings suggest that high-ranking males prolong their copulation time for mate guarding. In the present study, the duration of masturbation with ejaculation was not affected by male dominance rank or age. This suggests that copulation time remains the same as long as copulation patterns, such as the time interval between mounts and the number of thrusts per mount, do not vary.

12.4.2　Manual laterality

Several non-human primate species showed individual-level bias toward the left or right hand when performing particular behaviours, but none of them showed population-level bias in manual preference (McGrew and Marchant, 1997). Three male Japanese macaques in Arashiyama E troop used only their right hand when performing masturbation behaviour and one male used only his left hand. Another male used his left hand more often than his right hand. There was no hand preference at the group level. These results were consistent with previous findings obtained in macaques in general (McGrew and Marchant, 1997), and in the Arashiyama E troop of Japanese macaques when performing stone-handling behaviour in particular (Leca *et al.*, 2010). Since male masturbation can be found in various primate species, a cross-species comparison of hand preference in this behavioural pattern may contribute to the growing database of manual laterality and a better understanding of its evolution in non-human primates.

12.4.3   Effect of male dominance rank and age on the frequency of masturbation

The absence of correlation between male dominance rank and masturbation frequency was inconsistent with previous results obtained in the Japanese macaques of Yakushima (Thomsen and Soltis, 2004). This discrepancy was probably due to the fact that a positive correlation between dominance rank and copulation frequency was found in Yakushima (Soltis *et al.*, 2001), whereas no significant correlation was found in the Arashiyama E troop (Inoue and Takenaka, 2008). In this chapter, I showed that male masturbation frequency was correlated with male copulation frequency. In Yakushima, lower-ranking males masturbated more frequently due to their lower mating success.

Another factor that may affect male masturbation is age. Although I did not follow adolescent males (between 4 and 6 years of age) in this study, it should be noted that among adult males (i.e. over 7 years old), age did not appear to affect ejaculation frequency. In the Arashiyama E troop, there were several old males (around 20 years) and they ejaculated as frequently as younger adult males did.

12.4.4   Do males masturbate when they have less chance to copulate?

In each individual and in each month, ejaculation frequency by masturbation and ejaculation frequency by copulation were negatively correlated. Masturbation was more frequently observed on the days when copulation was not observed (Table 12.4). These data suggest that males masturbated when their copulation frequency was low.

Many studies have shown that females, and particularly oestrous females, may trigger male masturbation (Kollar *et al.*, 1968; Hanby *et al.*, 1971; Bielert and van der Walt, 1982; Nieuwenhuijsen *et al.*, 1986, 1987; Thomsen and Soltis, 2004). However, in the free-ranging Arashiyama-Kyoto Japanese macaques, oestrous females possibly triggered male masturbation in only seven of 47 cases (15%). As I showed in CASE 1 and CASE 2, males masturbated where they stayed in the vicinity of oestrous females for long periods of time, but could not copulate with them. These cases were rare in Arashiyama compared with reports from Yakushima, where oestrous females possibly triggered male masturbation in approximately 70% of cases (Thomsen and Soltis, 2004). This may be due to the difference in the number of females simultaneously in oestrus. In Arashiyama, when a male was unable to access an oestrous female, the best strategy in terms of mating success was to search for other oestrous females. The age of males may also account for this inter-group difference.

Young males (between 4 and 6 years of age) may masturbate more often in the presence of oestrous females because they may have less chance to access oestrous females than adult males even in a group containing a large number of oestrous females. However, this hypothesis could not be tested here because I did not collect data on young males.

In the Arashiyama E troop, male Japanese macaques masturbated under 'relaxed' conditions, such as when being groomed or alone. In seven cases (15%), males masturbated while they were groomed. Carpenter (1942) also reported that a captive rhesus macaque masturbated after being groomed. Male Japanese macaque sometimes exhibit erections while being groomed, even by close relatives. The duration of male masturbation with ejaculation in grooming contexts was shorter than that in other contexts. This suggests that grooming may lead to male sexual arousal. In 28 cases (60%), masturbation occurred when males were alone (i.e. no individuals within 5 metres). In most of these cases, it was difficult to determine what triggered male masturbation. Thomsen and Soltis (2004) reported that 19% of cases of male masturbation occurred when they were alone in the forest. I observed that males masturbated when resting alone after walking around and searching for oestrous females. Masturbation may be associated with the difficulty to find a mate and copulate. The time interval between ejaculations by masturbation was longer than that of ejaculation by copulation. Ejaculation by copulation preceded ejaculation by masturbation in only two cases, and in these cases, the time interval between ejaculations by masturbation was also longer than that between ejaculations by copulation. Overall, these results were consistent with the 'sexual outlet' hypothesis predicting that males would be more likely to masturbate when opportunities to copulate were rare (cf. Dixson and Anderson, 2004).

However, these findings contrast with those recently reported by Waterman (2010) about masturbation in male Cape ground squirrels (*Xerus inauris*). This author found that (1) masturbation frequency was higher in dominant individuals, which copulated more, (2) rates of masturbation increased with the number of mates accepted by the female and (3) masturbation usually occurred after a male had copulated with a female. Waterman's results are consistent with the hypothesis that masturbation in this species of squirrel may be a form of genital grooming, where males use their accessory gland fluids to cleanse their reproductive tract and reduce the transmission of sexually transmitted infections.

12.4.5 Masturbation and reproduction

In the Arashiyama E troop, male masturbation frequency and the number of offspring sired were not correlated. Therefore, male masturbation does not appear to negatively affect reproductive success. Masturbation could function to increase sperm fitness because the next ejaculate may contain fewer but quicker moving sperm (Baker and Bellis, 1993; Thomsen and Soltis, 2004). In the Arashiyama E troop, the time interval between ejaculations when masturbation preceded copulation was longer than that between ejaculations by copulation. This suggests that the quality of sperm in the next ejaculate after masturbation is probably not better than that in the next ejaculate after copulation. However, males with low copulation frequency, such as young males, might somehow increase the quality of their sperm by masturbation. In order to further discuss the function of masturbation, more information on the quantity and quality of sperm in different contexts of ejaculation is necessary.

Acknowledgements

I would like to thank the following people: Mr S. Asaba, Manager of the Iwatayama Monkey Park, Arashiyama; Mrs S. Kobatake, M. Morita and other staff at this park for their assistance in sampling and research at Arashiyama; Drs J.-B. Leca, M. A. Huffman and P. L. Vasey for their valuable comments on previous versions of the manuscript and their efforts in publishing this edited volume; Drs J. Yamagiwa, N. Nakagawa and other members of Human Evolution Studies, Kyoto University, for their advice during my research. The study was financed by JSPS global COE program (A06, Biodiversity) and by grants from MEXT, Japan (#19107007 to Dr J. Yamagiwa).

References

Altmann, J. (1974). Observational study of behavior: sampling methods. *Behaviour*, **49**, 227–267.

Baker, R. R. and Bellis, M. A. (1993). Human sperm competition: ejaculate adjustment by males and the function of masturbation. *Animal Behaviour*, **46**, 861–885.

Bielert, C. and van der Walt, L. A. (1982). Male chacma baboon (*Papio ursinus*) sexual arousal: mediation by visual cues from female conspecifics. *Psychoneuroendocrinology*, **7**, 31–48.

Carpenter, C. R. (1942). Sexual behavior of free ranging rhesus monkeys (*Macaca mulatta*). *Journal of Comparative Psychology*, **33**, 143–164.

Dixson, A. F. and Anderson, M. J. (2004). Sexual behavior, reproductive physiology and sperm competition in male mammals. *Physiology and Behavior*, **83**, 361–371.

Enomoto, T. (1974). The sexual behavior of Japanese monkeys. *Journal of Human Evolution*, **3**, 351–372.

Hanby, J. P., Robertson, L. T. and Phoenix, C. H. (1971). The sexual behavior of a confined troop of Japanese macaques. *Folia Primatologica*, **16**, 123–143.

Huffman, M. A. (1991). History of the Arashiyama Japanese macaques in Kyoto, Japan. In *The Monkeys of Arashiyama: Thirty-five Years of Research in Japan and the West*, eds. L. M. Fedigan and P. J. Asquith. Albany, NY: State University of New York Press, pp. 21–53.

Inoue, E. and Takenaka, O. (2008). The effect of male tenure and female mate choice on paternity in free-ranging Japanese macaques. *American Journal of Primatology*, **70**, 62–68.

Kollar, E. J., Beckwith, W. C. and Edgerton, R. B. (1968). Sexual behavior of the ARL colony chimpanzees. *Journal of Nervous and Mental Disease*, **147**, 444–459.

Leca, J. B., Gunst, N. and Huffman, M. A. (2010). Principles and levels of laterality in unimanual and bimanual stone handling patterns by Japanese macaques. *Journal of Human Evolution*, **58**, 155–165.

Linnankoski, I., Hytönen, Y., Leinonen, L. and Hyvärinen, J. (1981). Determinants of sexual behavior of *Macaca arctoides* in a laboratory colony. *Archives of Sexual Behavior*, **10**, 207–222.

Linnankoski, I., Grönroos, M. and Pertovaara, A. (1993). Eye contact as a trigger of male sexual arousal in stump-tailed macaques. *Folia Primatologica*, **60**, 181–184.

Manson, J. (1996). Male dominance and mount series duration in Cayo Santiago rhesus macaques. *Animal Behaviour*, **51**, 1219–1231.

McGrew, W. C. and Marchant, L. F. (1997). On the other hand: current issues in and metaanalyses of the behavioral laterality of hand function in nonhuman primates. *Yearbook of Physical Anthropology*, **40**, 201–232.

Mootnick, A. R. and Baker, E. (1994). Masturbation in captive *Hylobates* (Gibbons). *Zoo Biology*, **13**, 345–353.

Nieuwenhuijsen, K., de Neef, K. J. and Slob, A. K. (1986). Sexual behaviour during ovarian cycles, pregnancy and lactation in group-living stumptail macaques (*Macaca arctoides*). *Human Reproduction*, **1**, 159–169.

Nieuwenhuijsen, K., de Neef, K. J., van der Werff, J. J., Bosch, T. and Slob, A. K. (1987). Testosterone, testis size, seasonality, and behavior in group-living stump-tailed macaques (*Macaca arctoides*). *Hormones and Behavior*, **21**, 153–169.

Shively, C., Clarke, S., King, N., Schapiro, S. and Mitchell, G. (1982). Patterns of sexual behavior in male macaques. *American Journal of Physical Anthropology*, **2**, 373–384.

Soltis, J., Thomsen, R. and Takenaka, O. (2001). The interaction of male and female reproductive strategies and paternity in wild Japanese macaque, *Macaca fuscata*. *Animal Behaviour*, **62**, 485–494.

Sprague, D. S. (1998). Age, dominance rank, natal status, and tenure among male macaques. *American Journal of Physical Anthropology*, **105**, 511–521.

Strain, E. D. (2004). Masturbation observations in Temmnick's red colobus. *Folia Primatologica*, **75**, 114–117.

Takahata, Y. (1982). The socio-sexual behavior of Japanese monkeys. *Zeitschrift für Tierpsychologie*, **59**, 89–104.

Thomsen, R. and Soltis, J. (2004). Male masturbation in free-ranging Japanese macaques. *International Journal of Primatology*, **25**, 1033–1041.

Thomsen, R., Soltis, J. and Teltscher, C. (2003). Sperm competition and the function of male masturbation in non-human primates. In *Sexual Selection and Reproductive Competition in Primates: New Perspectives and Directions*, ed. C. B. Jones. ASP-Book Series, Special Topics in Primatology, Vol. 3, pp. 436–453.

Tokuda, K. (1961–62). A study on the sexual behavior in the Japanese monkey troop. *Primates*, **3**, 1–40.

Waterman, J. (2010). The adaptive function of masturbation in a promiscuous African ground squirrel. *PLoS ONE*, **5**, e13060.

Part III

Cultural behaviour, social interactions and ecology

13 Thirty years of stone handling tradition in Arashiyama-Kyoto macaques: implications for cumulative culture and tool use in non-human primates

JEAN-BAPTISTE LECA, NOËLLE GUNST
AND MICHAEL A. HUFFMAN

Glance-71–81, the oldest male (29 years) in the Arashiyama-Kyoto troop in October 2010, who still exhibits stone handling behaviour (photo by N. Gunst).

The Monkeys of Stormy Mountain: 60 Years of Primatological Research on the Japanese Macaques of Arashiyama, eds. Jean-Baptiste Leca, Michael A. Huffman and Paul L. Vasey. Published by Cambridge University Press. © Cambridge University Press 2012.

13.1   Animal cumulative culture: a debated topic

If culture (also termed 'tradition' by ethologists) is defined as a population-specific behavioural practice, persistent in several group members across generations or at least over a number of years, and dependent on social means for its transmission and maintenance (Perry and Manson, 2003), then culture is not limited to humans. There is increasing evidence for cultural variations in a wide range of behavioural patterns (e.g. interspecific interactions, communicatory, courtship, display, grooming, object play and social play behaviours, feeding habits, food processing techniques, medicinal plant use and tool use) and across various animal taxa (including fish, birds, rodents, cetaceans and non-human primates) (for reviews, see Lefebvre and Palameta, 1988; Avital and Jablonka, 2000; Fragaszy and Perry, 2003).

However, some authors argue that 'animal traditions' and 'human culture' should be distinguished and considered analogues rather than homologues on the basis of several major differences: (1) the content of what is transmitted (simple versus elaborate behavioural patterns); (2) the social learning mechanisms that support them (local enhancement and social facilitation versus imitation and teaching); (3) the stability and durability of the phenomenon (ephemeral animal tradition drifts or fads lasting from only a portion of an individual's life span up to a few generations versus stable human cultural traits enduring across centuries); and (4) the cumulativity of the process (no obvious improvement of behavioural patterns showing little if any change over generations versus progressive accumulation of cultural modifications over time leading to increasingly complex behaviours) (Galef, 1992; Tomasello *et al.*, 1993; Enquist and Ghirlanda, 2007; Caldwell and Millen, 2009; Hill, 2009).

Cumulative cultural evolution refers to situations in which 'the achievements of one pattern of behaviour form the basis for the selection of a modified and better-adapted descendent pattern' (Avital and Jablonka, 2000, p. 94). This process involves a 'ratchet-like effect' where a beneficial modification is retained until it can be improved upon, and results in behaviours or artefacts with cultural histories, i.e. that no individual could invent on their own (Tomasello, 1990; Tomasello *et al.*, 1993). On the one hand, human societies typically exhibit elaborate cumulative cultural evolution, with new patterns and methods building upon their predecessors', often leading to increasing diversity, complexity and efficiency of cultural or technological products (Tomasello, 1990; Boyd and Richerson, 1996; Caldwell and Millen, 2008a, 2008b, 2010). These accumulated adaptive knowledge and artefacts have allowed our species to occupy and exploit a far wider range of habitats than any other animal (Boyd and Richerson, 1996).

On the other hand, current evidence for cumulative culture and ratcheting in non-human species remains rare and controversial (Galef, 1992; Boesch and Tomasello, 1998; Tomasello, 1999; Laland and Hoppitt, 2003; Tennie *et al.*, 2009). There are only a few well-documented cases in which cultural changes seem to accumulate over generations, leading to the evolution of behavioural patterns that no single individual could invent. In New Caledonian crows, tool manufacture skills may partly be acquired through cumulative cultural evolution (Hunt and Gray, 2003). In killer whales, the foraging techniques consisting of briefly beaching in order to prey on sea-lion pups appear more diverse and complex across generations (Guinet and Bouvier, 1995). Some forms of ant-fishing and nut-cracking behaviours currently performed by particular chimpanzee communities indicate a step-by-step elaboration on earlier and simpler variants that may reflect accumulated modifications of socially transmitted behavioural patterns (Whiten *et al.*, 2003).

Finally, since Japanese researchers started providing food for the Japanese macaques living on Koshima island, this troop has gradually acquired a whole new lifestyle (Avital and Jablonka, 2000). Feeding the monkeys first with sweet potatoes, then with wheat grains, on the sandy seashore of Odomari beach, directly led to the appearance of two successive food-washing traditions: (1) potato-washing, with an original form described as dipping the potatoes into the fresh water of a nearby stream, thus washing off sand and dirt before eating them, and a subsequent elaboration of this behaviour consisting of biting the potatoes before dipping them into the shallow salty seawater, not only to wash them, but also presumably to season them before they were consumed; and (2) wheat-washing, defined as picking up a handful of mixed sand and wheat and throwing it into the seawater, which resulted in separating the heavier sand that sank from the lighter wheat that floated on the surface, allowing the monkeys to collect it easily (Kawai, 1965; Itani and Nishimura, 1973; Kawai *et al.*, 1992; Watanabe, 1994).

Moreover, the habit of spending more and more time on the beach, an unnatural habitat for Japanese macaques, also had ulterior indirect effects on the diffusion of additional behavioural innovations, through the influence of food provisioning on the troop's activity budget and sedentary lifestyle (cf. Huffman and Hirata, 2003; Leca *et al.*, 2008a). As young monkeys brought to the beach by their mothers (who had learned washing their food) became accustomed to the salty water, they started playing in it. Thus, sea-related subsistence traditions triggered the social traditions of using the sea for swimming, jumping and diving, as well as cooling in summer, newly acquired behaviours that became characteristic of the whole troop, including the adults, and had not been reported before in this troop or in other troops of Japanese macaques (Kawai, 1965; Kawai *et al.*, 1992; Watanabe, 1994). Another consequence of these beach activities occurred later: the monkeys started to eat raw fish, a feeding habit that is still present in the troop today (Watanabe, 1989; Leca *et al.*, 2007a). In sum, Koshima macaques have accumulated and elaborated over decades their food-related and social traditions in a ratcheted way by developing a new lifestyle associated with a new habitat, the sandy beach and the sea (Kawai *et al.*, 1992; Watanabe, 1994; Avital and Jablonka, 2000).

However, these few cases of cumulative cultural evolution in animals are still considered speculative and contentious. Some authors argue that cognitive constraints and contrasting social learning abilities make the evolutionary improvement of behaviours by the gradual accumulation of cultural adaptations much more likely in humans than in other animals (Galef, 1992; Heyes, 1993; Tomasello, 1999). As pointed out by Boyd and Richerson (1996), while social learning and culture are common in nature, cumulative cultural evolution is strikingly rare. Therefore, the pervasive human ability to accumulate socially learned behaviours over many generations poses an evolutionary puzzle: if cumulative culture is such an effective means of adaptation, why do non-human cultures not ratchet to any substantial degree? Lack of evidence for such a process does not mean its absence in nature (Danchin and Wagner, 2008). In order to tackle this issue, more 'provocative and intriguing instances of animal cumulative culture' based on systematic and long-term research are needed (Sapolsky, 2006). This report aims to show how the longitudinal study over 30 years (1979–2009) of one of the most thoroughly documented behavioural traditions in non-human primates, namely stone handling by the Japanese macaques living at Arashiyama, Japan, can contribute to the understanding of cumulative culture in animals, through the gradual transformation of stone-directed behavioural patterns that could be regarded as tool-use precursors.

13.2   Stone handling as a behaviour: structural and functional aspects

Stone handling (SH, hereafter) activity is typically defined as the spontaneous, solitary, non-instrumental and seemingly playful manipulation of stones, through the performance of multiple behavioural variants, also called SH patterns, with one or both hands, and occasionally in combination with the feet and mouth (Huffman, 1984; Leca *et al.*, 2010a, 2011). SH is typically categorised as a form of solitary object play, and differs both structurally and functionally from object exploration (Huffman and Quiatt, 1986; see also Candland *et al.*, 1978; Fagen, 1981; Hall, 1998). An individual engaged in SH activity can perform, for several minutes, a series of different SH patterns, often repeated and varied in sequence, while showing a relaxed facial expression and focusing most of its attention on the stones being manipulated (Huffman, 1984; Leca *et al.*, 2007b).

Like in other types of object play (cf. Fagen, 1981), some SH patterns are similar in form to those used during foraging activity, but the behaviours are performed out of context and modified in structure (Leca *et al.*, 2007c, 2008a, 2011). SH occurrence and frequency are largely dependent upon the time available for non-subsistence activities (Huffman and Hirata, 2003; Leca *et al.*, 2008a). SH is mainly practised by young individuals but is also continued into adulthood. In macaques, SH is probably the only example of routine object play among adults (Huffman and Hirata, 2003; Leca *et al.*, 2007b). Age appears to affect the diversity and type of SH patterns displayed. As they grow older, individuals tend to perform less varied and more simple patterns, such as gather, scatter or pick up stones (Huffman and Quiatt, 1986; Leca *et al.*, 2007b; Nahallage and Huffman, 2007a).

Although SH is primarily a solitary activity, the social aspects involved in the occurrence of this behaviour should not be overlooked. First, there is no doubt that it is socially transmitted (Huffman, 1984; Nahallage and Huffman, 2007b; Leca *et al.*, 2010b). Second, an inter-group comparative study showed that troop size was correlated with the proportion of troop members exhibiting SH simultaneously. The effect of troop size on the synchronised performance of SH may reveal the contagious nature of play (Leca *et al.*, 2007b). Third, SH is occasionally integrated with social interactions such as play wrestling and allogrooming (Huffman, 1984; Leca *et al.*, 2008a; Figure 13.1). Fourth, once particular stones are involved in a solitary SH episode, they appear to trigger great interest from other individuals who sometimes try to snatch them away from the handler as if they were the only stones available, and such supplanting interactions over the stones suggest the existence of a rudimentary form of 'possession' in monkeys (Huffman and Quiatt, 1986; Leca *et al.*, 2010b).

Regarding functional aspects, SH is largely considered a non-directly adaptive behaviour (Huffman, 1984; Huffman and Quiatt, 1986; Leca *et al.*, 2011). Most of the 45 SH patterns listed in the Japanese macaque repertoire do not seem to serve any immediate function (Leca *et al.*, 2007c; Nahallage and Huffman, 2007a). Despite the rare occurrence of percussive and complex SH patterns combining two stones, stones and substrates or objects, and stones and body parts (e.g. *flint, pound on surface* and *put/rub on fur*, cf. Table 13.1), and with the notable exception of unaimed stone-throwing, a SH pattern that may serve to augment the effect of agonistic displays in a captive troop housed at the Kyoto University Primate Research Institute (Leca *et al.*, 2008b), the stones handled are never used as tools to achieve an overt goal. Even complex combinatorial SH patterns did not meet the descriptive criteria of Beck's (1980) definition of tool use. The combination of stones with other objects, including food items, did not 'efficiently alter the form, position, or condition' (Beck, 1980, p. 10) of these objects (Leca *et al.*, 2011). Therefore, there is no local survival advantage in performing a particular SH pattern rather than another.

However, two proximate explanations for the performance of SH have been suggested. First, we believe that all monkeys, regardless of age, may simply enjoy manipulating stones, and pleasurable feedback potentially gained from the activity may be an immediate reinforcement (Huffman, 1996; Leca *et al.*, 2007c; Nahallage and Huffman, 2007a). Second, and at least in troops

Figure 13.1. Examples of stone handling (SH) patterns by Japanese macaques at Arashiyama. (a) cuddle, (b) rub stones together, (c) gather, (d) carry, (e) grasp with hands, (f) rub in mouth; (g) Glance-64–76 (the SH innovator) handling stones on 7 December 1979; Social influence of the mother in SH acquisition by infants, (h) Glance-64–76 and her infants in 1987, (i) Kusha-59–71–76–82 and her infant in 2008; Handling stones while involved in a social play interaction (j) and in a grooming interaction (k) (photos a, c, e, i, j, and k by J.-B. Leca; b and f by N. Gunst; d, g, and h by M. A. Huffman).

Figure 13.1. (*cont.*)

Figure 13.1. (*cont.*)

Figure 13.1. (*cont.*)

Figure 13.1. (*cont.*)

Figure 13.1. (*cont.*)

provisioned with cereal grains several times a day, like at Arashiyama, handling stones may be an extension of foraging-like behaviours, a continuation of manipulatory actions directed at alternative objects, while chewing food that does not require further food-processing behaviours (Huffman and Hirata, 2003; Leca *et al.*, 2008a).

13.3   Stone handling as a tradition: inter-group variation, social transmission and long-term maintenance

Japanese macaques are known for their cultural behaviours, among which is SH. The behaviour meets the set of criteria typically used to define a tradition. First a systematic comparative survey of SH in multiple populations of

Table 13.1. *Comprehensive list of the 35 stone handling (SH) patterns performed by Japanese macaques at Arashiyama between 1979 and 2009, and categorised according to general activity patterns*

Category	Name (code)	Definition
Investigative activities	Bite (B)	Bite a stone
	Hold (H)	Pick up a stone in one's hand and hold on to it, away from the body
	Lick (L)	Lick a stone
	Move inside mouth (MIM)	Make a stone move inside one's mouth with tongue or hands
	Pick (P)	Pick up a stone
	Put in mouth (PIM)	Put a stone in one's mouth and keep it some time
	Sniff (SN)	Sniff a stone
Locomotion activities	Carry (CA)	Carry a stone cuddled in hand from one place to another
	Carry in mouth (CIM)	Carry a stone in mouth while locomoting
	Grasp walk (GW)	Walk with one or more stones in the palm of one or both hands
	Move and push/pull (MP)	Push/pull a stone with one or both hands while walking forward/backward
	Toss walk (TW)	Toss a stone ahead (repeatedly) and pick it up while walking
Collection or gathering activities	Cuddle (CD)	Take hold of, grab or cradle a stone against the chest
	Gather (GA)	Gather stones into a pile in front of oneself
	Grasp with hands (GH)	Clutch a stone or a pile of stones gathered and placed in front of oneself
	Pick up (PU)	Pick up a stone and place it into one's hand
	Pick and drop (PUD)	Pick up a stone and drop it repeatedly
	Pick up small stones (PUS)	Pick up small stones and hold them between fingertips (like the picking up of wheat grains)
	Clack (CL)	Clack stones together (both hands moving in a clapping gesture)
Percussive or rubbing sound-producing activities	Combine with object (COO)	Combine (rub or strike) a stone with an object different from a stone (food item, piece of wood, metal, etc.)
	Flint (FL)	Strike a stone against another held stationary
	Grind with teeth (GWT)	Press and rub with a crushing noise one's teeth against a stone held in hand
	Pound on surface (POS)	Pound a stone on a substrate
	Rub in mouth (RIM)	Rub a stone against another held in mouth
	Rub/roll on surface (ROS)	Rub or roll a stone on a substrate
	Rub stones together (RT)	Rub stones together
	Scatter (SC)	Scatter stones about, on a substrate, in front of oneself
	Shake in hands (SIH)	Take stones in one's open palm hand and shake the stones with the hand moving back and forth
	Swipe (SW)	Swipe stones together (both hands moving in a sweeping gesture)

Table 13.1. (*cont.*)

Category	Name (code)	Definition
Other complex manipulative activities	Flip (FP)	Turn a stone over with both hands
	Put in water (PIW)	Put a stone in water
	Roll in hands (RIH)	Roll a stone in one's hands
	Rub/put on fur (ROF)	Rub or put a stone on one's fur while self-grooming
	Rub with hands (RWH)	Hold a stone in one hand and rub it with the other (like potato-washing)
	Wash (W)	Put a stone in water or pick up a stone from water and rub it with hands

Japanese macaques revealed substantial inter-group variation in the frequency and form of the behaviour, with a minor role of genetic determinants and environmental factors in explaining such differences (Leca *et al.*, 2007c, 2008c). Instead, the geographic distribution of clear troop-dependent clusters of SH variants was suggestive of the notion of cultural zones, based on inter-troop observation and possibly males transferring SH patterns when migrating from one troop to another (Leca *et al.*, 2007c). Second, longitudinal and experimental studies provided sound evidence for the role of social factors in the acquisition of the behaviour and the maintenance of the tradition, which may involve not only direct social influences through the observation by naïve infants of their mothers as SH demonstrators, but also indirect social inputs through the stimulating effect of SH artefacts, such as piles of stones left on the ground by previous stone handlers (Nahallage and Huffman, 2007b; Leca *et al.*, 2010b). Moreover, the pathways of intra-group diffusion of SH were in accordance with affiliated networks: the behaviour spread among social partners, along matrilineages, or within same-age classes (Huffman, 1984; Leca *et al.*, 2007b, 2008b). Third, transmitted over generations, SH behaviour persists over decades within several groups of Japanese macaques, where it occurs on a regular basis (Leca *et al.*, 2007c, 2010b).

13.4   What makes the Arashiyama-Kyoto troop 'special' for the study of the SH tradition?

When SH behaviour is mentioned in the primate culture literature, it is often associated with one particular location in Japan: Arashiyama, Kyoto Prefecture (e.g. Thierry, 1994, p. 98; de Waal, 2001, p. 230). However, it is not the place where SH was first noticed or reported. The very first observation of SH in Japanese macaques might have occurred around 1966 at Funakoshiyama,

Table 13.2. *The different periods of survey of stone handling at Arashiyama. MAH: Michael A. Huffman, DQ: Duane Quiatt, JBL: Jean-Baptiste Leca, NG: Noëlle Gunst*

Survey period	No. observation days	Main observers	Troop name	Troop size	No. stone handlers	% of stone handlers	No. SH patterns	Reference
Aug. 1979– Sept. 1980	170	MAH	B	243	1	0.4	4	Huffman (1984)
Nov. 1983– Jun. 1984	47	MAH	B	236	115	48.7	8	Huffman (1984)
Sept. 1984– Feb. 1985	113	MAH, DQ	B	236	142	60.2	8	Huffman and Quiatt (1986)
May– Jul. 1991	41	MAH	E	139	113	81.3	17	Huffman (1996)
May– Aug. 2004	96	JBL	E	141	131	92.9	32	Leca et al. (2007b, 2007c)
Jun.– Oct. 2008	66	JBL, NG	E	132	123	93.2	35	Leca et al. (2010a, 2010b)

Hyogo Pref. (cf. I. Narahara cited in Huffman and Hirata, 2003), and the first published study on SH was conducted at Takagoyama, Chiba Pref. (Hiraiwa, 1975). Moreover, Arashiyama is only one of the ten sites across the Japanese archipelago where SH behaviour has been observed, reported to occur or studied (Huffman and Hirata, 2003; Huffman et al., 2010; Leca et al., 2007c).

There are two reasons to account for the association between SH and Arashiyama in the public and scholars' minds. First, the initial research article written in English and providing original detailed descriptions about the conditions of appearance and initial diffusion of SH behaviour within a group of Japanese macaques was drawn from observations done at Arashiyama (cf. Huffman, 1984). Second, Arashiyama is the only study site where the prevalence of SH behaviour among individually identified group members and the diversity of SH patterns have been documented at several points in time for three decades (Figure 13.2, Table 13.2). As Perry (2006) pointed out, cultural primatology is a relatively new discipline and long-term databases that could bring a historical perspective on cultural modifications within the same populations and across multiple generations are lacking (but see Kawai et al., 1992; Perry et al., 2003; Nishida et al., 2009; this study for notable exceptions). In sum, Arashiyama is the first field site where a combination of longitudinal,

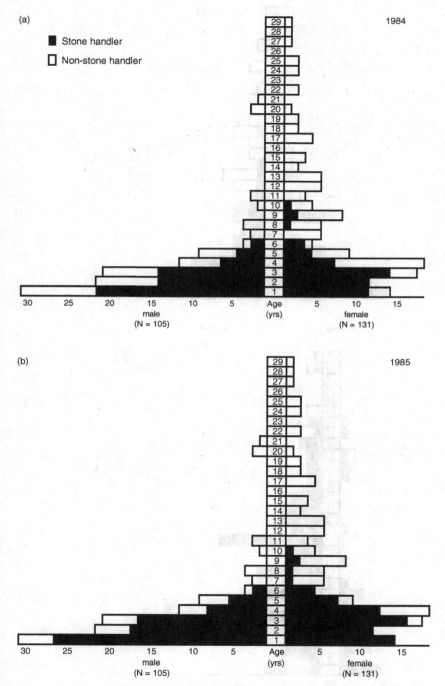

Figure 13.2. Distribution of stone handlers and non-stone handlers at Arashiyama, according to age and sex classes, and at several points in time: (a) 1984, (b) 1985, (c) 1991, (d) 2004, (e) 2008.

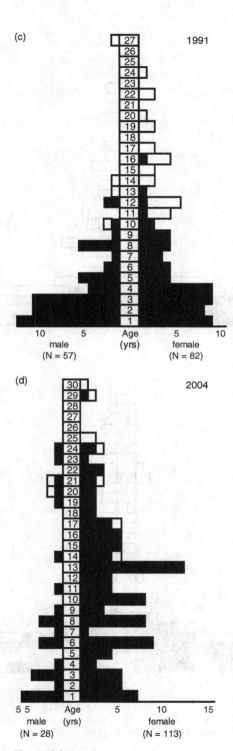

Figure 13.2. (*cont.*)

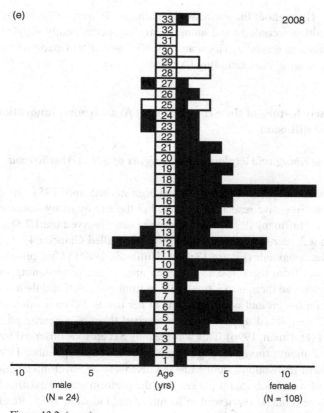

Figure 13.2. (*cont.*)

comparative and experimental approaches has provided sound evidence for the long-term maintenance, inter-troop variability and social transmission of a single cultural behaviour in Japanese macaques (reviewed in Huffman *et al.*, 2010).

For the present report, we conducted long-term analyses (1979–2009) on the free-ranging provisioned troop of Japanese macaques living at the Iwatayama Monkey Park, Arashiyama, Kyoto Prefecture. In 1986, a troop fission occurred at Arashiyama, splitting the original B troop into two sister troops named E and F; from then, only E troop stayed around the provisioning area and could be surveyed (cf. Huffman, 1991). According to the survey period, the group comprised between 132 and 243 members, of all age and sex classes, and the vast majority of them were sampled for SH behaviour (Table 13.2). Individual identities, exact age and kin relations through maternal lineages were known. The study subjects could be approached and observed within

3–5 metres. Throughout the entire study, data on SH were collected by using continuous video-recorded focal-animal sampling, occasionally supplemented with instantaneous group activity scan sampling, as well as video-recorded and pen-and-paper *ad libitum* sampling (Altmann, 1974).

13.5   Early history of the SH tradition at Arashiyama: innovation and diffusion

13.5.1   *Describing and explaining the origins of SH at Arashiyama*

Although the Arashiyama-Kyoto troop had been studied since 1954, and despite long-term and intensive research conducted at the site by many scientists successively (cf. Huffman, 1991), SH had never been observed until 7 December 1979, when a 3-year-old middle-ranking female, called Glance-64–76, started to exhibit the behaviour (Figure 13.1; cf. Huffman, 1984). After bringing several flat stones from the forest to the open area of the provisioning site, she repeatedly gathered them into a small pile in front of herself and then scattered them about on the ground with the palms of her hands. When another monkey approached, she picked up a few stones, carried them to a nearby place, and resumed SH (Huffman, 1996). This was the only SH episode observed by MAH during the 12-month survey lasting from August 1979 to September 1980.

Like in most innovations, defined as the discovery of novel information, the emergence of new behavioural patterns, or the performance of existing behaviours in a novel context (reviewed in Kummer and Goodall, 1985; Reader and Laland, 2003), we can only speculate about the factors that may have favoured the appearance of SH at Arashiyama, including the environmental context, the structural and functional aspects of the behaviour, and the individual characteristics of the innovator. First, food provisioning has undoubtedly affected the animals' activity budget, relaxed selective pressures on foraging, and created favourable environmental conditions under which various behavioural innovations by Japanese macaques may occur (Huffman and Hirata, 2003; Leca *et al.* 2007c, 2008a, 2010c). More specifically, attracting monkeys to the open space of feeding areas, where many stones occur, increases considerably their opportunities to encounter these objects. Feeding monkeys also gives them 'free time' since they can devote less time to foraging compared with their wild counterparts.

These proximate explanations are in agreement with the gradual disappearance of SH at Takagoyama after provisioning was stopped (Fujita, personal communication, cited in Huffman, 1984), and with the lack of observations of SH in wild, non-provisioned troops of Japanese macaques at other sites (e.g. Kinkazan: Shimooka, personal communication; Yakushima: Hanya, personal communication). In non-provisioned troops, foraging interspersed with travelling between food patches accounts for a large proportion of the daily activity budget (Hanya, 2004), and there may simply be less time available for non-subsistence activities such as SH (Huffman and Hirata, 2003; Leca *et al.*, 2008a). Therefore, food provisioning is likely to enhance the chances for SH to emerge. Although at Arashiyama (and other field sites too), there is now a strong temporal relationship between SH and feeding activities, with most SH episodes occurring within 20 minutes after food was distributed, it should be noted that the SH tradition emerged several decades after the onset of provisioning in these troops (Huffman and Hirata, 2003; Leca *et al.*, 2008a). The reasons for the late appearance of SH are not fully understood. Possibly, sporadic SH appeared earlier without spreading within the troop, and without being noticed by human observers (Huffman, 1984).

Second, the general behavioural predispositions of a species make behavioural innovation relatively predictable (Huffman and Hirata, 2003). Considering the natural propensity for Japanese macaques to manipulate stones (cf. Leca *et al.*, 2007c), and provided equivalent stone availability (cf. Leca *et al.*, 2008c), SH traditions are theoretically equally likely to emerge in all provisioned troops, although the relative rate of exposure to stones does not influence the latency of infants to acquire SH (Nahallage and Huffman, 2007b). Because chance may account for a good number of behavioural innovations (Reader and Laland, 2003), and SH is essentially a playful activity, we suggest that the SH innovation is an accidental by-product of object playing. Finally, individual characteristics of the SH innovator may partly account for the appearance of the novel behaviour. The fact that the first individual observed to perform SH at Arashiyama was a juvenile emphasises the playful nature of this behaviour (Huffman, 1984). Glance-64–76 might have temperamental traits that made her prone to behavioural innovation. This is consistent with previous research showing that most Japanese macaque innovators are juvenile females (Kawai, 1965; Itani and Nishimura, 1973; Kawai *et al.*, 1992; Leca *et al.*, 2010c).

13.5.2 *Analysing the diffusion of SH at Arashiyama*

During the time elapsed between the first two surveys (October 1980–October 1983), SH behaviour had spread to almost half of the Arashiyama-Kyoto troop and had become a daily occurrence (Table 13.2). Despite this 3-year gap in observation, a detailed analysis of the 1984 distribution of identified stone handlers according to age/sex classes and matrilineal membership allowed MAH to reconstruct, at least partially and a posteriori, the initial pathway of diffusion of SH within the troop (Figure 13.2; Huffman, 1984). In order to facilitate the comparison with other behavioural traditions, Huffman and Quiatt (1986) proposed that the diffusion of innovative behaviours could be chronologically divided into two distinct stages, namely the 'transmission phase' and the 'tradition phase' (after Itani, 1958; Kawamura, 1959; Kawai, 1965).

Transmission phase

This early period of behavioural diffusion is typically similar across groups and presumably species. The first individual(s) to display a novel behaviour may do so repeatedly and persistently, which facilitates its initial transmission to a network of close spatial-interactional associates of the innovator (Huffman and Quiatt, 1986; Nishida *et al.*, 2009; Leca *et al.*, 2010c). According to Coussi-Korbel and Fragaszy (1995), the spatial proximity and behavioural coordination exhibited by tolerant partners are expected to enhance opportunities for social learning, and therefore, the rate and speed of behavioural diffusion should be high within these subgroups.

Previous studies of subsistence traditions involving the diffusion of food-related innovations in Japanese macaques showed that most of these behaviours initially spread among young individuals, immediately followed by the upwardly vertical transmission to older kin members and to other adults regardless of kinship (Kawai, 1965; Itani and Nishimura, 1973; Kawai *et al.*, 1992; but see the special case of fish eating in Watanabe, 1989 and Leca *et al.*, 2007a). In contrast, it appeared that the transmission phase of SH behaviour occurred exclusively horizontally and among a particular cohort of young individuals, mainly peer playmates, starting with the innovator's cousins (Huffman, 1984). After a few years, as the first stone handlers grew older and their social networks extended, new and younger siblings and peers became stone handlers. Unlike food-washing behaviours, as no individuals over 5 years old were seen to perform SH behaviour during the transmission phase, there would be a critical period after which SH cannot be acquired (Huffman, 1984).

Most feeding and food-washing innovations found in Japanese macaques showed a wide and rapid intra-group diffusion – it took less than 4 years for most of these novel behaviours to be transmitted to at least a second group member – probably because information about food is critical to every individual (Itani, 1958; Kawai, 1965; Azuma, 1968; Itani and Nishimura, 1973; Watanabe, 1989; Kawai *et al.*, 1992; Nakamichi *et al.*, 1998). Likewise, the playful nature of SH behaviour could account for its fast transmission within the Arashiyama-Kyoto troop (Huffman, 1996; Leca *et al.*, 2007b). Seeing group members playing is a reliable cue for more individuals that the current environmental conditions are safe enough to engage in play (Spinka *et al.*, 2001). Although SH is primarily a solitary activity, the sight of nearby stone handlers and even the loud noise generated by percussive patterns may increase an individual's probability to start handling stones (Leca *et al.*, 2007b). This stimulation effect may be amplified by an increasing number of troop members and eventually result in a form of 'hysterical contagion' (Kerckhoff, 2002). This may help to explain the increase in number of SH individuals (synchronised occurrence) around feeding time at Arashiyama, as this is the only time when most troop members are all together in the same location (Leca *et al.*, 2008a). The rapid transmission of SH at this site may also have been enhanced by local construction projects when a large number of stones were left at the edge of the feeding area (Huffman and Hirata, 2003).

Tradition phase

In this later period of diffusion, the behaviour is passed down along multigenerational lines. At Arashiyama, when the first female stone handlers reached reproductive maturity, SH was mainly acquired vertically from mothers to offspring via observational learning (Huffman, 1984, 1996; see also Nahallage and Huffman, 2007b). During the tradition phase, the rate of SH diffusion was approximately equal to the birth rate: an infant primarily learnt SH from its mother, and complementarily from an infant playmate whose mother handled stones, or from an older sibling who had learned SH from a playmate (Huffman, 1996). However, it should be noted that the mother is the primary source of an infant's early exposure to SH (Huffman, 1984, 1996; see also Nahallage and Huffman, 2007b). From 1985, all infant macaques living at Arashiyama acquired SH behaviour within their first 6 months of life and thus, the increase in the number of new stone handlers was purely a function of new births (Huffman and Quiatt, 1986).

Since 1979, SH has spread gradually within the Arashiyama-Kyoto troop and across multiple generations of all matrilineages. Cross-sectional and longitudinal analyses on a 30-year time scale allowed us to assess the rate, speed and pathways of diffusion of this behaviour (Figure 13.2; Table 13.2; see also Huffman, 1996; Huffman and Hirata, 2003). In June 1984, 48.7% of the troop exhibited SH, and by February 1985, an additional 27 individuals (i.e. 60.2% of the troop) born before June 1984 were added to the list. In 1991, 12 years after the appearance of SH at Arashiyama, the diffusion rate increased to 81.3%, and every member of the E troop under the age of 10 was verified to have acquired SH (Huffman, 1996; Figure 13.2). Finally, during more recent surveys in 2004 and 2008, the percentages of stone handlers in the troop were 92.9% and 93.2%, respectively. In 2008, only nine individuals (eight females and one male) out of 132 troop members were qualified as verified non-stone handlers, i.e. they were sufficiently sampled but were not observed performing SH. They were all 25 years and older. Among them, the five youngest individuals (25–28 years old) were recorded as stone handlers in the 1991 or 2004 survey but had stopped engaging in this behaviour since then, whereas the four oldest individuals (28 years and older) had never acquired SH. At Arashiyama, as well as several other study sites, SH frequency was significantly lower in old adults than in younger troop members (Leca *et al.*, 2007b; Nahallage and Huffman, 2007a).

During the tradition phase, as long as mothers continue to practise SH, and provided the initial environmental conditions (in terms of food provisioning and stone availability) prevail, this behaviour will persist in young individuals and will thus become established in the troop across generations (Huffman and Hirata, 2003). However, the case of Takagoyama – where the SH tradition gradually disappeared after food provisioning was stopped and the monkeys began to feed solely on natural vegetation (Fujita, personal communication, cited in Huffman, 1984) – suggests that the persistence of the cultural practice of SH may be contingent on diet and foraging circumstances (Leca *et al.*, 2008a).

13.6   Cumulative transformation of the SH tradition

With a 30-year history, the SH tradition at Arashiyama has now reached its 'transformation phase', defined as the late period in which long-enduring practice with the behaviour and acquired familiarity with the properties of the stones are gained through the integration of SH with other daily activities by many age and sex classes (cf. Huffman and Quiatt, 1986; Huffman and Hirata, 2003). In 2004, we conducted a comparative survey of SH among multiple troops of Japanese macaques. We found that the Arashiyama-Kyoto troop presented a unique profile in terms of frequencies of SH patterns, i.e. its own SH tradition (Leca *et al.*, 2007c). However, a longitudinal study of SH in this troop showed that the emergence of this tradition was not an overnight process. By using similar methods of data collection for three decades of continued observation at Arashiyama, we found that the monkeys have gradually increased the size and the complexity of their SH repertoire and largely diversified the contexts in which SH activity was practised compared to earlier generations of stone handlers (Leca *et al.*, 2007c, 2008a).

13.6.1   *Gradual increase in the size and complexity of the SH repertoire*

The first aspect of the transformation of the SH tradition is an increase in the size and complexity of the SH repertoire over a number of years, that is an accumulation across generations of stone-related behavioural diversity

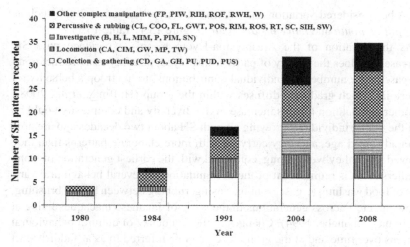

Figure 13.3. Accumulation over time and generations of stone handling (SH) patterns diversity and complexity (for categories, names and definitions of SH patterns, please see Table 13.1).

and sophistication (Figure 13.3). In 1980, the SH innovator, Glance-64–76, displayed only four SH patterns, namely *gather*, *pick up*, *scatter* and *carry*. In 1984, eight basic SH patterns were reported in the Arashiyama-Kyoto troop, including the original ones: *gather*, *pick up*, *scatter*, *carry*, *cuddle*, *roll in hands*, *rub stones together* and *clack* (Huffman, 1984). In 1991, an additional nine SH patterns were recognised – making a total of 17 patterns in the SH repertoire of the troop – with six of those patterns being obvious variations of the previous eight (*pick up and drop*, *pick up small stones*, *rub on surface*, *rub with hands*, *flint* and *grasp with hands*). The three new variants were *toss walk*, *move and push* and *grasp walk*, behavioural patterns considered to reflect an increasing familiarity with stones in general and their integration with locomotion activity, as the practice of SH spread and became a substantial part of the individual and the troop's daily activities (Huffman, 1996).

Between 1991 and 2004, the size of the SH repertoire almost doubled. During the 2004 survey, a total of 32 SH patterns were observed in the Arashiyama-Kyoto troop (Leca *et al.*, 2007c). The late emergence of SH patterns not recorded before involved percussive and complex manipulative actions, such as *pound on surface*, *combine with object*, *rub/put on fur* and *wash*, revealed an increased diversity in the combination of stones with other objects or substrates (Leca *et al.*, 2008a). Finally, in 2008, two new SH patterns were recorded – i.e. 34 patterns in the group SH repertoire – *grind with teeth* and *rub in mouth*, that could be considered variations of patterns already observed in 2004, such as *bite*, *put in mouth* and *move inside mouth* (Figure 13.1; Table 13.1).

As the duration of the Arashiyama-Kyoto troop's experience with SH increases, so does the variety of patterns displayed, possibly as a product of an increase in the number of 'individual contributions' to the troop's behavioural repertoire which gradually diffuses within the group (Huffman *et al.*, 2008). Another explanation for this increase in SH diversity and complexity·could be that the young individuals growing up with SH about two decades ago are now at an advanced age, and they carry on with more elaborate patterns than they showed when they were young, compared with the earliest generation of stone handlers. This is reminiscent of the accumulation of several behavioural variants of food washing (e.g. seasoning, rinsing, rubbing between hands, brushing, throwing, etc.) across generations in the troop of Japanese macaques living at Koshima (Watanabe, 1994). Changes in the frequency of cultural behavioural patterns over time and at the group level may be referred to as a 'faddish shift in the practice of certain behavioural sub-types' (Huffman and Quiatt, 1986, p. 413). In the case of SH, numerous behavioural patterns are accumulated modifications of earlier forms. For example, the recently appeared *flint* and *swipe* patterns can be considered slight variations of the original form *clack* (Leca *et al.*, 2007c). Despite such accumulation, it should be noted that all the original variants were also maintained in the Arashiyama-Kyoto troop over 30 years of continued observation at this site.

13.6.2   Diversification of the contexts of SH practice

The second aspect of the transformation of the SH tradition is the expansion of the contexts in which SH is practised, also referred to as 'mixed-activity SH'. Recently, SH was found to be integrated with social play or grooming interactions (Figure 13.1: j–k) and during inter-mount intervals taking place within the context of heterosexual and homosexual consortships (Leca, pers. obs.). In free-ranging troops where food provisioning plays a central role in the activity budget (e.g. at Arashiyama), our long-term study also revealed the integration of SH with food-related activities and the gradual emergence of food-directed SH patterns (e.g. rubbing stones and peanuts together on the ground). From 1985 to 1991, a few instances of expansion of SH practice to feeding context were reported at Arashiyama. For example, during the winter of 1985, there were two incidents of monkeys rubbing a stone on food items such as an acorn and a sweet potato (Huffman and Quiatt, 1986). However, these observations were too anecdotal to refer to them as a new SH pattern (Huffman, 1996).

By contrast, in the 2004 survey, various behavioural patterns combining provisioned and natural food items with stones (e.g. scattering stones mixed with chestnut shells on the ground, rolling a stone and pieces of peanut shell in one's hands, and the SH pattern called *combine with object*) were more frequently observed (mean = 1.3 bouts/hour of SH activity; cf. Leca *et al.*, 2008a). Moreover, the recent appearance of SH variants combining the use of hands and mouth (*put in mouth, carry in mouth, move inside mouth, bite, lick, flint in mouth* and *rub in mouth*) suggested that SH had become more integrated with foraging and feeding activities.

The intergroup comparative study showed that the integration of SH with food-related activities and the emergence of food-directed SH patterns were more frequent in free-ranging troops where food provisioning strongly influenced the activity budget (Leca *et al.*, 2008a). In troops frequently provisioned, the daily performance of SH was highly contingent on food provisioning: SH mainly occurred immediately after feeding on provisioned food (Huffman, 1984, 1996; Leca *et al.*, 2008a). Because they are provisioned with food several times a day, Arashiyama-Kyoto troop members have 'free time on their hands', and this opportunity could lead them to further explore various objects (including stones) and incorporate them into feeding activities (Huffman and Quiatt, 1986; Leca *et al.*, 2008a). Thus, food provisioning may be a key factor in the transformation phase of the SH tradition in Japanese macaques. Considering a troop's ranging conditions and its history in relation to feeding habits may be crucial in predicting the transformation phase of the SH tradition. This does not mean that a particular type of food provisioning is a necessary and sufficient condition for SH to appear and diffuse among group members. However, the way SH is practised by most group members on a daily basis, and its integration with other activities may differ from one troop to another, depending on the type of food provisioning (Leca *et al.*, 2008a).

13.6.3 Role of SH artefacts in the maintenance of the SH tradition

Recent field experiments conducted at Arashiyama aimed to simulate the context under which SH might be socially maintained in the wild, and infer which form(s) of social influence might support the persistence of the SH culture in Japanese macaques (Leca *et al.*, 2010b). Our main goal was to investigate experimentally how the physical traces typically left in the environment by previous stone handlers (such as piles of stones left on the ground) might help, through a stimulus enhancement process, trigger SH behaviour in individuals on a daily basis, and thus contribute to the long-term maintenance of the SH tradition at the group level. Our results supported the 'stimulus/local enhancement hypothesis' that individuals preferentially direct their SH behaviour toward typical physical traces of SH activity (piles of stones) over randomly scattered stones (Leca *et al.*, 2010b). In other words, encountering SH artefacts enhanced the subsequent use of these particular stones to perform SH activity in that particular part of the environment. Therefore, we provided the first experimental evidence for the role of indirect social influence in the daily performance of SH behaviour by most group members, and thus the maintenance of the SH tradition, through the stimulating effect of SH by-products. To some extent, our findings allowed us to reconstruct some elements of the environmental and social contexts underlying the SH culture. By supporting the view that SH is a socially influenced behaviour, this study contributes to validate the concept of SH culture (see also Huffman, 1984, 1996; Leca *et al.* 2007b, 2007c, 2008a, 2008c; Nahallage and Huffman, 2007b).

As they become more deeply ingrained into the behavioural landscape of the monkeys, these 'play stations' (sic Quiatt and Huffman, 1993) could ensure a baseline level of visual persistence of this form of material culture in Japanese macaques. This is particularly true for free-ranging provisioned troops, characterised by an increased sedentary lifestyle, with most group members staying around feeding grounds, i.e. open areas with stones (cf. Leca *et al.* 2008a, 2008c). Smaller home ranges are likely to increase individual probability to encounter SH artefacts, which in turn, may enhance SH activity. Moreover, we showed that piles of stones are frequently reused and constantly modified by the monkeys themselves through the transport of stones between and around SH artefacts. The frequent transport of randomly scattered stones to already gathered stones suggest cumulative environmental modifications. Therefore, through the ever-changing physical traces they leave in the environment, their subsequent stimulating effect on other group members and across generations, and their possible role on the maintenance of the SH tradition, we argue that stone handlers can be considered niche constructors. Our study suggests that a niche construction process could underlie the cultural maintenance of SH behaviour in Japanese macaques.

Similar indirect social influences are likely to occur in the acquisition and maintenance of tool-use behaviours in wild chimpanzees and brown capuchins, through the stimulating effect of nut-cracking by-products (nutshells, stones) left by skilled foragers around nut-cracking ateliers (Tomasello *et al.*, 1993; Visalberghi *et al.*, 2009). In general, conspecifics provide 'tools' (*sensu* socio-cultural learning theory: Forman *et al.*, 1993) for the individual acquisition, as well as the diffusion, and maintenance of behaviours at the group level.

From a developmental perspective, constant exposure to various artefacts could increase individual attention to some relevant environmental features, as suggested by Furlong *et al.* (2008) with respect to young chimpanzees reared in a human socio-cultural environment. For young Japanese macaques growing up in a troop where the SH tradition is well-established and has reached its transformation phase, resulting in a stimulating environment enriched in SH artefacts, a form of 'SH enculturation' process may facilitate their early acquisition of the behaviour. This argument is all the more relevant as we found a preferential use of piles of stones for SH across all age classes, including infants and yearlings (Leca *et al.*, 2010b).

13.6.4 Towards a stone-related cumulative culture in Japanese macaques?

Our results clearly show an accumulation over time and generations of SH diversity and complexity. However, in the light of the main definition of cumulative culture, which is based on the accumulation of *beneficial* modifications, this phenomenon will not be referred to as 'cumulative SH culture' because we could not demonstrate any direct benefits in the practice of SH (but see Leca *et al.*, 2008b). However, the transformation phase of the SH tradition is all the more likely and flexible since SH is currently acknowledged to be a non-adaptive behaviour with no obvious survival value (Huffman, 1984, 1996; Leca *et al.*, 2007c), as opposed to stone tool-use traditions for which an efficient behavioural pattern should be maintained unchanged (e.g. Sumita *et al.*, 1985; Boesch, 1991). The long-term cultural transformation of the SH tradition, associated with a generational increase in the diversity and complexity of SH patterns could ultimately result in future stone-tool use, as stone-related behaviours become more deeply ingrained into the behavioural landscape of Arashiyama-Kyoto macaques at the group level (Huffman and Quiatt, 1986; Leca *et al.*, 2008a).

13.7 Functional considerations: SH as a behavioural precursor to stone tool-use

13.7.1 Maintenance of a selectively neutral tradition

It has been argued that 'whether or not a particular pattern of behaviour persists obviously depends on its effects on the survival and reproductive success of its bearers' (Avital and Jablonka, 2000, p. 99). However, our findings show that even traditional behaviours with no obvious function and no apparent adaptive value, such as SH at Arashiyama, can not only be practised on a daily basis and maintained over several decades within a large proportion of group members, but can also be modified on the basis of a transgenerational accumulation (Huffman, 1996; Leca *et al.*, 2010b; see also 'games' as social conventions in white-faced capuchins: Perry *et al.*, 2003). How can we account for such a puzzling phenomenon?

Several reasons may partly explain the maintenance of the SH tradition at Arashiyama (and presumably at other sites). First, the original motivations underlying SH may be different from what they are today, both at the individual and group levels. Most Arashiyama-Kyoto monkeys observed handling stones in 2008 were born into troops with well-established SH traditions. Furthermore, individuals grew up into a troop with either a strong or a weak connection between SH and provisioning. The conformity-enforcing hypothesis, which proposes that culturally non-conforming individuals may be discriminated against (cf. Lachlan *et al.*, 2004), predicts that immature individuals should integrate the same type of connection between SH and feeding activities as most older group members (Leca *et al.*, 2008a). Individually, the immediate motivation to perform SH could be mere serendipity, as this behaviour appears to be self-rewarding (Huffman, 1984). As Avital and Jablonka (2000, p. 85) pointed out, animals may engage in 'apparently non-functional activities that seem like the luxurious by-products of extensive behavioural plasticity'. SH behaviour may also be maintained because of some internal (physiological and/or psychological) consequences that we cannot measure yet (Huffman and Hirata, 2003).

Second, although SH is not a subsistence activity, it should be noted that no SH pattern is deleterious and the SH tradition is not locally maladaptive but selectively neutral, at least under the favourable environmental conditions of food provisioning (Huffman and Hirata, 2003; Leca *et al.*, 2008a). Third, Huffman (1996) suggested that if SH persists sufficiently in a given troop, direct material benefits may be acquired in the future, provided some modifications of the behavioural patterns or the direct integration of SH with foraging activities (e.g. stone-tool-use) or social interactions (e.g. agonistic display) occur (Huffman and Quiatt, 1986; Huffman and Hirata, 2003; Leca *et al.*, 2008b). By relaxing selective pressure on foraging, food provisioning has created favourable environmental conditions under which SH may simply serve the function of maintaining in some troops (such as Arashiyama-Kyoto) a set of behaviours, involving a high level of behavioural complexity and familiarity with stones, that could evolve into tool-use provided particular environmental circumstances.

13.7.2   SH as an exaptive tradition?

Can the daily performance of SH with feeding activity by Arashiyama-Kyoto macaques lead by transformation to stone-tool use in a foraging context? If tool-use is defined as moving a detached object for the purpose of changing the condition and/or position of another object or organism (Beck, 1980), then SH behaviour as a whole and most SH patterns cannot be considered stone-tool use. However, there is a series of arguments suggesting that when practised on a daily basis and by most members of a group, the non-instrumental manipulation of stones could be considered as a behavioural precursor to the possible use of stones as tools (Huffman and Quiatt, 1986; Huffman, 1996; Leca *et al.*, 2008b).

First, the non-instrumental manipulation of objects, such as SH, has long been recognised as a behavioural precursor to tool-use, in terms of individual development and cross-species comparison (Beck, 1980; Huffman and Quiatt, 1986; Hayashi *et al.*, 2005). Second, at several sites such as Arashiyama, the SH tradition is undergoing a phase of transformation, including an increase in the diversity and complexity of the behavioural patterns and the integration of SH with foraging activities (Leca *et al.*, 2008a). Third, the occurrence of SH spots or 'play stations' revisited daily by Arashiyama-Kyoto macaques is likely to lead to an increased familiarity with SH artefacts that may result in the use of stones as tools (Huffman and Quiatt, 1986; Leca *et al.*, 2010b). Fourth, although macaques are not frequent tool-users (Beck, 1980; but see Weinberg and Candland, 1981; Sinha, 1997; Leca *et al.* 2008b, 2010c), long-tailed macaques have recently been reported to display oyster-cracking behaviour with stones (Malaivijitnond *et al.*, 2007). Finally, we recently witnessed a first case of tool-use probably derived from prolonged SH practice: spontaneous stone-throwing as an agonistic display (Leca *et al.*, 2008b).

Therefore, although most SH patterns do not currently meet the criteria used to define tool use, we hypothesised that the long-enduring practice of stone-related combinatorial behaviours by Arashiyama-Kyoto macaques could be considered a behavioural precursor to the use of stones as tools. This scenario is consistent with the 'perception-action' perspective on the development of tool-use and foraging competence in monkeys, apes and humans, postulating that skilled actions are acquired through the routine generation of species-typical exploratory actions, coupled with learning about the outcomes and affordances of each action that generates directly perceptible information (Lockman, 2000; Gunst *et al.*, 2010). As an unselected but eventually beneficial trait, the SH tradition would be an exaptation (cf. Gould and Vrba, 1982).

13.8   Conclusion and future directions

Arashiyama-Kyoto macaques largely contributed to make SH the best-known non-adaptive traditional behaviour in non-human primates. Three decades of continued observation at Arashiyama showed that the monkeys have largely extended and diversified their SH repertoire. Our findings have important implications for understanding cumulative cultural evolution, particularly the reasons for its rarity in non-humans. Research on SH as a tool-use precursor also provides new insights into the emergence of hominid material culture through stone-tool technology. We drew an overall picture of rich cultural diversity in a particular type of object-play behaviour in macaques, and suggest that multiple factors should be jointly considered to identify the mechanisms of emergence, diffusion and maintenance of a behavioural tradition in animals.

Acknowledgements

The authors' work was funded by: a Grant-In-Aid for scientific research (No. 1907421 to M.A. Huffman) sponsored by Ministry of Education, Science, Sports and Culture, Japan, by a Lavoisier postdoctoral Grant, Ministère des Affaires Etrangères, France and a JSPS (Japan Society for the Promotion of Science) postdoctoral fellowship to J.-B. Leca (No. 07421), and by travel funds from the HOPE Project, a core-to-core program sponsored by JSPS to M.A. Huffman and J.-B. Leca. We thank the researchers, students, staff and friends who provided permission to work, assistance, and valuable specific information about the Arashiyama-Kyoto macaque troop. We are grateful to the following people in Japan: S. Asaba, J. Hashiguchi, T. Kawashima, S. Kobatake and S. Tamada (Iwatayama Monkey Park, Arashiyama), as well as T. Matsuzawa, A. Mori, H. Takemoto and K. Watanabe (Kyoto University Primate Research Institute, Japan). For fruitful collaboration and discussion about stone handling and the Arashiyama E troop, we thank S. Hanamura (Kyoto University Laboratory of Human Evolution Studies, Japan), C.A.D. Nahallage (University of Sri Jayewardenepura, Sri Lanka), D. Quiatt (University of Colorado, USA), M. Shimada (Teikyo University of Science and Technology, Japan), and P.L. Vasey (University of Lethbridge, Canada). For occasional assistance with data collection, we thank C.A.D. Nahallage, K.J. Petrzelkova (Academy of Sciences of the Czech Republic) and N. Tworoski (University of Minnesota, USA). We are indebted to the Enomoto family for providing us with logistic assistance at Arashiyama. We thank C. Caldwell for fruitful comments on an earlier version of this paper.

References

Altmann, J. (1974). Observational study of behavior: sampling methods. *Behaviour*, **49**, 227–267.
Avital, E. and Jablonka, E. (2000). *Animal Traditions: Behavioural Inheritance in Evolution*. Cambridge: Cambridge University Press.

Azuma, S. (1968). Acquisition and propagation of food habit in a troop of Japanese monkeys. In *Social Regulators of Behavior in Primates*, ed. C. R. Carpenter. Lewisburg, PA: Bucknell University, pp. 284–292.

Beck, B. B. (1980). *Animal Tool Behavior: The Use and Manufacture of Tools by Animals*. New York, NY: Garland STPM Press.

Boesch, C. (1991). Teaching among wild chimpanzees. *Animal Behaviour*, **41**, 530–532.

Boesch, C. and Tomasello, M. (1998). Chimpanzee and human cultures. *Current Anthropology*, **39**, 591–614.

Boyd, R. and Richerson, P. J. (1996). Why culture is common, but cultural evolution is rare. *Proceedings of the British Academy*, **88**, 77–93.

Caldwell, C. A. and Millen, A. E. (2008a). Studying cumulative cultural evolution in the laboratory. *Philosophical Transactions of the Royal Society B*, **363**, 3529–3539.

(2008b). Experimental models for testing hypotheses about cumulative cultural evolution. *Evolution and Human Behavior*, **29**, 165–171.

(2009). Social learning mechanisms and cumulative cultural evolution: is imitation necessary? *Psychological Science*, **20**, 1478–1483.

(2010). Conservatism in laboratory microsocieties: unpredictable payoffs accentuate group-specific traditions. *Evolution and Human Behavior*, **31**, 123–130.

Candland, D. K., French, J. A. and Johnson, C. N. (1978). Object-play: test of a categorized model by the genesis of object-play in *Macaca fuscata*. In *Social Play in Primates*, ed. E. O. Smith. New York, NY: Academic Press, pp. 259–296.

Coussi-Korbel, S. and Fragaszy, D. M. (1995). On the social relation between social dynamics and social learning. *Animal Behaviour*, **50**, 1441–1453.

Danchin, E. and Wagner, R. H. (2008). Cultural evolution. In *Behavioural Ecology*, eds. E. Danchin, L.-A. Giraldeau and F. Cézilly. Oxford: Oxford University Press, pp. 693–726.

de Waal, F. B. M. (2001). *The Ape and the Sushi-Master: Cultural Reflections of a Primatologist*. New York: Basic Books.

Enquist, M. and Ghirlanda, S. (2007). Evolution of social learning does not explain the origin of human cumulative culture. *Journal of Theoretical Biology*, **246**, 129–135.

Fagen, R. (1981). *Animal Play Behavior*. New York, NY: Oxford University Press.

Forman, E., Minick, N. and Stone, C. A. (1993). *Contexts for Learning: Sociocultural Dynamics in Childrens' Development*. Oxford: Oxford University Press.

Fragaszy, D. M. and Perry, S. (2003). *The Biology of Traditions: Models and Evidence*. Cambridge: Cambridge University Press.

Furlong, E. E., Boose, K. J. and Boysen, S. T. (2008). Raking it in: the impact of enculturation on chimpanzee tool use. *Animal Cognition*, **11**, 83–97.

Galef, B. G. (1992). The question of animal culture. *Human Nature*, **3**, 157–178.

Gould, S. J. and Vrba, E. S. (1982). Exaptation – a missing term in the science of form. *Paleobiology*, **8**, 4–15.

Guinet, C. and Bouvier, J. (1995). Development of intentional stranding hunting techniques in killer whale (*Orcinus orca*) calves at Crozet Archipelago. *Canadian Journal of Zoology*, **73**, 27–33.

Gunst, N., Boinski, S. and Fragaszy, D. M. (2010). Development of skilled detection and extraction of embedded prey by wild brown capuchin monkeys (*Cebus apella apella*). *Journal of Comparative Psychology*, **124**, 194–204.

Hall, S. L. (1998). Object play by adult animals. In *Animal Play: Evolutionary, Comparative and Ecological Perspectives*, eds. M. Bekoff and J. A. Byers. Cambridge: Cambridge University Press, pp. 45–60.

Hanya, G. (2004). Seasonal variations in the activity budget of Japanese macaques in the coniferous forest of Yakushima: effects of food and temperature. *American Journal of Primatology*, **63**, 165–177.

Hayashi, M., Mizuno, Y. and Matsuzawa, T. (2005). How does stone-tool use emerge? Introduction of stones and nuts to naïve chimpanzees in captivity. *Primates*, **46**, 91–102.

Heyes, C. M. (1993). Imitation, culture and cognition. *Animal Behaviour*, **46**, 999–1010.

Hill, K. (2009). Animal 'culture'? In *The Question of Animal Culture*, eds. K. N. Laland and B. G. Galef. Cambridge, MA: Harvard University Press, pp. 269–287.

Hiraiwa, M. (1975). Pebble-collecting behavior by juvenile Japanese monkeys. *Monkey*, **19**, 24–25. (in Japanese)

Huffman, M. A. (1984). Stone-play of *Macaca fuscata* in Arashiyama B troop: transmission of a non-adaptive behavior. *Journal of Human Evolution*, **13**, 725–735.

   (1991). History of Arashiyama Japanese Macaques in Kyoto, Japan. In *The Monkeys of Arashiyama: Thirty-five Years of Research in Japan and the West*, eds. L. M. Fedigan and P. J. Asquith. Albany, NY: State University of New York Press, pp. 21–53.

   (1996). Acquisition of innovative cultural behaviors in non-human primates: a case study of stone handling, a socially transmitted behavior in Japanese macaques. In *Social Learning in Animals: The Roots of Culture*, eds. B. G. Galef and C. Heyes. Orlando, FL: Academic Press, pp. 267–289.

Huffman, M. A. and Hirata, S. (2003). Biological and ecological foundations of primate behavioral tradition. In *The Biology of Tradition: Models and Evidence*, eds. D. M. Fragaszy and S. Perry. Cambridge: Cambridge University Press, pp. 267–296.

Huffman, M. A. and Quiatt, D. (1986). Stone handling by Japanese macaques (*Macaca fuscata*): implications for tool use of stones. *Primates*, **27**, 413–423.

Huffman, M. A., Nahallage, C. A. D. and Leca, J.-B. (2008). Cultured monkeys, social learning cast in stones. *Current Directions in Psychological Science*, **17**, 410–414.

Huffman, M. A., Leca, J.-B. and Nahallage, C. A. D. (2010). Cultured Japanese macaques – a multidisciplinary approach to stone handling behavior and its implications for the evolution of behavioral tradition in nonhuman primates. In *The Japanese Macaques*, eds. N. Nakagawa, M. Nakamichi and H. Sugiura. Tokyo: Springer, pp. 191–219.

Hunt, G. R. and Gray, R. D. (2003). Diversification and cumulative evolution in New Caledonian crow tool manufacture. *Proceedings of the Royal Society B – Biological Sciences*, **270**, 867–874.

Itani, J. (1958). On the acquisition and propagation of a new food habit in the troop of Japanese monkeys at Takasakiyama. *Primates*, **1**, 84–98.

Itani, J. and Nishimura, A. (1973). The study of infra-human culture in Japan. In *Precultural Primate Behavior*, ed. E. Menzel. Basel: S. Karger, pp. 26–50.

Kawai, M. (1965). Newly acquired pre-cultural behavior of a natural troop of Japanese monkeys on Koshima Island. *Primates*, **6**, 1–30.

Kawai, M., Watanabe, K. and Mori, A. (1992). Pre-cultural behaviors observed in free-ranging Japanese monkeys on Koshima islet over the past 25 years. *Primate Report*, **32**, 143–153.

Kawamura, S. (1959). The process of sub-cultural propagation among Japanese macaques. *Primates*, **2**, 43–60.

Kerckhoff, A. C. (2002). A theory of hysterical contagion. *Human Nature and Collective Behavior*, **21**, 81–93.

Kummer, H. and Goodall, J. (1985). Conditions of innovative behaviour in primates. *Philosophical Transactions of the Royal Society, London B*, **308**, 203–214.

Lachlan, R. F., Janik, V. M. and Slater, P. J. B. (2004). The evolution of conformity-enforcing behaviour in cultural communication systems. *Animal Behaviour*, **68**, 561–570.

Laland, K. N. and Hoppitt, W. (2003). Do animals have culture? *Evolutionary Anthropology*, **12**, 150–159.

Leca, J.-B., Gunst, N., Watanabe, K. and Huffman, M. A. (2007a). A new case of fish-eating in Japanese macaques: implications for social constraints on the diffusion of feeding innovation. *American Journal of Primatology*, **69**, 821–828.

Leca, J.-B., Gunst, N. and Huffman, M. A. (2007b). Age-related differences in the performance, diffusion, and maintenance of stone handling, a behavioral tradition in Japanese macaques. *Journal of Human Evolution*, **53**, 691–708.

(2007c). Japanese macaque cultures: inter- and intra-troop behavioural variability of stone handling patterns across 10 troops. *Behaviour*, **144**, 251–281.

(2008a). Food provisioning and stone handling tradition in Japanese macaques: a comparative study of ten troops. *American Journal of Primatology*, **70**, 803–813.

Leca, J.-B., Nahallage, C. A. D., Gunst, N. and Huffman, M. A. (2008b). Stone-throwing by Japanese macaques: form and functional aspects of a group-specific behavioral tradition. *Journal of Human Evolution*, **55**, 989–998.

Leca, J.-B., Gunst, N. and Huffman, M. A. (2008c). Of stones and monkeys: Testing ecological constraints on stone handling, a behavioral tradition in Japanese macaques. *American Journal of Physical Anthropology*, **135**, 233–244.

(2010a). Principles and levels of laterality in unimanual and bimanual stone handling patterns by Japanese macaques. *Journal of Human Evolution*, **58**, 155–165.

(2010b). Indirect social influence in the maintenance of the stone handling tradition in Japanese macaques (*Macaca fuscata*). *Animal Behaviour*, **79**, 117–126.

(2010c). The first case of dental flossing by a Japanese macaque (*Macaca fuscata*): implications for the determinants of behavioral innovation and the constraints on social transmission. *Primates*, **51**, 13–22.

(2011). Complexity in object manipulation by Japanese macaques (*Macaca fuscata*): a cross-sectional analysis of manual coordination in stone handling patterns. *Journal of Comparative Psychology*, **125**, 61–71.

Lefebvre, L. and Palameta, B. (1988). Mechanisms, ecology, and population diffusion of socially-learned food-finding behavior in feral pigeons. In *Social Learning, Psychological and Biological Perspectives*, eds. T. Zentall and B. G. Galef, Jr. Hillsdale, NJ: Lawrence Erlbaum Asssociates, pp. 141–165.

Lockman, J. (2000). A perception-action perspective on tool use development. *Child Development*, **71**, 137–144.

Malaivijitnond, S., Lekprayoon, C., Tandavanitj, N. *et al.* (2007). Stone-tool usage by Thai long-tailed macaques (*Macaca fascicularis*). *American Journal of Primatology*, **69**, 227–233.

Nahallage, C. A. D. and Huffman, M. A. (2007a). Age-specific functions of stone handling, a solitary-object play behavior, in Japanese macaques (*Macaca fuscata*). *American Journal of Primatology*, **69**, 267–281.

(2007b). Acquisition and development of stone handling behavior in infant Japanese macaques. *Behaviour*, **144**, 1193–1215.

Nakamichi, M., Kato, E., Kojima, Y. and Itoigawa, N. (1998). Carrying and washing of grass roots by free-ranging Japanese macaques at Katsuyama. *Folia Primatologica*, **69**, 35–40.

Nishida, T., Matsusaka, T. and McGrew, W. C. (2009). Emergence, propagation or disappearance of novel behavioral patterns in the habituated chimpanzees of Mahale: a review. *Primates*, **50**, 23–36.

Perry, S. (2006). What cultural primatology can tell anthropologists about the evolution of culture. *Annual Review of Anthropology*, **35**, 171–190.

Perry, S. and Manson, J. H. (2003). Traditions in monkeys. *Evolutionary Anthropology*, **12**, 71–81.

Perry, S., Baker, M., Fedigan, L. *et al.* (2003). Social conventions in wild white-faced capuchin monkeys. *Current Anthropology*, **44**, 241–268.

Quiatt, D. and Huffman, M. A. (1993). On home bases, nesting sites, activity centers, and new analytic perspectives. *Current Anthropology*, **34**, 68–70.

Reader, S. and Laland, K. N. (2003). *Animal Innovation*. Oxford: Oxford University Press.

Sapolsky, R. M. (2006). Social cultures among nonhuman primates. *Current Anthropology*, **47**, 641–656.

Sinha, A. (1997). Complex tool manufacture by a wild bonnet macaque. *Folia Primatologica*, **68**, 23–25.

Spinka, M., Newberry, R. C. and Bekoff, M. (2001). Mammalian play: training for the unexpected. *Quarterly Review of Biology*, **76**, 141–168.

Sumita, K., Kitahara-Frisch, J. and Norikoshi, K. (1985). The acquisition of stone-tool use in captive chimpanzees. *Primates*, **26**, 168–181.

Tennie, C., Call, J. and Tomasello, M. (2009). Ratcheting up the ratchet: on the evolution of cumulative culture. *Philosophical Transactions of the Royal Society, B*, **364**, 2405–2415.

Thierry, B. (1994). Social transmission, tradition and culture in primates: from the epiphenomenon to the phenomenon. *Techniques & Culture*, **23/24**, 91–119.

Tomasello, M. (1990). Cultural transmission in chimpanzee tool use and signaling? In *'Language' and Intelligence in Monkeys and Apes*, eds. S. T. Parker and K. R. Gibson. Cambridge: Cambridge University Press, pp. 274–311.

(1999). *The Cultural Origins of Human Cognition*. Cambridge, MA: Harvard University Press.

Tomasello, M., Kruger, A. C. and Ratner, H. H. (1993). Cultural learning. *Behavioural and Brain Sciences*, **16**, 495–552.

Visalberghi, E., Spagnoletti, N., Ramos da Silva, E. D. *et al.* (2009). Distribution of potential suitable hammers and transport of hammer tools and nuts by wild capuchin monkeys. *Primates*, **50**, 95–104.

Watanabe, K. (1989). Fish: A new addition to the diet of Koshima monkeys. *Folia Primatologica*, **52**, 124–131.

(1994). Precultural behavior of Japanese macaques: longitudinal studies of the Koshima troops. In *The Ethological Roots of Culture*, eds. R. A. Gardner, B. T. Gardner, B. Chiarelli and F. X. Plooij. Boston, MA: Kluwer Academic Publishers, pp. 81–94.

Weinberg, S. M. and Candland, D. K. (1981). 'Stone-grooming' in *Macaca fuscata*. *American Journal of Primatology*, **1**, 465–468.

Whiten, A., Horner, V. and Marshall-Pescini, S. (2003). Cultural panthropology. *Evolutionary Anthropology*, **12**, 106–108.

14 Social object play among juvenile Japanese macaques: Comparison between the provisioned Arashiyama-Kyoto troop and the non-provisioned Kinkazan troop

MASAKI SHIMADA

Two juvenile Japanese macaques play wrestling at Arashiyama
(photo by M. Shimada).

The Monkeys of Stormy Mountain: 60 Years of Primatological Research on the Japanese Macaques of Arashiyama, eds. Jean-Baptiste Leca, Michael A. Huffman and Paul L. Vasey. Published by Cambridge University Press. © Cambridge University Press 2012.

14.1   Introduction

Since Hayaki (1983), Koyama (1985) and Imakawa (1990) studied social relationships through social play in juvenile Japanese macaques (*Macaca fuscata*) at Koshima, Arashiyama and Katsuyama, respectively, there have been few other studies of play behaviour in this species, with the notable exceptions of long-term research on stone-handling behaviour, a form of object play (cf. Huffman, 1984a; Huffman and Quiatt, 1986; Leca *et al.*: chapter 13), and recent studies of play-fighting (cf. Reinhart *et al.* 2010; VanderLaan *et al.*: chapter 11). In other study sites throughout Japan, the situation concerning the scientific study of play in Japanese macaques is similar: few expert primatologists have studied play behaviour in this species as a main research topic (cf. Shimada, 2010).

The free-ranging, habituated and individually identified Japanese macaques of the Arashiyama E troop, living at Iwatayama Monkey Park, Arashiyama, near Kyoto city, provide the ideal conditions for the ethological study of play behaviour, an activity that requires detailed observations on known individuals and from a short distance (cf. Symons, 1978; Reinhart *et al.*, 2010). All the macaques living at this site have been identified for several decades by researchers and park staff who have accumulated basic information on various attributes such as kinship, birth dates and social ranks (Fedigan and Asquith, 1991; Koyama *et al.*, 1992).

I have conducted research on play behaviour among juvenile macaques at Arashiyama since 1999. In this chapter, I will focus on two studies addressing a particular form of play: 'social object play' (SOP), i.e. social play while holding (a) portable object(s). First, I will introduce and describe some structural features of SOP (Study 1: from Shimada, 2006). Second, I will compare SOP among juvenile macaques in the provisioned Arashiyama-Kyoto troop and in the non-provisioned Kinkazan troop (Study 2). Finally, I will discuss whether variation in SOP between these two troops can be explained in terms of regional traditional or cultural differences.

14.2   Study 1: SOP in Arashiyama-Kyoto juvenile macaques

To date, there has been no single definition of play behaviour upon which all researchers can agree (e.g. Burghardt, 2005; Shimada, 2010). However, according to Burghardt (2005, p. 382), playful activities can be characterised as being: '(1) incompletely functional in the context expressed, (2) voluntary, pleasurable, or self-rewarding, (3) different structurally or temporally from related serious behaviour systems, (4) expressed repeatedly during at least some part of an animal's life span, and (5) initiated in relatively benign situations'. In this study, I use the term 'play' as the behavioural category meeting these five criteria.

Traditionally, ethologists have divided animal play behaviour into three categories: (1) locomotor (-rotational) play: solitary activities, such as running, leaping or rolling; (2) object play: manipulative activities using objects, including pushing, pulling or breaking objects; and (3) social play: playful interactions among two or more individuals, such as play-chasing or play-fighting (Fagen, 1981; Bekoff and Byers, 1998; Burghardt, 2005). Juvenile and adult Japanese macaques routinely engage in object play and while doing so, they handle various kinds of portable objects available in their environment, such as stones or pieces of glass (e.g. Huffman, 1984a). Juveniles also engage in social play by forming play groups (e.g. Hayaki, 1983; Shimada, 2009).

Now, let us think about the following two play patterns: (1) A pair of domestic dogs playing tug-of-war by biting a tree branch and pulling it at both ends; (2) Human children playing a game of baseball. Although these two examples seem totally different from the viewpoint of the internal rules and actions performed during the play, they incorporate characteristics of both social and object play at the same time. I suggest classifying them under a combined category called 'social object play (SOP)'. In its broadest sense, SOP is defined as an individual holding an object during social play activity. In its narrowest sense, it also includes games during which the object(s) is (are) manipulated according to strict rules. In this chapter, I use a broad interpretation of SOP in order to include the various patterns of SOP observed in nature.

Some authors described that juvenile individuals of *Macaca* species engage in SOP, although they do not use this term (*M. fuscata*: Itani, 1954; Hayaki, 1983; *M. mulatta*: Symons, 1978). For example, Symons (1978, p. 37) mentioned it within a general description of various forms of social play: 'Although free-ranging rhesus monkeys rarely manipulate or play with non-food objects, an object (usually a leaf or twig) is occasionally incorporated into chasing play, the chasee being the monkey holding the object'. However, few studies have examined SOP in detail.

Social object play should be of general interest to a wide range of researchers from the viewpoint of object possession rules in non-human primates. When a valuable object, whose availability is limited in the local environment, is unclaimed, a dominant individual will most likely gain access to it (Kummer, 1973; Thierry et al., 1989). However, according to the so-called 'prior possession rule' (Bakeman and Brownlee, 1982), individuals rarely compete directly for valuable portable objects, such as nutritious food items claimed by other individuals, even when the object holders are subordinates (e.g. *Papio hamadryas*: Sigg and Falett, 1985; *Cebus apella*: Thierry et al., 1989; *Macaca fascicularis*: Kummer and Cords, 1991).

If juvenile macaques automatically and strictly apply this prior possession rule during SOP, I predict that non-holders will keep their distance from a possessed object or its holder during play, which means that stealing an object from a holder should be rare, and SOP should not last for long periods of time. Taking this prediction into account, let us examine a particular observation of a playful interaction among juvenile macaques in the Arashiyama E troop.

14.2.1 Observation 1: SOP between two juveniles in Arashiyama

In May 2000, two juveniles belonging to the same matriline (Momo-lineage), namely 'Momo-59-78-98 (Mo-98)' (2-year-old high-ranking female) and 'Momo-59-78-92-99 (Mo-99)' (1-year-old low-ranking female and Mo-98's nephew), had been involved in social play activity intermittently since 13:00 in the woods away from the other troop members.

13:20:28 – Both of them sat on the ground 3 m away from each other. Mo-98 found an object in the bush where she sat, and picked it up. The object was a steel semicircular hoop, approximately 30 cm in diameter and 3 cm wide. Mo-98 handled the object.

13:20:56 – Mo-99 ran towards Mo-98, and Mo-98 ran in an attempt to escape with the object. Mo-99 chased Mo-98.

13:21:09 – Both of them sat 1 m away from each other.

13:21:30 – Mo-98 ran away from Mo-99 with the object and Mo-99 chased Mo-98.

13:21:47 – Mo-99 stole the object and escaped immediately from Mo-98. Now Mo-98 chased Mo-99.

13:22:09 – Both of them sat 1 m away from each other.

13:22:20 – Mo-98 stole the object and escaped immediately from Mo-99.

13:22:30 – Both of them grappled with each other on the ground, while Mo-98 kept holding the object.

13:22:37 – Mo-98 stepped 1 m away from Mo-99 and sat down.

13:23:04 – Mo-98 dropped the object, picked it up immediately and walked away from Mo-99, while Mo-99 kept sitting without reaction.

Because it meets all of the above-mentioned criteria, including play-fighting and play-chasing while holding an object, this interaction can be categorised as a SOP bout. These descriptions and features obviously contradict the prediction above, suggesting that SOP mechanisms differ from those applied in the prior possession rule. The question here is whether SOP was observed only rarely or frequently among juvenile macaques in Arashiyama. If frequently observed, did SOP present distinctive interactive patterns and how could SOP interactions last for a long period of time? First of all, I will further describe the structures and other characteristics of SOP among juvenile Japanese macaques, and discuss the underlying proximate mechanisms, by using the data collected at Arashiyama in 2000 (Shimada, 2006).

14.2.2   Methods of Study 1

The study subjects comprised all the immature individuals present in the troop in 2000, including three infants (< 1-year old), 28 juveniles (1- to 3-year-old), and ten 4-year-old subadults. During the study period, the troop consisted of 15 independent matrilines and about 160 individuals. Data were collected from 25 July to 17 October 2000 (38 days), for a total of 246.7 hours. Prior to the commencement of data collection, I learned to identify every immature individual in the troop.

To collect data on continuous interactions occurring around an object, I employed a modified version of the sequence sampling method (Altmann, 1974). When I observed a juvenile holding an object in the hand or in the mouth, I began to follow the object and the holder. If the focal object moved to a new holder, I recorded the time and continued following the object and the new holder until the object was finally abandoned. All the objects held during social play sequences were categorised as natural, artificial or others. Natural objects included plants (with species identification) and stones, whereas artificial objects included plastic, metallic, glass and paper.

I recorded all types of interactions between the holder and other individuals located within a 3-metre radius of the object. I operationally divided these interactions into nine mutually exclusive categories based on previous descriptions of social play among *Macaca* species (e.g. Symons, 1978; Fagen, 1981; Hayaki, 1983; Thierry *et al.*, 2000): (1) play-fighting (touching, grasping, biting, boxing, jumping on), (2) play-chasing (running at once within a 3-metre radius, leaping at the other from a distance), (3) tug-of-war with an object (multiple individuals grabbing an object), (4) taking an object from a holder, (5) walking together (moving at once touching each other), (6) other active interactions, (7) stopping (sitting, standing or lying down stationary while not touching each other within a 3-metre radius of the object), (8) clinging (sitting, standing or lying down stationary while touching each other), and (9) unknown (incomplete observation). Categories from (1) to (6) are active interactions, whereas categories (7) and (8) are inactive interactions.

In this study, I operationally defined a SOP bout as a continuous series of active interactions around one focal object with between-play intervals of less than 1 minute. That is, if the same individuals played socially with the same object after a pause of one or more than 1 minute, it was considered a subsequent bout. The duration of a bout was defined from the moment when I first observed socially playing object holders to the time the object was abandoned. Participants in a SOP bout were defined as individuals who actively interacted (categories from (1) to (6) defined above) at least once during the bout. A long bout was defined as a bout lasting at least 0.5 minute.

I noted the starting time of play-chasing interactions between an object holder and non-holder(s). For each of these events, I recorded the name of the holder and any non-holders within a 3-metre radius, along with information about which individuals were chased. These data were collected whenever it was possible to determine a single chaser and a single object holder. When two objects or more were used socially within a 3-metre radius, or when I was unable to determine the identity of the holder, I excluded these data from the analysis of play-chasing episodes. I analysed the correspondence between the role of an object holder and a chaser during play-chasing in SOP bouts.

14.2.3 Results of Study 1

Long SOP among immatures

I recorded 298 SOP bouts, of which 151 were long bouts, i.e. lasting at least 0.5 minutes. Since I did not use a focal animal sampling method, I was unable to assess individual frequencies of participation in SOP. However, it should be noted that 90.2% and 85.4% of juveniles participated in SOP bouts and long SOP bouts at least once, respectively. In the 141 bouts that I could observe completely, 62 bouts (44%) were long SOP bouts, and the longest bout lasted 15.2 minutes. In sum, I found an average of one SOP bout/h, and most juveniles participated in SOP at least once.

The video movies on typical examples of SOP interactions are available on the website 'Movie Archives of Animal Behaviour':

MOMO: http://zoo2.zool.kyoto-u.ac.jp/ethol/mov/01/0109/momo010930 mf03.mov

Objects held by immatures during long SOP

I identified a total of 135 objects held during long SOP bouts. Among them, 108 objects (80%) were natural. They consisted mostly of plants (n = 107), including 88 tree branches, 12 ferns, one herb and six other parts of plants. A stone was held once (n = 1). Twenty-three artificial objects were held (17%).

They consisted of 14 plastic materials (including six bags, two bottles and six other items), four metallic materials (including two aluminium cans and two steel cans), four pieces of glass and one piece of paper. Finally, four other objects (two pieces of processed lumber and two pieces of plywood) were held (3%). Immature macaques were never observed holding provisioned food items during long SOP bouts.

Among the tree branches held during SOP bouts, I identified 28 tree species from 17 families. Interestingly, according to the lists of Arashiyama-Kyoto macaque foods (Murata and Hazama, 1968; Huffman, 1984b; Huffman and MacIntosh: chapter 17), all of these tree species, except *Clethra barbinervis*, which was held once, contained some parts that were edible to Arashiyama-Kyoto macaques. At Arashiyama, immatures were frequently provisioned by staff and tourists with highly nutritious foods (e.g. wheat, soy beans, bananas and apples) and they seldom, if at all, fed on other edible food items around the park during the observation period.

Number of holders and objects

The total duration of long SOP bouts was 297.4 min, among which the following five states were recorded: (a) 'single holder': one individual held a focal object and others located within a 3-metre radius held no objects (90.2% of the total time of long SOP bout); (b) 'tug-of-war': two or more individuals held the same focal object simultaneously (1.9%); (c) 'one individual holding more than one object': an individual held two or more objects, including the focal object (0.7%); (d) 'multiple holders': an individual held a focal object and (an) other individual(s) within a 3-metre radius held different object(s) (5.5%); and (e) 'unknown': incomplete observation (1.7%).

Interactive features of SOP

Most long SOP bouts included one or more play-chasing episodes. Among all the long SOP bouts, I observed 373 play-chasing episodes between an object holder and non-holder(s). I classified these episodes according to the number of non-holders (from one to three). For each case, the observed and expected frequencies of play-chasing episodes in which the holder escaped and those in which one of the non-holders escaped are shown in Figure 14.1. In the case of play-chasing, there was a strong tendency for an object holder to assume the role of chasee, regardless of the number of non-holders around.

The tendency for the object holder to escape during play-chasing episodes may be attributed to particular attributes of the SOP protagonists, such as sex, age, relative rank or matriline. In order to investigate this possibility, I used data from 178 play-chasing episodes, for which the identities of the holder and

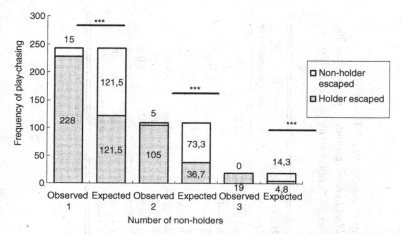

Figure 14.1. Frequency of a holder or non-holder escaping in play-chasing during SOP. For this analysis, one case in which the number of non-holders was unknown (n = 1) was excluded. Each expected frequency was calculated based on the assumption that all participants had an equal chance of being chased. ***: $p < 0.001$ (binominal test).

the non-holder were recorded, and I examined whether each holder escaped more often than was chased in each dyadic play-chasing episode (i.e. where the number of non-holders was equal to one) during long SOP bouts. Table 14.1 shows that the tendency for the object holder to escape was not significantly affected by factors such as sex, age, relative rank or kinship.

Holder-reversal of a focal object occurred, on average, more than once per long SOP bout which I observed completely. In most cases, the new holder assumed the role of chasee, and the former holder chased the new holder or just stopped playing. Juveniles rarely exhibited serious aggression or submission around objects during SOP. Only two cases of serious competition were observed during long SOP. Two aggressive interactions were considered to have happened accidentally during play-fighting and there was no serious competition about objects.

14.2.4   Discussion of Study 1

The results showed that SOP was a common activity for immature macaques in Arashiyama and that they often engaged in SOP for long periods of time. Immature macaques held various kinds of portable objects available in their environment, except provisioned food items. However, it is interesting to note that most of the natural plants that the immatures integrated with social

Table 14.1. *Tendency of the holder to escape and chase with regard to the attributions of non-holders.* *Twenty-six individuals were observed as holders at least once. However, the total number of holders in each attribute is higher than 26 because data for some individuals were included in different combinations of attributes. **Insufficient number of holders for the statistical test*

Key attribute	Attributes of holders and non-holders	No. of holders*	Holder escaped: mean ($\pm$s.d.)	Holder chased: mean ($\pm$s.d.)	Wilcoxon signed-ranks test	
Sex	Holder is male, non-holder is male	7	2.4 (1.8)	0.3 (0.5)	−2.20	<0.05
	Holder is male, non-holder is female	5	3.6 (1.8)	0.0 (0.0)	**	**
	Holder is female, non-holder is male	8	2.1 (1.0)	0.3 (0.5)	−2.52	<0.05
	Holder is female, non-holder is female	18	6.2 (4.3)	0.6 (1.0)	−3.62	<0.001
Age	Holder and non-holder are the same age	16	5.1 (4.0)	0.4 (0.6)	−3.41	<0.001
	Holder is older than non-holder	15	2.5 (2.0)	0.4 (0.7)	−2.86	<0.01
	Holder is younger than non-holder	14	3.3 (2.8)	0.1 (0.4)	−3.18	<0.01
Relative rank	Holder is ranked higher than non-holder	17	5.7 (3.6)	0.4 (0.8)	−3.52	<0.001
	Holder is ranked lower than non-holder	19	3.5 (2.8)	0.4 (0.7)	−3.53	<0.001
Matrilines	Holder's and non-holder's matriline is the same	19	3.2 (2.0)	0.4 (0.8)	−3.62	<0.001
	Holder's and non-holder's matriline differs	24	4.3 (3.6)	0.3 (0.6)	−4.15	<0.001

play activity contained edible parts. Because Arashiyama-Kyoto macaques were provisioned with plenty of nutritious food, they seldom fed on these wild plants during the study period. Thus, the nutritional values of the natural objects held during SOP were considered to be very low for immature macaques in Arashiyama. Moreover, artificial objects, such as plastic, metallic or glass materials, held during long SOP had no nutritional value.

Most of the long SOP bouts among immature macaques in Arashiyama had two distinctive interactive features, independent from the attributes of the participants: (1) multiple immature macaques treated only one object as a target of play and (2) the holder of the target object escaped from non-holder(s) during play-chasing. Hereafter, I define repetitive play-chasing episodes including these two features as 'play-chasing with a target object' (PCT). Thus, Observation 1 can be regarded as a typical example of PCT.

Although it was expected that SOP would not last for long periods of time if immatures applied the prior possession rule, the opposite was found. Perception of the value of the objects as food for immature macaques at Arashiyama seems to explain the reason why SOP among them could last for long periods. Even though participants apparently perceived the ownership of the object, they did not seem to respect the possession of object, because they simultaneously perceived the nutritional value of the object as being negligible. In other words, because the nutritional value of the objects held during SOP was very low or null, the prior possession rule hardly operated among immatures. In strong support of this interpretation, I never observed juveniles playing socially with any food items provisioned by humans.

However, when considering how macaques may perceive the value of (non-food) objects handled by conspecifics during playful activities, it should be noted that long-term research on stone handling behaviour in Japanese macaques suggested the existence of a rudimentary form of 'possession' in monkeys. Although stone handling behaviour is not a form of SOP, it is a form of solitary object play – specifically directed to stones, i.e. objects with no nutritional value – that can occasionally trigger social interactions around these natural objects. Several reports on Arashiyama-Kyoto macaques showed that once particular stones, or sets of stones, were involved in a stone handling episode, they seemed to become valuable objects for the handler who may pick them up and carry them to different places rather than leave them behind, and they appeared to trigger great interest from others who tried to snatch them away from the handler as if they were the only stones available (Huffman and Quiatt, 1986; Leca *et al.*, 2008a, 2010). Therefore, my perspective about the prior possession rule and how it may apply or not in macaques may largely depend on the type of playful activities and contexts under study.

14.3   Study 2: Comparison of SOP between a provisioned and a non-provisioned troop

Among Arashiyama-Kyoto immature macaques in 2000, PCT was the most frequently observed interactive pattern during long SOP. This suggests that PCT was a common play pattern established in this troop, although PCT is only one of the many patterns possible during long SOP. As reported above, other possible forms of SOP included 'tug-of-war', situations where a socially playing individual held two or more objects, an object holder chasing a non-holder, or even many object holders socially playing together. In addition, PCT is a regulative interaction in the sense that immature macaques need to restrict their own behaviour in order to match their partners' behaviours when playing for long periods.

What remains unknown is whether socially demanding patterns of SOP, such as PCT, are common in juvenile Japanese macaques within any troops, and if not, what factors may account for a possible inter-troop variation. Indeed, there is evidence for PCT-like forms of SOP in other species, especially in humans and some apes (e.g. Tanner and Byrne, 2010). Play patterns similar to the PCT observed at Arashiyama in 2000 have been reported to occur in several Japanese macaque troops living in different geographical regions: Takasakiyama (Itani, 1954), Koshima (Mori and Kudo, 1986), Jigokudani (Shimada, unpublished data), Funakosiyama (Hayaki, pers. comm.) and Kinkazan (Sugiura, pers. comm.; Shimada, unpublished data). However, due to the informal nature of these reports, it remains unclear whether PCT is typical of all these populations of Japanese macaques or restricted to juveniles of particular troops.

For example, if there is no water pool of size and depth suitable for diving into in their habitat, the absence of swimming play can be ascribed to an absence of opportunity, not to a lack of capacity for such a form of play. Thus, such environmental differences may account for some of the variation reported in play behaviour across troops of the same and different species (e.g. McGrew, 1992; Whiten *et al.*, 1999).

As I showed in Study 1, immature macaques at Arashiyama held various kinds of detached natural and artificial objects available in their environment during PCT. When considering the objects used during social play, the environmental conditions affecting PCT seem similar among all troops. Indeed, twigs, branches or other detached and portable materials are widespread and readily available in the habitat of any troops of wild Japanese macaques. Thus, such obvious environmental reasons are not sufficient to account for possible regional variations in the frequency of PCT among juveniles.

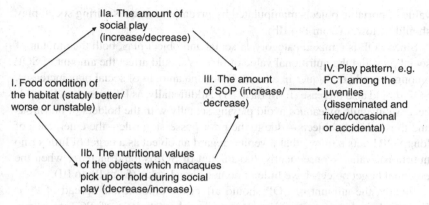

Figure 14.2. A causal model about the effect of food condition on the dissemination of a particular play pattern. Each factor is connected by arrows as a causal relation.

14.3.1 A model explaining differences in the dissemination of PCT between troops

Even though local availability in the objects typically used during play activities is probably similar, inter-troop differences in food conditions may partly be responsible for variation in play patterns. Huffman and Hirata (2003) and Leca *et al.* (2008b) showed that the type of food provisioning had a strong influence on the emergence and expression of stone-handling behaviour in Japanese macaques, as well as its maintenance as a behavioural tradition. The occurrence or non-occurrence of provisioning may also give rise to regional variations in PCT. Here, I propose a model to explain how local food conditions may influence PCT in juvenile Japanese macaques among different troops (Figure 14.2).

If the nutritional status of a given troop is positive and stable, the birth rate in this troop should increase and mortality should decrease. Consequently, the number of juvenile macaques in the troop should increase (e.g. Sugiyama and Ohsawa, 1982). At the same time, the nutritious food items provided by humans should reduce the time spent by the group foraging for natural foods (e.g. Baulu and Redmond, 1980). This in turn should lead to surplus time for non-essential activities such as play (cf. Southwick, 1967; Loy, 1970; Huffman and Hirata, 2003). Furthermore, as the number of peer immature individuals with more time to play increases, there should be more social play at the group level (cf. Hayaki, 1983; Mori and Kudo, 1986) (from I to IIa in Figure 14.2). Juvenile macaques with reduced foraging requirements should be more likely to pick up and hold objects with low or no nutritional values in their habitat. Thus, when the nutritional status of a troop is higher, the relative nutritional value of portable objects manipulated by juvenile macaques during social play should be lower (from I to IIb).

Since SOP is a mixed category of social and object play, both the amount of social play and the nutritional value of objects should affect the amount of SOP among juvenile macaques in a troop. When the amount of social play increases, SOP should also increase (from IIa to III). Additionally, as I discussed in Study 1, because juvenile macaques avoid playing socially with the holders of nutritious and thus valuable objects – due to the prior possession rule – the emergence of long SOP bouts requires that juveniles regard an object as a target of little or no nutritional value. Consequently, the amount of SOP should increase when the potential target objects have little or no nutritional value (from IIb to III).

Finally, the amount of SOP should affect the amount and spread of PCT. For the long-lasting repetitive interactions characteristic of PCT to occur, play partners must regard the object in the holder's possession as the target of play. As each juvenile macaque participates in more SOP, it should have more opportunities to learn socially and consolidate the relevant patterns of such playful interactions. Consequently, PCT should be both present and widely disseminated among juveniles of provisioned troops (from III to IV).

The purpose of Study 2 is to examine whether PCT was commonly observed among juveniles in different troops of Japanese macaques, and if not, how to account for such inter-troop variation. If PCT is generally observed in juveniles, it should also be seen even in a troop where social play is relatively infrequent. If PCT occurs only among the juvenile macaques of particular troops, then the model presented in Figure 14.2 should lead to the following predictions: (1) The duration and frequency of social play sequences of juvenile macaques in a provisioned troop should be higher than in a non-provisioned troop; (2) The nutritional values of objects held during SOP by juvenile macaques in a provisioned troop should be lower than in a non-provisioned troop; (3) Interactions in which multiple individuals regard an object as a target should be more frequent among juvenile macaques in a provisioned troop than in a non-provisioned troop; and (4) The duration of SOP bouts in a provisioned troop should be higher than in a non-provisioned troop.

14.3.2   Methods of Study 2

To test the model, I chose two free-ranging troops of Japanese macaques: Arashiyama (E troop) and Kinkazan (A troop). Like the provisioned Arashiyama-Kyoto macaques, the non-provisioned Kinkazan macaques have been perfectly habituated to humans. All the members of the Kinkazan A troop have been identified since 1983, and long-term records of their dates of birth, kin relationships and social ranks have been systematically collected (Izawa, 2009). In addition, it is reasonable to assume that juveniles of both troops had similar access to typical play target objects (e.g. twigs, branches and other detached and portable materials) in their respective environments.

The Arashiyama E troop ranges in an area of about 0.7 km² around a feeding site located in the Arashiyama Mountains, Kyoto prefecture (Fedigan and Asquith, 1991), whereas the Kinkazan A troop ranges in an area of 4–5 km² in the north-west area of Kinkazan Island, Miyagi prefecture (Izawa, 2009). Although the sizes of their home ranges were different, a large part of the habitats of both troops were covered by deciduous broad-leaved forests, or coniferous forests (Arashiyama: Murata and Hazama, 1968; Huffman, 1984b; Kinkazan: Yoshii and Yoshioka, 1949; Tsuji and Takatsuki, 2004). Since 80% of the objects held during long-lasting PCT bouts among Arashiyama-Kyoto immatures were plant materials, one can assume that juveniles of both troops had similar opportunities to encounter such portable objects in their respective habitats. Thus, it is reasonable to consider that the main environmental difference between the Arashiyama E troop and the Kinkazan A troop consisted of food provisioning by humans, present in the former and absent in the latter.

I compared social play, SOP and PCT, between juveniles of these two troops. Data were collected from 5 June to 16 July 2004 in Arashiyama (21 days) for a total of 131.9 h, and from 24 July to 25 August 2004 in Kinkazan (23 days) for a total of 149.8 h, after I had identified every juvenile in each troop. During the study period, the Arashiyama E troop consisted of about 160 individuals, including 26 juveniles (1- to 3-year-old), whereas the Kinkazan A troop consisted of 35 individuals, including eight juveniles.

I used the focal animal sampling method (Altmann, 1974), focusing on juvenile macaques in both troops, and continuously recording their behaviours to the nearest second. Because the number of juvenile macaques differed between troops, I tried to follow each focal subject for at least 2 hours in Arashiyama (n = 26), and 12 hours in Kinkazan (n = 8), respectively. In order to collect as much data on SOP as possible, I also used an *ad libitum* sampling method (Altmann, 1974), recording SOP whenever it occurred as long as the focal individual being simultaneously followed was inactive.

Each day, before collecting data, I decided which juvenile individual to follow according to the list of accumulated observation time for each juvenile. Whenever possible, I followed one juvenile for about an hour in Arashiyama, and for about two hours in Kinkazan. After collecting data on one juvenile, I followed the next listed juvenile, as far as possible. For the analyses, I only used data that were continuously collected for at least 10 minutes. The object held in each bout of SOP was recorded in field notes. If the object was part of a tree, I identified the species. I also collected data on food items that the juvenile macaques fed on, including species or parts of plants, the number of juveniles that fed on each item and how long they fed on them each day.

Social play and SOP were defined as in the Methods of Study 1, and PCT was defined as in the Discussion of Study 1. A sequence of social play was defined as a continuous series by multiple juvenile macaques of interactions categorised in Study 1, including a sampled individual, with between-play intervals of less than 1 minute. The duration of each sequence of social play was defined from start to end of the interactions involving the target individual. The frequency of social play for each juvenile was defined as the total number of observed sequences of social play divided by the total focal observation time. I defined the target of SOP as the portable object that multiple juvenile macaques appeared to be trying to possess or retain during any phase of social play. For example, when a non-holder individual reached for an object which was held by the holder, and the latter withdrew in order to retain the object, the object was regarded as a target of SOP for both individuals.

The value of food items on each observation day was operationally defined as follows: (a) food items which at least two juveniles ate for at least 10 minutes; (b) food items which at least two juveniles ate for at least 1 minute but not (a); (c) food items which juveniles ate only rarely and not (a) or (b). Moreover, each food item for juveniles in each troop was attributed to one of the following categories of objects: Main food: objects which were scored as (a) for at least 2 days; Supplemental food: objects which scored (a) or (b) for at least 2 days, and not main food; Rare food: objects which scored (a), (b) or (c) at least 1 day, and not main or supplemental food. Objects that juvenile macaques never ate were defined as non-food. All objects juvenile macaques held during SOP were noted and classified as main, supplemental, rare, or non-food, at least in part.

14.3.3   Results of Study 2

SOP bouts and PCT in both troops

I observed 241 sequences of social play (324.8 min, cumulatively) in Arashiyama and 116 sequences (71.9 min) in Kinkazan. Twenty out of 26 juveniles (76.9%) and five out of eight juveniles (62.5%) were observed to participate in SOP at least once in Arashiyama and Kinkazan, respectively. Fifty-five SOP bouts were observed in Arashiyama, and seven bouts in Kinkazan.

The mean duration of SOP bouts was 0.64 ± 1.19 min (mean value ± 1 s.d.; n = 50; median = 0.17; range: 0.05–6.20) in Arashiyama, and 0.08 ± 0.03 min (n = 7; median = 0.08; range: 0.05–0.13) in Kinkazan. SOP bouts tended to be longer in Arashiyama than in Kinkazan, although this result was not statistically significant (Mann–Whitney U test; $U = 253.5$; $p = 0.053$). In

Arashiyama, 17 out of the 50 SOP bouts observed lasted at least 0.5 minute and the longest bout lasted 6.20 minutes, whereas all SOP bouts lasted less than 0.5 minute in Kinkazan. PCT occurred in 13 SOP bouts in Arashiyama, whereas no PCT was observed in Kinkazan. Of the 20 juveniles participating in SOP bouts at Arashiyama, 12 (41.4%) took the role of both chasee and chaser at least once.

In Arashiyama, 29 of 55 SOP bouts involved interactions in which multiple individuals regarded an object as a target, compared with only one out of seven in Kinkazan. In this particular case, two 1-year-old females, whose abbreviations were KN and SM, were playing socially, when KN sat and started to eat small mushrooms on the surface of a piece of rotten wood. SM approached to grasp KN's back and reached toward the wood. KN turned around and kept eating. SM did not touch the wood and ran away, and the sequence of social play ended.

Comparison of frequencies and durations of social play

I observed 17 juveniles in Arashiyama for at least 2 hours and all eight juveniles in Kinkazan for at least 12 hours, recording the exact duration of 210 sequences of social play in Arashiyama and 112 in Kinkazan. The mean frequency of sequences of social play among 17 juveniles was 2.79 ± 1.49 times/min (mean value ± 1 s.d.; range: 0.00–5.45) in Arashiyama, whereas that among eight juveniles was 0.91 ± 0.28 times/min (range: 0.50–1.38) in Kinkazan. The frequencies of social play sequences were significantly different between the two sites (Mann–Whitney U test; n = 17 and 8, respectively; $U = 115, p < 0.01$).

The mean duration of social play sequences observed was 0.53 ± 2.53 min (mean value ± 1 s.d.; n = 210; range: 0.05–25.6) in Arashiyama, whereas it was 0.23 ± 0.88 min (n = 112; range: 0.05–4.55) in Kinkazan. The durations of sequences in both sites were also significantly different (n = 210 and 112, respectively; $U = 14015, p < 0.01$). Percentages of social play in the daily activities of juveniles were 6.9 ± 5.2% (mean value ± 1 s.d.) in Arashiyama and 0.9 ± 0.6% in Kinkazan.

Food items and objects held in SOP bouts

Table 14.2 shows a list of food items consumed in both troops. The more frequently a food item was consumed by juveniles, the higher it appears on the list. The main food items of juveniles in Arashiyama were mixed wheat and soybeans provided regularly by the park staff, peanuts provided occasionally by tourists or staff, and natural herbaceous vegetation of unidentified species. Other natural plants were eaten only as supplemental or rare food items. On the other hand, the main food of juveniles in Kinkazan consisted of 14 natural plants.

Table 14.2. *List of food items eaten by juveniles in Arashiyama and Kinkazan. The number indicates the days counted at each level (a–c).* *Species names were unknown, including multiple species. ** Provisioned food

Arashiyama E troop

Food items	Part	Level a	b	c	Category
Mixed food (wheat + soy beans)**		19			Main
Herbs*		10	6	2	Main
Peanuts**		2	6	1	Main
Sweet potatoes**		1	4	1	Supplemental
Acer palmatum	bark	1	3	2	Supplemental
Apples**			4	1	Supplemental
Clod			3	7	Supplemental
Prunus jamasakura	bark		3	2	Supplemental
Swida macrophylla	flower		3		Supplemental
Swida macrophylla	leaf		3		Supplemental
Styrax shiraiana	calyx		3		Supplemental
Fungi*			2	2	Supplemental
Styrax shiraiana	leaf		2	1	Supplemental
Rhododendron macrosepalum	leaf		2		Supplemental
Insecta*			1	4	Rare
Acer palmatum	leaf		1	1	Rare

Kinkazan A troop

Food items	Part	Level a	b	c	Category
Fungi*	fruit	17	4	2	Main
Benthamidia japonica	fruit	6	6		Main
Aralia elata	bark/sap	6	4	3	Main
Berberis thunbergii	leaf	5	13	1	Main
Zanthoxylum piperitum	leaf	5	4	4	Main
Prunus jamasakura	fruit	5		1	Main
Berchemia racemosa	fruit	4	4	2	Main
Prunus jamasakura	bark/sap	4	3	1	Main
Magnolia obovata	leaf	3	1	3	Main
Zanthoxylum piperitum	fruit	3	1		Main
Torreya nucifera	seedling	2	4	6	Main
Vitis flexuosa	leaf	2	3	1	Main
Zanthoxylum piperitum	bark	2	2		Main
Zelkova serrata	leaf	2		2	Main
Herbs*		1	7	8	Supplemental
Smilax china*	fruit	1		1	Supplemental
Insecta*			5	8	Supplemental
Rubus microphyllus	leaf		3	7	Supplemental
Viburnum dilatatum	fruit		3		Supplemental
Symplocos chinensis	leaf		1	3	Rare
Torreya nucifera	fruit			3	Rare
Carpinus tschonoskii	stem			3	Rare
Stephanandra incisa	leaf			2	Rare
Ilex macropoda	fruit			2	Rare
Magnolia obovata	fruit			2	Rare

Table 14.3 shows the list of objects that juveniles held during bouts of SOP and which category of food items the objects included. Most objects held during SOP bouts in Arashiyama were non-food items, and when food items were involved they were mostly of the rare or supplemental categories. In Kinkazan, two of the objects held in SOP were fungi, which were main food objects. All the other five objects were pieces of rotting wood of unidentified species. I found fungi on the surface of all of these rotting woods, and juveniles fed on them.

14.3.4   Discussion of Study 2

Difference in PCT between Arashiyama and Kinkazan

Among the juvenile macaques at Arashiyama and Kinkazan, more than half of them participated in SOP bouts at least once. This suggests that SOP is a general play category among juveniles in both troops, despite inter-group differences in food conditions and in the frequency of social play. By contrast, SOP bouts including PCT were observed only among Arashiyama-Kyoto juveniles. In Kinkazan, social play accounted for 71.9 min of 149.8 hours of total observation time, but no PCT was observed. However, PCT was seen among Kinkazan juveniles in another season (Sugiura, pers. comm.; Shimada, unpublished data), so the behaviour does occur in this troop. Thus the results of the present study do not suggest that juveniles in Kinkazan are unable to participate in PCT, but rather that PCT is a play pattern not as common in Kinkazan as it is in Arashiyama.

At least 41.4% of juvenile macaques in Arashiyama participated in PCT during the 2004 study period. Study 1 showed that 85.4% of immature individuals participated in continuous SOP bouts during 38 observation days in 2000, and suggested that PCT was a more common play pattern of long SOP than the other possible play patterns, such as tug-of-war. The reason for the lower percentage of juveniles participating in PCT in 2004 than in 2000 might be due to the shorter duration of this study. However, PCT has been intermittently observed in this troop since 1999 when I first visited Arashiyama, and thus, it appears to be a stable phenomenon.

Test of predictions

The frequently provisioned macaques in Arashiyama ate sufficient amounts of nutritious food every day so that time was freed up for juveniles to engage in more social play. In contrast, as Nakagawa (1997) pointed out, in Kinkazan,

Table 14.3. *List of objects held by juveniles in social object play (SOP). The number of each sample is shown in parentheses. 'BL' indicates a branch with leaves. 'B' indicates a branch without leaves. 'L' indicates a leaf. 'P' indicates a part of a tree*

Category	Arashiyama	Total no. of objects	Kinkazan	Total no. of objects
Main food	Herbs (1), a peanut with shell (1)	2	A fungus (2), B with fungi (3)	5
Supplemental food	BL of *Swida macrophylla* (3), BL of *Styrax japonica* (3), BL of *Acer palmatum* (2), BL of *Cleyera japonica* (1), B of *Prunus jamasakura* (1), L of *Swida macrophylla* (2), L of *Styrax japonica* (1), a mushroom (1), a piece of *Prunus jamasakura* (1)	15	No data	0
Rare food	BL of *Lyonia ovalifolia* (3), BL of *Eurya japonica* (1), L of *Quercus crispula* (1), L of *Camellia japonica* (2), a fern (1)	8	No data	0
No food	L of *Quercus glauca* (4), stone (4), a peanut shell (3), BL of *Ilex pedunculosa* (2), a glass piece (2), a plastic bag (2), BL of *Thujopsis dolabrata* (1), a plastic toy (1)	19	No data	0
Unidentified	B of unidentified wood (5), a piece of unidentified wood (5), L of unidentified wood (1)	11	A piece of unidentified rotting wood (2)	2
Total		55		7

summer is the second hardest season of the year in terms of food availability, and this is when the study was conducted. Thus, juveniles in Kinkazan probably needed to spend more time foraging for relatively low nutritious food items, leaving less time available for play.

As a result, juvenile macaques in Arashiyama played socially much more than those in Kinkazan. Previous authors have reported percentages of play behaviour in the daily activities of juvenile macaques between 1% and 8% (wild rhesus macaques: 1–6%, Fagen, 1981; provisioned pigtailed macaques, *Macaca nemestrina*: 4–8%, Bernstein, 1970; provisioned Japanese macaques: 5.4%, Mori and Kudo, 1986). Although the present study dealt with only social play, the percentage obtained in Arashiyama (6.9 ± 5.2%) appears equal to or higher than in the provisioned troops of previous studies. For the Kinkazan troop, 0.9 ± 0.6% appears equal to or lower than that reported in wild macaques. This result suggests that the amount of social play was positively affected by the food supplied to the troop, and therefore supports prediction (1).

In the study period, most objects held by juvenile macaques in Kinkazan during or at the onset of social play were relatively valuable nutritionally. Although pieces of rotting wood themselves were not food items, small fungi adhered to their surface, which were edible and main food items. Thus, all seven objects held in social play were considered to have high nutritional value. On the contrary, most of the objects held by juvenile macaques in Arashiyama had no or low nutritional value: only two out of 55 objects were main food items, while half were rare or non-food items. In addition, Study 1 showed that Arashiyama-Kyoto juvenile macaques in 2000 held only low-value food objects and never held nutritional provisioned food items during SOP (see first Results of Study 1). The present study showed a similar pattern, thereby supporting prediction (2).

Because the objects that juveniles in Kinkazan held during SOP were nutritionally valuable, the prior possession rule (cf. Bakeman and Brownlee, 1982) appears to apply strictly. Consequently, they declined interactions in which multiple individuals regarded an object as a target of play. On the contrary, juveniles in Arashiyama did not decline such interactions as the nutritional value of objects was relatively low and the prior possession rule applied weakly, if at all (Study 1).

Interactions in which multiple individuals regarded an object as a target were more frequent in Arashiyama (29 of 55 SOP bouts) than in Kinkazan (one of seven). In one case in Kinkazan, which I described in the first Results of Study 2, the juveniles ceased playing socially immediately after this interaction. Although this particular interaction was defined as SOP which happened at the end of social play, and the object was defined as a target of KN and SM, the case can be interpreted as a disruption to continue their playful interaction; KN regarded the piece of wood as a food item, whereas SM regarded the same wood as a target of play since SM played socially with KN until this moment. The data from the two troops suggest that interactions in which multiple individuals regarded an object as a target were frequent in Arashiyama, but only rare or incomplete in Kinkazan, and thus, support prediction (3).

SOP bouts observed in Arashiyama tended to last longer than in Kinkazan. All SOP bouts observed in Kinkazan lasted 0.13 min or less, whereas 34% (17 of 50) of SOP bouts in Arashiyama continued for at least 0.5 min. Study 1 showed that 44% of SOP bouts observed in Arashiyama-Kyoto juvenile macaques in 2000 continued for at least 0.5 min, generally similar to the present study. These findings support prediction (4).

Food provisioning and dissemination of PCT

Inter-troop difference in the duration of SOP bouts may also explain the differential dissemination of play patterns among juveniles in the two troops. Because of the repetitive interactive features of PCT shown in Study 1 – i.e. a holder of the target object always escaped from the non-holders whenever play-chasing occurred in SOP – each juvenile needs to restrict and regulate its behaviour according to its partners' play behaviour. However, the association between the role of chasee and being the object holder cannot be an innate propensity, because not all juvenile macaques, such as those in Kinkazan, played socially with objects in the same way as those in Arashiyama did. It is suggested that the modification of patterned playful interactions, thus PCT here, is achieved by a social learning process through the repetitive SOP.

In Kinkazan, juvenile macaques avoided interacting with an object holder at least during the study period. I observed brief SOP bouts, and the objects held during SOP were rarely or only incompletely regarded as targets. Each SOP bout was probably too short and too infrequent to learn the association. Consequently, juvenile macaques participated in PCT only occasionally, if at all. In contrast, during continuous SOP bouts, juveniles in Arashiyama had many chances to play with the object holder, regarding the nutritionally invaluable object as a target of play. Through repetitive SOP interactions, juveniles could learn the above association. Consequently, PCT as a form of SOP would be disseminated, standardised and established among juvenile macaques in Arashiyama.

The results of Study 2 supported all four predictions. Therefore, the hypothesis about the causal chain between food condition and dissemination of a particular play pattern in a troop was also supported. However, the causation was not simple because there were at least two possible channels (Figure 14.2). That is, provisioning affects not only the quantity of social play but also the possibility of interactions which constitute a particular type of social play. Why PCT has not been disseminated and established among juvenile macaques in Kinkazan may be explained by the same mechanism underlying the dissemination and establishing of PCT among juvenile macaques in Arashiyama. It is not individual differences or habitat differences between troops that matter, but a difference in the local food condition.

Is PCT a behavioural tradition?

In this chapter, I pointed out that, during the surveys conducted in 2000 and 2004, continuous SOP among Arashiyama-Kyoto juvenile macaques included repetitive PCT, i.e. regulative play-chasing interactions while holding objects with no or low nutritional value. In addition, PCT was suggested to be an established play pattern among juvenile macaques in Arashiyama but not in Kinkazan. The final question is whether the regional difference in the establishment of PCT between two troops of Japanese macaques can be regarded as an example of a cultural difference.

Primatologists have asked whether regional differences in various primate behaviours reflect traditions (or cultural behaviour) or not (e.g. Whiten *et al.*, 1999). On the one hand, some of them claim that a tradition does exist when a regional difference in behaviour cannot be explained by ecological or genetic differences (Whiten *et al.*, 1999; van Schaik, 2003). Thus, when the habitats of two particular troops were obviously different, researchers have simply regarded inter-troop behavioural differences as the product of these different environmental conditions and not as cultural (McGrew, 1992; Whiten *et al.*, 1999; Perry and Manson, 2003). If I follow their definition of tradition as a residual concept which cannot be explained by other possible explanations, the present findings cannot be considered an example of tradition. Indeed, I compared two troops living under obviously different food conditions. In addition, this study pointed out that food provisioning represented an environmental variation that could actually explain regional variation in the dissemination and establishment of PCT in Arashiyama but not in Kinkazan.

On the other hand, Perry and Manson (2003, p. 71) defined a tradition as 'a behavioral practice that is relatively long-lasting and shared among members of a group in part through social learning'. In other words, a traditional or cultural behavioural pattern should be persistent over years at the group level, and most of all, it should be acquired, transmitted and maintained by social learning (e.g. Fragaszy and Perry, 2003; Perry and Manson, 2003). From this definition, they nominated 12 behavioural patterns which can be considered traditions in Japanese macaques (see Table 1 in Perry and Manson, 2003). 'Sweet-potato washing' behaviour in Koshima (Kawai, 1965) and 'stone handling' behaviour in Arashiyama (Huffman, 1984a) are famous examples illustrating the role of food provisioning, as a human-induced environmental change, in promoting the innovation and dissemination of novel behaviours, leading to traditions in each troop (McGrew, 1992; Huffman and Hirata, 2003; Perry and Manson, 2003).

From this perspective, I claim that PCT observed in Arashiyama should be added to the list of traditions in Japanese macaques. Although only anecdotal data were available, PCT has been observed in this troop from 1999 to the present. Therefore, PCT has been maintained among Arashiyama-Kyoto juvenile macaques for several years. In addition, in order to prolong the repetitive interactions of PCT, each participant needs to achieve a cognitive association between the role of chasee and that of the holder of a non-nutritious object, which cannot be an innate propensity. Because PCT is the regulative interaction generated by multiple juvenile macaques, this association cannot be achieved only by individual learning, but also by social learning through repetitive SOP.

Detailed future studies will need to provide direct evidence of social learning in order to claim that PCT is actually a traditional interactive pattern, that may be influenced by local food conditions. However, the present study is important to better understand the role of environmental factors in the emergence of intergroup behavioural differences, and further discuss the occurrence of culture in non-human primates.

Acknowledgements

I thank Mr Shinsuke Asaba and the late Nobuo Asaba, the present and the former park wardens of Iwatayama Monkey Park, Arashiyama, respectively, and many park staff for support officially and privately. I thank Dr Kosei Izawa, Mr Takeharu Uno and Ms Hiroko Fujita for support in Kinkazan. I would like to thank the following fieldworkers for exciting discussions in both the study sites, that inspired me: Dr Yuji Takenoshita, Koichiro Zamma, Eiji Inoue, Mariko Fujimoto, James V. Wakibara, Jean-Baptiste Leca and Michael A. Huffman in Arashiyama; Dr Yamato Tsuji, Ms Nobuko Kazahari and Ms Naoko Higuchi in Kinkazan. I also thank Dr Juichi Yamagiwa, Naofumi Nakagawa, Michio Nakamura, Hideki Sugiura (Kyoto University), Hitoshige

Hayaki (Kobe Gakuin University) and Toshisada Nishida (Japan Monkey Center) for their comments on this paper. This study was partially supported by the Ministry of Education, Science, Sports and Culture, Grant-in-Aid for JSPS fellows, and the Cooperation Research Program of Primate Research Institute, Kyoto University.

References

Altmann, J. (1974). Observational study of behavior: sampling methods. *Behaviour*, **49**, 227–267.

Bakeman, R. and Brownlee, J. R. (1982). Social rules governing object conflicts in toddlers and preschoolers. In *Peer Relationships and Social Skills in Childhood*, eds. K. H. Rabin and H. S. Ross. New York: Springer, pp. 99–111.

Baulu, J. and Redmond, D. E. Jr. (1980). Social and nonsocial behaviours of sex- and age-matched enclosed and free-ranging rhesus monkeys (*Macaca mulatta*). *Folia Primatologica*, **34**, 239–258.

Bekoff, M. and Byers, J. A. (1998). *Animal Play: Evolutionary, Comparative, and Ecological Perspectives*. Cambridge: Cambridge University Press.

Bernstein, I. S. (1970). Activity patterns in pigtail monkey groups. *Folia Primatologica*, **10**, 1–19.

Burghardt, G. M. (2005). *The Genesis of Animal Play: Testing the Limits*. Cambridge, MA: MIT Press.

Fagen, R. (1981). *Animal Play Behavior*. Oxford: Oxford University Press.

Fedigan, L. M. and Asquith, P. J. (1991). *The Monkeys of Arashiyama: Thirty-five Years of Research in Japan and the West*. Albany, NY: State University of New York Press.

Fragaszy, D. M. and Perry, S. (2003). *The Biology of Traditions: Models and Evidence*. Cambridge: Cambridge University Press.

Hayaki, H. (1983). The social interactions of juvenile Japanese monkeys on Koshima Islet. *Primates*, **24**, 139–153.

Huffman, M. A. (1984a). Stone-play of *Macaca fuscata* in Arashiyama B troop: transmission of a non-adaptive behavior. *Journal of Human Evolution*, **13**, 725–735.

    (1984b). Plant foods and foraging behavior of the Arashiyama Japanese macaques. In *Arashiyama Japanese Monkeys: Arashiyama Natural History Research Station Report, Volume 3*, ed. N. Asaba. Osaka: Osaka Seihan Printers, pp. 55–65. (in Japanese)

Huffman, M. A. and Hirata, S. (2003). Biological and ecological foundations of primate behavioral tradition. In *The Biology of Traditions: Models and Evidence*, eds. D. M. Fragaszy and S. Perry. Cambridge: Cambridge University Press, pp. 267–296.

Huffman, M. A. and Quiatt, D. (1986). Stone handling by Japanese macaques (*Macaca fuscata*): implications for tool use of stones. *Primates*, **27**, 413–423.

Imakawa, S. (1990). Playmate relationships of immature free-ranging Japanese monkeys at Katsuyama. *Primates*, **31**, 509–521.

Itani, J. (1954). *Japanese Monkeys in Takasakiyama.* Tokyo: Kobunsha. (in Japanese)

Izawa, K. (2009). *The Study of Wild Japanese Monkeys.* Tokyo: Dobutsusha. (in Japanese)

Kawai, M. (1965). Newly-acquired pre-cultural behavior of the natural troop of Japanese monkeys on Koshima islet. *Primates*, **6**, 1–30.

Koyama, N. (1985). Playmate relationships among individuals of the Japanese monkey troop in Arashiyama. *Primates*, **26**, 390–406.

Koyama, N., Takahata, Y., Huffman, M. A., Norikoshi, K. and Suzuki, H. (1992). Reproductive parameters of female Japanese macaques: thirty years data from the Arashiyama troops, Japan. *Primates*, **33**, 33–47.

Kummer, H. (1973). Dominance versus possession: an experiment on hamadryas baboon. In *Precultural Primate Behavior, Vol. 1,* ed. E. W. Menzel. London: Krager, pp. 226–231.

Kummer, H. and Cords, M. (1991). Cues of ownership in long-tailed macaques (*Macaca fascicularis*). *Animal Behaviour*, **42**, 529–549.

Leca, J. B., Gunst, N. and Huffman, M. A. (2008a). Of stones and monkeys: testing ecological constraints on stone handling, a behavioral tradition in Japanese macaques. *American Journal of Physical Anthropology*, **135**, 233–244.

(2008b). Food provisioning and stone handling tradition in Japanese macaques: a comparative study of ten troops. *American Journal of Primatology*, **70**, 803–813.

(2010). Indirect social influence in the maintenance of the stone handling tradition in Japanese macaques (*Macaca fuscata*). *Animal Behaviour*, **79**, 117–126.

Loy, J. (1970). Behavioral responses of free-ranging rhesus monkeys to food shortage. *American Journal of Physical Anthropology*, **33**, 263–271.

McGrew, W. C. (1992). *Chimpanzee Material Culture: Implications for Human Evolution.* Cambridge: Cambridge University Press.

Mori, U. and Kudo, H. (1986). *Social Development and Social Relationship among Female Japanese Monkeys.* Tokyo: Tokai University Press. (In Japanese)

Murata, G. and Hazama, N. (1968). Flora of Arashiyama, Kyoto, and plant food of Japanese monkeys. In *Iwatayama Shizenshi Kenkyujo Chosa Kenkyu Hokoku,* **2**, 1–59. (in Japanese)

Nakagawa, N. (1997). Determinants of the dramatic seasonal changes in the intake of energy and protein by Japanese monkeys in a cool temperate forest. *American Journal of Primatology*, **41**, 267–288.

Perry, S. and Manson, J. (2003). Tradition in monkeys. *Evolutionary Anthropology*, **12**, 71–81.

Reinhart, C. J., Pellis, V. C., Thierry, B. *et al.* (2010). Targets and tactics of play fighting: Competitive *versus* cooperative styles of play in Japanese and Tonkean macaques. *International Journal of Comparative Psychology*, **23**, 166–200.

Shimada, M. (2006). Social object play among young Japanese macaques (*Macaca fuscata*) in Arashiyama, Japan. *Primates*, **47**, 342–349.

(2009). An ethnography of play of Japanese macaques: juveniles of Kinkazan, Arashiyama, Koshima, and Shigakogen. In *Anthropology of Children*, ed. N. Kamei. Kyoto: Showado, pp. 81–133. (in Japanese)

(2010). Social object play among juvenile Japanese macaques. In *The Japanese Macaques*, eds. N. Nakagawa, M. Nakamichi and H. Sugiura. Tokyo: Springer, pp. 375–385.

Sigg, H. and Falett, J. (1985). Experiments on the respect of possession and property in hamadryas baboon (*Papio hamadryas*). *Animal Behaviour*, **33**, 978–984.

Southwick, C. H. (1967). An experimental study of intragroup agonistic behavior in rhesus monkeys (*Macaca mulatta*). *Behaviour*, **28**, 182–209.

Sugiyama, Y. and Ohsawa, H. (1982). Population dynamics of Japanese monkeys with special reference to the effect of artificial feeding. *Folia Primatologica*, **39**, 238–263.

Symons, D. (1978). *Play and Aggression: A Study of Rhesus Monkeys*. New York, NY: Columbia University Press.

Tanner, J. E. and Byrne, R. W. (2010). Triadic and collaborative play by gorillas in social games with objects. *Animal Cognition*, **13**, 591–607.

Thierry, B., Wunderlich, D. and Gueth, C. (1989). Possession and transfer of objects in a group of brown capuchins (*Cebus apella*). *Behaviour*, **110**, 294–305.

Thierry, B., Bynum, E. L., Baker, S. *et al.* (2000). The social repertoire of Sulawesi macaques. *Primate Research*, **16**, 203–226.

Tsuji, Y. and Takatsuki, S. (2004). Food habits and home range use of Japanese macaques on an island inhabited by deer. *Ecological Research*, **19**, 381–388.

van Schaik, C. P. (2003). Local traditions in orangutans and chimpanzees: social learning and social tolerance. In *The Biology of Traditions: Models and Evidence*, eds. D. M. Fragaszy and S. Perry. Cambridge: Cambridge University Press, pp. 297–328.

Whiten, A., Goodall, J., McGrew, W. C. *et al.* (1999). Culture in chimpanzees. *Nature*, **399**, 682–685.

Yoshii, Y. and Yoshioka, K. (1949). Plant communities on Kinkazan Island. *Ecological Review, Sendai*, **12**, 84–105. (in Japanese)

14 Box essay 1 *Play fighting in Japanese macaques: A comparative perspective*

SERGIO M. PELLIS AND VIVIEN C. PELLIS

14 Box 1.1 Introduction

Play fighting involves competition for access to a play target, which if contacted is gently bitten or otherwise contacted (Aldis, 1975; Pellis, 1988). These play targets are protected by using various defensive manoeuvres and successful defence is often followed by counterattacks against the partner's play target, leading to complex and prolonged interactions (Pellis and Pellis, 1998a). Moreover, for play fighting to remain playful, it has to be reciprocal (Dugatkin and Bekoff, 2003), and this is achieved by the participants following rules of competition that involve some form of restraint during play fights (Pellis and Pellis, 1998b) – a restraint that is not present during serious fighting (Geist, 1978; Pellis, 1997). Thus, play fighting, unlike serious fighting, has a cooperative component as well as a competitive component (Pellis and Pellis, 1998b), although the degree of competitiveness can vary across species (Biben, 1998; Thompson, 1998; Bauer and Smuts, 2007).

Some species, such as rats, organise play fighting in a way that ensures that when attacking a play target, the attacker does so without simultaneously defending against the partner's retaliation (Pellis *et al.*, 2005). Other species, such as the degu, a South American rodent, organise their play fighting in a way in which attack is coupled with defence, which increases the likelihood of successfully gaining an advantage over the partner. In play fighting, restraint is shown once the tactic succeeds, where the winner does not take advantage of its success; this allows the loser to regain its stability. In contrast, in serious fighting, the successful degu will take advantage of its off-balance partner and lunge to deliver a bite (Pellis *et al.*, 2010). That is, irrespective of the species-typical manner in which reciprocity is integrated into play fighting (Dugatkin and Bekoff, 2003), all species that engage in play fighting incorporate an element of cooperation into the competition for accessing play targets (Pellis and Pellis, 2009).

Studies of play fighting in various species of monkeys (Pellis and Pellis, 1997; Biben, 1998) suggest that they are more like rats, in that they tend to

Box 1. Play fighting in Japanese macaques   285

refrain from incorporating defensive measures in their attacks and use tactics that facilitate counterattacks by the partner (Pellis and Pellis, 1998b). Viewed from this perspective, it would be expected that, in play fighting, Japanese macaques would show restraint in their execution of the tactics of offence and defence. Although we will show that this expectation is partly met, it is also critical to understand that such restraint is a 'matter of degree' in that different species show quantitative and qualitative differences in the cooperative aspect of play fighting (see above). As Japanese macaques are no exception to this, it is necessary to examine the play fighting of this species within the context of related species, to determine what aspects of its play are unique and which are variations on a broader theme.

14 Box 1.2 Play fighting in macaques

The play fighting of Japanese macaques is comparable to that described in other species of macaques (Symons, 1978), and, indeed, that of other Old World monkeys (Emory, 1975; Owens, 1975; Pellis and Pellis, 1997). Typically, among Japanese macaques, other macaques and many Old World monkeys, play fighting involves attack and defence of the upper arm, shoulder and the lateral edge of the neck, which if contacted is bitten, albeit gently (Reinhart *et al.*, 2010). The defender uses manoeuvres to block access to the play target and then to launch its own attack (Figure 14 Box 1.1). While macaques and other Old World monkeys appear to use a common suite of defensive tactics, there are species differences, with some tactics being more frequently used by certain species than others. For example, gelada baboons and mandrills are more likely to rear upright onto their hind legs and use their arms to grapple with their partners (Emory, 1975) than are baboons, which, while often doing the same thing, will also frequently roll over onto their backs (Cheney, 1978).

Similarly, there are many species differences in play fighting across macaques, such as differences in the frequency with which adults play not only with juveniles, but also with other adults. Among juvenile macaques, there are differences in the degree to which dominance and familial relationships influence the selection of play partners (Caine and Mitchell, 1979). For example, in Japanese macaques, juveniles preferentially play with partners from the same matriline and of comparable social status (Koyama, 1985). Even with regard to the attack and defence behaviour during play fighting itself, there are subtle species differences (Thierry *et al.*, 2000). For example, compared with macaque species from Sulawesi, such as the crested macaque and the Tonkean macaque, Japanese macaques tend to be more competitive in their play fighting (Petit *et al.*, 2008). One spectacular difference is that the Sulawesi species

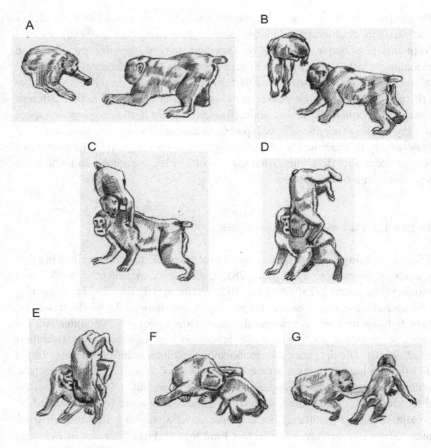

Figure 14 Box 1.1. A sequence of play fighting by two juvenile Japanese macaques drawn from videotaped records shows a complete encounter starting from the face-to-face orientation (A). The smaller individual (on the left) attacks by jumping up and twisting its body (B) while orienting to the side of the defender's neck (C). As the attacker moves through the air, the defender rotates to face its attacker and delivers a retaliatory bite to its neck (C–F). As the original attacker lands on the ground, the defender continues to bite the back of its attacker's neck (F) until the attacker swerves away (G). Note that both the attacker and the defender are targeting the neck/shoulder/upper arm region of the body. (From Reinhart *et al.*, 2010. Targets and tactics of play fighting: Competitive *versus* cooperative styles of play in Japanese and Tonkean macaques. *International Journal of Comparative Psychology*, **23**, 166–200. © International Society for Comparative Psychology, reprinted with permission.)

frequently engage in play fights involving three or more participants and these multi-animal interactions break the pattern typical of dyadic play fighting. That is, they bite their partners' bodies more opportunistically and do not adopt the typical defensive tactics when playing with one partner. In contrast, in

Box 1. Play fighting in Japanese macaques   287

Japanese macaques, multi-animal interactions, when they occur, resemble the targets and tactics of dyadic play fights (Reinhart *et al.*, 2010).

Even dyadic play fights differ between Japanese macaques and the Sulawesi macaques – the play fights of Tonkean macaques, for instance, involve rolling over onto the back as a defensive tactic more often, which leads to longer-lasting interactions. Moreover, when defending themselves, retaliatory bites by many species of monkeys are often directed at the side of the face as opposed to the play target (i.e. shoulder area) (Pellis and Pellis, 1997). The relative use of the side of the face bites versus play target bites can be used as an index of the degree of competition in play fights (Pellis and Pellis, 1998a). Japanese macaques have a higher proportion of the side of the face bites both during initiating and retaliatory bites than do Tonkean macaques (Reinhart *et al.*, 2010).

Thus, even though they use the same targets and tactics, Japanese macaques and Tonkean macaques differ in some of the ways that those targets and tactics are used, with the main difference being that Tonkean macaques tend to emphasise *cooperation* relative to Japanese macaques and Japanese macaques tend to emphasise *competition* relative to Tonkean macaques (Reinhart *et al.*, 2010; see also Petit *et al.*, 2008). As indicated above, this contrast is most strikingly brought out in the multi-animal play fights so typical of the Sulawesi species. For these, the multi-animal play fights involve animals piling onto one another, often lying on their backs on top of other animals and grabbing and biting whatever limbs or other body parts are accessible – in these instances, the whole competitive component of play fighting seems to have evaporated (Reinhart *et al.*, 2010). It is important to note that, whatever the origin of the troops observed, these species-typical patterns persist, and so Japanese macaques from Arashiyama show the same behaviour as those derived from other populations (Petit *et al.*, 2008; Reinhart *et al.*, 2010). Given the consistency within the species, what may account for these species-level differences?

14 Box 1.3 Macaque social systems

A clue to answering this question arises from the finding that some other monkey species, such as the rhesus macaques, are more similar in their play fighting to Japanese macaques than they are to the Sulawesi species (Symons, 1978). Similarly, while detailed studies are yet to be conducted, the few comparative data available (see Caine and Mitchell, 1979, for a review) seem to show that some other species, such as pig-tailed macaques, have play that is more reminiscent of that of Japanese and rhesus macaques (Evans, 1967), whereas others, such as stump-tailed macaques, have play that appears more like that of

Sulawesi macaques (Rhine, 1973). A common thread in these species differences in play relates to species differences in their social systems.

The genus *Macaca* comprises about 20 species that cluster into patterns of social organisation, ranging from 'despotic' to 'egalitarian' (Matsumura, 1999; Thierry, 2000). Furthermore, these differences in social styles reflect species differences in temperament that influence a wide range of behaviours, including dominance relationships (Thierry, 2000, 2004). Thus, while all these species are female philopatric, with female dominance hierarchies within and between matrilines and a separate dominance hierarchy for the males, there is considerable variation in how rigidly these hierarchies are enforced. For example, in the more despotic species, within troop disputes are more likely to lead to biting by the more dominant partner, whereas for the more egalitarian species, not only is biting less frequent, but it is also more likely to be reciprocated by the recipient (Thierry, 1985). It should, however, be borne in mind that the 20 species fall on a gradient ranging from more tolerant (egalitarian) to more intolerant (despotic) that probably reflects small changes in traits rather than there being distinct categories (Thierry, 2007), so it is more accurate to consider species as being more or less tolerant. Such differences in tolerance influence a variety of traits such as dominance (Thierry, 1985), the likelihood of reconciliation after conflict (Thierry *et al.*, 2008) and play fighting (Thierry *et al.*, 2000).

With regard to social tolerance, Japanese macaques are at the far end of the gradient, and so, along with the closely related rhesus macaques form a cluster of those macaques that show the least tolerance. In contrast, the Sulawesi macaques are in the cluster formed of the most tolerant macaques. Therefore, it may not be surprising that the play fighting of Japanese macaques tends to be more competitive than that of the Sulawesi species (Petit *et al.*, 2008; Reinhart *et al.*, 2010) – the species-typical social system influences the balance of competition and cooperation in play fighting.

An underlying variable that differs along the tolerance–intolerance gradient is temperament, which results in a number of psychological and social traits co-varying (Thierry, 2004), with the degree of competition–cooperation in play fighting being one of those traits (Petit *et al.*, 2008; Reinhart *et al.*, 2010). Thus, species-typical temperaments may explain a large amount of the behavioural variation across closely related species (Clarke and Boinski, 1995; Capitanio, 2004), and these temperament differences, even though of large consequence, may themselves arise from small genetic differences that influence the brain mechanisms involved in impulse control (Barr *et al.*, 2003; Wendland *et al.*, 2006).

That temperamental differences among species may account for differences in social traits, such as play fighting style, suggests that it may be unproductive to seek adaptive explanations for species differences in play (Thierry, 2005).

Box 1. Play fighting in Japanese macaques   289

Indeed, the degree of relatedness among species seems to be a better predictor of play style than differences in mating systems and ecology (Thierry *et al.*, 2000; Thierry, 2004). That is, the ancestors of Japanese macaques are more important in accounting for the species-typical pattern of competition and cooperation in play fighting than are current ecological factors. However, there are other peculiarities of play in Japanese macaques that may well require more specific explanations that are related to the unique history and geography of Japanese macaques.

14 Box 1.4 Sex rears its head

Japanese macaques have some features of play fighting that are not shared with species with similar social systems. Compared with other macaques, whether from the intolerant end of the spectrum like themselves, or from the more tolerant end of the distribution like Tonkean macaques, Japanese macaques incorporate mounting in their play to an exaggerated degree (see VanderLaan *et al.*: chapter 11). Male–male mounting is prevalent in the social interactions of Japanese macaques (Hanby, 1974), and this prevalence extends to stages of development prior to sexual maturity (Hanby and Brown, 1974). Mounting also occurs at an exaggerated prevalence among adult females in this species (Vasey, 1998), and so, compared with other species of macaques, Japanese macaques appear to be hyper-sexualised (Vasey, 2006). Because of the sexual idiosyncrasies of Japanese macaques, they may have had a unique opportunity to put elements of sexual behaviour, such as mounting, to novel uses in play fighting. Indeed, it would appear that, in juvenile males, mounting is used in several ways to facilitate playful contact and to promote prolonged wrestling when initiating play fighting (see VanderLaan *et al.*: chapter 11).

These sexual oddities that pervade play cannot be accounted for by clade-specific patterns of temperament, as is the case for the relative balance of competition and cooperation in play fighting (see above). Rather, the questions that arise are whether these sexual peculiarities in the play fighting of Japanese macaques are by-products of the hyper-sexual social behaviour of this species, or whether they are merely an exaggeration of what occurs in other species of macaques, and are simply more evident and empirically accessible in Japanese macaques because of that exaggeration. Finally, while the competitive–cooperative balance in play fighting is consistent across troops of Japanese macaques irrespective of their origin (Reinhart *et al.*, 2010), it remains to be determined whether the involvement of mounting in play, especially with regard to its play-promoting properties, is present across all populations of this species. With regard to the female homosexual behaviour, it does appear to be more prevalent in monkeys from Arashiyama and troops derived from that population than among other populations of Japanese macaques (Vasey and Jiskoot, 2010), although the full range of variation in homosexual behaviour has yet to be fully described (Vasey and Reinhart, 2009). The issue thus becomes one of determining whether the play-promoting mounting of Japanese macaques is unique to this species, and, if so, whether the monkeys from Arashiyama are unique among this species.

References

Aldis, O. (1975). *Play Fighting*. New York, NY: Academic Press.

Barr, C. S., Newman, T. K., Becker, M. L. *et al.* (2003). The utility of non-human primate models for studying gene by environment interactions in behavioral research. *Genes, Brain and Behavior*, **2**, 336–340.

Bauer, E. B. and Smuts, B. B. (2007). Cooperation and competition during dyadic play in domestic dogs, *Canis familiaris*. *Animal Behaviour*, **73**, 489–499.

Biben, M. (1998). Squirrel monkey play fighting: making the case for a cognitive training function for play. In *Animal Play: Evolutionary, Comparative, and Ecological Perspectives*, eds. M. Bekoff and J. A. Byers. Cambridge: Cambridge University Press, pp. 161–182.

Caine, N. and Mitchell, G. (1979). A review of play in the genus *Macaca*: social correlates. *Primates*, **20**, 535–546.

Capitanio, J. P. (2004). Personality factors between and within species. In *Macaque Societies: A Model for the Study of Social Organization*, eds. B. Thierry, M. Singh, and W. Kaufmanns. Cambridge: Cambridge University Press, pp. 13–33.

Cheney, D. L. (1978). The play patterns of immature baboons. *Animal Behaviour*, **26**, 1038–1050.

Clarke, A. S. and Boinski, S. (1995). Temperament in nonhuman primates. *American Journal of Primatology*, **37**, 103–125.

Dugatkin, L. A. and Bekoff, M. (2003). Play and the evolution of fairness: a game theory model. *Behavioural Processes*, **60**, 209–214.

Emory, G. R. (1975). The patterns of interaction between the young males and group members in captive groups of *Mandrillus sphinx* and *Theropithecus gelada*. *Primates*, **16**, 317–334.

Evans, C. S. (1967). Methods for rearing and social interaction in *Macaca nemestrina*. *Animal Behaviour*, **15**, 263–266.

Geist, V. (1978). On weapons, combat and ecology. In *Advances in the Study of Communication and Affect. Volume 4, Aggression, Dominance and Individual Spacing*, eds. L. Krames, P. Pliner and T. Alloway. New York: Plenum Press, pp. 1–30.

Hanby, J. P. (1974). Male-male mounting in Japanese monkeys (*Macaca fuscata*). *Animal Behaviour*, **22**, 836–849.

Hanby, J. P. and Brown, C. E. (1974). The development of sociosexual behaviours in Japanese macaques (*Macaca fuscata*). *Behaviour*, **49**, 152–196.

Box 1. Play fighting in Japanese macaques 291

Koyama, N. (1985). Playmate relationships among individuals of the Japanese monkey troop in Arashiyama. *Primates*, **26**, 390–406.

Owens, N. W. (1975). Social play behaviour in free-living baboons, *Papio anubis*. *Animal Behaviour*, **23**, 387–408.

Matsumura, S. (1999). The evolution of 'egalitarian' and 'despotic' social systems among macaques. *Primates*, **40**, 23–31.

Pellis, S. M. (1988). Agonistic versus amicable targets of attack and defense: consequences for the origin, function, and descriptive classification of play-fighting. *Aggressive Behavior*, **14**, 85–104.

(1997). Targets and tactics: The analysis of moment-to-moment decision making in animal combat. *Aggressive Behavior*, **23**, 107–129.

Pellis, S. M. and Pellis, V. C. (1997). Targets, tactics, and the open mouth face during play fighting in three species of primates. *Aggressive Behavior*, **23**, 41–57.

(1998a). Play fighting of rats in comparative perspective: a schema for neurobehavioral analyses. *Neuroscience & Biobehavioral Reviews*, **23**, 87–101.

(1998b). The structure-function interface in the analysis of play fighting. In *Animal Play: Evolutionary, Comparative, and Ecological Perspectives*, eds. M. Bekoff and J. A. Byers. Cambridge: Cambridge University Press, pp. 115–140.

(2009). *The Playful Brain. Venturing to the Limits of Neuroscience*. Oxford: Oneworld Press.

Pellis, S. M., Pellis, V. C. and Foroud, A. (2005). Play fighting: aggression, affiliation and the development of nuanced social skills. In *Developmental Origins of Aggression*, eds. R. Tremblay, W. W. Hartup and J. Archer. New York, NY: Guilford Press, pp. 47–62.

Pellis, S. M., Pellis, V. C. and Reinhart, C. J. (2010). The evolution of social play. In *Formative Experiences: The Interaction of Caregiving, Culture, and Developmental Psychobiology*, eds. C. Worthman, P. Plotskyn and D. Schechter. Cambridge: Cambridge University Press, pp. 406–433.

Petit, O., Bertrand, F. and Thierry, B. (2008). Social play in crested and Japanese macaques: testing the covariation hypothesis. *Developmental Psychobiology*, **50**, 399–407.

Symons, D. (1978). *Play and Aggression: A Study of Rhesus Monkeys*. New York, NY: Columbia University Press.

Reinhart, C. J., Pellis, V. C., Thierry, B. *et al.* (2010). Targets and tactics of play fighting: competitive *versus* cooperative styles of play in Japanese and Tonkean macaques. *International Journal of Comparative Psychology*, **23**, 166–200.

Rhine, R. J. (1973). Variation and consistency in the social behavior of two groups of stumptail macaques (*Macaca arctoides*). *Primates*, **14**, 21–35.

Thierry, B. (1985). Patterns of agonistic interactions in three species of macaque (*Macaca mulatta, M. fascicularis, M. tonkeana*). *Aggressive Behavior*, **11**, 223–233.

(2000). Covariation of conflict management patterns across macaque species. In *Natural Conflict Resolution*, eds. F. Aureli and F. B. M. de Waal. Berkeley, CA: University of California Press, pp. 106–128.

(2004). Social epigenesis. In *Macaque Societies: A Model for the Study of Social Organization*, eds. B. Thierry, M. Singh and W. Kaufmanns. Cambridge: Cambridge University Press, pp. 267–290.

(2005). Integrating proximate and ultimate causation: just one more go! *Current Science*, **89**, 1180–1183.

(2007). Unity in diversity: lessons from macaque societies. *Evolutionary Anthropology*, **16**, 224–238.

Thierry, B., Iwaniuk, A. N. and Pellis, S. M. (2000). The influence of phylogeny on the social behavior of macaques (Primates: Cercopithecidae, genus *Macaca*). *Ethology*, **106**, 713–728.

Thierry, B., Aureli, F., Nunn, C. L., Petit, O., Abegg, C. and de Waal, F. B. M. (2008). A comparative study of conflict resolution in macaques: Insights into the nature of trait co-variation. *Animal Behaviour*, **75**, 847–860.

Thompson, K. V. (1998). Self assessment in juvenile play. In *Animal Play: Evolutionary, Comparative, and Ecological Perspectives*, eds. M. Bekoff and J. A. Byers. Cambridge: Cambridge University Press, pp. 183–204.

Vasey, P. L. (1998). Female choice and inter-sexual competition for female sexual partners in Japanese macaques. *Behaviour*, **135**, 579–597.

(2006). The pursuit of pleasure: Homosexual behaviour, sexual reward and evolutionary history in Japanese macaques. In *Homosexual Behaviour in Animals: An Evolutionary Perspective*, eds. V. Sommer and P. L. Vasey. Cambridge: Cambridge University Press, pp. 191–219.

Vasey, P. L. and Jiskoot, H. (2010). The biogeography and evolution of female homosexual behavior in Japanese macaques. *Archives of Sexual Behavior*, **39**, 1439–1441.

Vasey, P. L. and Reinhart, C. J. (2009). Female homosexual behavior in a new group of Japanese macaques: evolutionary implications. *Laboratory Primate Newsletter*, **48**, 8–10.

Wendland, J. R., Lesch, K. P., Newman, T. K. *et al.* (2006). Differential functional variability of serotonin transporter and monoamine oxidase A genes in macaque species displaying contrasting levels of aggression-related behavior. *Behavior Genetics*, **36**, 163–172.

14 Box essay 2 *Eye-covering play in Japanese macaques and orangutans*

ANNE E. RUSSON AND PAUL L. VASEY

14 Box 2.1 Introduction

Eye-covering play, deliberately closing or covering one's eyes during a play sequence, occurs in various non-human primates. Some suggest it involves pretending to be blind – acting *as if* one can't see, doesn't exist or exists in some altered form (Hahn, 1982; de Waal, 1986). However, only one systematic study has been made of eye-covering play in non-human primates (Thierry, 1984) and no systematic attempts have been made to investigate the mental processes involved. We systematically studied eye-covering play in captive Japanese macaques (*Macaca fuscata*) and orangutans (*Pongo abelii*) to assess its relation to pretending.

Eye-covering play has been reported in all great apes and several other primates, including gibbons, guenons (*Cercopithecus nictitans, C. pogonias, C. neglectus*), mangabeys (*Cercocebus albigenal, C. galeritus*), talapoins (*Miopithecus talapoin*), rhesus macaques (*Macaca mulatta*), olive baboons (*Papio anubis*), brown capuchins (*Cebus apella*) and douc langurs (*Pygathrix nemaeus*) (Harrison, 1962; Gautier-Hion, 1971; Lawick-Goodall, 1971; Kavanagh, 1978; Hahn, 1982; de Waal, 1986). It involves stumbling or staggering about in 'drunken' fashion with eyes closed or covered, sometimes also with arms spread out or groping about and occasionally using objects to cover the eyes. For example, douc langurs will sometimes keep their eyes closed while they move towards a potential play partner or during a play session. Although they were never harmed, this sometimes resulted in their falling or reaching out in the wrong place for support or for a conspecific (Kavanagh, 1978).

Whether eye-covering play involves pretence is not easily determined. The essence of pretence is deliberately behaving as if something were real when it is not, as if doing one thing while in fact doing something else, or projecting an imagined situation onto an actual one (Lillard, 1993; Mitchell, 1994). It requires simultaneously entertaining *two* representations of a situation, real versus imagined, deliberately and without confusion, plus recognising and 'marking' the imagined representation as different from the real one (Leslie, 1987; Lillard, 1993). Some consider symbolic 'metarepresentational' cognition essential (e.g. Leslie, 1987), where metarepresentation means re-representing

293

existing representations; others see a graded range of simpler forms guided by pre-symbolic cognition (e.g. McCune, 1995).

Cognitive criteria are especially interesting because they may differentiate primate species in their capacity for pretending. The simplest metarepresentations are secondary representations that re-represent primary (sensory–motor, or 'real') representations of the world. Secondary representations may represent great apes' peak cognitive achievements and distinguish their cognition from that of other non-human primates (e.g. Parker and McKinney, 1999; Russon, 2004). A few cases of language-trained great apes playing with non-existent objects appear to satisfy criteria for meta-representational level pretending (e.g. pulling an invisible pull toy; Hayes, 1951). We know of no accepted evidence for pretending at this level in other non-human primates.

We undertook a systematic study of spontaneous eye-covering play in captive Japanese macaques and orangutans to assess what cognitive processes may be involved, whether macaque eye-covering play qualifies as pretence and how it compares to orangutan eye-covering.

14 Box 2.2 Methods

14 Box 2.2.1 Subjects and settings

Japanese macaque subjects were housed at the University of Montréal's Laboratory of Behavioural Primatology, in a mixed-sex group of 23 adults (18 females, five males) and 14 immatures (seven females, seven males) (for details see Vasey, 1998). The group's founding members originated from the Arashiyama West colony of Japanese macaques that were translocated from Arashiyama, Japan, to southern Texas in 1972 (Fedigan and Asquith, 1991). Their enclosure was furnished with climbing and swinging devices plus loose enrichment items like coconut shells and pails. Orangutan subjects lived in a mixed-sex group at the Toronto Zoo; during observations in 1995, the group comprised three adults (two females, one male), three adolescents (two males, one female), a juvenile male and a 2-year-old infant female; the same individuals were observed in 1998. Their enclosure was an open area surrounded by a moat and plexiglass walls, and furnished with climbing structures, and changing arrays of ropes, nets, logs and loose behavioural enrichment items (e.g. boxes, clothes, scoops).

14 Box 2.2.2 Sampling and data collection

We defined *eye-covering play* as an actor deliberately covering their eyes as part of a play sequence, where *play* was intentional activity performed for

Box 2. Eye-covering play in Japanese macaques and orangutans 295

its own sake, for amusement or to simulate another end-directed activity for benign ends (Mitchell, 1990). In eye-covering play, *events* were periods when eyes were continuously covered and *bouts* were sequences of eye-covering events that were related but separated by brief peeking (i.e. uncovering eyes for a few seconds). For each event we recorded actor (name/age/sex), component behaviours and play type (stationary/locomotory/positional, arboreal/terres-trial, social/non-social). When we detected bouts, we also recorded behaviour during intervening eyes-uncovered periods.

For Japanese macaques, we recorded all spontaneous eye-covering play during observational sessions for other projects from March 1993 to March 1995 (1264 hours). Eye-covering was rare, so more systematic procedures were not feasible. For orangutans, we recorded all eye-covering play during weekly 2-hour observational sessions (1) from October 1995 to March 1996 (*c.* 50 hours, minimum six events per orangutan: Hickerson, 1996) and (2) from February to July 1998 (*c.* 15 hours, focused on two females who engaged in eye-covering play in that period, Ramai, 12 years, and Sekali, 5 years). We facilitated orangutan eye-covering by providing several items of the sort they used spontaneously to cover their eyes (e.g. cloth, containers) at the start of each session.

14 Box 2.3 Results

In Japanese macaques, eye-covering play was rare (n = 75 events/69 bouts by 13 monkeys: 3.1 bouts/month; mean = 5.8 events/actor, range 1–14), brief (mean = 4.4 seconds, range 1–20), solitary more than social (84% versus 16%), terrestrial more than arboreal (56% versus 44%), seen only in juve-niles 1–3.5 years old (82.7%) and adolescents 3.5–4.5 years old (17.3%), and more frequent in males (n = 3, mean = 7 events/male) than females (n = 10, mean = 5.4 events/female). Types of eye-covering play included simple locomotion (walk, run), positional behaviour (spin, hang from swing, somer-sault), object manipulation (rub surfaces, manual exploration) and sit (alone and inactive).

In orangutans, eye-covering play was frequent (1995/6: 100 events/50 hours; 1998: 71 events/15 hours; Ramai: 38 events, 2.73 events/hour; Sekali: 33 events, 2.2 events/hour), perhaps due to facilitation. In 1998, eye-covering events were brief (mean = 12 seconds, range 1–45) but many occurred within longer bouts (mean = 41.2 seconds, range 2–199), solitary more than social (1995/6: 79% versus 21%; Hickerson, 1996; 1998: 97% versus 3%), and arboreal more than terrestrial (1998: 100% arboreal). In 1995, all orangutans performed eye-covering play but juveniles and adolescents did so more than infants or adults (Hickerson, 1996). Types of eye-covering play included arboreal locomotion, positional behaviour (spin, hang from bar or rope), wrestle, chase, object manipulation and sit (social or solitary, 'hiding').

Eye-covering play did not suggest a unified purpose or cognitive processing for either species, so we analysed only events that involved complex behaviour or 'cheating' (deliberately using vision or touch to obtain extra information in the midst of an eye-covering bout). Both species cheated by seeking extra-sensory information when they could not continue their activity without breaking their apparent 'rule' for the game, guiding behaviour from memory. Both used two forms of cheating: *peek* (deliberately uncover eyes for a few seconds, look around, then re-cover them) and *grope* (feel around for an object, or fumbling manipulation). Cheating showed that their vision was impaired when their eyes were covered but they were probably not blind. It also offered hints of the cognitive processes guiding eye-covering play, e.g. when and why actors used stored versus experiential representations and whether higher-level cognitive processes might be involved (e.g. if cheating occurred in anticipation of difficulties).

14 Box 2.4 Cheating as a cognitive indicator

Japanese macaques peeked and/or groped in 16 eye-covering events (21.3%). All were females (n = 8); they did so mostly in non-social contexts (87.5%), arboreally versus terrestrially (68.7% versus 31.3%) and while locomoting versus stationary (75% versus 25%). Locomotion was along continuous structures in all but one event, when a female covered her head with a pail and ran towards a fence; when near the fence, she paused, groped for the fence, grasped it, and climbed a few feet before uncovering her eyes.

Seven females groped in nine eye-covering events for objects (e.g. platform edge, swing) or conspecifics. They did not sweep their hands widely through space as if they had no sense of what was there; instead, their groping seemed directed. In five of the nine cases, the actor positioned herself before covering her eyes so that she was in proximity to and oriented towards a specific object or individual; then, still stationary, she covered her eyes and groped in a directed manner towards that target. In all five cases, groping appeared to be the aim of the eye-covering play. One female positioned herself beside a tyre-swing before covering her eyes with a deflated plastic ball and then groped for the swing with both hands in a well-directed manner. In another case, a nulliparous female tried repeatedly to touch a newborn with her eyes open but failed because the mother moved to avoid contact. She then tried a new tactic, covering her eyes and groping for the newborn (the mother still avoided contact). In the other four cases, groping was one component of a behavioural sequence: actors covered their eyes while travelling then groped for nearby objects (e.g. fence, wall, platform edge), apparently to avoid travel errors. Twice they slowed and moved more cautiously as they neared barriers, then groped for the barrier with an outstretched arm, and twice they groped along the edge of a continuous arboreal structure while travelling on it.

Four females peeked in four eye-covering play bouts, all while travelling on continuous substrates. In three cases, they appeared to peek to avoid errors. One covered her face with a coconut shell that blocked all but her periph- eral vision then travelled on an arboreal platform while repeatedly turning her head to the side. This way, she could peek in the direction she was moving and correct her position. In the fourth case, a female appeared to be checking her position in space. She covered her eyes, spun rapidly in circles, stopped, uncovered her eyes, looked around, covered her eyes again and spun for a few more seconds.

None of the macaques cheated to correct errors. On four occasions females made mistakes while travelling blind by bumping into objects like walls, swing- ing doors or poles. Each time, they simply uncovered their eyes and stopped locomoting.

Orangutans were assessed from 24 events of 'blind' travel, i.e. travel with eyes covered (34% of 1998 events; Ramai: 14/24, Sekali: 10/24). These two females typically travelled blind in their climbing structures, making several complex transfers through a discontinuous array of arboreal supports (poles, ropes, etc.). Ramai once climbed blind for 15 seconds through a complex tan- gle of bars, poles and ropes, along a route that required three transfers between discontinuous structures; the only hints of difficulty were occasional slight hesitations. Despite the complex trajectories, their errors were minor (e.g. a fumbling grab for a support); they never backtracked from impassible or awk- ward gaps or grabbed then dropped unsuitable supports.

These females groped in 14/24 blind travel events (Ramai: 9, Sekali: 5). Like the macaques, they never groped as if they had no sense of what was there. Their groping was always one component of a complex behavioural sequence rather than the focus of their eye-covering play. Both *only* groped when travel- ling across discontinuous or unstable supports, e.g. transferring onto ropes or curved poles, crossing a junction (e.g. log to platform), or approaching obstruc- tions (e.g. a rope blocking travel over a straight log). They *never* groped in transferring from one straight pole to another or travelling on continuous sub- strates. They groped to correct (Ramai: 5/9, Sekali: 4/5) as well as to prevent small errors (Ramai: 5/9, Sekali: 1/5). To prevent errors, they typically slowed locomotion as they neared a discontinuity, reached directly but slowly toward the next support object, then cautiously grabbed hold. Ramai, for example, approached a rope that dangled across her path while walking blind on a log. Before contacting the rope, she paused her stride and *then* reached towards the rope; only *after* touching the rope did she grope a little to grab it, and immediately *after* grabbing it she resumed her stride. To correct errors, they groped *after* their reach missed its mark to correct reaching details. Once, as Ramai was climbing blind up a straight pole, her uppermost hand bumped a rope and a log attached to the pole; she then groped around the obstructions for the pole, grabbed the pole just above the rope, and resumed climbing.

Ramai and Sekali peeked in 12/24 eye-covering play events (Ramai: 8, Sekali: 4). They peeked just *before* attempting a difficult transfer (Ramai: 6/8 cases, Sekali: 2/4 cases), i.e. they anticipated an upcoming transfer difficulty while blind, obtained new visual information, then transferred without error. Sekali also peeked just *after* groping in 2/4 of the events in which she peeked. Sometimes, when Ramai and Sekali peeked, they seemed to look at a specific location in the enclosure and then reoriented themselves in that direction; when they re-covered their eyes and resumed travelling, they travelled until they reached the location they had just looked at and oriented towards.

14 Box 2.5 Discussion

Our aim was to assess whether eye-covering play involved pretence. If pretence means acting on mental images deliberately at odds with reality, the Japanese macaques and orangutans were probably not pretending in their eye-covering play. In blind travel, especially climbing, acting on imaginary features of space is dangerous. Their eye-covering play did not appear to concern pretending about a self that cannot see or exists in another state: they *really* could not see and *really* existed in another state. Eye-covering play could involve pretending not to exist if it reduces stimulation or contact, as when children think they can't be seen if they can't see, but most of our events involved stimulation seeking where the point is that actors *do* exist, not that they don't.

To explore pretence in cognitive perspective, we assessed whether eye-covering play involved: two mental representations of the situation held simultaneously and without confusion; an imagined mental representation, marked as such, that directed eye-covering play; and metarepresentation.

Both species showed signs of two mental representations of space during eye-covering play. They probably had stored primary or 'real' representations of stable features of their enclosures (structures, objects). They also took in sensory information from their experience of moving about blind; whether it produced a new primary representation, a perceptual representation, or was simply assimilated to the stored representation is unclear. What is clear, from

Box 2. Eye-covering play in Japanese macaques and orangutans 299

their errors, is that the stored and immediate representations could differ and both species recognised discrepancies: they cheated to avoid making errors at points of reduced predictability and seemed to enjoy playing with this discrepancy (electing to travel blind). At least at error-prone points, stored and immediate representations must have co-existed simultaneously and without confusion. The Japanese macaques never corrected their errors on the spot, suggesting limited abilities to manipulate the discrepancy voluntarily.

The stored representations could be considered 'imagined' in that they were retrieved from memory versus generated from immediate experience, but we saw no signs that they deliberately flouted real-world rules. They must have been marked as distinct from immediate experiential representation because subjects voluntarily chose which to use. During eye-covering they chose to use the imagined representation to guide their movement, deliberately blocking the visual input that would normally enhance it, but then chose the experiential representation (by peeking or groping) when the imagined representation was probably imprecise. They also appeared to check the match between representations: Japanese macaques, for example, peeked after blind travel as if checking whether their actual position matched their imagined one. That they distinguished imagined and experiential representations is shown by the fact that they were reasonably good at choosing when to use each: they managed to travel and anticipate the position of objects using an imagined representation, and switched to the experiential representation when they anticipated difficulty or to check their success.

Whether metarepresentation was involved was suggested by several indicators. The macaques did not seem to plan destinations and travel routes in advance, unless groping at a wall or fence qualifies as a travel goal and moving a few metres in a straight line qualifies as a route. They positioned themselves relative to a target before eye-covering and groped from a stationary position, suggesting some attempt to 'see ahead' but not planning routes. Their trajectories and the travel problems they tackled blind were also simple and accordingly, their solutions were of low complexity; they often peeked and groped as an end in itself, not to handle problems that arose during eye-covering; and they did not correct their travel errors 'on line' but instead stopped dead.

These features are important because planning travel routes can require more than primary representations when it involves the relations between objects in space. Spatial cognition about object–object relations is considered to require secondary representation, i.e. rudimentary metarepresentation (Case, 1985; Langer, 1996; Byrne and Russon, 1998). Some of the orangutan evidence supports this interpretation over the alternative of routes defined by primary representations generated from past travel experiences, including: *handling errors online* then resuming travel along the same trajectory, using peeking and groping as *tactics in the service of* elaborate eye-covering travel, and *facultative use of* cheating tactics (i.e. peeking and groping in one bout) all suggest hierarchisation, the generative cognitive process considered responsible for metarepresentation. Japanese macaques' eye-covering play did not show these features; instead, it more closely resembles the linear, chained organisation characteristic of primary representations and association-based cognitive processes (e.g. Langer, 1996; Byrne & Russon, 1998).

Other patterns in the Japanese macaques' eye-covering play suggest cognitive limitations. Relatively little of their blind travel was arboreal or through discontinuous structures, so the routing and locomotor problems they chose were relatively simple. Their eye-covering events and bouts were relatively short, suggesting they were unable to sustain the mental images without frequently refreshing them with new experiential input. They peeked to avoid errors but rarely, and over very short distances in straight line arboreal travel. They seemed to have difficulty following a straight line with their eyes covered, drifting to the edge of the structure. They peeked after coming near enough to the edge to feel it with their hands so they handled difficulties by detecting them at an early stage experientially rather than by anticipating them using a stored representation.

The cognitive limitations suggested by Japanese macaques' eye-covering play stand out when compared with the orangutans' eye-covering play. The orangutans similarly showed signs of maintaining two distinct representations of space, but were more proficient in handling discrepancies between them (e.g. predicting object locations from imagined representations versus detecting them early, switching to experiential representations in anticipation of an upcoming difficulty). The orangutans also showed evidence of planning specific destinations and travel routes in advance, planning more complex trajectories for blind travel and anticipating problems during blind travel – all suggesting that mental representations guided their travel routes. They also peeked and groped to solve problems 'on line' along their trajectory.

In sum, our findings suggest that Japanese macaques' and orangutans' eye-covering play did not involve pretence. It did not flout the constraints of reality and did not suggest the requisite cognitive processes. It could, however, involve relatively sophisticated cognition. Both the macaques and the orangutans seemed to play with discrepancies between several mental representations of the world, to recognise and test differences between their stored representation and the representation generated by the experience of acting blind, and to have voluntary control over which representation to use. This could be one way in which actors discover the possibility of pretence – by discovering the fallibility of mental representations when tested against the real world. In the macaques, all representations were probably primary but in the orangutans,

Box 2. Eye-covering play in Japanese macaques and orangutans 301

secondary representations also appeared to have been involved. This raises the question of what other cognitive exercises might be involved.

Pretending is only one of several ways of playing with normally occurring discrepancies among multiple representations of a situation. Secondary representations of a situation, for example, are derived from relevant primary representations but are not tied to real-world constraints in the same way. In pretence, actors voluntarily play with that distinction in a particular way: they elaborate and behave in line with a secondary representation so as to deliberately flout rules of the primary representation, the situation 'as it really is' (Leslie, 1987). Another way of playing with these discrepancies is to reduce them, to come closer in the mind to meeting real constraints. The difference could be likened to that between representational and abstract painters or scientific and literary writers: both sides generate images inspired by 'reality' and both recognise discrepancies between what they imagine and reality. One side aims to come closer to reality and the other to escape it, but both could be seen as engaging in the same game of manipulating discrepancies between representations.

This second game is the one that both orangutans and macaques appeared to be playing. Their eye-covering play seemed to involve some element of imagination in that their actions were guided by images held in memory, so it could be considered a form of imaginative play. The orangutans, but probably not the Japanese macaques, appeared to be functioning at the same cognitive level involved in pretending, so they were engaging in an equally sophisticated form of imaginative play. All may have been playing with the awareness that when they can't see with their eyes, they can see with their minds.

Acknowledgements

P.L.V. thanks Bernard Chapais, Carole Gauthier, Annie Gautier-Hion and James G. Pfaus. A.E.R. thanks Pam Hickerson, Colleen O'Connell, Sheryl Parks and Juan-Carlos Gomez.

References

Byrne, R. W. and Russon, A. E. (1998). Learning by imitation: a hierarchical approach. *Behavioral and Brain Sciences*, **21**, 667–721.

Case, R. (1985). *Intellectual Development: Birth to Adulthood.* New York, NY: Academic Press.

de Waal, F. B. M. (1986). Imaginative bonobo games. *Zoonooz*, **59**, 6–10.

Fedigan, L. M. and Asquith, P. J. (1991). *The Monkeys of Arashiyama: Thirty-five Years of Research in Japan and the West.* Albany, NY: State University of New York Press.

Gautier-Hion, A. (1971). Répertoire comportemental du talapoin (*Miopithecus talapoin*). *Biologia Gabonica*, **7**, 295–391.

Hahn, E. (1982). Annals of Zoology: Gorillas – Part I. *New Yorker*, **58**, 39–62.

Harrison, B. (1962). *Orangutan*. Singapore: Oxford University Press.

Hayes, C. (1951). *The Ape in our House*. New York, NY: Harper & Row.

Hickerson, P. (1996). *Do Orang-utans Engage in Pretence?* Unpublished BA thesis, Glendon College, York University, Toronto, Canada.

Kavanagh, M. (1978). The social behaviour of doucs (*Pygathrix nemaeus nemaeus*) at San Diego Zoo. *Primates*, **19**, 101–114.

Langer, J. (1996). Heterochrony and the evolution of primate cognitive development. In *Reaching into Thought: The Minds of the Great Apes*, eds. A. E. Russon, K. A. Bard and S. T. Parker. Cambridge: Cambridge University Press, pp. 257–277.

Lawick-Goodall, J. (1971). *In the Shadow of Man*. New York, NY: Houghton Mifflin.

Leslie, A. M. (1987). Pretense and representation. the origins of theory of mind. *Psychological Review*, **94**, 412–426.

Lillard, A. S. (1993). Pretend play skills and the child's theory of mind. *Child Development*, **64**, 348–371.

McCune, L. (1995). A normative study of representational play at the transition to language. *Developmental Psychology*, **31**, 198–206.

Mitchell, R. W. (1990). A theory of play. In *Interpretation and Explanation in the Study of Animal Behavior. Volume 1: Interpretation, Intentionality and Communication*, eds. M. Bekoff and D. Jamieson. Boulder, CO: Westview Press, pp. 197–227.

   (1994). The evolution of primate cognition: simulation, self-knowledge, and knowledge of other minds. In *Hominid Culture in Primate Perspective*, eds. D. Quiatt and J. Itani. Boulder, CO: University Press of Colorado, pp. 177–232.

Parker, S. T. and McKinney, M. L. (1999). *Origins of Intelligence: The Evolution of Cognitive Development in Monkeys, Apes, and Humans*. Baltimore, MD: Johns Hopkins University Press.

Russon, A. E. (2004). Great ape cognitive systems. In *The Evolution of Great Ape Intelligence*, eds. A. E. Russon and D. R. Begun. Cambridge: Cambridge University Press, pp. 76–100.

Thierry, B. (1984). Descriptive and contextual analysis of eye-covering behavior in captive rhesus macaques (*Macaca mulatta*). *Primates*, **25**, 62–77.

Vasey, P. L. (1998). Female choice and inter-sexual competition for female sexual partners in Japanese macaques. *Behaviour*, **135**, 579–597.

15 Behavioural sequences involved in grooming interactions in adult female Japanese macaques: How do participants change roles and maintain interactions?

MARIKO FUJIMOTO

Two adult females engaged in a grooming interaction at Arashiyama (photo by M. Fujimoto).

The Monkeys of Stormy Mountain: 60 Years of Primatological Research on the Japanese Macaques of Arashiyama, eds. Jean-Baptiste Leca, Michael A. Huffman and Paul L. Vasey. Published by Cambridge University Press. © Cambridge University Press 2012.

15.1 Introduction

Social grooming represents the most common affiliative social behaviour in non-human primates, and numerous studies have examined this behaviour in a variety of primate species (reviewed in Goosen, 1987; Dunbar, 1996). Grooming provides a practical and beneficial way of removing ectoparasites (Tanaka and Takefushi, 1993) whereby the recipient (hereafter the 'groomee') receives a benefit in terms of hygiene, and the cost to the performer (hereafter the 'groomer') is the time and effort spent removing such parasites. Noë and Hammerstein (1994) proposed a 'biological market' theory wherein social grooming was interpreted as a tradeable commodity similar to a currency. Social grooming in non-human primates has also been studied from the perspective of reciprocal altruism (Seyfarth, 1977; Silk, 1982; Seyfarth and Cheney, 1984; Hemelrijk and Ek, 1991; Henzi and Barrett, 1999; Manson *et al.*, 2003; Schino *et al.*, 2003, 2005; Schino and Aureli, 2009, 2010) or on the measurement of existing affiliations between participants (Kaplan and Zucker, 1980; Cheney, 1992; Nakagawa, 1992; Cords, 1997). These studies have addressed the exchange of grooming for other social benefits, such as support in agonistic interactions (Schino and Aureli, 2009), allomothering behaviour (Muroyama, 1994) and co-feeding (Ventura *et al.*, 2006). These social grooming interactions are analysed 'at a group level' (Hemelrijk, 1990a, 1990b).

Earlier studies of social grooming in Japanese macaques showed that: (1) females engaged in grooming interactions more often than males (Kudo, 1986; Koyama, 1991); (2) females tended to choose kin-related females, especially dependent offspring, as grooming partners (Oki and Maeda, 1973; Koyama, 1991); and (3) among non-kin females, subordinates groomed dominants more often than the reverse (Sade, 1972; Oki and Maeda, 1973; Fairbanks, 1980; Sambrook *et al.*, 1995). In addition, Nakamichi and Shizawa (2003) found that some females groomed dominant partners more frequently, whereas others groomed subordinate partners more often, and, in other cases, some individuals groomed subordinates and dominants equally. They concluded that females in high-ranking and large kin-groups tended to choose related females as grooming partners, whereas females in middle- or low-ranking and smaller kin-groups tended to choose unrelated females as grooming partners, and their grooming interactions were directed both down and up the hierarchy. The presence of the latter 'egalitarian females' may contribute to strengthen female-bonded groups.

Several studies have focused on the behavioural patterns involved in social grooming on the 'dyadic interactional level' (Tsukahara, 1990; Muroyama, 1991a, 1991b, 1996; Dunbar, 1996). When Japanese macaques initiated social grooming, they often uttered a particular sound (Mori, 1975; Shizawa, 2001). In this species, Tsukahara (1990) reported that males tended to display solicitation behaviours toward females, whereas females tended to approach the alpha male to initiate grooming interactions. Muroyama (1991a, 1991b) reported that in affiliated pairs, one partner tended to approach and solicit grooming from the other, whereas in unaffiliated pairs, one partner tended to groom the other when approached. Only a few studies have focused on the patterns of role choice in grooming interactions, i.e. groomer versus groomee (Muroyama, 1996; Manson *et al.*, 2003; Frank and Silk, 2009).

Chimpanzees have a unique grooming pattern in which both partners groom each other simultaneously (Goodall, 1968; Boesch and Boesch-Achermann, 2000). This pattern is known as 'Groom mutually (GM)' in the ethogram of Mahale chimpanzees (Nishida *et al.*, 1999). Furthermore, chimpanzees exhibit social grooming clusters in which a gathering of ten or more individuals participate in complex grooming activity with one individual grooming another and being simultaneously groomed by one or more individuals (Goodall, 1986; Nakamura, 2003). In such polyadic grooming interactions, one chimpanzee can play both roles simultaneously, groomer and groomee. In contrast, most cases of social grooming among Japanese monkeys consist of a dyadic interaction where one individual plays only one role – groomer – and the other plays the other role – groomee – and they sometimes change roles in turns (Oki and Maeda, 1973; Goosen, 1987). In this continuation of role reversals between the groomer and the groomee, an agreement between the participants is required for roles to be exchanged (Muroyama, 1991b).

Another example of affinitive social activity whose episodes have been studied on the interactional level is social play (Fagen, 1981; Curtin, 1984; Hayaki, 1985; Fontaine, 1994; de Oliveira *et al.*, 2003; Shimada, 2006). Role choice also exists in social play, with one participant chasing and the other being chased (Shimada, 2006). In social play among mammals, 'self-handicapping' plays an important role in maintaining the playful interactions (Pereira and Preisser, 1998; Petrů *et al.*, 2009). Self-handicapping consists for the animals in restricting their physical strength, skills or social potential within social play episodes. In order to achieve and maintain social play for a prolonged time, the elder or larger participant needs to restrict its power, skills and kinematic ability. This can be regarded as a kind of cooperation among participants to achieve the same goal (Axelrod and Hamilton, 1981; Bauer and Smuts, 2007).

In Japanese macaque groups, all members are characterised by a strict matrilineal social rank order, with each dyad being defined by a clear dominance/subordination relationship; this hierarchical linearity severely constrains many aspects of their social behaviours (Kawamura, 1958; Koyama, 1967; Nakamichi *et al.*, 1995). How does such social asymmetry influence grooming episodes? In this chapter, my objective is to investigate how adult female Japanese macaques change grooming roles and maintain grooming interactions, through a systematic analysis of the behavioural sequences involved in their social grooming episodes.

15.2   Methods

15.2.1   Data collection

This study was carried out between October 2002 and September 2003. The author was the only observer. In October 2002, the Arashiyama E group consisted of 156 individuals including 30 adult males (at least 4 years old), 102 adult females (at least 4 years old), seven juvenile males (between 1 and 3 years old), seven juvenile females (between 1 and 3 years old), and ten infants (less than 1 year old). Of these, ten adult females between 10 and 12 years old and from various kin groups and social ranks were selected as study subjects. Table 15.1 provides the targeted animals' names, ages, family composition and the social ranks of their respective kin-groups.

Subjects were observed from 7:30 to 16:30 by using the focal-animal sampling method (Altmann, 1974). The total observation time was 373 hours. The mean observation time per focal subject was 37.1 hours, ranging from 34.3 to 43.5 hours. The study period incorporated the mating season for Japanese macaques, which lasted from September 2002 to February 2003. Focal animals were not sampled when they showed any physical or behavioural signs of oestrus characteristic of the mating season. Physical signs include redness of face and sexual skin, smelling of sexual skin or vaginal secretion, and swelling of sexual skin, whereas behavioural signs include female-to-male approach patterns, female receptivity to male mounts, and female–female homosexual interactions (Enomoto, 1974; Wolfe, 1979; Takahata, 1980; Vasey and VanderLaan: chapter 10). From these physical and behavioural features of female oestrus, I employed three criteria to identify whether the females were in oestrus: (1) skin colour of their face and anogenital region, (2) receptive behaviours toward males attempting to consort and (3) female-to-male active approach aimed at attracting males.

The term 'bout' was defined as a continuous grooming interaction between two participants without role reversal. When grooming roles changed, this was counted as a separate bout. The term 'episode' referred to a social grooming sequence between two participants, from initiation to termination, possibly including a number of bouts. When grooming was interrupted for more than 2 minutes or both participants moved more than 5 metres apart, subsequent grooming interaction was counted as a separate episode. Soliciting behaviours were recorded only when they occurred within 3 metres of a prospective grooming partner. I systematically recorded the names of all grooming partners, the

Table 15.1. *Focal subjects and lineage composition.*

Focal animal	Age (years)	Lineage rank	Own offspring (< 4 years)	Number of members	Adult male (≥ 4 years)	Adult female (≥ 4 years)	Infant and Juvenile (< 4 years)
						Lineage composition	
Mino-63–69–74–92	10	1	0	11	2	8	1
Mino-63–75–91	11	2	0	24	3	17	4
Kojiwa-62–74–79–91	11	5	0	11	3	8	0
Cooper-65–71–90	12	8	2	8	1	5	2
Yun-76–81–91	11	10	2	9	1	6	2
Chonpe-69–79–92	10	12	2	5	0	2	3
Blanche-59–64–75–91	11	14	0	6	2	4	0
Ai-61–72–91	11	17	1	5	0	3	2
Glance-60–71–83–90	12	20	1	2	0	1	1
Rakushi-59–79–92	10	21	1	5	1	2	2

duration and nature of the behaviours involved in initiating grooming interactions, whether the focal individual acted as groomer or groomee, the possible changes of role, and the time when a bout was terminated.

15.2.2   Social rank among group members

To assess dominance order among participants of grooming interactions, I referred to the social rank index provided by the staff of Iwatayama Monkey Park, Arashiyama. Dominance relationships among all group members were periodically measured by using the peanut test, consisting of throwing a peanut between two individuals, and considering the one who takes it as dominant over the other. However, in large groups of Japanese macaques, and particularly among middle- or low-ranking females, there are some dyads whose dominance relationships are difficult to predict from the linearity of the dominance hierarchy established at the group level (cf. Takahata, 1988; Nakamichi *et al.*, 1995; Kutsukake, 2000). Therefore, in addition to the data provided by the staff, I also used my personal data on the direction of approach/avoidance and dyadic unidirectional agonistic interactions collected during provisioning time among my focal subjects. It should be noted that my data strictly corresponded to the social rank index provided by the park staff.

15.2.3   Data analysis

All grooming episodes were categorised according to three events, namely initiation, role reversal and termination. Initiation was referred to as two types of event, I1 and I2. I1 consisted of initiation by the initial groomee via soliciting behaviour, and I2 consisted of initiation by the initial groomer with no soliciting behaviour displayed by the groomee. Both I1 and I2 were further defined through the detailed descriptions of subcategories of behavioural sequences, namely I11–I14 and I21–I22 (Table 15.2). Role reversal and termination were classified into different subcategories of behavioural sequences depending on whether the event was triggered by the groomer or the groomee (Table 15.2). The third type of termination, labelled T3, included grooming interactions that were terminated by external factors (e.g. agonistic behaviours involving third parties, food provisioning). Grooming episodes that were terminated due to external factors were excluded from the study.

15.3   Results

15.3.1   Number of grooming episodes

The total number of grooming episodes observed in the present study was 1179. Of these, 734 episodes (62.3%) concerned adult female–female dyadic interactions, and 19 episodes (1.6%) involved participation of three individuals. In all of the 19 triadic episodes, a kin relationship existed among all three participants, and one was the groomee while the other two acted as groomers. Such triadic grooming was exceptional and often brief; therefore, this type of interaction was also excluded from the data analysis.

15.3.2   Frequency of each type of grooming event

Of the 734 female–female dyadic grooming episodes, 332 (45.2%) involved kin females, and 402 (54.8%) involved non-kin females. Table 15.3 indicates the

Table 15.2. *Definition of events: Initiation, Role reversal and Termination*

Event		Subcategories of event		Definition
Initiation				
I1	by initial groomee	I11	Lie down (included Roll over)	Groomee approaches partner or stays on the spot and lies down or rolls over
		I12	Stand on all fours	Groomee approaches partner or stays on the spot and stands on all fours
		I13	Present its back	Groomee approaches partner or stays on the spot and sits presenting its back to the partners
		I14	Few seconds groom	Participant begins to groom but stops within 3 seconds and then solicits partner to groom
I2	by initial groomer	I21	Groom partner	Groomer approaches partner or reaches arms and starts grooming partner
		I22	Reply with solicitation	Participant solicits partner to groom but partner responds with solicitation to groom
Role reversal				
R1	by groomee	R1	Turn and raise hands	Groomee turns and raises hands toward partner before groomer withdraws hands
R2	by groomer	R21	Withdraw hand → Solicit grooming	Groomer stops grooming and withdraws hands and then solicits partner for grooming
		R22	Withdraw hand → stay still	Groomer stops grooming and withdraws hands and then maintains proximity with the partner
Termination				
T1	by groomee	T11	Ignore	Groomee ignores groomer's signal to change roles
		T12	Leave	Groomee leaves before groomer withdraws hands
		T13	Scare partner away	Groomee scares partner away before groomer withdraws hands
T2	by groomer	T2	Withdraw hand → Leave	Groomer stops grooming and leaves
T3	External factors	T31	Outsider's approach	Outsider approaches the dyad
		T32	Food provisioning time	Food provisioning time, both artificial and natural foods
		T33	Incidental affairs	Participants disturbed by agonistic interactions involving third parties, visitors approaching, etc.

Table 15.3. *Number of cases categorised into each subcategory for three types of event: Initiation, Role reversal and Termination*

(a) Number of each type of Initiation

Event		Kin		Non-kin		Total
I1	I11	53	80	45	80	160
	I12	6		7		
	I13	20		25		
	I14	1		3		
I2	I21	249	I212	319	I212	574
	I22	3		3		
Total			332		402	734

(b) Number of each type of Role reversal

Event		Kin		Non-kin		Total
R1	R1	14	14	33	**33	47
R2	R21	128	**164	43	57	221
	R22	36		164		
Total			178		90	168

(c) Number of each type of Termination

Event		Kin		Non-kin		Total
T1	T11	19		12		
	T12	67	T12	63	T12	172
	T13	5		6		
T2	T2	128	128	189	189	317
T3	T31	61		77		
	T32	17	113	19	T32	245
	T33	35		36		
Total			332		402	734

Event subcategories (e.g. I11, I12 ...). See Table 15.2 (**: $p < 0.01$).

frequency of each type of grooming event. Table 15.3(a) shows the number of each type of Initiation. During the initiation phase of grooming episodes involving both kin and non-kin females, I21 (in which the initial groomer initiated grooming with no soliciting behaviour displayed by the groomee) was more frequently observed than I1 (in which the initial groomee initiated grooming via soliciting behaviour) (kin: 75.0%, 249/332; non-kin: 79.4%, 319/402). I compared the frequency of I1 and I2 between kin and non-kin groups by using the Chi-square test for independence. There were no significant differences between kin and non-kin groups (Chi-square test for independence, $\chi^2 = 1.88$, df = 1, n.s.). Table 15.3(b) shows the number of each type of Role reversal.

Similarly I compared the frequency of R1 and R2 between kin and non-kin groups by using the Chi-square test for independence. In kin-group, R2 was frequently observed and in non-kin group, R1 was frequently observed (Chi-square test for independence, $\chi^2 = 34.29$, df = 1, $p < 0.001$). Among kin pairs, R1 (in which the groomee changed posture and began grooming without being solicited) was only observed in a few cases (7.9%, 14/178), whereas R21 (in which the groomer ceased grooming and began displaying solicitation behaviour) was the most frequent type of role reversal (71.9%, 128/178). Among non-kin pairs, R21 was also the most frequent behavioural sequence performed during role reversal (47.8%, 43/90).

Table 15.3(c) shows the number of each type of Termination. I compared the frequencies of T1, T2 and T3 between kin and non-kin dyads by using the Chi-square test for independence. There were no significant differences in the frequencies of each types of termination between the two types of dyads (Chi-square test for independence, $\chi^2 = 7.18$, df = 2, n.s.). In both kin and non-kin female dyads, T2 (in which the groomer stopped grooming and moved away) was the most frequently observed form of termination (kin; 38.6%, 128/332, non-kin; 47.0%, 189/402).

15.3.3   Number of bouts and duration of grooming episodes

To reveal how participants maintain their grooming interactions, those episodes terminated due to external factors (T3) were excluded from the following analyses, leaving 219 episodes that occurred between kin females and 270 episodes between non-kin females. Of these, 146 episodes (66.7%) involving kin females and 235 episodes (87.0%) involving non-kin females were single-bout episodes that did not include role reversal. In grooming episodes among kin females, no significant difference in frequency was found between single-bout episodes and multi-bout episodes (Wilcoxon signed-rank test, $z = -2.10$, n = 10, n.s.). However, in the case of non-kin females, single-bout episodes were significantly more frequent than multi-bout episodes (Wilcoxon signed-rank test, $z = -2.70$, n = 10, $p < 0.001$). Kin dyads showed significantly more grooming episodes with multiple role reversals than non-kin dyads.

The maximum duration of single-bout episodes was 36 minutes in non-kin dyads and 18 minutes in kin dyads. The former was a I2–T1 episode where the alpha female's daughter groomed the nearly lowest ranking female unidirectionally. The dominant female started grooming the subordinate female and left without soliciting role reversal. At some point during the grooming episode, the subordinate female heard her dependent offspring screaming, and she stood quadrupedally for a while as if she was about to leave. But when the dominant female uttered a threat vocalisation repeatedly, the subordinate female lay down again and the grooming episode resumed within a minute. This example illustrates how some grooming episodes may last for periods of time without role reversals.

Generally, one may think that the more role reversal events occur, the longer the grooming episodes should last. I analysed the correlation between the duration of grooming episodes and the number of grooming bouts. I found statistically significant correlations for both kin (Spearman's correlation coefficient by rank test, $r_s = 0.68$, n = 219, $p < 0.001$) and non-kin dyads ($r_s = 0.55$, n = 270, $p < 0.001$). I also found a significant difference between kin and non-kin dyads in the frequency distribution of the number of grooming bouts (Kolmogorov–Smirnov test, $p < 0.001$).

15.3.4 *Frequency of transition between grooming events*

Each event (namely Initiation, Role reversal and Termination) appears to be the result of a particular behavioural sequence during grooming interactions. In Table 15.4, I compared the frequencies of transitions for each event by using the Mantel–Haenszel chi-square test. In Table 15.4(a), there was no significant difference in the frequencies of transition from initiation to termination between kin and non-kin groups in single-bout episodes (Mantel–Haenszel's chi-square test, $\chi^2 = 0.01$, df = 1, n.s.). Table 15.4(b) presents the frequency of transition from initiation to the first role reversal. The transition from I2 to R2 was the most frequently observed in both kin and non-kin dyads (kin: 65.8%, 48/73, non-kin: 51.4%, 18/35). No significant differences were found between kin and non-kin dyads ($\chi^2 = 1.94$, df = 1, n.s.). Table 15.4(c) shows the frequency of transition between the last role reversal and termination. In both kin and non-kin dyads, the R2–T2 transition was the most frequently observed (i.e. the previous groomer solicited role reversal, and the next groomer stopped grooming and left). No significant differences were found in the frequencies of transitions from the last role reversal to termination between kin and non-kin dyads ($\chi^2 = 1.11$, df = 1, n.s.).

Table 15.4(d) shows a significant difference between kin and non-kin dyads in the frequency of the transition between two adjacent role reversals from multi-bout grooming episodes ($\chi^2 = 19.16$, df = 1, $p < 0.0001$). In kin females, the R1–R1 transition was not observed, and the R2–R2 combination was frequently observed (90.3%, 65/72), whereas in non-kin females, R1–R1 was the most frequently observed form of transition (59.3%, 16/27). This raises the question as to whether differences might exist between those episodes with only a few role reversals and those with a greater number of role reversals. Figure 15.1 presents the complete sequence of role reversals for grooming episodes with more than two role reversals. For kin females, 29 episodes included

Table 15.4. *Number of cases recorded in 16 types of transition between events*

(a) Transition from Initiation to Termination

		T1	T2	Total
Kin	I1	12 (8.2)	18 (12.3)	30
	I2	55 (37.7)	61 (41.8)	116
		67 (45.9)	79 (54.1)	146
Non-kin	I1	13 (5.6)	23 (9.8)	36
	I2	58 (24.7)	141 (60.0)	199
		71 (30.2)	164 (69.8)	235

(b) Transition from Initiation to Role reversal

		R1	R2	Total
Kin	I1	2 (2.7)	18 (24.7)	20
	I2	5 (6.9)	48 (65.8)	53
		7 (9.6)	66 (86.3)	73
Non-kin	I1	1 (2.9)	10 (28.6)	11
	I2	6 (17.1)	18 (51.4)	24
		7 (20.0)	28 (80.0)	35

(c) Transition from Role reversal to Termination

		T1	T2	Total
Kin	R1	3 (4.1)	4 (5.5)	7
	R2	21 (28.8)	45 (61.6)	66
		24 (32.9)	49 (67.1)	73
Non-kin	R1	3 (8.6)	4 (11.4)	7
	R2	7 (20.0)	21 (60.0)	28
		10 (28.6)	25 (71.4)	35

(d) Transition between two Role reversals

		R1	R2	Total
Kin	R1	0	4 (5.56)	4
	R2	3 (4.2)	65 (90.3)	68
		3 (4.2)	69 (95.8)	72
Non-kin	R1	16 (59.3)	0	16
	R2	1 (3.7)	10 (37.0)	11
		17 (63.0)	10 (37.0)	27

I1, I2, R1, R2, T1, T2: See Table 15.2.

more than two role reversals, with a maximum of 12 reversals. For non-kin females, 13 episodes had more than two role reversals, with a maximum of seven. In kin dyads, only R21 was observed following the fourth role reversal, and in non-kin dyads, only R1 was observed following the third role reversal.

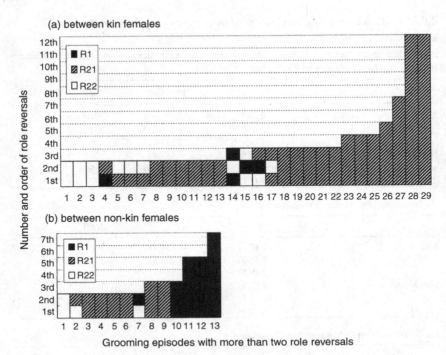

Figure 15.1. Sequence of Role reversals in each grooming episode with more than three bouts. R1, R21 and R22 are defined in Table 15.2.

15.4   Discussion

The present study set out to examine the behavioural sequences involved in grooming interactions among adult female Japanese macaques. Many grooming interactions showed no role reversals, and in a large number of these, the initial groomer began to groom the partner without being solicited, and then stopped and left without soliciting grooming. This description applied to grooming episodes involving both kin and non-kin dyads. Solicitation was not frequently observed in such cases. The finding that a large proportion of grooming interactions observed in Japanese macaques are unidirectional and without role reversals has been reported previously (Shizawa, 2001; Schino *et al.*, 2003).

I found that the more frequently Japanese macaques changed roles during grooming, the longer the grooming episodes, a result which is consistent with a previous study of grooming interactions among male Japanese macaques (Sugawara, 1975). In several multi-bout episodes, role reversals were occasionally observed. Grooming episodes including more than three role reversals continued over a long period of time. The present study identified role reversals as the factor that maintained grooming interactions over an extended period.

In grooming episodes involving numerous role reversals, both participants displayed the same type of behaviour with regard to changing roles. Among kin females, when the roles changed, the former groomer would display soliciting behaviour toward the partner. After the roles had changed, the next groomer would stop grooming and proceed to display soliciting behaviour to the partner again. Conversely, when the roles changed in non-kin female dyads, the former groomee would change posture and begin to groom the partner before the former groomer had withdrawn its hands. Subsequently, the groomee would change posture and begin to groom the partner again without being solicited. Among non-kin females, grooming roles changed without any solicitation behaviour on the part of any participants. Non-kin dyads exhibited the same behavioural patterns regardless of the relative social rank of the partners.

A notable result was that the participants adopted different behavioural patterns depending on whether the grooming partner was a kin or a non-kin individual. Moreover, it is important to point out that the participants changed grooming roles during interactions in a symmetrical way. Bateson (1972) suggested that various social interactions, including play, greeting and fighting, could be divided into two types: the first ones involved symmetrical interactions, such as shaking hands, kissing and hugging, and the others consisted of more complementary behavioural patterns involving give and take or touching and being touched. Bateson (1972) also emphasised that the behaviour of giving a gift to one's friend involves complementary interactions if completed as a single act. If, however, the recipient provides a gift in return to the sender and such an action is repeated, this becomes a 'symmetrical' interaction. The former category of symmetrical acts including shaking hands or kissing, and the latter, which results in exchanging complementary behaviours such as to groom and to be groomed repeatedly, are different in their logical types.

In the present study, social grooming was treated as an interaction involving complementary behavioural patterns that occurred between the groomer and the groomee. If grooming terminated without role reversal taking place, the interaction was not symmetrical but remained complementary. Many of the grooming episodes presented here were single bouts, and yet some continued over a prolonged period. Multi-bout grooming episodes, however, were characterised by both participants exchanging roles in a symmetrical way. Interestingly, Sugawara (1980) also reported that mounting interactions in adolescent male

Japanese macaques involved repeated symmetrical role reversals, referred to as mounting and being mounted.

Similarly, the more often role reversals occurred, the longer and continuous the interactions tended to be – a phenomenon that has also been observed in humans. For example, in conversations, humans regulate utterances to avoid problems caused by overlap, a practice referred to as 'turn taking' (Sacks *et al.*, 1974). Moreover, Dunbar (1996) has proposed that humans acquired language in order to create affiliative social bonds and human speech may have eventually replaced what social grooming represents in non-human primates. This hypothesis is significant in that social grooming in primates appears to show some parallels with human conversation (Dunbar, 1996; Nakamura, 2000).

The current study focused on the behavioural sequences involved in grooming interactions and revealed how Japanese macaques exchanged roles and how grooming interactions were maintained over an extended period. When grooming interactions were prolonged and included several role reversals, the participants eventually achieved meta-level symmetry, even when the complementary behaviour of grooming and being groomed had been the norm. It can therefore be surmised that such symmetrical role reversals in grooming are to some extent similar to the turn-taking that occurs in human conversation.

In the genus *Macaca*, some species have strict dominance hierarchies and asymmetrical and kin-biased agonistic and affinitive social interactions, whereas others have more relaxed dominance relationships and more symmetrical and less kin-biased social interactions (de Waal, 1989; de Waal and Luttrell, 1989; Hemelrijk, 1990a; Matsumura, 1999; Majolo *et al.*, 2010). The former are called 'despotic' species and include Japanese macaques (*Macaca fuscata*) and rhesus macaques (*M. mulatta*), whereas the latter are called 'egalitarian' species and include Barbary macaques (*M. sylvanus*), bonnet macaques (*M. radiata*) and tonkean macaques (*M. tonkeana*) (de Waal and Luttrell, 1989; Moore, 1992; Thierry *et al.*, 1994). In non-kin female Japanese macaques, social rank order is paramount (Kawamura, 1958; Nakamichi *et al.*, 1995). Their asymmetrical social interactions, examined at the group level, have been explained by the biological market theory (Noë and Hammerstein, 1994; Henzi and Barrett, 1999).

In this study, I described the structure of social grooming interactions through their behavioural sequences and elements, namely initiation, role reversal and termination, from the perspective of communicatory patterns between both participants, as has been done for the behavioural analysis of social play behaviour (cf. Fagen, 1981; Smith, 1982; Pereira and Preisser, 1998; Nishida and Wallauer, 2003; Shimada, 2006).

In order to prolong social play interactions, both protagonists need to display a temporal equality in their skills and power, otherwise play may stop quickly or turn into aggression (Fagen, 1981; Shimada, 2006). Thus, it is important for strong participants to restrict their physical strength or kinematic skills, a phenomenon called 'self-handicapping' (Pereira and Preisser, 1998; Petrů *et al.*, 2009). Likewise, in social grooming interactions including repeated role reversals, the participants seem to show temporal equality in taking the groomer role alternately. Despite the strict dominance relationships exhibited within all non-kin female dyads, those showing prolonged grooming interactions with several role reversals may be able to disguise temporarily their social asymmetry, just like the display of temporal equality of skills and strength in social play. From the perspective of communicative patterns, the 'meta-symmetrical' practice shown in certain grooming interactions may be the underlying mechanism of other social interactions representative of group membership in social mammals (Connor, 1995; Špinka *et al.*, 2001; Bauer and Smuts, 2007).

Acknowledgements

This study received financial support thanks to a grant toward the project 'The ecological and behavioural study of Japanese monkeys ranging around the Hozu valley, Kyoto' provided by the Science and Engineering Research Institute, Doshisha University. I would like to thank Professor Akisato Nishimura for supervising this study and Professor Akira Takada for useful suggestions on an earlier version of this manuscript. This paper was financed by JSPS Grant-in-Aid for Young Scientists (S) (#19672002, to Akira Takada). I wish also to express special thanks to Shinsuke Asaba, the director of Iwatayama Monkey Park, Arashiyama, and Shuhei Kobatake, the staff there for a range of support provided during the data collection.

References

Altmann, J. (1974). Observational study of behaviour: Sampling methods. *Behaviour*, **49**, 227–265.
Axelrod, R. and Hamilton, W. D. (1981). The evolution of cooperation. *Science*, **211**, 1390–1396.
Bateson, G. (1972). *Steps to an Ecology of Mind*. New York, NY: Ballantine Books.
Bauer, E. B. and Smuts, B. B. (2007). Cooperation and competition during dyadic play in domestic dogs, *Canis familiaris*. *Animal Behaviour*, **73**, 489–499.
Boesch, C. and Boesch-Achermann, H. (2000). *The Chimpanzees of the Taï Forest: Behavioural Ecology and Evolution*. New York, NY: Oxford University Press.

Cheney, D. L. (1992). Intragroup cohesion and intergroup hostility – the relation between grooming distributions and intergroup competition among female primates. *Behavioral Ecology*, **3**, 334–345.

Connor, R. C. (1995). Impala allogrooming and the parceling model of reciprocity. *Animal Behaviour*, **49**, 528–530.

Cords, M. (1997). Friendships, alliances, reciprocity and repair. In *Machiavelian Intelligence II*, eds. A. Whiten and R. W. Byrne. Cambridge: Cambridge University Press, pp. 24–49.

Curtin, R. A. (1984). Play, practice and predictability in nonhuman primates: a study of the langur *Presbytis entellus*. In *Current Primate Research*, eds. M. L. Roonwal, S. M. Monhot and N. S. Rathore. Jodhpur: University of Jodhpur, pp. 287–293.

de Waal, F. B. M. (1989). *Peacemaking among Primates*. Cambridge, MA: Harvard University Press.

de Waal, F. B. M. and Luttrell, L. M. (1989). Toward a comparative socioecology of the genus *Macaca*: different dominance styles in rhesus and stumptail monkeys. *American Journal of Primatology*, **19**, 83–109.

Dunbar, R. I. M. (1996). *Grooming, Gossip and Evolution of Language*. Cambridge: Harvard University Press.

Enomoto, T. (1974). The sexual behavior of Japanese monkeys. *Journal of Human Evolution*, **3**, 351–372.

Fagen, R. M. (1981). *Animal Play Behavior*. New York, NY: Oxford University Press.

Fairbanks, L. (1980). Relationships among adult females in captive vervet monkeys: testing a model of rank-related attractiveness. *Animal Behaviour*, **28**, 853–859.

Fontaine, R. P. (1994). Play as physical flexibility training in five ceboid primates. *Journal of Comparative Psychology*, **108**, 203–212.

Frank, R. E. and Silk, J. B. (2009). Impatient traders or contingent reciprocators? Evidence for the extended time-course of grooming exchanges in baboons. *Behaviour*, **146**, 1123–1135.

Goodall, J. (1968). The behavior of free-living chimpanzees in the Gombe Stream Reserve. *Animal Behaviour Monograph*, **1**, 161–311.

Goodall, J. (1986). *The Chimpanzees of Gombe*. Cambridge, MA: Harvard University Press.

Goosen, C. (1987). Social grooming in primates. In *Comparative Primate Biology, Volume 2b: Behavior, Cognition, and Motivation*, eds. G. Mitchell and J. Erwin. New York, NY: Alan R. Liss, pp. 107–131.

Hayaki, H. (1985). Social play of juvenile and adolescent chimpanzees in the Mahale Mountains National Park, Tanzania. *Primates*, **26**, 343–360.

Hemelrijk, C. K. (1990a). Models of, and tests for, reciprocity, unidirectionality and other social interaction patterns at a group level. *Animal Behaviour*, **39**, 1013–1029.

(1990b). A matrix partial correlation test in investigations of reciprocity and other social interaction patterns at group level. *Journal of Theoretical Biology*, **143**, 405–420.

Hemelrijk, C. K. and Ek, A. (1991). Reciprocity and interchange of grooming and 'support' in captive chimpanzees. *Animal Behaviour*, **41**, 923–935.

Henzi, S. P. and Barrett, L. (1999). The value of grooming to female primates. *Primates*, **40**, 47–59.

Kaplan, J. R. and Zucker, E. (1980). Social organization in a group of free-ranging patas monkeys. *Folia Primatologica*, **34**, 196–213.

Kawamura, S. (1958). Matriarchal social ranks in the Minoo-B troop: A study of the rank system of Japanese monkeys. *Primates*, **1**, 148–156.

Koyama, N. (1967). On dominance rank and division of a wild Japanese monkey troop at Arashiyama. *Primates*, **8**, 189–216.

   (1991). Grooming relationships in the Arashiyama group of Japanese monkeys. In *The Monkeys of Arashiyama: Thirty-five Years of Research in the Arashiyama in Japan and the West*, eds. L. M. Fedigan and P. J. Asquith. Albany, NY: State University of New York Press, pp. 211–226.

Kudo, H. (1986). Social behavior of adolescent females. In *Social Development and Social Relationships of Female Japanese Monkeys*, eds. U. Mori and H. Kudo. Tokyo: Tokaidaigaku-shuppankai, pp. 94–129. (in Japanese)

Kutsukake, N. (2000). Matrilineal rank inheritance varies with absolute rank in Japanese macaques. *Primates*, **41**, 321–335.

Majolo, B., Venture, R. and Schino, G. (2010). Asymmetry and dominance of relationship quality in the Japanese macaque (*Macaca fuscata yakui*). *International Journal of Primatology*, **40**, 736–750.

Manson, J. H., Navarrete C. D., Silk, J. B. and Perry, S. (2003). Time-matched grooming in female primates: new analyses from two species. *Animal Behaviour*, **67**, 493–500.

Matsumura, S. (1999). The evolution of 'egalitarian' and 'despotic' social systems among macaques. *Primates*, **40**, 23–31.

Moore, J. (1992). Dispersal, nepotism, and primate social behavior. *International Journal of Primatology*, **13**, 361–378.

Mori, A. (1975). Signals found in the grooming interactions of wild Japanese monkeys of the Koshima troop. *Primates*, **16**, 107–140.

Muroyama, Y. (1991a). Mutual reciprocity of grooming in female Japanese macaques (*Macaca fuscata*). *Behaviour*, **119**, 161–170.

   (1991b). Role choice in the sequence of grooming interaction of Japanese monkeys. In *Primatology Today*, eds. A. Ehara, T. Kimura, O. Takenaka and M. Iwamonto. Amsterdam: Elsevier, pp. 159–161.

   (1994). Exchange of grooming for allomothering in female patas monkey. *Behaviour*, **128**, 103–119.

   (1996). Decision making in grooming by Japanese macaques (*Macaca fuscata*). *International Journal of Primatology*, **17**, 817–830.

Nakagawa, N. (1992). Distribution of affiliative behaviours among adult females within a group of wild patas monkeys in a non-mating, nonbirth season. *International Journal of Primatology*, **13**, 73–93.

Nakamichi, M. and Shizawa, Y. (2003). Distribution of grooming among adult females in a large, free-ranging group of Japanese macaques. *International Journal of Primatology*, **24**, 607–625.

Nakamichi, M., Itoigawa, N., Imakawa, S. and Machida, S. (1995). Dominance relations among adult females in a free-ranging group of Japanese monkeys at Katsuyama. *American Journal of Primatology*, **37**, 241–251.

Nakamura, M. (2000). Is human conversation more efficient than chimpanzee grooming? Comparison of clique size of chimpanzee grooming and of human conversation. *Human Nature*, **11**, 281–297.

    (2003). 'Gatherings' of social grooming among wild chimpanzees: implications for evolution of sociality. *Journal of Human Evolution*, **44**, 59–71.

Nishida, T. and Wallauer, W. (2003). Leaf-pile pulling: an unusual play pattern in wild chimpanzees. *American Journal of Primatology*, **60**, 167–173.

Nishida, T., Kano, T., Goodall, J., McGrew, W. C. and Nakamura, M. (1999). Ethogram and ethnography of Mahale chimpanzees. *Anthropological Science*, **107**, 141–188.

Noë, R. and Hammerstein, P. (1994). Biological markets: supply and demand determine the effect of partner choice in cooperation, mutualism and mating. *Behavioral Ecology and Sociobiology*, **35**, 1–11.

Oki, J. and Maeda, Y. (1973). Grooming as a regulator of behavior in Japanese macaques. In *Behavioral Regulator of Behavior in Primates*, ed. C. R. Carpenter. Lewisburg, PA: Bucknell University Press, pp. 149–163.

de Oliveira, C. R., Ruiz-Miranda, C. R., Kleiman, D. G. and Beck, B. B. (2003). Play behavior in juvenile golden lion tamarins (*Callitrichidae*: Primates): organization in relation to costs. *Ethology*, **109**, 593–612.

Pereira, M. E. and Preisser, M. C. (1998). Do strong primate players 'self-handicap' during competitive social play? *Folia Primatologica*, **69**, 177–180.

Petrů, M., Spinka, M., Charvatova, V. and Lhota, S. (2009). Revisiting play elements and self-handicapping in play: a comparative ethogram of five Old World monkey species. *Journal of Comparative Psychology*, **123**, 250–263.

Sacks, H., Schegloff, E. A. and Jefferson, G. (1974). A simplest systematics for the organization of turn-taking for conversation. *Language*, **50**, 696–735.

Sade, D. S. (1972). Sociometrics of *Macaca mulatta*. I: Linkages and cliques in grooming matrices. *Folia Primatologica*, **18**, 196–223.

Sambrook, T. D., Whiten, A. and Strum, S. C. (1995). Priority of access and grooming patterns of females in a large and small group of olive baboons. *Animal Behaviour*, **50**, 1667–1682.

Schino, G. and Aureli, F. (2009). Reciprocal altruism in primates: partner choice, cognition, and emotions. *Advances in the Study of Behavior*, **39**, 45–69.

    (2010). The relative roles of kinship and reciprocity in explaining primate altruism. *Ecology Letters*, **13**, 45–50.

Schino, G., Ventura, R. and Troisi, A. (2003). Grooming among female Japanese macaques: distinguishing between reciprocation and interchange. *Behavioral Ecology*, **14**, 887–891.

    (2005). Grooming and aggression in captive Japanese macaques. *Primates*, **46**, 207–209.

Seyfarth, R. (1977). A model of social grooming among adult female monkeys. *Journal of Theoretical Biology*, **65**, 671–698.

Seyfarth, R. and Cheney, D. L. (1984). Grooming, alliances and reciprocal altruism in vervet monkeys. *Nature*, **308**, 541–542.

Shimada, M. (2006). Social object play among young Japanese macaques (*Macaca fuscata*) in Arashiyama, Japan. *Primates*, **47**, 342–349.

Shizawa, Y. (2001). Vocalization before grooming interactions in Japanese macaques. *Japanese Journal of Animal Psychology*, **51**, 39–46. (in Japanese)

Silk, J. B. (1982). Altruism among female *Macaca radiata*: explanations and analysis of patterns of grooming and coalition formation. *Behaviour*, **79**, 162–188.

Smith, P. K. (1982). Does play matter? Function and evolutionary aspects of animal and human play. *Behavioral and Brain Sciences*, **5**, 139–184.

Špinka, M., Newberry, R. C. and Bekoff, M. (2001). Mammalian play: Training for the unexpected. *Quarterly Review of Biology*, **76**, 141–168.

Sugawara, K. (1975). Analysis of the social relations among adolescent males of Japanese monkeys (*Macaca fuscata fuscata*) at Kohshima island. *Journal of Anthropological Society of Nippon*, **83**, 330–354. (in Japanese)

  (1980). Structure of social encounter among non-troop males of Japanese macaques. *Kikan Jinruigaku*, **11**, 3–76. (in Japanese)

Takahata, Y. (1980). The reproductive biology of a free-ranging troop of Japanese monkeys. *Primates*, **21**, 303–329.

  (1988). Dominance rank order of Adult female Japanese monkeys of the Arashiyama B troop. *Primate Research*, **4**, 19–32. (in Japanese with English summary)

Tanaka, I. and Takefushi, H. (1993). Elimination of external parasites (lice) is the primary function of grooming in free-ranging Japanese macaques. *Anthropological Science*, **101**, 187–193.

Thierry, B., Anderson, J. R., Demaría, C., Desportes, C. and Petit, O. (1994). Tonkean macaque behavior from the perspective of the evolution of Sulawesi macaques. In *Current Primatology, Vol. 2*, eds. J. J. Roeder, B. Thierry, J. R. Anderson and N. Herrenschmidt. Strasbourg: Louis Pasteur University Press, pp. 103–117.

Tsukahara, T. (1990). Initiation and solicitation in male-female grooming in a wild Japanese macaque troop on Yakushima island. *Primates*, **31**, 147–156.

Ventura, R., Majolo, B., Koyama, N. N., Hardie, S. and Schino, G. (2006). Reciprocation and interchange in wild Japanese macaque: grooming, and agonistic support. *American Journal of Primatology*, **68**, 1138–1149.

Wolfe, L. (1979). Behavioral patterns of estrous females of the Arashiyama West troop of Japanese macaques (*Macaca fuscata*). *Primates*, **20**, 525–534.

15 Box essay *Dental flossing behaviour as a grooming-related innovation by a Japanese macaque*

JEAN-BAPTISTE LECA

15 Box 1 Evolutionary significance of animal innovation

Innovation is typically defined as the discovery of novel information, the emergence of new behavioural patterns or the performance of established behaviours in a novel context (Reader and Laland, 2003). Various animal taxa – including fish, birds, rodents and non-human primates – are known to innovate in a wide range of behavioural domains such as travel routes, song acoustic patterns, food selection, food-processing techniques, agonistic display and tool use (see Reader and Laland, 2003 for a review).

In many ways, innovation is related to cultural diversity and evolution. First, when a novel behavioural practice, initially invented by an individual – the innovator – spreads to other group members, and is dependent on social means for its diffusion and maintenance, it becomes a tradition (Fragaszy and Perry, 2003). Second, similar factors may affect the likelihood of innovation and subsequent propagation within a group, including individual attributes, social relationships, group size and structural, contextual and functional aspects of the new behaviour (Huffman and Hirata, 2003). Third, several comparative research programmes linked innovation rate, cultural transmission and the relative size of brain associative areas in birds and primates (Lefebvre *et al.*, 2004). Therefore, investigating the determinants of the spontaneous appearance of new behavioural patterns in non-human animals, and particularly tool-use innovations, is of special interest to understanding the evolution of material culture in hominids (cf. McGrew, 1992).

However, little is known about the initial process by which novel behavioural patterns emerge spontaneously (but see Kummer and Goodall, 1985; Huffman and Hirata, 2003 for a few notable exceptions). It is not clear whether innovators are exceptionally creative individuals, subjects with a particular motivational state, or individuals that simply adjust their behavioural responses to novel,

322

appropriate or stressful environmental circumstances. Despite its significance to a broad range of research disciplines, the topic of behavioural innovation in animals is widely neglected in the evolutionary literature (Sol, 2003).

15 Box 2   Tool-use innovations in non-human primates

A multitude of tool-use innovations have been reported in the primate literature. Wild chimpanzees (*Pan troglodytes*) are known to innovate at a relatively high rate in the domain of spontaneous tool-use (sensu Beck, 1980), including aimed object throwing as an agonistic display, leaf-clipping as a courtship display and leaf-spooning/folding/sponging as ways to get access to water sources located in the hollows of trees (see Nishida *et al.*, 2009 for a review). Health maintenance, a sub-division of self-medicative behaviour (Huffman, 2007), in the form of the use of twigs as 'toothpicks' has been reported in great apes, including chimpanzees (McGrew and Tutin, 1973), bonobos, *Pan paniscus* (Ingmanson, 1996) and orang-utans, *Pongo pygmaeus* (Russon *et al.*, 2009). Although macaques are not frequent tool-users (Beck, 1980; but see Malaivijitnond *et al.*, 2007; Leca *et al.*, 2008a), dental flossing behaviour has been reported in long-tailed macaques, *Macaca fascicularis* (Watanabe *et al.*, 2007). The spontaneous use of tools in hygienic contexts, such as stones in allo-grooming and sticks as vaginal probes, has been occasionally reported in this genus (Weinberg and Candland, 1981; Sinha, 1997).

Despite the numerous examples of socially transmitted tool-use innovations in several non-human primate species, it should be noted that only a subset of such innovations become traditions. Many new tool-use behaviours have been reported to appear in primate troops, but were either idiosyncratic, or independently adopted by very few individuals, or their performance was restricted to a small class of the population, and for some reason, they never widely spread within the group by social means, and disappeared after the death of their few performers (Nishida *et al.*, 2009).

However, the role of the importance of these behaviours to the performers themselves in the likelihood to innovate has received little attention (Kummer and Goodall, 1985). Surprisingly, few attempts have been made to address the factors that could favour or constrain the spontaneous appearance of novel tool-use patterns (Reader and Laland, 2003). As Huffman (1996) pointed out, it is critical to know the history of a behavioural innovation in order to assess how environmental factors and social influences may enhance or limit its propagation within a group.

15 Box 3   Dental flossing behaviour as a tool-use and grooming-related innovation

In an effort to encourage the compilation of relevant data on the determinants of tool-use innovations in non-human primates, I report the first case of dental flossing (DF) behaviour in Japanese macaques (*Macaca fuscata*). From June to October 2008, I collected video-recorded focal data on a 14-year-old middle-ranking female (named Chonpe-69–85–94, hereafter 'the innovator') living in the free-ranging Arashiyama E troop at the Iwatayama Monkey Park, Arashiyama, Kyoto. I systematically documented her frequent and spontaneous use of hair as dental floss to remove food remains stuck between her teeth, and I examined the initial conditions that may have favoured this tool-use innovation (cf. Leca *et al.*, 2010).

I distinguished three different DF techniques or variants. First, the 'stretching with mouth' technique consisted of stretching its own hair or another individual's hair by clenching its lips onto the basal part of the hair, inserting the hair between the front teeth by slightly pulling its head downwards, and pulling the head backwards while gradually moving the lips to the distal end of the hair and performing repeated teeth-chattering. Second, the 'stretching with hand' technique consisted of stretching its own hair or another individual's hair by grasping and pulling the tip(s) of the hair between the thumb and forefinger of one hand, moving the mouth to the hair, and inserting the hair between the front teeth by performing repeated teeth-chattering (Figure 15 Box.1a). Third, the 'plucking' technique consisted of pulling out its own hair with one hand, holding the hair horizontally by grasping and pulling the tips of the hair between the thumb and forefinger of both hands, taking the hair to the mouth, and inserting the hair between the front teeth by performing repeated teeth-chattering (Figure 15 Box.1b).

The innovator's dependence on this form of tool use was shown by the fact that in all DF episodes, she used hair to floss her teeth and never used her fingers alone. Although DF may serve to remove food remains stuck between the teeth, there was no significant difference in the DF frequency between post-feeding and non-feeding focals. Regarding the context of occurrence, the DF behaviour was always associated with self- or allo-grooming activity.

15 Box 4   Determinants of the dental flossing innovation

Because they are provisioned with food several times a day, Arashiyama E troop members have 'free time on their hands', and this environmental opportunity

Figure 15 Box 1. Chonpe-69–85–94 performing two dental flossing techniques:
(a) 'Stretching with hand' technique during self-grooming; (b) 'Plucking' technique
(photos by J.-B. Leca).

could lead them to further explore various objects and incorporate them into
feeding activities (Leca *et al.*, 2008b). These artificial conditions are likely to
enhance the appearance of food-related tool use, such as the use of dental floss.
Provisioning has relaxed selective pressures on foraging, and created favourable environmental conditions under which various behavioural innovations by
Japanese macaques may occur (Leca *et al.*, 2007c, 2008c).

Since DF is always grooming-related, the behaviour is more likely to appear in frequent groomers, i.e. central group members, than in individuals less involved in grooming interactions, i.e. peripheral group members (cf. Nakamichi and Shizawa, 2003). Thus, it is not surprising that the innovator was a central individual. As an adult female, the current age of the innovator was consistent with most studies, whereas the sex was not. It has been found that innovators were more frequently males and adults than females and non-adults (see Reader and Laland, 2001 for a review). However, previous observations suggest that the innovator may have started performing the DF behaviour from the age of 2 years (Zamma, pers. comm.). This is consistent with previous research showing that most Japanese macaque innovators are juvenile females (Itani and Nishimura, 1973; Huffman, 1984; Kawai *et al.*, 1992).

Finally, Chonpe-69–85–94 might have temperamental traits that made her prone to behavioural innovation. Despite extensive observation of most troop members in a study of stone-handling behaviour, she has been the only individual observed repeatedly rolling small stones on the palm of her hand while intensively grooming her palm and presumably trying to remove a spine stuck into it (Leca, pers. obs.). This first report in Japanese macaques of an apparent attempt to use stones in a health-maintenance context is suggestive of a generally inventive temperament. Sinha (2005) suggested the role of temperament in bipedal begging innovation by bonnet macaques (*Macaca radiata*).

Since chance may account for a good number of behavioural innovations (Reader and Laland, 2003), and DF was always associated with grooming activity, I suggest that the DF innovation is an accidental by-product of grooming. Thus, the following is a reasonable scenario: during regular grooming episodes, Japanese macaques sometimes bite into hair or pull it through their mouths to remove external parasites, such as louse eggs (Tanaka and Takefushi, 1993). Due to particular anatomical constraints such as diastema (i.e. gaps between incisors), pieces of hair may accidentally have stuck between the innovator's teeth, and as she drew them out, she may have noticed the presence of food remains attached to them. The immediate reward consisting of licking the food remains off the hair may have encouraged her to repeat the behaviour for the same effect in the future, by actively inserting the hair between her teeth.

Therefore, the DF innovation could be a transformation of grooming patterns via the running of hair between the teeth to remove louse eggs. These scenarios are consistent with the 'perception–action' perspective on the development of tool use and foraging competence in monkeys, apes and humans, postulating that skilled actions are acquired through the routine generation of species-typical exploratory actions, coupled with learning about the outcomes and affordances of each action that generate directly perceptible information (Lockman, 2000).

Besides possible proximate causes, the problem arises as to why the DF behaviour has been maintained by its innovator for several years. A first parsimonious explanation for this behaviour is that the flosser may simply enjoy the interaction between the hair and its teeth, and apparent pleasurable feedback potentially gained from the activity may be an immediate reinforcement (cf. Leca *et al.*, 2007b). Second, DF could alleviate the possible physical annoyance caused by a piece of food stuck between the teeth. A third beneficial consequence of DF could be an improvement in the teeth condition. Flossing one's teeth would be a form of health maintenance, which is considered a level of self-medication (Huffman, 2007). Since all self-medicative behaviours are driven by some quest for comfort, these interpretations are congruent with the classification of the tooth-pick behaviour in orang-utans as a 'comfort innovation' (Russon *et al.*, 2009). Since DF was not more frequent in post-feeding than non-feeding periods, we doubt that the DF innovation had a significant or even any survival value, through the very small amount of extra food the flosser can obtain from its behaviour.

15 Box 5 A scenario for the emergence of new dental-flossing variants

Although the present data did not allow us to accurately determine the order of appearance of the different DF techniques, past observations and a comparative analysis of the behavioural patterns support the view that the 'plucking' technique was acquired later than the two other DF variants. First, previous long-term behavioural observations of this group showed that Chonpe-69–85–94 had been using the 'stretching with hand' and the 'stretching with mouth' techniques for at least four years (Leca, unpublished data; Vasey, pers. comm.; Zamma, pers. comm.). Although the 'plucking' technique was not noticed before 2008, its absence remains speculative. Second, several elements show a higher level of complexity of the 'plucking' technique relative to the two other variants: (1) the former consists of the manipulation of a detached object (pluck hair) whereas hair is attached to the skin in the latter, (2) the former requires both hands to be used whereas only one hand is used in the latter, (3) hair selection was more frequent in the former than in the latter, and (4) since the former necessitates hair to be pulled out, it is more invasive than the latter.

Likewise, the 'stretching with hand' technique, that requires the use of hand and mouth can be considered more complex, in terms of sequence of actions, than the 'stretching with mouth' technique, in which only the mouth is used. Although we cannot determine the exact timeline, we propose the following order in the emergence of DF techniques: first the 'stretching with mouth' technique, then the 'stretching with hand' technique, and last the 'plucking' technique. In long-tailed macaques, there was a generalisation of the DF behaviour with hair to the use of coconut-shell fibres for the same purpose (Watanabe *et al.*, 2007). In Japanese macaques, even non-instrumental object manipulation can undergo a major 'transformation' process over time, with an increase in the diversity and complexity of the behavioural patterns exhibited (Leca *et al.*, 2007a, 2008a).

This is one of the rare studies to document the spontaneous appearance of tool-use behaviour in Japanese macaques under natural conditions (see also Leca *et al.*, 2008a). The lack of report on DF in other troops of Japanese macaques and the idiosyncratic presence of the behaviour in this troop may reflect possible intra- and inter-troop variations in (1) the likelihood of behavioural innovation, (2) the social constraints on the early dissemination and long-term maintenance of such inventions and (3) appropriate social and/or environmental reinforcement for the emergence, propagation and continued practice of this behaviour. When a behaviour is restricted to one or very few group members, it is likely to disappear at the group level (Leca *et al.*, 2007c; Nishida *et al.*, 2009). Further investigation, including experimentally elicited DF, may help to determine more accurately the conditions of appearance of this novel behaviour, and to elucidate why it has not spread within the group (cf. Watanabe *et al.*, 2007; Leca *et al.*, 2008a).

Acknowledgements

This work was funded by a Grant-In-Aid for scientific research (No. 1907421 to M.A. Huffman) sponsored by the Ministry of Education, Science, Sports and Culture, Japan, a JSPS (Japan Society for the Promotion of Science) postdoctoral fellowship to J.-B. Leca (No. 07421), and by travel funds from the HOPE Project, a core-to-core program sponsored by JSPS to J.-B. Leca. I thank the researchers, students and staff who provided permission to work, assistance and valuable specific information about the Arashiyama macaque troop. I am particularly grateful to S. Asaba, J. Hashiguchi, S. Kobatake and S. Tamada (Iwatayama Monkey Park, Arashiyama). For fruitful discussion, I thank M. A. Huffman (Kyoto University Primate Research Institute), P. L. Vasey (University of Lethbridge, Canada) and K. Zamma (Great Ape Research Institute, Okayama). For assistance with data collection, I thank N. Gunst (University of Lethbridge, Canada) and N. Tworoski (University of Minnesota, USA).

References

Beck, B. B. (1980). *Animal Tool Behavior: The Use and Manufacture of Tools by Animals*. New York, NY: Garland Press.

Fragaszy, D. M. & Perry, S. (2003). Towards a biology of traditions. In *The Biology of Traditions: Models and Evidence*, eds. D. M. Fragaszy and S. Perry. Cambridge: Cambridge University Press, pp. 1–32.

Huffman, M. A. (1984). Stone-play of *Macaca fuscata* in Arashiyama B troop: transmission of a non-adaptive behavior. *Journal of Human Evolution*, **13**, 725–735.

(1996). Acquisition of innovative cultural behaviors in non-human primates: A case study of stone handling, a socially transmitted behavior in Japanese macaques. In *Social Learning in Animals: The Roots of Culture*, eds. B. Galef and C. Heyes. Orlando, FL: Academic Press, pp. 267–289.

(2007). Primate self-medication. In: *Primates in Perspective*, eds. C. J. Campbell, A. Fuentes, K. C. MacKinnon, M. Panger and S. K. Bearder. New York, NY: Oxford University Press, pp. 677–690.

Huffman, M. A. and Hirata, S. (2003). Biological and ecological foundations of primate behavioral tradition. In *The Biology of Traditions: Models and Evidence*, eds. D. M. Fragaszy and S. Perry. Cambridge: Cambridge University Press, pp. 267–296.

Ingmanson, E. J. (1996). Tool-using behavior in wild *Pan paniscus*: social and ecological considerations. In *Reaching into Thought: The Minds of the Great Apes*, eds. A. E. Russon, K A. Bard and S. T. Parker. New York, NY: Cambridge University Press, pp. 190–210.

Itani, J. and Nishimura, A. (1973). The study of infra-human culture in Japan. In *Precultural Primate Behaviour*, ed. E. Menzel. Basel: S. Karger, pp. 26–50.

Kawai, M., Watanabe, K. and Mori, A. (1992). Pre-cultural behaviors observed in free-ranging Japanese monkeys on Koshima islet over the past 25 years. *Primate Report*, **32**, 143–153.

Kummer, H. and Goodall, J. (1985). Conditions of innovative behaviour in primates. *Philosophical Transactions of the Royal Society London B*, **308**, 203–214.

Leca, J.-B., Gunst, N. and Huffman, M. A. (2007a). Japanese macaque cultures: inter- and intra-troop behavioural variability of stone handling patterns across 10 troops. *Behaviour*, **144**, 251–281.

Leca, J.-B., Gunst, N. and Petit, O. (2007b). Social aspects of fur rubbing in *Cebus capucinus* and *C. apella*. *International Journal of Primatology*, **28**, 801–817.

Leca, J.-B., Gunst, N., Watanabe, K. and Huffman, M. A. (2007c). A new case of fish-eating in Japanese macaques: implications for social constraints on the diffusion of feeding innovation. *American Journal of Primatology*, **69**, 821–828.

Leca, J.-B., Nahallage, C. A. D., Gunst, N. and Huffman, M. A. (2008a). Stone-throwing by Japanese macaques: form and functional aspects of a group-specific behavioral tradition. *Journal of Human Evolution*, **55**, 989–998.

Leca, J.-B., Gunst, N. and Huffman, M. A. (2008b). Variability of food provisioning regimes and stone handling tradition in Japanese macaques: a comparative study of ten troops. *American Journal of Primatology*, **70**, 803–813.

(2008c). Of stones and monkeys: testing ecological constraints on stone handling, a behavioral tradition in Japanese macaques. *American Journal of Physical Anthropology*, **135**, 233–244.

(2010). The first case of dental flossing by a Japanese macaque (*Macaca fuscata*): implications for the determinants of behavioral innovation and the constraints on social transmission. *Primates*, **51**, 13–22.

Lefebvre, L., Reader, S. M. and Sol, D. (2004). Brains, innovations and evolution in birds and primates. *Brain, Behavior, and Evolution*, **63**, 233–246.

Lockman, J. (2000). A perception-action perspective on tool use development. *Child Development*, **71**, 137–144.

Malaivijitnond, S., Lekprayoon, C., Tandavanitj, N. *et al.* (2007). Stone-tool usage by Thai long-tailed macaques (*Macaca fascicularis*). *American Journal of Primatology*, **69**, 227–233.

McGrew, W. C. (1992). *Chimpanzee Material Culture: Implications for Human Evolution*. Cambridge: Cambridge University Press.

McGrew, W. C. and Tutin, C. E. G. (1973). Chimpanzee tool use in dental grooming. *Nature*, **241**, 477–478.

Nakamichi, M. and Shizawa, Y. (2003). Distribution of grooming among adult females in a large, free-ranging group of Japanese macaques. *International Journal of Primatology*, **24**, 607–625.

Nishida, T., Matsusaka, T. and McGrew, W. C. (2009). Emergence, propagation or disappearance of novel behavioral patterns in the habituated chimpanzees of Mahale: a review. *Primates*, **50**, 23–36.

Reader, S. M. and Laland, K. N. (2001). Primate innovation: sex, age and social rank differences. *International Journal of Primatology*, **22**, 787–805.

(2003). Animal innovation: An introduction. In *Animal Innovation*, eds. S. Reader and K. N. Laland. Oxford: Oxford University Press, pp. 3–35.

Russon, A. E., van Schaik, C. P., Kuncoro, P. *et al.* (2009). Innovation and intelligence in orangutans. In *Orangutans: Geographic Variation in Behavioral Ecology and Conservation*, eds. S. A. Wich, S. S. Utami Atmoko and T. Mitra Setia. New York, NY: Oxford University Press, pp. 279–298.

Sinha, A. (1997). Complex tool manufacture by a wild bonnet macaque. *Folia Primatologica*, **68**, 23–25.

(2005). Not in genes: phenotypic flexibility, behavioural traditions and cultural evolution in wild bonnet macaques. *Journal of Bioscience*, **30**, 51–64.

Sol, D. (2003). Behavioural innovation: a neglected issue in the ecological and evolutionary literature? In *Animal Innovation*, eds. S. Reader and K. N. Laland. Oxford: Oxford University Press, pp. 63–82.

Tanaka, I. and Takefushi, H. (1993). Elimination of external parasites (lice) is the primary function of grooming in free-ranging Japanese macaques. *Anthropological Science*, **101**, 187–193.

Watanabe, K., Urasopon, N. and Malaivijitnond, S. (2007). Long-tailed macaques use human hair as dental floss. *American Journal of Primatology*, **69**, 940–944.

Weinberg, S. M. and Candland, D. K. (1981). 'Stone-grooming' in *Macaca fuscata*. *American Journal of Primatology*, **1**, 465–468.

16 The impact of kinship, defence cost and priority of access on food competition

PATRICK BÉLISLE, JEAN PRUD'HOMME
AND CONSTANCE DUBUC

B4 male (6 months old), member of the Arashiyama-Montréal troop (photo by Jean Prud'homme).

The Monkeys of Stormy Mountain: 60 Years of Primatological Research on the Japanese Macaques of Arashiyama, eds. Jean-Baptiste Leca, Michael A. Huffman and Paul L. Vasey. Published by Cambridge University Press. © Cambridge University Press 2012.

16.1   Introduction

Several studies have shown the marked effects of female kinship on positive behaviours such as contact, proximity, grooming, protective interactions
in conflicts and various forms of coalitions and alliances (Gouzoules, 1984;
Gouzoules and Gouzoules, 1987; Walters, 1987; Bernstein, 1991; Kapsalis
and Berman, 1996; Chapais *et al.*, 2001; Bradley *et al.*, 2007; Charpentier
et al., 2008; Fraser *et al.*, 2008; Perry *et al.*, 2008; Mitani, 2009). Kinship
also promotes selection for active food sharing, especially the act of giving
from mothers or other adults to dependent immatures. Although this behaviour
is especially well documented in some *Callithricidae* species (Feistner and
McGrew, 1989; Saito *et al.*, 2008) and apes (Teleki, 1973; Nowell and Fletcher,
2006; Jaeggi *et al.*, 2008), active sharing of food is uncommon in non-human
primates (Feistner and McGrew, 1989).

Dominance rank regulates priority of access to food resources (Whitten,
1983; Barton, 1993; Saito, 1996; Sterk *et al.*, 1997). However, this does not
rule out the possibility of a dominant individual tolerating a subordinate one
when feeding simultaneously at a common food site. Indeed, although active
sharing is rare, 'co-feeding' or passive sharing resulting from inter-individual
tolerance has been observed in several primate species and in different contexts
(Teleki, 1973; Brown and Mack, 1978; Boesch and Boesch, 1989; Feistner
and McGrew, 1989; Price and Feistner, 1993; de Waal, 1989, 1993, 1997).
Here also, kinship seems to influence feeding behaviours by extending inter-
individual tolerance at the food site during feeding (Yamada, 1963; Furuichi,
1983; Ihobe, 1989).

Generally, studies on nepotism in the context of food competition have
taken a dichotomous approach by comparing two categories of individuals,
kin and non-kin. This dichotomous approach has at least two drawbacks.
First, where does one draw the line between 'kin' and 'non-kin' and on what
basis? As such, these dichotomous categories of relatedness are arbitrarily
determined and this may obscure subtle differences. Second, such a binary
categorisation will conceal any potential effect of degree of genetic kinship
on the behaviour of dominant and subordinate individuals during co-feeding
episodes.

Even when the genetic degree of relatedness of two co-feeding individuals
is precisely known, it will remain difficult to ascertain without any doubt that
the apparent nepotism is true altruism among kin. This is, because in particular contexts, co-feeding tolerated by a dominant animal might be the optimal
'selfish' strategy by which it avoids wasteful energy expenditure at defending
a resource, rather than true altruism (Bélisle, 2002). Following this argument,
one would expect that the more a food patch is dispersed, the more probable

any observed passive food sharing among kin reflects a selfish optimal strategy on the part of the 'tolerant' dominant animal. Some studies have in fact revealed a correlation between the degree of food concentration and the frequency of aggressive behaviours (for a review, see Janson, 1988).

Another factor that may result in the erroneous attribution of nepotism to a particular bout of co-feeding is the order of access that individuals have to a food patch. In general, the direction of agonism and displacement at a coveted resource site can be accurately predicted on the basis of dominance relationships. However, when subordinate individuals arrive at the food patch first, this could theoretically induce inhibition of aggression in more dominant individuals. Some studies on non-human primates have shown such a first-access advantage in relation to resources such as territories (van Schaik *et al.*, 1992; Perry, 1996), sexual partners (Kummer *et al.*, 1973) or coveted objects (Torii, 1975; Thierry *et al.*, 1989). However, few studies have tested this strategy in a context of feeding competition, let alone in a despotic species such as the Japanese macaque (*Macaca fuscata*) where dominance relationships are rigid (Kummer *et al.*, 1973; Dubuc and Chapais, 2007).

In this chapter, we aim to test how co-feeding in a group of Japanese macaques derived from the Arashiyama population is influenced by: (1) various degrees of genetic relatedness, (2) different costs of resource defence by dominant individuals and (3) priority of access to a resource. For each of the three variables tested we carried out a distinct set of experiments. The research was conducted in the Laboratory of Behavioural Primatology at the University of Montréal. Age and degree of matrilineal kinship were known for all members of the group. A coveted food resource (raisins) was distributed in a translucent Plexiglas box with two apertures set in it allowing two animals to feed simultaneously in close proximity. Dominance hierarchy in Japanese macaques is rigid or despotic, so any co-feeding episode between dominant and subordinate females was considered to be a potential manifestation of unilateral altruism from the dominant kin.

16.2   Hamilton's classic rule and the deployment of altruism among kin

Kin favouritism among primates, or nepotism, is generally understood in the light of the theory of kin selection (Hamilton, 1964). Altruistic behaviours directed towards kin are the most patent expression of kin favouritism. Hamilton's equation states that altruism among kin can be profitable to the donor as long as the benefits to the recipient (*b*) once weighed by the genetic degree of relatedness between the two (*r*) are higher than the costs (*c*) associated with the behaviour for the donor (i.e. when $br > c$). Stated otherwise, altruistic behaviour that is profitable to both kin can evolve when the level of genetic relatedness between two individuals is superior to the c/b ratio.

Although this formula defines the relatedness threshold at which altruism becomes theoretically unprofitable, it is of little help when it comes to predicting the distribution of altruistic behaviours among different potential kin recipients. To illustrate this point, imagine a situation in which a dominant individual can relinquish part of a coveted, easily defensible food to kin. The benefits to the receiver and the costs to the donor might vary according to different factors. For example, during passive food-sharing, these two values can vary according to the absolute amount of food coveted, its level of concentration, the energetic needs of the two co-feeders in relation to the intrinsic energetic value of the resource, and respective health status of the two co-feeders, the general density of food on the territory, the probability of aggression from the dominant, the cost of aggression in terms of energetic expenditure and so on. Accordingly, it becomes almost impossible to weigh the two variables in an attempt to predict the behaviours of kin. Moreover, while the values of these variables are expected to fluctuate for the same behaviour when exhibited in different contexts, they are also expected to differ from one behaviour to another (e.g. co-feeding versus support in a conflict).

In short, Hamilton's equation allows us, at best, to predict that there will be altruism among kin as long as br is greater than c and that this propensity should subside as the genetic distance between the kin increases. Thus, the limit at which altruistic behaviours among kin will cease should match the kinship threshold at which altruism becomes non-profitable to the donor (when $bc < c$) or should match the limit of kin discrimination, or both.

16.3   The distribution of nepotism and its confounding factors

As stated previously, there is an abundant literature on the impact of genetic matrilineal relatedness on positive behaviours in primates such as contact grooming or interventions in conflicts. It is well known that the frequencies of affinitive behaviours decrease more or less linearly with degree of relatedness, while in contrast, the frequency of interventions in conflicts plummets past the $r = 0.50$ degree of kinship (Chapais *et al.*, 2001). Although kinship is fundamental to understanding nepotism, other factors are known to predict nepotism among various categories of kin (Altmann, 1979; and for review see Chapais and Bélisle, 2004). We know, for example, that differences in the reproductive value of various kin can affect the choice of a female beneficiary (Hamilton,

1964; Chapais and Schulman, 1980; Combes and Altmann, 2001). Similarly, differing intensities of competition among various categories of kin also affect the distribution of kin favouritism (Chapais *et al.*, 1994).

Three factors are of particular importance to fully understand the limits of nepotism, because these variables can negatively influence the deployment of nepotistic behaviours toward distant genetic kindred beyond the mother–offspring genetic bond (i.e. those whose level of relatedness is less than $r = 0.50$). This can lead one to mischaracterise the real motivation of individuals and to mistakenly conclude that they have reached the limit of their kin discrimination.

The first factor we are concerned with has to do with the variability in spatial distribution of kindred. On average, close kin tend to be nearer in proximity to each other than to distant kin and this is explained due to the common genetic link they share with the same individual: their mother. Thus, by spending more time around their mother, close kin happen to spend more time around each other than around more distant kin. Various studies have revealed a positive correlation between degree of kinship and average time spent in proximity (Kurland, 1977; Kapsalis and Berman, 1996). For this reason, close kin have more opportunities to be nepotistic towards each other. This is reflected in the sudden decrease in the frequency of nepotism in relation to genetic distance. In other words, the observable distribution curve of nepotistic behaviour correlated to genetic distance cannot be explained only by kin selection per se, but also by one of its correlates, that is differential availability of kin.

The second factor affecting the distribution of nepotism relates to individual time budgeting. Postulating that group members spend, on average, most of the day fulfilling their needs and those of their kin through nepotism, one might argue that low frequencies of nepotistic interactions in favour of distant kin would not be caused by a low motivation to favour one's distant kin, as much as the consequence of a lack of time to fulfil the latter's needs.

Thirdly, in natural habitats individuals will rely on various kin donors of nepotism to fulfil daily needs such as grooming, cuddling and passive food sharing. However, in as much as these needs will generally be fulfilled by close kin due to the biases discussed above, opportunities for nepotistic behaviours among distant kin are expected to be significantly restricted. This reasoning is based on the assumption that daily needs are not unlimited and can attain saturation. Supplemental nepotism from distant kin would then become unnecessary. However, it is worth noting that this does not apply to all types of nepotistic behaviours. For example, agonistic support received from kin during conflicts cannot attain saturation.

In conclusion, one can expect the joint and cumulative effects of these three factors to limit the deployment of nepotistic behaviours towards distant kin,

obscuring the fact that these animals might recognise each other as kin not-
withstanding the weak frequency of observable nepotism. In nature, there are
a number of reasons why an individual might never experience passive co-
feeding with distant kin. First, distant kin might rarely be in close proximity
while foraging. Second, distant kin may have spent most of their time with
their own close kin, or have co-fed with their close kin and, as such, might have
satiated their appetites or expended their foraging time budgets. Finally, the
energetic needs of distant kin might be fulfilled if, by chance, an opportunity
for co-feeding occurred.

16.4   The value of kin selection in the understanding of nepotism

The reader may have noticed that we preferred the terms nepotism and favouritism instead of altruism when describing positive interactions between kin.
This was done purposely, as it is often very difficult to ascertain the altruistic
nature of a given behaviour, or more specifically if such an interaction incurs
costs to the donor. For example, kin may sometimes engage in cooperative
interactions because the immediate benefits derived are superior to the alternative of non-cooperative behaviour. Such mutualistic behaviours (Pusey and
Packer, 1997; Chapais, 2006) can be explained without recourse to the genetic tie between the individuals involved; cooperative hunting, defence against
predators, huddling, aggressive interventions in conflicts, all fall in that category (see Gouzoules and Gouzoules, 1987).

Kin may also engage in reciprocity of nepotistic behaviours, which may be
misinterpreted as pure kin altruism (Schino and Aureli, 2010). It is possible,
in contexts in which help is costly to the donor, to understand the selection of
such behaviours in terms of probability of reciprocation in a near future rather
than in terms of genetic benefits to the latter. Such 'bilateral altruism' or reciprocal altruism, as defined by Trivers (1971), is difficult to identify in nature but
it should be selected when certain conditions prevail (Axelrod and Hamilton,
1981).

Thus, although kinship appears to be important both in the distribution of
mutualistic interactions (see Gouzoules and Gouzoules, 1987) and reciprocal
altruism (Kurland, 1977; Kaplan, 1978; Janus, 1989), the observed kin bias
is wholly insufficient proof of the presence of pure kin altruism via kin selection. A more direct means to test kin selection theory consists of studying the
deployment of unilateral altruistic behaviours only (i.e. pure altruism) within
a given population of related individuals (Bélisle and Chapais, 2001; Chapais
et al., 2001; Bélisle, 2002). This category of behaviour is distinct from other

nepotistic, cooperative or mutualistic interactions in that the benefits to the donor can only ensue from the genetic bond shared by both individuals and, as such, possible confounding factors are eliminated.

Bearing this in mind, we designed an experimental setting allowing various kin dyads to engage in clear unilateral altruistic interactions, while controlling for the previously described confounding variables. Moreover, during the experimental tests, the kin dyads interacted in isolation from the rest of the group. This was done in order to control for conspecific influences that could potentially induce tolerance from the dominant animal and, in turn, result in food sharing to avoid potential high risk conflicts. In such instances, co-feeding episodes would reflect a selfish optimal strategy benefiting the dominant animal, which Blurton-Jones (1984) has referred to as '*tolerated theft*' and Wrangham (1975) has referred to as '*buying peace while eating*'.

16.5 Experiment 1: Effect of matrilineal kinship on passive food sharing

The founding members of the study group originated from the Arashiyama-Kyoto group of Japanese macaques and were translocated to Texas in 1972 (Fedigan, 1991). In 1984, the study group was moved from Texas to the Laboratory of Behavioural Primatology at the University of Montréal, which was founded and directed by Bernard Chapais. The initial group of monkeys consisted of 15 animals belonging to three distantly related matrilines named A, B and C. At the time of the experiments, the group numbered 40 individuals. Age and degree of matrilineal relatedness were known for all individuals. The group lived in five indoor rooms and two outdoor pens covering 230 m².

In Experiment 1, we aimed at analysing the influence of degree of kinship on the propensity of dominant monkeys to behave altruistically towards a subordinate kin in a context of food competition. We reasoned that in this type of context, the benefits ensuing from kinship would be clearly discernible. It is worth noting that with this experiment, we did not aim at proving primates systematically behave as to improve their fitness. We rather aimed at testing the postulate that kinship, through the proximate mechanism of familiarity, should have a positive impact so as to promote passive food sharing.

In order to do so, we presented a limited quantity of highly prized food (50 dry raisins) in a compartmented translucid Plexiglas box (46 × 20 × 17 cm) to a dyad of females isolated in a room. Two openings, one at each end of the box, constrained the females' spatial distribution and required them to stand face to face at less than 1 metre from each other when feeding simultaneously (Figure 16.1). This degree of proximity is greater than the average distance at

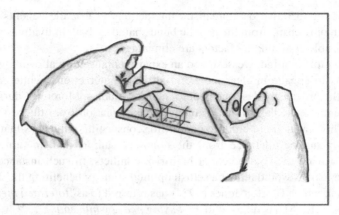

Figure 16.1. Two adult females eating simultaneously at the foodbox.

which kin females forage in the wild (Mori, 1977; Ihobe, 1989) and is associated with aggressive behaviours in a feeding context (Masataka and Fujii, 1980; Furuichi, 1983).

We tested 18 different adult female dyads accounting for five degrees of genetic relatedness (i.e. four mother–daughter dyads, two grandmother–granddaughter dyads, four sister dyads, four aunt–niece dyads, and four non-kin dyads). Each female dyad was tested 12 times for a total of 216 tests. *Co-feeding duration* was defined as the period of time during which the two subjects ate simultaneously at the box. The *proportion of co-feeding* time was calculated in relation to the period of time during which the dominant female had been present at the box and, therefore, could have monopolised food.

Japanese macaques practise a very despotic form of dominance and, in the experiment, food dispersion was highly concentrated. Consequently, we assumed that any tolerance from the dominant animal that resulted in the subordinate female being able to co-feed simultaneously was an act of unilateral altruism on the part of the former. Kinship had a positive effect on co-feeding. The percentage of time the subjects spent co-feeding increased significantly in relation to their degree of relatedness (Jonckheere test, n = 18, $J^* = 3.12$, $p = 0.001$; Figure 16.2). We also measured the exact percentage of raisins eaten by the subordinate subject. To obtain figures that would vary between 0 and 100, we took as denominator the utmost number of raisins the subordinate subject could have eaten assuming equal sharing (i.e. one-half of 50). The feeding success of the subordinate subject increased significantly with degree of relatedness (n = 18, $J^* = 3.08$, $p = 0.001$; Figure 16.1).

To test if kinship had any effect on the inhibition of the subordinate females to co-feed, we calculated the mean distance between the dominant female and the

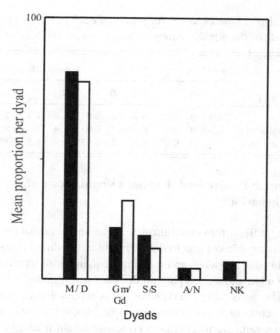

Figure 16.2. Effect of degree of kinship on the percentage of time spent
co-feeding (black) and the percentage of food eaten by the subordinate female
(open). M/D: mother–daughter dyads ($r = 0.50$), Gm/Gd: grandmother–granddaughter
dyads ($r = 0.25$), S/S: sister dyads ($r = 0.25$), A/N: aunt–niece dyads ($r = 0.125$),
NK: non-kin dyads ($r < 0.0004$).

subordinate female when the former was feeding. The results showed a positive
correlation between degree of kinship and mean distances observed ($n = 18$,
$J^* = 3.08$, $p = 0.001$). Put differently, the more closely related the partners were,
the closer the subordinate female stood to the box when the dominant female was
eating. These three results (co-feeding time, feeding success and mean distance)
held unchanged whether the tests were carried out with the 0.25 relatedness category restricted to the sister dyads ($p = 0.004$, 0.003 and 0.003, respectively), or
to the grandmother–granddaughter dyads ($p = 0.001$, 0.002 and 0.001).

The finding that kinship has a positive effect on co-feeding does not tell
which categories of kin were responsible for this effect. To address this issue,
we considered the scores of non-kin dyads as a baseline against which to compare the scores of each of the three categories of relatedness. The 0.5 and 0.25
categories of dyads scored significantly higher than non-kin dyads for all three
behavioural measures, but the 0.125 category did not (Table 16.1). Hence,
aunt–niece dyads behaved as non-kin dyads in these respects and kinship only
exerted a positive effect on co-feeding among closer kin.

Table 16.1. *Comparison of the behavioural scores of each category of relatedness with those of the non-kin category, using Mann–Whitney tests*

	$r = 0.5$			$r = 0.25$			$r = 0.125$		
	$N_{1,2}$	U	p	$N_{1,2}$	U	p	$N_{1,2}$	U	p
Co-feeding time	4, 4	0	0.02	6, 4	3	0.05	4, 4	5.5	0.29
Feeding success of subordinate	4, 4	0	0.02	6, 4	3	0.05	4, 4	6	0.55
Mean distance between females	4, 4	0	0.02	6, 4	2	0.03	4, 4	6.5	0.39

16.6   Experiment 2: Passive food sharing: a singular case of altruistic behaviour

Passive food sharing differs from other altruistic behaviours, in that refraining from sharing in a favourable context to do so, might entail direct costs to the 'would be donor'. In contrast, when an individual abstains from intervening in a conflict involving a kin and a third party, it does not result in direct immediate costs to itself. The decision to intervene or not is essentially taken according to the potential costs or benefits brought about by the intervention. In a food-sharing context, refusing to tolerate a co-feeder might mean, in certain circumstances, having to defend the resource aggressively. Such defence might even entail higher energetic expenditure than the cost of losing a portion of the resource to the subordinate co-feeder.

In keeping with this line of reasoning, food sharing between kin could sometimes be purely altruistic (i.e. the product of kin selection, exclusively), sometimes a selfish strategy to avoid the costs of defence or sometimes a combination of both possibilities. Thus, it is possible that the rates of sharing observed in the first experiment did not only reflect the altruistic nature of dominant female kin, but also reflected their selfish interest in avoiding energy expenditure involved in the defence of the resource in question.

Janson (1988) showed that there is a positive correlation between increased food concentration and increased intra-specific competition for access to the resource. With respect to co-feeding, this correlation exacerbates the relative importance of dominance rank between individuals (Whitten, 1983; Saito, 1996), as well their degree of genetic relatedness (de Waal, 1986; van Schaik, 1989). In Experiment 1, the spatial arrangement of the two apertures at either end of the food box required that the dominant female had to shift position and actively drive the subordinate away from the box if she sought to monopolise the feeding site. Thus, in Experiment 1, the dominant female had to expend energy to monopolise the feeding site. In contrast, in Experiment 2, we tested food sharing between kin in a context where costs of food defence would be almost nil. In order to do so, we applied the protocol of the first experiment except the feeding box was modified. Specifically, the two apertures were relocated to the forefront of the feeding box. This new configuration maximised the proximity between the females, allowing the dominant animal to control the food without having to move around the box to drive the subordinate female away (as was the case in Experiment 1). In this context, any passive food sharing would constitute 'pure' altruism by the dominant female.

During Experiment 2, only subordinate daughters and grand-daughters risked approaching the box when a dominant female was feeding (i.e. mother or grandmother). As such, it was not possible to test the costs of defence on the same five degrees of kinship dyads analysed in Experiment 1. Nonetheless, the results from the six kin dyads tested (i.e. mother–daughter and grandmother–granddaughter) revealed a significant decrease in the rates of aggression by the dominant females (i.e. lungeing and chasing with the subordinate fleeing), compared with that of the first experiment (mean rate of aggression: Experiment 1 $= 1.22$; Experiment 2 $= 0.01$; Wilcoxon test; n $= 6$ pairs; $T^* = 21$; $p = 0.03$). Thus, there was a clear drop in the need to actively defend the food box in Experiment 2 and, by extension, a drop in the cost of defence.

As expected, by lowering the costs of defence at the feeding box, dominant females were less inclined to tolerate passive co-feeding by subordinate ones, even though they were kin. For example, compared to Experiment 1, dominant females exhibited more low-intensity, non-contact threats (i.e. open-mouth stare threats) from a greater inter-individual distance when the subordinate female attempted to approach the feeding box (n $= 6$ pairs; $T^* = 20$; $p = 0.05$). Compared with Experiment 1, when high-intensity aggression did occur, dominant females exhibited more contact aggression given that they were at close range with the subordinate target of aggression during co-feeding (n $= 6$ pairs; $T^* = 21$; $p = 0.03$).

The lower tolerance threshold of dominant females resulted in a drastic reduction in the average time spent co-feeding (n $= 6$ pairs, $T^* = 21$, $p = 0.03$; Figure 16.3). However, it also caused subordinate females to pick the raisins more rapidly (n $= 5$ pairs; $T^* = 15$; $p = 0.05$). Accordingly, there was no significant difference in the average number of raisins gathered by the subordinate females between the two experiments (n $= 6$ pairs, $T^* = 16$, $p = 0.31$; Figure 16.3).

In the light of the above results and taking into account that the second experiment replicated the protocol of the first one except for the box apertures' location, it is reasonable to conclude that the tolerance manifested by the dominant females in Experiment 1 cannot be entirely explained through indirect genetic benefits defining unilateral altruism. The relatively greater proportion of co-feeding observed in Experiment 1, compared to Experiment 2,

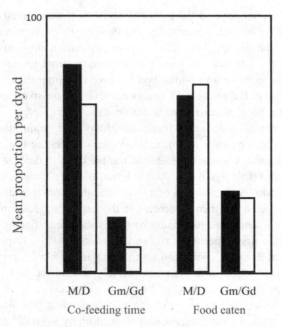

Figure 16.3. Effect of costs of defence on the proportion of time spent co-feeding and the proportion of raisins eaten by the subordinate female: when inter-individual distance is 1 metre (black: experiment I) and when individuals are in contact (open: experiment 2). M/D: mother–daughter dyads ($r = 0.50$), Gm/Gd: grandmother–granddaughter dyads ($r = 0.25$).

can be explained, at least in part, by the benefits dominant females selfishly obtained from conserving energy that they would have otherwise expended by directing aggression at subordinate females. Following this reasoning, it seems reasonable to conclude that the proportion of co-feeding observed in Experiment 2 represents 'pure' altruism, which is concentrated mainly within mother–daughter dyads. As such, the deployment of altruism in the context of passive food-sharing for this species is more limited than was suggested by the results of Experiment 1.

16.7   Priority of access as an alternative strategy

The order of access at the food patch can also influence how competition for food is manifested. Simply arriving first at a resource is sometimes sufficient to inhibit a competitive response from a more dominant individual. The advantage ensuing from priority of access has been observed in different species of non-human primates whether it be a territory (van Schaik *et al.*, 1992; Perry, 1996), a sexual partner (Kummer *et al.*, 1973; Bachmann and Kummer, 1980), a food resource (Sigg and Fallet, 1985) or even simply a coveted object (Thierry *et al.*, 1989).

Different proximate factors have been proposed to explain the 'passive' reaction of a dominant competitor arriving second at the resource, including (1) small rank distance between two individuals (Sigg and Falett, 1985), (2) the ease with which an owner can carry away a coveted object (Torii, 1975; Kummer and Cords, 1991) or (3) the presence of the subordinate individual's potential allies (Bachmann and Kummer, 1980). From a functional perspective, game theory (Maynard Smith and Parker, 1976; Hammerstein, 1981; Maynard Smith, 1982) provides an explanation for this effect. According to that model, when the costs of a potential agonistic escalation (in terms of risk of injury and energetic expenditure) are high relative to the value of a coveted resource, mere possession of the resource acts as an arbitrary rule settling the conflict in the initial possessor's favour. Note that to be considered a true arbitrary rule, respect of possession must not simply ensue from an asymmetry in the capacity to withhold the resource (resource holding power or RHP) due to relative weight, dominance rank or fighting ability (Parker, 1974; Maynard Smith and Parker, 1976). Likewise the value that each competitor allots to the coveted resource cannot be asymmetrical because theoretically this would result in the possessor having greater interests in initiating or increasing the aggressive escalation (Davies and Houston, 1981; Krebs, 1982).

16.8   Experiment 3: Impact of priority of access on food competition

We tested the effect of priority of access by repeating the protocol of Experiment 2 (apertures on the front of the box maximising inter-individual proximity) and by setting priority of access at the feeding box to each member of a given dyad, alternately. Thus, for each dyad, two tests were performed, in Test 1 the dominant female accessed the box first, and in Test 2, the subordinate female was given priority of access. For distant kin dyads the subordinate females were too inhibited to come forward to the box first. So, Experiment 3 was carried out on seven mother–daughter dyads and one grandmother–granddaughter dyad. Each dyad was tested 20 times, for a total of 160 tests.

None of the proximate factors discussed above that might explain respect of possession was an issue here. Japanese macaques form despotic matrilineal hierarchies, with clear unidirectional submissiveness (Kawamura, 1965; Chapais, 1988). Females arriving first at the box during the experiment could not move it away from the competitor because the box was attached to a wall and the raisins were placed within compartments allowing just a fraction of the resource to be manipulated at any one time. Moreover, although female dyads were not isolated from other group members during the experiments, the females tested were always the two most dominant members of their lineage, thus excluding the possibility of a potential third party intervention influencing the interactions.

In such a context, complete inhibition from the dominant intruder (i.e. dominant female second to the box) was not expected. However, we considered that respect of possession accounted for those instances of co-feeding in which a subordinate female arrived first at the food box followed by a dominant female who engaged in co-feeding with the subordinate for a longer duration of time, compared with when the dominant female was first to arrive at the food box.

Priority of access had a significant impact on time spent in co-feeding. For five of the eight dyads tested, there was an increase in the average time spent at co-feeding, ranging from 44% to 539%, when the subordinate female was first at the feeding box (Wilcoxon test, n = 5 pairs, $T^* = 15$, $p = 0.03$; Figure

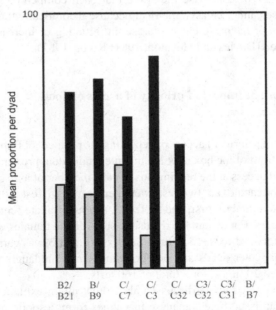

Figure 16.4. Proportion of co-feeding time observed for each dyad, when the dominant female was first at the feeding box (grey) and when the subordinate female had priority of access (black).

16.4). The average number of raisins ingested by the subordinate females also increased in similar proportions (n = 5 pairs, $T^* = 15$, $p = 0.03$; Figure 16.4). These results indicate that the dominant females showed respect of possession towards the subordinate ones. For the three other dyads in which no co-feeding was observed when the dominant female arrived second at the box, it is worth noting that co-feeding also did not occur when she was first. Thus, co-feeding seems to be a prior condition necessary for respect of possession to occur. Neither the rank distance, the coefficient of relatedness among the females, nor the number of individuals per matriline can explain the lack of tolerated co-feeding by these three dominant females.

It is worth noting that although passive co-feeding increased dramatically in five of eight dyads, 'absolute' respect from the dominant female arriving second at the feeding box never took place. Thus in fact, it would be more accurate to talk about impact of priority of access, rather than respect of possession. Note also, that the inversion of the priority of access at the feeding box, in relation to dominance, had an impact on the behaviour of both the dominant and the subordinate females. On the one hand, the reversal of priority of access increased the period of latency before an aggression from the dominant female occurred (n = 8 pairs, $T^* = 33$, $p = 0.04$) and it also increased significantly the frequency and rapidity with which subordinate females returned to the feeding box after being targeted with aggression (n = 8 pairs, $T^* = 34$, $p = 0.02$; and n = 8 pairs, $T^* = 33$, $p = 0.04$, respectively). On the other hand, the tolerance expressed by the dominant female extended only to the point at which the subordinate female was driven away from the feeding box. This latter finding suggests that the 'notion' of possession is closely linked to the physical proximity between the possessor and the coveted object (see Torii, 1975; Thierry *et al.*, 1989; Kummer and Cords, 1991).

In sum, when the subordinate females had priority of access to the food box, there was a significant increase in the passive co-feeding time observed. These results can be explained in terms of temporarily elevated levels of tolerance from the dominant females combined with a more prolonged reduction in inhibition on the part of the subordinate female.

16.9   Impact of dominance despotism on priority of access

The significant increase in co-feeding time following the reversal of normal priority of access for five of the eight dominant Japanese macaques is somewhat surprising and rather counter-intuitive considering the despotic type of dominance demonstrated by this species. It is doubtful that dominant females would be tolerant simply due to the potential risk of aggressive escalation. Indeed, retaliation from a subordinate female following aggression from a dominant individual never occurred.

The effect of a rigid dominance hierarchy was clear on distant kin dyads ($r < 0.50$). Given the manner in which the food box was designed, close physical proximity between two individuals was an inescapable condition for co-feeding to occur. This clearly inhibited the subordinate females during trial tests. In fact, inhibition was such that only mother–daughter dyads and one grandmother–granddaughter dyad could be tested. Assuming that the subordinate females' inhibition is a precise indication of the tolerance level of dominant individuals, then one is forced to conclude that in a species ruled by despotism, respect of possession will occur only among close kin.

In order to better understand the impact of the hierarchy's level of rigidity on co-feeding, we tested whether respect of possession would be more effective in a species ruled by a more egalitarian system of dominance relationships and in which conflicts are more symmetrical than in Japanese macaques. Thus, we conducted a similar experiment on a group of Tonkean macaques living at the Centre de Primatologie of l'Université Louis Pasteur, Strasbourg, France, which is under direction of Bernard Thierry. Compared with Japanese macaques, aggression is more symmetrical in this species. Moreover, in Tonkean macaques, conflicts are rare but risks of retaliation from subordinate targets are high (Thierry, 1986, 1999; Petit et al., 1997; Schino et al., 1998).

We could not replicate the protocol carried on the Japanese macaques because the infrastructure at the Centre de Primatologie of l'Université Louis Pasteur did not allow for it. In lieu of this, a limited amount of coveted food was placed in a concentrated area just outside the fence of the enclosure (i.e. 50 raisins dispersed on a 30×30 cm surface). The monkeys accessed the food by passing their arm through the mesh of the fence. To co-feed, the individuals had to stand at a similar distance as during Experiment 2 and 3 on Japanese macaques, as described above. The group consisted of 29 individuals (12 males and 17 females). We were able to test 46 adult dyads (17 male–male, 22 male–female and 7 female–female dyads) representing five different genetic bonds, including non-kin. Because the group was living in semi-captive conditions, we could not isolate the dyads from the rest of the group during the tests. In order to minimise the potential influence of a third party during co-feeding, when a third animal was within 5 metres' distance, the test was abandoned. Also, priority of access to the food spot was not induced but varied randomly. We carried out a total of 622 tests. It is worth noting that few agonistic behaviours took place in proximity to the resource, and when agonistic behaviour did occur, it was always of mild intensity.

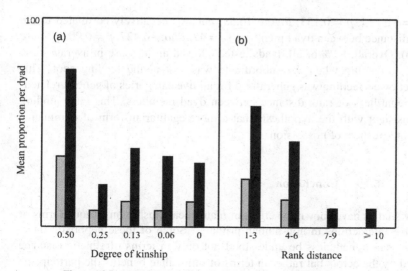

Figure 16.5. (a) Proportion of co-feeding time observed in relation to genetic distance, when the dominant was first at the feeding box (grey) and when the subordinate had priority of access (black). (b) Proportion of co-feeding time observed in relation to rank distance, when the dominant was first at the feeding box (grey) and when the subordinate had priority of access (black).

The results revealed that kinship had a positive impact on co-feeding only for those dyads who were closely related. The average time spent co-feeding was nearly twice as long for closely related dyads ($r = 0.50$) as for any other more distantly related or unrelated ones, regardless of whether a dominant or subordinate individual arrived at the feeding site first (dominant first: Mann–Whitney U test, n = 6 and 41 respectively, $Z = 3.328$, $p = 0.001$; dominant second: n = 6 and 41 respectively, $Z = 2.598$; $p = 0.009$; Figure 16.5a). Rank distance also had an impact on co-feeding for the Tonkean group, which was not the case for the Japanese macaques. For all five categories of genetic relatedness, there was a negative correlation between rank distance and time spent in co-feeding, regardless of the order of arrival at the feeding spot. In other words, the closer individuals were in rank, the more tolerant the dominant individual was towards the subordinate, regardless of who arrived first at the feeding site (dominant first: Spearman rank correlation, n = 47, $Rho = -0.415$, $p = 0.004$, dominant second: n = 47, $Rho = -0.555$, $p < 0.001$; Figure 16.5b).

Finally, when the subordinate individuals had priority of access to the feeding site, mean time of co-feeding more than doubled compared with when dominant individuals were first (Wilcoxon, n = 47 pairs, $Z = 4.279$, $p \leq 0.001$). This finding held regardless of the degree of relatedness (Spearman, n = 47,

Rho = −0.212, *p* = 0.153; Figure 16.5a) but was negatively correlated to the rank distance between dyad members (n = 47, *Rho*: −0.427, *p* = 0.003; Figure 16.5b). Overall, 82% of all dyads tested showed an increase in average time spent co-feeding when the subordinate was first at the feeding spot. This impact on co-feeding was generalised for all five categories of genetic relatedness regardless of rank distance between dyad members. This latter finding is consistent with the hypothesis that a more egalitarian form of dominance promotes respect of possession.

16.10   Conclusion

Many studies have shown the effect of matrilineal kinship on various forms of positive interactions in non-human primates. Some of these nepotistic interactions can nonetheless be understood not only in terms of genetic distance shared by the actors but rather in terms of immediate benefits the participants derive from such interactions. In order to test the real effect of kinship on nepotistic interactions, we designed an experimental protocol in which genetic relatedness would be the only possible explanation for an act of unilateral altruism by excluding the possibility of any immediate benefit to the donor. In Experiment 1, we showed that for two given individuals, time spent in co-feeding varied significantly with degree of kinship. Aunt–niece dyads showed the same tolerance at co-feeding as did non-kin. Consequently, we concluded that the threshold for co-feeding in a context of food competition is at $r = 0.25$ (sister dyads and grandmother–granddaughter dyads). These results demonstrated the role of kin selection in the evolution of altruism in primates beyond the mother–daughter relationship.

Hamilton's equation predicts the maximal deployment of all forms of altruism among kin according to the cost–benefit ratio distinctive to each situation. In an average group, however, the mere presence of a number of different kin generates certain constraints that will substantially limit the deployment of altruistic acts. The experimental protocol we designed controlled for what we believe to be the three most important constraints. The first one has to do with differential availability of kin, the second one relates to time constraints with which individuals have to deal, and finally the theoretical saturation of one's needs per day. The deployment curve of altruism observed in Experiment 1 plummeted abruptly past the $r = 0.50$ threshold (mother–daughter dyads) which is very similar to what has been observed in nature. This similarity is all the more striking given that the co-feeding observed in Experiment 1 could be explained, at least in part, in terms of the dominant individual's selfish interests, coupled with the cost associated with defence of the resource. This leads us to presume that there would be almost no expression of unilateral altruism beyond the 0.50 level of genetic relatedness, which is consistent with the results of Experiment 2.

At least three non-mutually exclusive hypotheses can be put forward to explain this 'drop' in altruism past $r = 0.50$ (i.e. mother–daughter dyads). First, since the cost–benefit ratio of altruistic behaviours varies according to the context in which they occur, it is quite possible that the genetic bond of $r = 0.25$ represented the limit of 'profitability' associated to passive food sharing in these experiments. A second hypothesis relates to the various levels of familiarity any given distant kin dyad might share from one matriline to another. The familiarity among distant kin is likely to be affected by different factors, such as a matriline's demography. Thus, for any particular distant kin bond, familiarity might fluctuate from one lineage to another whereas the mother–daughter bond is expected to remain constant. According to this reasoning, the co-feeding times observed, or more generally, the propensity to behave altruistically, could simply reflect the degree of familiarity between two individuals. Thirdly, the competitive background that sisters share could act as an inhibitory factor among them.

Experiment 3 tested whether arriving first at a coveted resource granted any advantage to a subordinate animal in the context of food competition with a dominant individual. Priority of access proved to be an effective strategy for subordinate females. For five of the eight subordinate females tested, their mean level of co-feeding increased significantly when they were first at the feeding box, compared with when dominant individuals arrived first. This finding was rather unexpected given that Japanese monkeys have a very despotic dominance system. One hypothesis to explain these results is that the dominant animals benefit over the long run from sporadic respect of possession towards subordinate animals. By not doing so, the limited coveted resource might become valuable enough, from the perspective of the subordinate animals, to risk aggression. In other words, dominant animals may refrain from being 'too' despotic in order to 'buy peace'. Sporadic respect of possession would be a mixed evolutionary stable strategy (ESS) that could not be invaded through time by any alternative (in contrast to a pure EES in which dominant animals would always, or never, show respect of possession).

In Experiment 3, with the exception of one grandmother–granddaughter dyad, only subordinate daughters were assertive enough to approach the feeding box. On the one hand, close genetic relatedness appeared to be a prerequisite for priority of access to have a positive impact on co-feeding, but on the other hand it cannot account for the three dyads where priority of access did not lead to co-feeding. From a proximate qualitative perspective, the positive impact of priority of access appeared to be highly related to the low level of inhibition and the higher level of assertiveness shown by certain subordinate Japanese macaques at the feeding box (i.e. absence of hesitation to collect the raisins, body position, frequency of monitoring, etc.).

Variation in subordinate females' attitudes during the experiments may relate to the familiarity shared between the members of a given dyad. In species that have a more relaxed mode of dominance, assertiveness might also act as an honest signal of the subordinate's motivation, and even their willingness to counter-attack, should dominant individuals attack them. As such, more assertive subordinates might elicit more respect of possession from dominant individuals, at least temporarily. In keeping with these ideas, subordinate Tonkean macaques showed greater assertiveness during the co-feeding episodes. In this egalitarian species, priority of access was a highly successful strategy. It increased the average co-feeding times in dyads regardless of genetic distance (including non-kin dyads) or rank distance between competing individuals.

The positive effect of priority of access observed in the Tonkean macaques' group can be understood in terms of this species' dominance system. In this species, the individual's resource holding power (RHP) is more symmetrical than in Japanese macaques. Consequently, the probability of a dispute initiated by a subordinate individual is greater in Tonkean macaques. In practical terms, it is probably very safe for dominant female Japanese macaques not to respect the priority of access of subordinate females. In contrast, the same is not true for dominant individuals in more egalitarian species like the Tonkean macaques. In species with more egalitarian patterns of dominance, dominant individuals should acknowledge respect of possession of subordinate individuals, because by doing so, they 'buy peace' over the long run, and minimise the potential risk of an aggressive escalation. Although aggressive conflicts in egalitarian species, like Tonkean macaques, are generally 'mild' and biting is neither frequent nor injurious, the energy spent in such altercations might be costly when weighted against the intrinsic value of the resource at stake. Thus, as a general rule, one would expect that, compared with despotic species, egalitarian ones should use priority of access more often as an arbitrary rule for evaluating possession.

Acknowledgements

We thank Paul Vasey and Jean-Baptiste Leca for their offer to contribute to this volume. Our thanks also go to Bernard Chapais for his useful and challenging comments, to Bernard Thierry, Pierre Uhlrich, Ana Ducoing and Odile Petit for their warm welcome to Strasbourg. Finally, I am very grateful to Océane and Nathalie Diamond for their boundless patience.

References

Altmann, J. (1979). Altruistic behaviour: the fallacy of kin deployment. *Animal Behaviour*, **27**, 958–962.

Axelrod, R. T. and Hamilton, W. D. (1981). The evolution of cooperation. *Science*, **211**, 1390–1396.

Bachmann, C. and Kummer, H. (1980). Male assessment of female choice in hamadryas baboons. *Behavioral Ecology and Sociobiology*, **6**, 315–321.

Barton, R. A. (1993). Sociospatial mechanisms of feeding competition in female olive baboons. *Animal Behaviour*, **46**, 791–802.

Bélisle, P. (2002). *Apparentement et co-alimentation chez le macaque japonais*. PhD dissertation, University of Montréal.

Bélisle, P. and Chapais, B. (2001). Tolerated co-feeding in relation to degree of kinship in Japanese macaques. *Behaviour*, **138**, 487–509.

Bernstein, L. S. (1991). The correlation between kinship and behaviour in non-human primates. In *Kin Recognition*, ed. P. G. Hepper. Cambridge: Cambridge University Press, pp. 7–29.

Blurton-Jones, N. G. (1984). A selfish origin for human food sharing: tolerated theft. *Ethology and Sociobiology*, **5**, 1–3.

Boesch, C. and Boesch, H. (1989). Hunting behavior of wild chimpanzees in the Taï National Park. *American Journal of Physical Anthropology*, **78**, 547–573.

Bradley, B. J., Doran-Sheehy, D. M. and Vigilant, L. (2007). Potential for female kin associations in wild western gorillas despite female dispersal. *Proceedings of the Royal Society B: Biological Sciences*, **274**, 2179–2185.

Brown, K. and Mack, D. S. (1978). Food sharing among captive *Leontopithecus rosalia*. *Folia Primatologica*, **29**, 268–290.

Chapais, B. (1988). Experimental matrilineal inheritance of rank in female Japanese macaques. *Animal Behaviour*, **36**, 1025–1037.

(2006). Kinship, competence and cooperation in primates. In *Cooperation in Primates and Humans: Mechanism and Evolution*, eds. P. M. Kappeler and C. P. van Shaik. Berlin: Springer-Verlag, pp. 47–66.

Chapais, B. and Bélisle, P. (2004). Constraints on kin selection in primate groups. In *Kinship and Behavior in Primates*, eds. B. Chapais and C. M. Berman. New York, NY: Oxford University Press, pp. 365–386.

Chapais, B. and Schulman, S. (1980). An evolutionary model of female dominance relationships in primates. *Journal of Theoretical Biology*, **82**, 47–89.

Chapais, B., Prud'homme, J. and Teijeiro, S. (1994). Dominance competition among siblings in Japanese macaques: constraints on nepotism. *Animal Behaviour*, **48**, 1335–1347.

Chapais, B., Savard, L. and Gauthier, C. (2001). Kin selection and the distribution of altruism in relation to degree of kinship in Japanese macaques (*Macaca fuscata*). *Behavioral Ecology and Sociobiology*, **49**, 493–502.

Charpentier, M. J. E., Deubel, D. and Peignot, P. (2008). Relatedness and social behaviors in *Cercopithecus solatus*. *International Journal of Primatology*, **29**, 487–495.

Combes, S. L. and Altmann, J. (2001). Status change during adulthood: life-history by-product or kin selection based on reproductive value? *Proceedings of the Royal Society of London, B*, **268**, 1367–1373.

Davies, N. B. and Houston, A. I. (1981). Owners and satellites: the economics of territory defence in the pied wagtail. *Journal of Animal Ecology*, **50**, 157–180.

de Waal, F. B. M. (1986). Class structure in a rhesus monkeys group: the interplay between dominance and tolerance. *Animal Behaviour*, **34**, 1033–1040.

  (1989). Food-sharing and reciprocal obligations in chimpanzees. *Journal of Human Evolution*, **18**, 433–459.

  (1993). Preliminary data on voluntary food sharing in brown capuchin monkeys. *American Journal of Primatology*, **29**, 73–78.

  (1997). Food transfers through mesh in brown capuchins. *Journal of Comparative Psychology*, **111**, 370–378.

Dubuc, C. and Chapais, B. (2007). Feeding competition in *Macaca fascicularis*: an assessment of the early arrival tactic. *International Journal of Primatology*, **28**, 357–367.

Fedigan, L. M. (1991). History of the Arashiyama West Japanese macaques in Texas. In *The Monkeys of Arashiyama: Thirty-five Years of Research in Japan and the West*, eds. L. M. Fedigan and P. J. Asquith. Albany, NY: State University of New York Press, pp. 54–73.

Feistner, A. T. and McGrew, W. C. (1989). Food-sharing in primates: a critical review. In *Perspectives in Primate Biology, Volume 3*, eds. P. K. Seth and S. Seth. New Delhi: Today and Tomorrow's Publishers, pp. 21–36.

Fraser, O. N., Schino, G. and Aureli, F. (2008). Components of relationship quality in chimpanzees. *Ethology*, **114**, 834–843.

Furuichi, T. (1983). Interindividual distance and influence of dominance on feeding in a natural Japanese macaque troop. *Primates*, **24**, 445–455.

Gouzoules, S. (1984). Primate mating systems, kin association, and cooperative behaviour: evidence for kin recognition? *Yearbook of Physical Anthropology*, **27**, 99–134.

Gouzoules, S. and Gouzoules, H. (1987). Kinship. In *Primate Societies*, eds. B. B. Smuts, D. L. Cheney, R. M. Seyfarth, R. W. Wrangham and T. T. Struhsaker. Chicago, IL: The University of Chicago Press, pp. 299–305.

Hamilton, W. D. (1964). The genetical theory of social behavior. *Journal of Theoretical Biology*, **7**, 1–52.

Hammerstein, P. (1981). The role of asymmetries in animal contest. *Animal Behaviour*, **29**, 193–205.

Ihobe, H. (1989). How social relationships influence a monkey's choice of feeding sites in a troop of Japanese macaques on Koshima Islet. *Primates*, **30**, 17–25.

Jaeggi, A. V., van Noordwijk, M. A. and van Schaik, C. P. (2008). Begging for information: mother-offspring food sharing among wild Bornean orangutans. *American Journal of Primatology*, **70**, 533–541.

Janson, C. H. (1988). Intra-specific food competition and primate social structure: a synthesis. *Behaviour*, **29**, 493–505.

Janus, M. (1989). Reciprocity in play, grooming, and proximity in sibling and non sibling young rhesus monkeys. *International Journal of Primatology*, **10**, 243–261.

Kaplan, J. R. (1978). Fight interference and altruism in rhesus monkeys. *American Journal of Physical Anthropology*, **47**, 241–249.

Kapsalis, E. and Berman, C. M. (1996). Models of affiliative relationship among free-ranging rhesus monkeys, *Macaca mulatta*. I. Criteria for kinship. *Behaviour*, **133**, 1209–1234.

Kawamura, S. (1965). Matriarchal social ranks in the Minoo-B group: a study of Japanese monkeys. In *Japanese Monkeys: A Collection of Translations*, eds. K. Imanishi and S. A. Altmann. Atlanta: Emory University Press, pp. 105–112.

Krebs, J. R. (1982). Territorial defence in the great tit: Do residents always win? *Behavioral Ecology and Sociobiology*, **11**, 185–194.

Kummer, H. and Cords, M. (1991). Cues of ownership in long-tailed macaques. *Animal Behaviour*, **42**, 529–549.

Kummer, H., Götz, W. and Angst, W. (1973). Triadic differentiation: an inhibitory process protecting pair bonds in baboons. *Behaviour*, **49**, 62–87.

Kurland, J. A. (1977). *Kin Selection in the Japanese Monkey*. Basel: S. Karger.

Masataka, N. and Fujii, H. (1980). An experimental study on facial expressions and interindividual-distance in Japanese macaques. *Primates*, **21**, 340–349.

Maynard Smith, J. (1982). *Evolution and the Theory of Games*. Cambridge: Cambridge University Press.

Maynard Smith, J. and Parker, G. A. (1976). The logic of asymmetric contests. *Animal Behaviour*, **24**, 159–175.

Mitani, J. C. (2009). Male chimpanzees form enduring and equitable social bonds. *Animal Behaviour*, **77**, 633–640.

Mori, A. (1977). Intra-troop spacing mechanism of the wild Japanese monkeys of the Koshima troop. *Primates*, **18**, 331–357.

Nowell, A. A. and Fletcher, A. W. (2006). Food transfers in immature wild western lowland gorillas. *Primates*, **47**, 294–299.

Parker, G. A. (1974). Assessment strategy and the evolution of fighting behaviour. *Journal of Theoretical Biology*, **47**, 223–243.

Perry, S. (1996). Female-female social relationships in wild white-faced capuchin monkeys. *American Journal of Primatology*, **40**, 167–182.

Perry, S., Manson, J. H., Muniz, L., Gros-Louis, J. and Vigilant, L. (2008). Kin-biased social behaviour in wild adult female white-faced capuchins. *Animal Behaviour*, **76**, 187–199.

Petit, O., Abegg, C. and Thierry, B. (1997). A comparative study of aggression and conciliation in three cercopithecine monkeys. *Behaviour*, **134**, 415–431.

Price, E. C. and Feistner, A. T. C. (1993). Food sharing in lion tamarins: tests of three hypotheses. *American Journal of Primatology*, **31**, 211–221.

Pusey, A. E. and Packer, C. (1997). The ecology of relationships. In *Behavioral Ecology: An Evolutionary Approach*, 4th edn, eds. J. R. Krebs and N. B. Davies. Oxford: Blackwell, pp. 254–283.

Saito, C. (1996). Dominance and feeding success in female Japanese macaques: effects of food patch size and inter-patch distance. *Animal Behaviour*, **51**, 967–980.

Saito, A., Izumi, A. and Nakamura, K. (2008). Food transfer in common marmosets: parents change their tolerance depending on the age of offspring. *American Journal of Primatology*, **70**, 999–1002.

Schino, G. and Aureli, F. (2010). The relative roles of kinship and reciprocity in explaining primate altruism. *Ecology Letters*, **13**, 45–50.

Schino, G., Rosati, L. and Aureli, F. (1998). Intragroup variation in conciliatory tendencies in captive Japanese macaques. *Behaviour*, **135**, 897–912.

Sigg, H. and Falett, J. (1985). Experiments on the respect of possession and property in hamadryas baboons. *Animal Behaviour*, **33**, 978–984.

Sterck, E. H. M., Watts, D. P. and van Schaik, C. P. (1997). The evolution of female social relationships in nonhuman primates. *Behavioral Ecology and Sociobiology*, **4**, 291–309.

Teleki, G. (1973). *The Predatory Behavior of Wild Chimpanzees*. Lewisburg, PA: Bucknell University Press.

Thierry, B. (1986). A comparative study of aggression and response to aggression in three species of macaque. In *Primate Ontogeny, Cognition, and Social Behaviour*, eds. J. G. Else and P. C. Lee. Cambridge: Cambridge University Press, pp. 307–313.

(1999). Covariation of conflict management patterns across macaque species. In *Natural Conflict Resolution*, eds. F. Aureli and F. B. M. de Waal. Berkeley, CA: University of California Press, pp. 106–128.

Thierry, B., Wunderlich, D. and Gueth, C. (1989). Possession and transfer of objects in a group of brown capuchins. *Behaviour*, **110**, 294–305.

Torii, M. (1975). Possession by non-human primates. In *Contemporary Primatology*, eds. S. Kondo, M. Kawai and A. Ehara. Basel: S. Karger, pp. 310–314.

Trivers, R. L. (1971). The evolution of reciprocal altruism. *Quarterly Review of Biology*, **46**, 35–57.

van Schaik, C. P. (1989). The ecology of social relationships amongst female primates. In *Comparative Socioecology*, eds. V. Standen and R. A. Foley. London: Blackwell Scientific Publications, pp. 195–218.

van Schaik, C. P., Assink, P. R. and Salafsy, N. (1992). Territorial behavior in Southeast Asian langurs: resource defense or mate defense? *American Journal of Primatology*, **26**, 233–242.

Walters, J. R. (1987). Kin recognition in nonhuman primates. In *Kin Recognition in Animals*, eds. D. F. Fletcher and C. D. Michener. New York, NY: John Wiley, pp. 359–393.

Whitten, P. L. (1983). Diet and dominance among female vervet monkeys. *American Journal of Primatology*, **5**, 139–159.

Wrangham, R. W. (1975). *The Behavioural Ecology of Chimpanzees in the Gombe National Park, Tanzania*. PhD thesis, Cambridge University.

Yamada, M. (1963). A study of blood-relationship in the natural society of the Japanese macaques. *Primates*, **4**, 45–65.

17 *Plant-food diet of the Arashiyama-Kyoto Japanese macaques and its potential medicinal value*

MICHAEL A. HUFFMAN AND
ANDREW J. J. MACINTOSH

Female Japanese macaque feeding on flowers of *Lindera umbellata* (Kuromoji) at Arashiyama (photo by M. A. Huffman).

The Monkeys of Stormy Mountain: 60 Years of Primatological Research on the Japanese Macaques of Arashiyama, eds. Jean-Baptiste Leca, Michael A. Huffman and Paul L. Vasey. Published by Cambridge University Press. © Cambridge University Press 2012.

17.1   Studies on feeding ecology at Arashiyama

During the long history of research on Arashiyama-Kyoto Japanese macaques, only limited attention has been paid to their feeding ecology. One of the first studies conducted, however, was an extensive botanical survey of the area by Kyoto University researchers Gen Murata and Nanosuke Hazama. They catalogued all the vascular plant species, including ferns and grasses, and noted which of them were eaten by the monkeys (Murata and Hazama, 1968). Their study was conducted between 1954 and 1958 and published as the second research report of the Iwatayama Natural History Research Station (Murata and Hazama, 1968). Before their study was published, only a very general description of the flora of Arashiyama was available (Hayashi, 1941).

In their survey, Murata and Hazama recorded 815 species from 119 plant families and 442 genera. From this diverse flora, the monkeys were observed to ingest 192 species, from 51% of these families and 34% of these genera. The food items or time of year ingested were not given. In the early stages of provisioning, the group's home range gradually shifted toward its present location in the Iwatayama area, where it now lives at the western edge of Kyoto city (Huffman, 1991). Murata and Hazama's survey coincided with the early period of provisioning but they did not consider it to be a complete list of the group's natural diet. They concluded this by the fact that the Mt. Hiei group, which is located on the eastern side of Kyoto city and was later studied by Hazama from the early 1960s, was found to have a natural diet of over 370 plant species (Hazama, unpublished records cited in Murata and Hazama, 1968). The Arashiyama-Kyoto group's proximity to human habitation is reflected in the fact that 25 cultivar species were recorded as food sources, including garden fruits and vegetables, ornamental trees and some exotic species whose seeds were likely serendipitously carried by the wind or dispersed by animals from nearby temples or household gardens (Murata and Hazama, 1968). Two such interesting species in the diet mentioned by Murata and Hazama are *Firmiana platanifolia* (Styracaceae), a coastal species native to China, Taiwan, Okinawa and parts of southern Kyushu, and *Poncirus trifoliata* (syn. *Citrus trifoliata*) (Rutaceae), native to central China and introduced to Japan. Both species were found in substantial numbers in relatively remote parts of the forest.

After this extensive survey, little attention was paid to the feeding habits of this group again until 21 years later when Huffman conducted a year-long survey in 1979 of Arashiyama B troop's plant food diet and feeding habits (Huffman, 1984), followed by a similar study by Wakibara between 1998 and 1999 (Wakibara *et al.*, 2001). By this time the group's home range had decreased noticeably from Hazama's original estimation of 8 km^2 in the 1950s to a core area of approximately 1 km² in 1979 (Huffman, 1984, 1991). The provisioning of food three to four times daily is believed to be responsible for this marked decrease in habitat use. Nonetheless, the group continues to forage on natural vegetation to this day. Huffman (1984) recorded 111 food items from 67 plant-food species (34 families, 55 genera), including bark, fruits, buds, flowers, leaves, roots, seeds, shoots, stems and twigs at various stages of development. Wakibara *et al.* (2001) identified an additional nine new food items from seven new food species, including three new families and three new genera.

Provisioned foods such as wheat, soybeans, chestnuts and various fruits and vegetables are excluded from these two lists. In total, the latter two studies added 24 new species to the food list originally assembled by Murata and Hazama (1968). Five of these new species, including tea olive (*Osmanthus fragrans* var. *auranticus*: Oleaceae), Fortune's osmanthus (*O. fortunei*), tea (*Camellia sinensis*: Theaceae), jasmine (*Gardinia jasmioides*: Rubiaceae) and Japanese photinia (*Photinia glabra*: Rosaceae), are cultivars planted around the provisioning grounds and visitor station for the enjoyment of the public.

Murata and Hazama (1968) and Huffman (1984) were published in Japanese in limited numbers and have remained largely inaccessible to the larger primatological community. For this reason, we decided to compile a comprehensive list of the diet of Arashiyama-Kyoto macaques from records collected over a 45-year period of research. Combining these three studies, this Arashiyama plant food list consists of 219 species (Table 17.1, Figure 17.1).

17.2   Bioactivities in the diet and their potential importance for health maintenance

Primate food choices are also influenced by a number of factors, including nutrients and plant secondary metabolites with feeding deterrent properties, such as tannins, saponins and phenolics (e.g. Glander, 1982; Iwamoto, 1982; Nakagawa *et al.*, 1996; Worman and Chapman, 2005; Fashing *et al.*, 2007; Hanya *et al.*, 2007; Felton *et al.*, 2009; Jaman *et al.*, 2010). The possible role of these secondary metabolites and their potential medicinal benefits in the primate diet has recently begun to receive greater attention (e.g. Huffman, 2007; Forbey *et al.*, 2009; MacIntosh and Huffman, 2010). As an initial step toward understanding the possible relationship between secondary metabolites and health maintenance in Japanese macaques at Arashiyama and across Japan, we compiled a database listing those plant items containing such chemical compounds, and the ways in which they might benefit their consumers (Table 17.2).

Table 17.1. *Plant food list of the Arashiyama Japanese macaques*

Family	Species	Food item[1] (month eaten)[2]	Japanese common name
Aceraceae カエデ科	*Acer palmatum* Thunb.		Irohamomiji イロハモミジ
	Acer palmatum var. *amoenum* Carr.	n:lf (4)	Oomomiji オオモミジ
	Acer palmatum var. *matsumurae* CULT.		Yamamomiji ヤマモミジ
Actinidiaceae マタタビ科	*Actinidia arguta* (Siebold & Zucc.) Planch. ex Miq.		Sarunashi サルナシ
Amaranthaceae ヒユ科	*Achyranthes japonica* (Miq.) Nakai		Inokotsuji イノコヅチ
Anacardiaceae ウルシ科	*Rhus javanica* L.		Nurude ヌルデ
	Rhus sylvestris S. et Z.		Yamahaze ヤマハゼ
	Rhus trichocarpa Miq.	lf	Yamaurushi ヤマウルシ
Apocynaceae キョウチクトウ科	*Trachelosperum asiaticum* Thunb.	lf (12–2)	Teikakazura テイカカズラ
Aquifoliaceae モチノキ科	*Ilex chinensis* Sims	lf	Nanaminoki ナナミノキ
	Ilex integra Thunb.		Mochinoki モチノキ
	Ilex pedunculosa Mia.	lf (1–4), bk (1–)	Soyogo ソヨゴ
Araliaceae ウコギ科	*Aralia cordata* Thunb.		Udo ウド
	Aralia elata var. *subinermis* Ohwi.	bk (12–2)	Medara メダラ
	Euodiopanax innovans Nakai.	bk (9–)	Takanotsume タカノツメ
Aspidiaceae オシダ科	*Polystichum polyblempharum* Pr.	st (2)	Inode イノデ
Berberidacea メギ科	*Nandina domestica* Thunb. CULT.	lf (1–3)	Nanten ナンテン
Boraginaceae ムラサキ科	*Cynoglossum asperrimum* Nakai		Onirurisou オニルリソウ
Campanulaceae キキョウ科	*Codonopsis lanceolata* (Siebold. & Zucc.) Trautv.		Tsuruninjin ツルニンジン
Caprifoliaceae スイカズラ科	*Lonicera gracilipes* Miq. var. *glabra* Miq.		Uguisukagura ウグイスカグラ
	Sambucus sieboldiana Blume.	bk (1–), bd (1–), n:l (3/10)	Niwatoko ニワトコ
	Viburnum erosum Thunb.		Kobanogamazumi コバノガマズミ
	Viburnum sieboldii Miq.		Gomagi ゴマギ

Table 17.1. (*cont.*)

Family	Species	Food item[1] (month eaten)[2]	Japanese common name
Caryophyllaceae ナデシコ科	*Stellaria aquatica* (L.) Scop.		Ushihakobe ウシハコベ
	Stellaria neglecta (Weihe.) Murr.		Hakobe ハコベ
Cephalotaxaceae イヌガヤ科	*Cephalotaxus harringtonia* var. *nana* Rehder	lf (2–3)	Hai-inugaya ハイイヌガヤ
Clethraceae リョウブ科	*Clethra barvinervis* Sieb. et Zucc.	lf	Ryobu リョウブ
Commelinaceae ツユクサ科	*Commelina communis* L.		Tsuyukusa ツユクサ
	Pollia japonica Thunb.		Yabumyouga ヤブミョウガ
Compositae キク科	*Bidens frondosa* L.	st (12–1)	America sendangusa アメリカ　センダングザ
	Cirsium microspicatum Nakai.		Oharameazami オハラメアザミ
	Cirsium japonicum DC.		Noazami ノアザミ
	Erechtites hieraciifolia (L.) Raf. Ex DC.		Dandorobogiku ダンドロボロギク
	Erigeron annuus (L.) Pers.		Himejoon ヒメジョオン
	Eupatorium chinense L.		Hiyodoribana ヒヨドリバナ
	Hemisteptia lyrata (Bunge) F. E. L. Fischer et C. A. Meyer		Kitsuneazami キツネアザミ
	Ixeris dentata Nakai		Nigana ニガナ
	Ixeris stolonifera A. Gray		Iwanigana イワニガナ
	Lactuca indica L.		Akinogeshi アキノゲシ
	Lactuca sororia Miq.		Murasakinigana ムラサキニガナ
	Petasites japonicus (Siebold et Zucc.) Maxim.		Fuki フキ
	Picris hieracioides L.		Kozorina コウゾリナ
	Sonchus oleraceus L.		Nogeshi ノゲシ
	Sonchus asper (L.) Hill		Oninogeshi オニノゲシ
	Taraxacum japonicum Koidz.		Kansaitanpopo カンサイタンポポ
	Youngia japonica (L.) DC.		Onitabirako オニタビラコ

Table 17.1. (*cont.*)

Family	Species	Food item[1] (month eaten)[2]	Japanese common name
Cornaceae ミズキ科	*Aucuba japonica* Thunb.	lf (9–), st.lf (12–2)	Aoki アオキ
	Helwingia japonica Thunb.		Hanaikada ハナイカダ
Cruciferae アブラナ科	*Raphanus sativus* L. CULT.		Daikon ダイコン
Cupressaceae ヒノキ科	*Chamaecyparis obtusa* (Sieb. et Zucc.) Endl. CULT.		Hinoki ヒノキ
Cyperaceae カヤツリグサ科	*Carex gibba* Wahlenb.		Masukusa マスクサ
	Carex stenostachys Franch. et Sav.		Nitsunohonmonjisuge ニシノホンモンジスゲ
Dioscoreaceae ヤマノイモ科	*Dioscorea japonica* Thunb.	lf (6–8)	Yamanoimo ヤマノイモ
	Dioscorea quinqueloba Thunb.		Kaedokoro カエデドコロ
Ebenaceae カキノキ科	*Diospyrus kaki* Thunb. CULT.	b (7–), fr (7–), lf (7–)	Kaki カキ
	Diospyrus kaki var. *sulvestris* Makino.	lf (7–)	Yamakaki ヤマカキ
Elaeagnaceae グミ科	*Elaeagnus pungens* Thunb.	lf (1–2)	Nawashirogumi ナワシログミ
Ericaceae ツツジ科	*Lyonia ovalifolia* var. *elliptica* Hand. – Mzt.	lf (–1)	Nejiki ネジキ
	Pieris japonica D. Don.	fl (3–5)	Asebi アセビ
	Rhododendron indicum Sweet.	fl (4–6)	Satsukitsutsuji サツキツツジ
	Rhododendron kaempferi Planch.	fl (4–6)	Yamatsutsuji ヤマツツジ
	Rhododendron macrocepalum Maxim.	fl (5)	Mochitsutsuji モチツツジ
	Rhododendron metternichii var. *hondoense* Nakai CULT.		Honshakunage ホンシャクナゲ
	Rhododendron reticulatum D. Don		Kobanomitsutsuji コバノミツバツツジ
	Vaccinium bracteatum Thunb.	lf (2–4)	Shashanbo シャシャンボ
	Vaccinium hirtum Thub.		Kakuminosuki カクミノスノキ
	Vaccinium oldhamii Miq.		Natsuhaze ナツハゼ
Euphorbiaceae トウダイグサ科	*Daphniphyllum macropodum* Miq. CULT.		Yuzuriha ユズリハ

Table 17.1. (*cont.*)

Family	Species	Food item[1] (month eaten)[2]	Japanese common name
Fagaceae ブナ科	*Castanea crenata* Sieb et Zucc.	lf (9–11)	Kuri クリ
	Castanopsis cuspidata (Thunb. ex Murray)	gn (9–11)	Tsuburaji (Sudajii) ツブラジイ(スダジイ)
	Quercus acutissima Carruth.	lf	Kunugi クヌギ
	Quercus acuta Thunb.	lf	Akagashi アカガシ
	Quercus glauca Thunb.	tw (4–5), n:lf (4–5), gn (9–), lf (11–)	Arakashi アラカシ
	Quercus serrata Thunb.	lf (11–12)	Konara コナラ
	Quercus variabilis Blume.	lf (11–)	Abemaki アベマキ
Gramineae (Poaceae) イネ科	*Agropyron tsukushiense* (Honda) Ohwi.		Kamojigusa カモジグサ
	Arundinaria pygmaea (Miq.) Mitf. var. *glabra* (Makino) Ohwi	lf (9–4), sh (9–4)	Azumazasa アズマザサ
	Calamagrostis arundinacea (L.) Roth		Nogariyasu ノガリヤス
	Digitaria timorensis Balansa.	rt (4–5), n:lf (4–5), lf (6–10),	Komehishiba コメヒシバ
	Echinochloa crusgalli (L.) Beauv		Inubie イヌビエ
	Echinochloa crusgalli Beauv. var. *caudata* Kitagawa		Keinubia ケイヌビエ
	Isachne globosa O. Ktze.	lf	Chigosasa チゴササ
	Melica nutans L.		Komegaya コメガヤ
	Melica vimineum (Trin.)		Ashibon アシボソ
	Miscanthus sinensis Anderss.	lf (5–11), rt (5–11)	Susuki ススキ
	Oryza sativa L. CULT.		Ine イネ
	Paspalum thunbergii Kunth ex Steud.		Suzumenobie スズメノヒエ
	Pennisetum alopecuroides Spreng.	lf (4–11), n:rt (3–8), sh (3–8), rt (12)	Chikarashiba チカラシバ

Table 17.1. (*cont.*)

Family	Species	Food item[1] (month eaten)[2]	Japanese common name
	Phalaris arundinacea L.		Kusayoshi クサヨシ
	Phyllostachys bambusoides Sieb. et Zucc. CULT.		Madake マダケ
	Phyllostachys heterocycla var. *pubscens* Ohwi CULT.		Kikkou-chikku キッコウチク
	Poa acroleuca Stued.		Mizoichigotsunagi ミゾイチゴツナギ
	Poa annual L.	n:lf (11–3), lf (4–5)	Zuzumenokatabira ズズメノカタビラ
	Setaria viridis P. Beauv.		Enokorogusa エノコログサ
	Sporobolus indicus R. Br.	cl:fl (5–10)	Nezuminoo ネズミノオ
	Zea mays L. CULT.		Toumorokoshi トウモロコシ
Iridacea アヤメ科	*Iris japonica* Thunb.	lf (2–3)	Shaga シャガ
Juncaceae イグサ科	*Luzula plumosa* E. Meyer, Linn.		Nukaboshisou ヌカボシソウ
	Luzula multiflora (Ehrh.) Lej.		Yamasuzumenohie ヤマスズメノヒエ
Labeataceae シソ科	*Rhadosia inflexa* (Thunb.) Hara	lf, sd, st	Yamahakka ヤマハッカ
Lardizabalaceae メギ科	*Akebia quinata* Decne.	lf (10–11), fr (10–11)	Akebi アケビ
	Akebia trifoliata Koidz.		Mitsubaakebi ミツバアケビ
	Stauntonia hexaphylla (Thunb.) Decaisne.		Mube ムベ
Lauraceae クスノキ科	*Actinodaphne lancifolia* (Sieb. et Zucc.) Meisn.		Kagonoki カゴノキ
	Lindera umbellata Thunb.	fl (4)	Kuromoji クロモジ
	Cinamonium camphora Sieb. CULT.	fr (11–)	Kusunoki クスノキ
	Cinamonium japonicum Sieb.	fr (11–)	Yabunikkei ヤブニッケイ
	Neolistea sericea Koidz.	n:lf (3–)	Shirodamo シラダモ
Leguminosae (Fabaceae) マメ科	*Albizzia julibrissin* Durazz.		Nemunoki ネムノキ
	Desmodium oldhamii Oliver		Fujikanzou フジカンゾウ
	Desmodium oxyphyllum DC.		Nusubitohagi ヌスビトハギ

Table 17.1. (*cont.*)

Family	Species	Food item[1] (month eaten)[2]	Japanese common name
	Desmodium paniculatum DC.		Arechinuzubitohagi アレチヌスビトハギ
	Dumasia truncata Sieb. et Zucc.		Nosasage ノササゲ
	Glycine max (L.) Merrill CULT.		Daizu ダイズ
	Lespedeza bicolor Turcz.		Yamahagi ヤマハギ
	Lespedeza cyrtobotrya Miq.		Marubahagi マルバハギ
	Lespedeza pilosa Sieb. et Zucc.		Nekohagi ネコハギ
	Milletia japonica A. Gray.		Natsufuji ナツフジ
	Pueraria lobata (Willd.) Ohwi		Kuzu クズ
	Trifolium repens L.		Shirotsumekusa シロツメクサ
	Vicia angustifolia L.		Karasunodendou カラスノエンドウ
	Vicia hirsuta (L.) Gray		Susumenoendou スズメノエンドウ
	Wisteria floribunda DC.	bk (11–12), bd (11–12), bd (4)	Fuji フジ
Liliaceae ユリ科	*Allium fistulosum* L. CULT.		Negi ネギ
	Liriope platyhylla Wang et Tang.	lf (12–1), fr (12–1)	Yaburan ヤブラン
	Lilium japonicum Thunb.		Sasayuri ササユリ
	Ohiopogen japonicus Ker Gawler.	lf (12–1), fr (12–1)	Janohige ジャノヒゲ
	Polygonatum macranthum (Maxim.) Koidz.		Oonarukoyuri オオナルコユリ
	Smilax china L.	lf (11–)	Sarutoribara サルトリバラ
Loranthaceae オオバヤドリギ科	*Taxillus kaempferi* (DC) Danser		Matsugumi マツグミ
Magnoliaceae モクレン科	*Illicium religiosum* Seib. et Zucc.		Shikimi シキミ
	Magnolia kobus DC. var. borealis Sarg.		Kobushi コブシ
	Magnolia obovata Thunb.		Hoonoki ホオノキ
Moraceae クワ科	*Broussonetia kazinoki* Sieb.		Kouzo コウゾ

Table 17.1. (*cont.*)

Family	Species	Food item[1] (month eaten)[2]	Japanese common name
	Cannabis sativa Linné CULT.		Asa アサ
	Ficus carica L. CULT.		Ichijiku イチジク
	Morus bombycis Koidz.		Yamaguwa ヤマグワ
Myricacea ヤマモモ科	*Myrica rubra* Seib et Zucc.	b (6–7)	Yamamomo ヤマモモ
Oleaceae モクセイ科	*Ligustrum japonicum* Thunb.		Nezumimochi ネズミモチ
	Osmanthus fragrans var. *auranticus* Makino CULT.	lf (10–1), fl (10)	Kinmokusei キンモクセイ
	Osmanthus fortunei CULT.	lf	Hiragimokuse ヒラギモクセイ
	Osmanthus heterophyllus P. S. Green.	bk (1–4), lf (1–3), fl:bd (9), fl (10–11)	Hiiragi ヒラギ
Papaveraceae ケシ科	*Corydalis incisa* (Thunb.) Pers.		Murasakikeman ムラサキケマン
	Macleaya cordata (Willd.) R. Br.	st (4–5)	Takenigusa タケニグサ
Pinaceae マツ科	*Pinus densiflora* Seib et Zucc.	lf:sp (10–11), lf (12–3)	Akamatsu アカマツ
	Pinus thunbergii Parl. CULT.		Kuromatsu クロマツ
Polygoaceae タデ科	*Polygonum caespitosum* L.		Hanatade ハナタデ
	Polygonum cuspidata Sieb et Zucc.	lf (6–7)	Itadori イタドリ
	Polygonum filiforme Thunb.		Mizuhiki ミズヒキ
	Polygonum longisetum D. Bruyn.	st (12–2)	Inutaji イヌタジ
	Polygonum nipponense Makino		Yanonegusa ヤノネグサ
	Polygonum pubescens Blume		Bontokudade ボントクタデ
	Polygonum thunbergii H. Gross		Mizosoba ミゾソバ
	Rumex acetosa L. (*Polygonum acetosa* syn.)		Suiba スイバ
Pteridaceae ワラビ科	*Coniogramme intermedia* Hieron.	stalk (2–3)	Iwaganezenmai イワガネゼンマイ
Ranunculaceae キンポウゲ科	*Ranunculus quelpaertensis* DC.		Kitsunenobotan キツネノボタン
Rhamanaceae クロウメモドキ科	*Rhamnus crenata* Siebold et Zucc.	lf	Isonoki イソノキ

Table 17.1. (*cont.*)

Family	Species	Food item[1] (month eaten)[2]	Japanese common name
Rosaceae バラ科	*Duchesnea indica* (Andrews) Th.Wolf	b (4–), fl:b (4–)	Hebiichigo ヘビイチゴ
	Duchesnea major Major		Yabuhebiichigo ヤブヘビイチゴ
	Eriobotrya japonica (Thunb.) Lindl.		Biwa ビワ
	Kerria japonica DC.	nlf (4–)	Yamabuki ヤマブキ
	Malus sieboldii Rehder CULT.		Zumi ズミ
	Photinia glabra Maxim. CULT.	lf (11–3)	Kanamemochi カナメモチ
	Pourthiaea villosa Thunb.		Kamatsuka カマツカ
	Prunus ansu (Maxim.) Kom. CULT.		Anzu アンズ
	Prunus grayana Maxim.		Uwamizuzakura ウワミズザクラ
	Prunus jamasakura Sieb.	bk (1–), fl:bd (3–)	Yamazakura ヤマザクラ
	Prunus mume Siebold & Zucc.		Ume ウメ
	Prunus persica (L.) Batsch CULT.		Momo モモ
	Prunus yedoensis Matsum. CULT.		Soumeiyoshino ソメイヨシノ
	Pyrus pyrifolia (Burm.) Nak. Var. *culta* CULT.		Nashi ナシ
	Rubus buergeri Miq.		Fuyuichigo フユイチゴ
	Rubus corchorifolius L. f.	lf (12–2)	Biroodoichigo ビロウドイチゴ
	Rubus crataegifolius Bunge		Kumaichigo クマイチゴ
	Rubus hakonensis Fr. et. Sav.	lf (11–), b (12–)	Miyama fuyuichigo ミヤマフユイチゴ
	Rubus hirsutus Thunb.	lf (12–), st (12–)	Kusaichigo クサイチゴ
	Rubus illecebrosus Focke		Baraichigo バライチゴ
	Rubus microphyllus		Nigaiichigo ニガイチゴ
	Rubus palmatus Thunb.		Nagabamomijiichigo ナガバモミジイチゴ
	Rubus parvifolius L.		Nawashiroichigo ナワシロイチゴ

Table 17.1. (*cont.*)

Family	Species	Food item[1] (month eaten)[2]	Japanese common name
	Rubus sorbifolius D.		Kojikiichigo コジキイチゴ
Rubiaceae アカネ科	*Gardinia jasmioides* var. *frandiflora* Nakai. CULT.	tw (10–11), nL. (10), lf (10)	Kuchinashi クチナシ
	Galium spurium L.		Yaemugura ヤエムグラ
	Paederia scandens (Lour.) Merrill		Hekusokazura ヘクソカズラ
Rutaceae ミカン科	*Orixa japonica* Thunb.		Kokusagi コクサギ
	Poncirus trifoliata Rafin. (syn. *Citrus trifoliata*)		Karatachi カラタチ
	Poncirus junos (syn. *Citrus junos* Sieb. ex Tanaka) CULT.		Yuzu ユズ
	Zanthoxylum ailanthoides (Sieb. et Zucc.)		Karasusanshou カラスザンショウ
	Zanthoxylum piperitum DC.		Sanshou サンショウ
Saxifragaceae ユキノシタ科	*Hydrangea hirta* (Thunb. ex Murray) Sieb. et Zucc.		Koajisai コアジサイ
Solanaceae ナス科	*Physalis alkekengi* var. *francheti* Hort CULT.		Hoozuki ホオズキ
	Solanum japonense Nakai		Yamahoroshi ヤマホロシ
	Tubocapsicum anomalum (Franch. et Savat.) Makino		Hadakahoozuki ハダカホオズキ
Sterculiaceae アオギリ科	*Firmiana platanifolia* (L.f.) Marsili		Aogiri アオギリ
Styracaceae エゴノキ科	*Styrax japonica* Sieb. et Zucc.	nL (4), fl (5), sd (9–10), bk (11), tw (11)	Egonoki エゴノキ
Symplocaceae ハイノキ科	*Symplocos prunifolia* Sieb. et Zucc.		Kurobai クロバイ
Taxaceae イチイ科	*Torreya nucifera* Sieb. et Zucc.	lf (2–4)	Kaya カヤ
Taxodiaceae スギ科	*Cryptomeria japonica* D. Don.	lf (1–)	Kitayamaugi キタヤマスギ
Theaceae ツバキ科	*Camellia japonica* L.	lf (11–4), bd.fl (10–11), fl (1–3)	Tsubaki ツバキ
	Camellia sinensis O. Kuntze. CULT.	lf (1–2)	Cha チャ
	Cleyera japonica Thunb.	bk (1–2), nlf (3–)	Sakaki サカキ

Table 17.1. (*cont.*)

Family	Species	Food item[1] (month eaten)[2]	Japanese common name
	Eurya japonica Thunb.	lf (1–3)	Hisakaki ヒサカキ
Ulmaceae ニレ科	*Aphananthe aspera* Planch.	fr (5–6/9–), sd (5–6/9–), lf (10–11)	Mukunoki ムクノキ
	Celtis sinensis Planch.		Enoki エノキ
Umbelliferae セリ科	*Daucus carota* Linn. CULT.		Ninjin ニンジン
	Heracleum lanatum Michx.		Hanaudo ハナウド
Urticaeae イラクサ科	*Boehmeria longispica* Steud.		Yabumao ヤブマオ
	Boehmeria niponivea Koidz.		Kusamao (Karamushi) クサマオ(カラムシ)
	Boehmeria platanifolia Fr. et Sav.		Meyabumao メヤブマオ
	Urtica thunbergiana Siebold. & Zucc.		Irakusa イラクサ
Verbenaceae クマツヅラ科	*Callicarpa japonica* Thunb.	b (11)	Murasakishikibu ムラサキシキブ
	Callicarpa mollis Sieb. et Zucc.	b (11)	Yabumurasakishikibu ヤブムラサキシキブ
	Clerodendron trichotomum Thunb.	st.lf (11–12)	Kusagi クサギ
Violaceae スミレ科	*Viola grypoceras* A. Gray		Tachitsubosumire タチツボスミレ
Vitaceae ブドウ科	*Ampelopsis brevipedunculata* Trautv.		Nobudou ノブドウ
	Cayratia japonica (Thunb.) Gagnep.		Yabugarashi ヤブガラシ
	Parthenocissus tricuspidata (Siebold & Zucc.) Planch.		Tsuta ツタ
	Vitis flexuosa Thunb.		Sankakudsuru サンカクヅル
Zingiberaceae ショウガ科	*Zingiber mioga* (Thunb.) Roscoe		Myouga ミョウガ

CULT. = cultivars

1. Food items: bk (bark), b (berry), bd (bud of leaf), fl (florescence), n:lf (new leaf), lf (leaf), st:l (leaf stem), lf:sp (leaf sap), n (nut), gn (green nut), rt (root), fl:bd (flower bud), cl:fl (composite florescence), fr (fruit), n:rt (new root), sd (seed), sh (shoot), st (stem), tw (twig).
2. Months: Those periods (months) of use by monkeys are available only for those species reported in Huffman (1984).

(a)

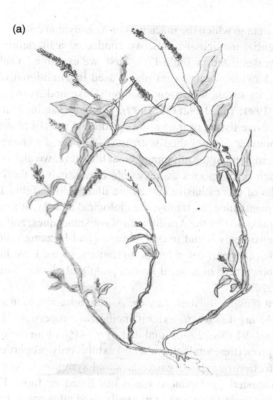

(b)

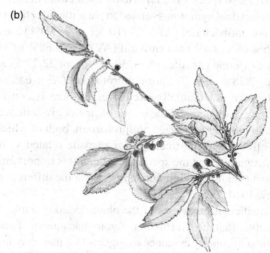

Figure 17.1. (a) *Penisetum longisetum*; (b) *Eurya japonica* (drawings by M. A. Huffman).

To determine the extent to which the macaques of Arashiyama consume plant items which are potential medicinal foods, we conducted a literature search for all of the plant species listed in Table 17.1. First, we examined a subset of the ethnomedicinal literature, focusing on plants used in Traditional Chinese Medicine (TCM) and its Japanese progeny, Kanpo (Duke and Ayensu, 1985; Kimura and Kimura, 1991; Yen, 1992; Ou, 1999). We then examined an online database with an extensive listing of the various medicinal uses of plants, with citations to a large number of published sources (*Plants for a Future*: http://www.pfaf.org/). To complement the ethnomedicinal literature, we also used the scientific name of each plant species as keywords in a search of the *ISI Web of Science*. The results of this extensive survey are illustrated in Table 17.2, in which we report the plant name, its family, the biological activity or medicinal use of each species consumed by the Arashiyama-Kyoto macaques, and a list of selected chemical constituents found in each. Murata and Hazama (1968) did not report the specific plant part eaten by the macaques, so we have included information on all known activities of all known plant parts found during our investigation.

From the 219 plant species consumed by the Arashiyama-Kyoto macaques, 135 species (61.64%) are known to exhibit medicinal properties. Of these medicinal plant species, 91 (38.72% of total diet or 67.41% of medicinal diet) exhibit antiparasitic properties, while the other 44 exhibit only properties unrelated to parasitic infection or related gastrointestinal symptoms. Among the antiparasitic items ingested by Japanese macaques listed in Table 17.2, 61 (27.85% or 45.18%) of these species are reportedly used ethnomedicinally by humans to treat gastrointestinal symptoms related to parasitic infection (GI), 48 (21.92% or 35.56%) are antibacterial (AB), 23 (10.5% or 17.04%) are anthelmintic (AH), 23 (10.5% or 17.04%) are antiviral (AV), 19 (8.68% or 14.07%) display activity against external parasites (XP), 30 (13.7% or 22.22%) are antifungal (AF), and five (2.28% or 3.7%) are antiprotozoan (AP) (e.g. antimalarial). These properties are not mutually exclusive, as there is considerable overlap in activity among plant items of the same species. A given item can, for example, be concurrently anthelmintic and antiprotozoan, both of which make it more likely to be effective in the treatment of parasite-related symptoms. Furthermore, our investigation is at the species level, but it is important to note that chemical constituents can vary significantly between the different parts of a single plant (McKey, 1974).

The bioactive and medicinal properties of the plants listed in Table 17.2 are not evidence in themselves that the Arashiyama-Kyoto macaques self-medicate. However, there is some preliminary evidence to suggest that they may indeed be benefiting from the bioactive properties of some of these plants. For example, *Macleaya cordata* (Papaveraceae) is an herbaceous perennial eaten by Japanese macaques at several sites (see Table 3 in MacIntosh and Huffman, 2010). Its ingestion was first noted by Junichiro Itani during his early observations at Takasakiyama in the 1950s, and he was struck by the fact that it turned the faeces of the monkeys to a gelatinous dark brown colour (J. Itani, pers. comm.). Itani and Huffman together later observed the use of this plant by monkeys at Funakoshiyama (Hyogo Prefecture). What is particularly interesting about *takenigusa* – the plant's common name in Japan – is that the pith, which is so voraciously eaten by monkeys in the spring, is extremely bitter to the taste and contains highly bioactive properties. A main ingredient of the plant, the milky yellowish alkaloid sanguinarine, is toxic and has been demonstrated to readily kill animal cells through its action on Na, K, ATPase transmembrane protein (Pitts and Meyerson, 1981; Dalvi, 1985). It is known for its anti-inflammatory and antioxidant properties, and is an active ingredient in many dental products for its beneficial antimicrobial properties (Yaniv and Bachrach, 2005).

During their communications, Itani told Huffman that he thought the monkeys might be using takenigusa to purge themselves of parasites in the spring. Recently, we have begun to investigate this possibility. Preliminary data suggest that the crude extract of *M. cordata* may exhibit potent nematodicidal activity against the common Japanese macaque parasites *Oesophagostomum aculeatum* (Strongylida) and *Strongyloides fuelleborni* (Rhabditida) in vitro (MacIntosh, unpublished data). On Yakushima Island in southern Japan, infection with these parasites appears to be most intense during spring and summer, respectively (MacIntosh *et al.*, 2010). Considering the patterns of climatic seasonality throughout most of Japan, infections are presumably most intense at Arashiyama during these seasons as well. The macaques could thus benefit from the ingestion of *M. cordata* if it played even a minor role in controlling these infections in vivo during periods of highest infection risk, as Itani had previously hypothesised.

On the other hand, primates also avoid ingesting certain potentially toxic plant items (Freeland and Janzen, 1974). Indeed, the Japanese macaques at Arashiyama have learned to remove or avoid toxic parts of some plants while still managing to obtain nutritional benefit from them. For example, the seeds of *Styrax japonica* (Styracaceae) are commonly eaten at Arashiyama despite the fact that the plant, including the seeds, is known to be highly toxic (see Table 17.2). The concentration of triterpene saponins was found to be particularly high in the seeds and was shown to be capable of complete haemolysis of erythrocytes at even low concentrations (Wakibara *et al.*, 2001). Interestingly, the concentration of this secondary metabolite varies according to seed part, being highest in the seed case and lowest in the seed itself (Wakibara *et al.*, 2001). The monkeys are able to exploit this potentially hazardous food item by carefully removing the thin layer of casing between the shell and the seed.

Table 17.2. *Plant items containing secondary metabolites and health maintenance benefits in Japanese macaques at Arashiyama and across Japan. PFAF refers to Plants for a Future: (http://www.pfaf.org/).*

Family	Latin name	Biological activity	Chemical constituents
Actinidiaceae	*Actinidia arguta* (Siebold & Zucc.) Planch. ex Miq.	*Fruit*: Antioxidant & cancer-related (Yoshizawa *et al.*, 2000)	
Alliaceae	*Allium fistulosum* L.	*Plant*: Antifungal (Pyun and Shin, 2006), *Seed*: Sex & pregnancy, renal disorders & calculi, general analgesic, gastrointestinal or digestive disorders (Yen, 1992), *Leaf*: Antioxidant & cancer-related (Stajner *et al.*, 2006), *Bulb*: Antibacterial, burns/lesions & skin disorders, antiperspirant, renal disorders & calculi, sex & pregnancy, gastrointestinal or digestive disorders, anthelmintic, metabolic and circulatory disorders (PFAF: Yeung, 1985; Duke and Ayensu, 1985), Inflammation & inflammatory disorders, burns/lesions & skin disorders (PFAF: Chevallier, 2001), Antioxidant & cancer-related (Stajner *et al.*, 2008), Metabolic and circulatory disorders (Chen *et al.*, 1999), *Root*: sedative (PFAF: Duke and Ayensu, 1985)	*Plant*: allicin (Pyun and Shin, 2006), *Seed*: alkaloids, saponins (Yen, 1992), *Leaf*: flavonoids, carotenoids, chlorophylls (Stajner *et al.*, 2006), *Bulb*: sulfur compounds (PFAF: Dan and Nhu, 1989), flavonoids, carotenoids, chlorophylls (Stajner *et al.*, 2008)
Amaranthaceae	*Achyranthes japonica* (Miq.) Nakai	*Plant*: Antifungal (Kim *et al.*, 2004a), *Root*: Sex & pregnancy, general analgesic, inflammation & inflammatory disorders, muscle-related, renal disorders & calculi, metabolic and circulatory disorders (PFAF: WHO, 1998), Antibacterial (Jung *et al.*, 2008), Inflammation & inflammatory disorders (Han *et al.*, 2005), *Leaf*: Antidote (Minakami and Jitouzono, 2002)	*Root*: triterpenoid saponins, protocatechuic acid (PFAF: WHO, 1998)
Anacardiaceae	*Rhus javanica* L.	*Bark*: Astringent, anthelmintic (PFAF: Duke and Ayensu, 1985), *Fruit*: Colic (PFAF: Chopra *et al.*, 1986), Cold/flu & respiratory diseases, antidote, gastrointestinal or digestive disorders (Kimura and Kimura, 1991), Liver diseases (Abbasi *et al.*, 2009), *Sap*: Cold/flu & respiratory diseases, gastrointestinal or digestive disorders,	*Fruit*: gallotannins, gallotanic acid, gallic acid, m-digallic acid (Abbasi *et al.*, 2009), *Root*: gallotannins, gallotanic acid, gallic acid, m-digallic acid (Abbasi *et al.*, 2009), *Stem*: benzofuranone lactones: rhuscolide

Family	Species	Uses	Constituents
		antiperspirant, blood purification & clotting, burns/lesions & skin disorders, eye-related (Yen, 1992), **Seed**: Cold/flu & respiratory diseases, gastrointestinal or digestive disorders, febrifuge, liver diseases, inflammation & inflammatory disorders, antiprotozoan (PFAF: Duke and Ayensu, 1985), Liver diseases (Abbasi et al., 2009), **Flower**: Cold/flu & respiratory diseases (Kimura and Kimura, 1991), **Leaf**: Metabolic and circulatory disorders (PFAF: Fogarty, 1977), Blood purification & clotting, inflammation & inflammatory disorders, cold/flu & respiratory diseases, antidote, gastrointestinal or digestive disorders, bone-related (PFAF: Duke and Ayensu, 1985), Cold/flu & respiratory diseases, antidote, gastrointestinal or digestive disorders (Kimura and Kimura, 1991), **Stem**: Antiviral (Wang et al., 2008; Gu et al., 2007), **Gall**: Astringent, gastrointestinal or digestive disorders, metabolic and circulatory disorders (PFAF: Duke and Ayensu, 1985), antiviral (Chang et al., 2010), Antifungal (Ahn et al., 2005), metabolic and circulatory disorders (Shim et al., 2003; Doo et al., 2003), **Root**: Inflammation & inflammatory disorders, cold/flu & respiratory diseases (Kimura and Kimura, 1991), Metabolic and circulatory disorders (PFAF: Fogarty, 1977), Blood purification & clotting, inflammation & inflammatory disorders, cold/flu & respiratory diseases, antidote, gastrointestinal or digestive disorders, bone-related (PFAF: Duke and Ayensu, 1985)	A (Gu et al., 2007), **Gall**: tannins (WHO, 1998), moronic acid (Chang et al., 2010), methyl gallate, gallic acid (Ahn et al., 2005)
Anacardiaceae	*Rhus sylvestris* S. et Z.	**Leaf & Stem**: Inflammation & inflammatory disorders (Ding et al., 2009b), bone-related (Ding et al., 2009a)	**Leaf, Stem**: triterpenes, sterol, chalcone (Ding et al., 2009b), megastigmane glycoside: rhusonoside A (Ding et al., 2009a)
Apiaceae	*Heracleum lanatum* Michx.	**Plant**: Burns/lesions & skin disorders, inflammation & inflammatory disorders, muscle-related, gastrointestinal & digestive disorders, febrifuge, tonic & stimulant, oral hygiene & disease (PFAF: Moerman, 1998), Neurological & stress-related disorders (PFAF: Weiner 1972), Antioxidant & cancer-related (Baba et al., 2002),	**Plant**: coumarins (Baba et al., 2002), **Root**: psoralen (PFAF: Foster and Duke, 2000)

Table 17.2. (cont.)

Family	Latin name	Biological activity	Chemical constituents
		Leaf: Tonic & stimulant, cold/flu & respiratory diseases, oral hygiene & disease, burns/lesions & skin disorders, muscle-related (PFAF: Moerman, 1998), **Stem**: Gastrointestinal or digestive disorders, antifungal (PFAF: Moerman, 1998), **Root**: Gastrointestinal or digestive disorders, cold/flu & respiratory diseases, inflammation & inflammatory disorders, oral hygiene & disease, antibacterial, burns/lesions & skin disorders (PFAF: Moerman, 1998; Foster and Duke, 2000), Burns/lesions & skin disorders, antioxidant & cancer-related, antiviral (PFAF: Foster and Duke, 2000), Antifungal (McCutcheon et al., 1994), **Seed**: General analgesic (PFAF: Moerman, 1998), **Flower**: External parasites (PFAF: Moerman, 1998)	
Aquifoliaceae	*Ilex chinensis* Sims	**Leaf**: Antibacterial, burns/lesions & skin disorders, metabolic and circulatory disorders, antioxidant & cancer-related, febrifuge (Ou, 1999), Burns/lesions & skin disorders, antidote, gastrointestinal & digestive disorders, tonic & stimulant (PFAF: Duke and Ayensu, 1985)	**Leaf**: protocatechuic acid, protocatechuic aldehyde (Ou, 1999)
Aquifoliaceae	*Ilex integra* Thunb.	**Fruit**: Antifungal, antibacterial (Haraguchi et al., 1999)	**Fruit**: triterpenes: rotundic acid, ulsolic acid, peduncloside (Haraguchi et al., 1999)
Aquifoliaceae	*Ilex pedunculosa* Mia.	**Fruit**: Gastrointestinal & digestive disorders, burns/lesions & skin disorders, tonic & stimulant (PFAF: Stuart, 1911)	**Fruit**: saponins (PFAF)
Araliaceae	*Aralia cordata* Thunb.	**Root**: General analgesic, inflammation & inflammatory disorders, gastrointestinal & digestive disorders, renal disorders & calculi, febrifuge, tonic & stimulant (PFAF: Duke and Ayensu, 1985; WHO, 1998), General analgesic, inflammation & inflammatory disorders (Yen, 1992), Antiperspirant, general analgesic, renal disorders & calculi, antibacterial, antiviral, inflammation & inflammatory disorders, neurological & stress-related disorders (Kimura and Kimura, 1991), Inflammation & inflammatory disorders,	**Root**: essential oil, sesquiterpenes, diterpenes (PFAF: WHO, 1998), diterpene acids (Cho et al., 2010), coumarins: angelical, bergapten, umbelliferone, angelol, columbianetin, columbianentin acetate, columbianidin (Yen, 1992),

Family	Species	Ethnomedicinal uses	Chemical constituents
		immuno-modulation & antiallergic, general analgesic, febrifuge, muscle-related (Cho et al., 2010), General analgesic, inflammation & inflammatory disorders (Kim et al., 2010c), Neurological & stress-related disorders (Cho et al., 2009b), bone-related (Baek et al., 2006), antibacterial (Ambrosio et al., 2008), *Aerial Parts*: Neurological & stress-related disorders, antioxidant & cancer-related (Cho et al., 2009b), Antioxidant & cancer-related (Lee et al., 2008b), *Stem*: Antiperspirant, general analgesic, renal disorders & calculi, antibacterial, antiviral, inflammation & inflammatory disorders, neurological & stress-related disorders (Kimura and Kimura, 1991), *Pith*: Antiperspirant, general analgesic, renal disorders & calculi, antibacterial, antiviral, inflammation & inflammatory disorders, neurological & stress-related disorders (Kimura and Kimura, 1991)	*Aerial Parts*: oleanolic acid (Cho et al., 2009b), triterpenes, diterpenes, saponins, sterols, cerebrosides (Lee et al., 2008b)
Araliaceae	*Aralia elata* var. *subinermis* Ohwi.	*Plant*: Inflammation & inflammatory disorders, cold/flu & respiratory diseases, metabolic and circulatory disorders, liver diseases, gastrointestinal or digestive disorders, antioxidant & cancer-related (PFAF), *Extract*: Metabolic and circulatory disorders (Jin et al., 2006), Metabolic and circulatory disorders, antioxidant & cancer-related, eye-related (Chung et al., 2005), Inflammation & inflammatory disorders, cold/flu & respiratory diseases, metabolic and circulatory disorders, gastrointestinal or digestive disorders, antioxidant & cancer-related (PFAF: Fogarty, 1977), *Bark*: Tonic & stimulant, sedative, renal disorders & calculi, metabolic and circulatory disorders (Kimura and Kimura, 1991), Alcoholism (Xu et al., 2005), *Root*: Inflammation & inflammatory disorders, renal disorders & calculi (Kimura and Kimura, 1991), *Root/Stem*: General analgesic, antioxidant & cancer-related, gastrointestinal & digestive disorders (PFAF: Fogarty, 1977), Inflammation & inflammatory disorders, cold/flu & respiratory diseases, metabolic and circulatory disorders, gastrointestinal or digestive disorders, antioxidant & cancer-related (PFAF: Fogarty, 1977; Duke and Ayensu, 1985),	*Plant*: saponins, glycosides (Chung et al., 2005), *Bark*: triterpenoids (Yagi-Chaves et al., 2006), saponins & glycosides: elatosides (Yoshikawa et al., 1993), *Root Bark*: saponins (Xu et al., 2005), *Root Bark*: araloside A (Lee et al., 2005a), *Bud*: triterpenoids (Zhang et al., 2006), *Shoot*: glycoprotein: aralin (Tomatsu et al., 2004), saponins: elatocides (Yoshikawa et al., 1995), *Leaf*: triterpenoids (Saito et al., 1993; Kim et al., 2005e)

Table 17.2. (cont.)

Family	Latin name	Biological activity	Chemical constituents
		Root Bark: Neurological & stress-related disorders, inflammation & inflammatory disorders, liver diseases, antiviral, metabolic and circulatory disorders, antioxidant & cancer-related (Yagi-Chaves *et al.*, 2006), Gastrointestinal or digestive disorders: stomach ulcer (Lee *et al.*, 2005a), Metabolic and circulatory disorders: diabetes (Yoshikawa *et al.*, 1993), Inflammation & inflammatory disorders, cold/flu & respiratory diseases, metabolic and circulatory disorders: diabetes, gastrointestinal or digestive disorders, antioxidant & cancer-related (PFAF: Fogarty, 1977; Duke and Ayensu, 1985), ***Bud***: Tonic & stimulant, inflammation & inflammatory disorders, metabolic and circulatory disorders: diabetes, antioxidant & cancer-related (Zhang *et al.*, 2006), Inflammation & inflammatory disorders, cold/flu & respiratory diseases, metabolic and circulatory disorders: diabetes, gastrointestinal or digestive disorders, antioxidant & cancer-related (PFAF: Fogarty, 1977; Duke and Ayensu, 1985), ***Shoot***: Antioxidant & cancer-related (Tomatsu *et al.*, 2004), Metabolic and circulatory disorders (Yoshikawa *et al.*, 1995), Inflammation & inflammatory disorders, cold/flu & respiratory diseases, metabolic and circulatory disorders, gastrointestinal or digestive disorders, antioxidant & cancer-related (PFAF: Fogarty, 1977; Duke and Ayensu, 1985), ***Leaf***: Gastrointestinal or digestive disorders (Kimura and Kimura, 1991), Liver diseases (Saito *et al.*, 1993), Inflammation & inflammatory disorders, cold/flu & respiratory diseases, metabolic and circulatory disorders, gastrointestinal or digestive disorders, antioxidant & cancer-related (PFAF: Fogarty, 1977; Duke and Ayensu, 1985)	
Berberidacea	*Nandina domestica* Thunb.	***Leaf***: Tonic & stimulant (PFAF: Duke and Ayensu, 1985), Antiviral, cold/flu & respiratory diseases, eye-related, renal disorders & calculi (Kimura and Kimura, 1991), ***Root***: Febrifuge, general analgesic, inflammation & inflammatory disorders (Kimura and Kimura, 1991), ***Fruit***: Cold/flu & respiratory diseases (Kimura and Kimura, 1991)	

Family	Species	Uses	Constituents
Campanulaceae	*Codonopsis lanceolata* (Siebold.&Zucc.) Trautv.	**Root**: Antioxidant & cancer-related, sex & pregnancy, sedative, metabolic and circulatory disorders, burns/lesions & skin disorders, cold/flu & respiratory diseases, inflammation & inflammatory disorders (PFAF: Fogarty, 1977; Duke and Ayensu, 1985), Antioxidant & cancer-related (Lee *et al.*, 2005b), liver diseases (Kim *et al.*, 2009b), Inflammation & inflammatory disorders (Xu *et al.*, 2008a; Byeon *et al.*, 2009)	**Root**: codonoposide 1c (Lee *et al.*, 2005b), saponins (Byeon *et al.*, 2009; Kim *et al.*, 2009b), triterpenoid saponins: codonolaside & codonolasides I-III (Xu *et al.*, 2008a)
Cannabidaceae	*Cannabis sativa* Linné	**Plant**: General analgesic, eye-related, renal disorders & calculi, anthelmintic, gastrointestinal or digestive disorders, inflammation & inflammatory disorders, muscle-related, burns/lesions & skin disorders, metabolic and circulatory disorders, neurological & stress-related disorders, sedative (PFAF: Stuart, 1911; Uphof, 1968; Grieve, 1971; Fogarty, 1977; Chiej, 1984; Yeung, 1985; Bown, 1995; Lust, 2001;), Burns/lesions & skin disorders, inflammation & inflammatory disorders (PFAF: Duke and Ayensu, 1985; Bown, 1995), Antibacterial, general analgesic (PFAF: Chopra *et al.*, 1986), Antibacterial, immuno-modulation & antiallergic, blood purification & clotting, burns/lesions & skin disorders, eye-related, sex & pregnancy, muscle-related, neurological & stress-related disorders, gastrointestinal or digestive disorders, febrifuge, antibacterial, inflammation & inflammatory disorders, sedative, liver diseases, antiprotozoan, general analgesic, antidote, oral hygiene & disease (PFAF: Duke, 1983)	**Plant**: THC: Δ9-tetrahydrocannabinol
Caprifoliaceae	*Lonicera gracilipes* Miq. var. glabra Miq.	**Plant**: Gastrointestinal or digestive disorders (PFAF: Reid, 1977)	
Caryophyllaceae	*Stellaria aquatica* (L.) Scop.	**Leaf**: Sex & pregnancy (PFAF: Chopra *et al.*, 1986), **Sprout**: Antioxidant & cancer-related (Chon *et al.*, 2009)	
Cephalotaxaceae	*Cephalotaxus harringtonia* var. nana Rehder	**Seed**: Antioxidant & cancer-related, antiviral, antifungal (Evanno *et al.*, 2008)	**Seed**: norditerpene: harringtonolide (Evanno *et al.*, 2008)

Table 17.2. (cont.)

Family	Latin name	Biological activity	Chemical constituents
Clethraceae	*Clethra barvinervis* Sieb. et Zucc.	**Root**: Antifungal, antibacterial (Furumai *et al.*, 2003)	**Root**: clethramycin (Furumai *et al.*, 2003)
Commelinaceae	*Commelina communis* L.	**Plant**: Metabolic and circulatory disorders, gastrointestinal or digestive disorders, febrifuge, antibacterial (PFAF: Duke and Ayensu, 1985), **Aerial Parts**: Antioxidant & cancer-related (Shibano *et al.*, 2008), **Leaf**: Renal disorders & calculi, febrifuge (PFAF: Fogarty, 1977), Cold/flu & respiratory diseases (PFAF: Flora of China, http://www.efloras.org/flora_page.aspx?flora_id=2),	**Aerial Parts**: flavonoid glycosides (Shibano *et al.*, 2008)
Compositae	*Bidens frondosa* L.	**Plant**: Antioxidant & cancer-related (Venkateswarlu *et al.*, 2004)	**Plant**: aurones (Venkateswarlu *et al.*, 2004)
Compositae	*Cirsium japonicum* DC.	**Plant**: Antioxidant & cancer-related (Kim *et al.*, 2010a; Liu *et al.*, 2006b, 2007), Metabolic and circulatory disorders: diabetes (Liao *et al.*, 2010), Metabolic and circulatory disorders (Kwon *et al.*, 2008), Neurological & stress-related disorders (Yeo *et al.*, 2007), **Leaf**: Blood purification & clotting, renal disorders & calculi (PFAF: Duke and Ayensu, 1985), Antioxidant & cancer-related, inflammation & inflammatory disorders (Lee *et al.*, 2008c), antifungal (Zhou *et al.*, 2007b), **Stem**: Blood purification & clotting (PFAF: Duke and Ayensu, 1985), **Root**: Inflammation & inflammatory disorders, renal disorders & calculi, blood purification & clotting (PFAF: Stuart, 1911; Usher, 1974; Fogarty, 1977; Yeung, 1985), Burns/lesions & skin disorders, inflammation & inflammatory disorders, sex & pregnancy, blood purification & clotting (PFAF: Fogarty, 1977), Blood purification & clotting, metabolic and circulatory disorders, antibacterial, inflammation & inflammatory disorders, renal disorders & calculi (Ou, 1999)	**Plant**: flavones (Liu *et al.*, 2006b, 2007)
Compositae	*Erigeron annuus* (L.) Pers.	**Flower**: Antioxidant & cancer-related, renal disorders & calculi (Kim *et al.*, 2009c), Metabolic and circulatory disorders (Jang *et al.*, 2008; Yoo *et al.*, 2008), Metabolic and circulatory disorders: cholesterol (Kim *et al.*, 2005a), **Leaf**, **Stem**: Metabolic and circulatory disorders: diabetes (Jang *et al.*, 2010)	**Flower**: erigeroflavanone (Yoo *et al.*, 2008; Kim *et al.*, 2009c)

Compositae	*Eupatorium chinense* L.	*Aerial Parts*: Antioxidant & cancer-related (Itoh *et al.*, 2009; Liaw *et al.*, 2008), *Leaf*: General analgesic, gastrointestinal & digestive disorders, neurological & stress-related disorders, renal disorders & calculi, anthelmintic (PFAF: Duke and Ayensu, 1985), Antibacterial, cold/flu & respiratory diseases, inflammation & inflammatory disorders (PFAF: Bown, 1995), *Seed*: Sex & pregnancy (PFAF: Duke and Ayensu, 1985), *Root*: Sex & pregnancy, metabolic and circulatory disorders (PFAF: Duke and Ayensu, 1985), Inflammation & inflammatory disorders: goitre (PFAF: Manandhar, 2002)	*Aerial Parts*: sesquiterpene lactones (Itoh *et al.*, 2009), sesquiterpene lactones: eupatozansins (Liaw *et al.*, 2008)
Compositae	*Hemisteptia lyrata* (Bunge) F. E. L. Fischer et C. A. Meyer	*Plant*: Febrifuge, metabolic and circulatory disorders, antioxidant & cancer-related, antibacterial, inflammation & inflammatory disorders (Peoples Press, 1977 cited in Ha *et al.*, 2003), Antioxidant & cancer-related (Jang *et al.*, 1999), *Flower*: Antioxidant & cancer-related (Ha *et al.*, 2003)	*Plant*: flavonoids, polyacetylenes, sesquiterpene lactones (Ha *et al.*, 2003), hemistepsins (Jang *et al.*, 1999), *Flower*: sesquiterpene lactones (Ha *et al.*, 2003)
Compositae	*Ixeris dentata* Nakai	*Plant*: Inflammation & inflammatory disorders, immuno-modulation & antiallergic (Chung *et al.*, 2002), Immuno-modulation & antiallergic (Park *et al.*, 2008), *Root*: Antioxidant & cancer-related (Ahn *et al.*, 2006), *Sap*: Immuno-modulation & antiallergic (Yi *et al.*, 2002)	*Root*: sesquiterpene lactones (Ahn *et al.*, 2006)
Compositae	*Lactuca indica* L.	*Plant*: Gastrointestinal or digestive disorders, tonic & stimulant (PFAF: Usher 1974), Antioxidant & cancer-related (Chen *et al.*, 2007), Metabolic and circulatory disorders (Hou *et al.*, 2003a), Antioxidant & cancer-related (Wang *et al.*, 2003), *Sap*: General analgesic, muscle-related, gastrointestinal or digestive disorders, renal disorders & calculi, sedative (PFAF: Launert, 1981; Uphof, 1968; Weiner, 1972; Bown, 1995), sedative, neurological & stress-related disorders, cold/flu & respiratory diseases, inflammation & inflammatory disorders, general analgesic, tonic & stimulant, sedative (PFAF: Bown, 1995), antifungal (PFAF: Foster and Duke, 2000), *Aerial Parts*: Liver diseases, antiviral (Kim *et al.*, 2007a), Inflammation & inflammatory disorders, antibacterial, gastrointestinal or digestive disorders (Kan, 1986 cited in Kim *et al.*, 2007a)	*Plant*: phenolic compounds: quercetin, caffeic acid, rutin, chlorogenic acid (Chen *et al.*, 2007), dimeric guianolides, lignan glycosides (Hou *et al.*, 2003a), *Sap*: lactucarium (PFAF: Launert, 1981; Uphof,1968; Weiner, 1972; Bown, 1995), *Aerial Parts*: quinic acid derivatives, flavonoids (Kim *et al.*, 2007a)

Table 17.2. (cont.)

Family	Latin name	Biological activity	Chemical constituents
Compositae	*Petasites japonicus* (Siebold et Zucc.) Maxim.	***Plant***: Immuno-modulation & antiallergic, cold/flu & respiratory diseases, burns/lesions & skin disorders, antibacterial (PFAF: Fogarty, 1977), ***Leaf***: Neurological & stress-related disorders (Jun et al., 2007), Antioxidant & cancer-related, neurological & stress-related disorders (Min et al., 2005), ***Root***: Immuno-modulation & antiallergic (Tobinaga et al., 1983), ***Flower***: Cold/flu & respiratory diseases (Yen, 1992)	***Leaf***: quercetin, luteolin (Jun et al., 2007), furofuran lignans (Min et al., 2005), ***Root***: eremophilenolides (Tobinaga et al., 1983), ***Flower***: tannoid, flavonoid (Yen, 1992)
Compositae	*Picris hieracioides* L.	***Leaf***: Febrifuge (PFAF: Chopra et al., 1986), General analgesic (PFAF: Manandhar, 2002)	
Compositae	*Sonchus asper* (L.) Hill	***Plant***: Burns/lesions & skin disorders, inflammation & inflammatory disorders (PFAF: Chopra et al., 1986; Manandhar, 2002), Antioxidant & cancer-related (Alpinar et al., 2009)	
Compositae	*Sonchus oleraceus* L.	***Plant***: Sex & pregnancy, liver diseases (PFAF: Usher, 1974; Moerman, 1998), Gastrointestinal or digestive disorders (PFAF: Moerman, 1998), External parasites: *Aromatictemia salina* (Lima et al., 2009), Antioxidant & cancer-related (El and Karakaya, 2004; Yin et al., 2008; Alpinar et al., 2009), ***Aerial Parts***: Gastrointestinal or digestive disorders, general analgesic, liver diseases, inflammation & inflammatory disorders, tonic & stimulant (Agra et al., 2007), Antioxidant & cancer-related, general analgesic, neurological & stress-related disorders: anxiety (Schaffer et al., 2005; Vilela et al., 2009a, 2009b), Febrifuge, inflammation & inflammatory disorders (Vilela et al., 2010a), Neurological & stress-related disorders (Vilela et al., 2010b), ***Stem***: Gastrointestinal or digestive disorders (PFAF: Duke and Ayensu, 1985), General analgesic (Vilela et al., 2009a), Antioxidant & cancer-related (Yin et al., 2008; Alpinar et al., 2009), ***Leaf***: Inflammation & inflammatory disorders (PFAF: Grieve, 1971), Febrifuge, tonic & stimulant (PFAF: Chopra et al., 1986), General analgesic (Vilela et al., 2009a; Alpinar et al., 2009), ***Latex***: Antifungal, antioxidant & cancer-related (PFAF: Duke and Ayensu, 1985), ***Root***: Febrifuge, tonic & stimulant (PFAF: Chopra et al., 1986), Antioxidant & cancer-related, antibacterial (Saad, 2009)	***Aerial Parts***: flavonoids, terpenes (Yin et al., 2008), ***Leaf***: triterpenoids, alkaloids, coumarins, flavonoids, saponins (Vilela et al., 2009a, b), ***Root***: sesquiterpene lactones, glycosides, loliolide 1 (Saad, 2009)

Compositae	*Taraxacum japonicum* Koidz.	***Plant***: Renal disorders & calculi, burns/lesions & skin disorders, tonic & stimulant, gastrointestinal or digestive disorders, sex & pregnancy (PFAF: Kariyone, 1971), ***Root***: Antioxidant & cancer-related, antiviral (Takasaki *et al.*, 1999)	***Root***: triterpenoids: taraxasterol, taraxerol (Takasaki *et al.*, 1999)
Compositae	*Youngia japonica* (L.) DC.	***Plant***: Antiviral, antioxidant & cancer-related (Ooi *et al.*, 2004), Cold/flu & respiratory diseases, febrifuge, burns/lesions & skin disorders, antidote (PFAF: Duke and Ayensu, 1985)	***Plant***: sesquiterpene lactones, triterpenoids (Ooi *et al.*, 2004)
Cornaceae	*Aucuba japonica* Thunb.	***Leaf***: Burns/lesions & skin disorders, inflammation & inflammatory disorders (PFAF: Duke and Ayensu, 1985). Antioxidant & cancer-related (Hung *et al.*, 2008), Burns/lesions & skin disorders (Kimura and Kimura, 1991), Inflammation & inflammatory disorders (Shimizu and Tomoo, 1994)	***Leaf***: E-phytol, phytone, bis(2-ethylhexyl) phthalate and friedelin (Shimizu and Tomoo, 1994)
Cruciferae	*Raphanus sativus* L.	***Plant***: Anthelmintic (PFAF: Fogarty, 1977), Antibacterial, antifungal (PFAF: Duke and Ayensu, 1985; Bown, 1995), Antioxidant & cancer-related (PFAF: Duke and Ayensu, 1985), ***Root & Aerial Parts***: Antioxidant & cancer-related (Ben Salah-Abbes *et al.*, 2009), ***Root***: Metabolic and circulatory disorders, gastrointestinal or digestive disorders, tonic & stimulant (PFAF: Chevallier, 2001), Metabolic and circulatory disorders, muscle-related, astringent, gastrointestinal or digestive disorders, renal disorders & calculi (PFAF: Lust, 2001; Duke and Ayensu, 1985), Burns/lesions & skin disorders, inflammation & inflammatory disorders, antiperspirant (PFAF: Duke and Ayensu, 1985), Renal disorders & calculi, metabolic and circulatory disorders (PFAF: Grieve, 1971), ***Leaf***: Cold/flu & respiratory diseases (PFAF: Duke and Ayensu, 1985), Renal disorders & calculi, gastrointestinal or digestive disorders (PFAF: Chopra *et al.*, 1986), Gastrointestinal or digestive disorders, sex & pregnancy (Gilani and Ghayur, 2004; Ghayur and Gilani, 2005), ***Seed***: Gastrointestinal & digestive disorders, renal disorders & calculi, cold/flu & respiratory diseases (PFAF: Duke and Ayensu, 1985; Yeung, 1985; Chopra *et al.*, 1986),	***Plant***: raphanin (PFAF: Duke and Ayensu, 1985; Bown, 1995), ***Leaf***: saponins, alkaloids (Ghayur and Gilani, 2005)

Table 17.2. (*cont.*)

Family	Latin name	Biological activity	Chemical constituents
		gastrointestinal or digestive disorders, cold/flu & respiratory diseases (PFAF: Bown, 1995), Antifungal (Games et al., 2008), Metabolic and circulatory disorders (Ghayur and Gilani, 2006), Antioxidant & cancer-related (Pop et al., 1998), **Sprout**: Metabolic and circulatory disorders (Taniguchi et al., 2006), **Bark**: Renal disorders & calculi (Vargas et al., 1999)	
Cupressaceae	*Chamaecyparis obtusa* (Sieb. et Zucc.) Endl.	**Leaf**: Neurological & stress-related disorders (Kuroyanagi et al., 2008), Antibacterial, antifungal (Yang et al., 2007; Hong et al., 2004), Antiviral (Kuo et al., 2006), **Heartwood**: Antioxidant & cancer-related (Koyama et al., 1997; Tanaka et al., 1999; Hsieh et al., 2010)	**Leaf**: lignan & sesquiterpene derivatives (Kuroyanagi et al., 2008), **Heartwood**: chamaecypanones (Hsieh et al., 2010), yoshixolTR (Koyama et al., 1997; Tanaka et al., 1999)
Daphniphyllaceae	*Daphniphyllum macropodum* Miq.	**Bark**: External parasites (Li et al., 2009)	**Bark**: alkaloids (Li et al., 2009)
Dioscoreaceae	*Dioscorea japonica* Thunb.	**Bulb**: Gastrointestinal or digestive disorders, sex & pregnancy, renal disorders & calculi, tonic & stimulant (PFAF: Duke and Ayensu, 1985), Cold/flu & respiratory diseases, inflammation & inflammatory disorders, sex & pregnancy (PFAF: Foster and Duke, 2000), **Tuber**: Gastrointestinal or digestive disorders, sex & pregnancy, renal disorders & calculi, tonic & stimulant (PFAF: Duke and Ayensu, 1985), Cold/flu & respiratory diseases, inflammation & inflammatory disorders, sex & pregnancy (PFAF: Foster & Duke, 2000), **Root**: Tonic & stimulant, sex & pregnancy, cold/flu & respiratory diseases, inflammation & inflammatory disorders (Kimura and Kimura, 1991), Tonic & stimulant, gastrointestinal or digestive disorders, cold/flu & respiratory diseases (Japanese Medicinal Plant Research Center Online Database, http://wwwts9.nibio.go.jp: http://wwwts9.nibio.go.jp/)	**Tuber**: diosgenin (PFAF: Foster and Duke, 2000), sesquiterpenoids, Stilbenes, steroids, amenes, amylase (Yen, 1992), saponins (Ou, 1999)

Ebenaceae	*Diospyros kaki* Thunb.	*Calyx*: Astringent, cold/flu & respiratory diseases, gastrointestinal or digestive disorders (Yen, 1992), Gastrointestinal or digestive disorders (Ou, 1999; Kimura and Kimura, 1991), *Fruit*: Cold/flu & respiratory diseases, astringent, gastrointestinal or digestive disorders, metabolic and circulatory disorders, febrifuge, anthelmintic, tonic & stimulant, gastrointestinal or digestive disorders (PFAF: Duke and Ayensu, 1985), Gastrointestinal or digestive disorders, burns/lesions & skin disorders, cold/flu & respiratory diseases, astringent, metabolic and circulatory disorders (PFAF: Bown, 1995), *Leaf*: Cold/flu & respiratory diseases, blood purification & clotting, antibacterial (Kimura and Kimura, 1991)	*Calyx*: triterpenoids, tannoids (Yen, 1992)
Elaeagnaceae	*Elaeagnus pungens* (Andrews) Th. Wolf	*Fruit*: Antioxidant & cancer-related (PFAF: Matthews, 1994), *Leaf*: Cold/flu & respiratory diseases, gastrointestinal or digestive disorders (PFAF: Duke and Ayensu, 1985; Fogarty, 1977)	
Ericaceae	*Lyonia ovalifolia* var. *elliptica* Hand. – Mzt.	*Leaf*: Burns/lesions & skin disorders, external parasites (PFAF: Chopra *et al.*, 1986; Manandhar, 2002)	
Ericaceae	*Pieris japonica* D. Don.	*Plant*: External parasites (Kimura and Kimura, 1991)	
Ericaceae	*Rhododendron metternichii* Sieb. et Zucc.	*Plant*: Renal disorders & calculi, tonic & stimulant (PFAF: Kariyone, 1971), *Leaf*: Renal disorders & calculi, tonic & stimulant, general analgesic (Kimura and Kimura, 1991), *Petiole*: Renal disorders & calculi, tonic & stimulant, general analgesic (Kimura and Kimura, 1991)	
Ericaceae	*Vaccinium bracteatum* Thunb.	*Plant*: Antioxidant & cancer-related (PFAF: Duke and Ayensu, 1985)	
Fagaceae	*Castanea crenata* Sieb et Zucc.	*Flower*: Gastrointestinal or digestive disorders, blood purification & clotting (Kimura and Kimura, 1991), *Root*: Oral hygiene & disease, inflammation & inflammatory disorders (Kimura and Kimura, 1991)	

Table 17.2. (cont.)

Family	Latin name	Biological activity	Chemical constituents
Fagaceae	*Castonopsis cuspidata* (Thunb. ex Murray)	*Stem*: Antioxidant & cancer-related (Kang *et al.*, 2007)	
Fagaceae	*Quercus acuta* Thunb.	*Gall*: Astringent, gastrointestinal or digestive disorders, blood purification & clotting (PFAF: Grieve, 1971), External parasites (PFAF: Grieve, 1971), *Wood*: Antibacterial (Serit *et al.*, 1991)	*Gall*: tannins (PFAF: Grieve, 1971), *Wood*: protoquercitols (Serit *et al.*, 1991)
Fagaceae	*Quercus acutissima* Carruth.	*Gall*: Astringent, gastrointestinal or digestive disorders, blood purification & clotting (PFAF: Grieve, 1971), External parasites (PFAF: Grieve, 1971)	*Gall*: tannins (PFAF: Grieve, 1971)
Fagaceae	*Quercus glauca* Thunb.	*Gall*: Astringent, gastrointestinal or digestive disorders, blood purification & clotting (PFAF: Grieve, 1971), External parasites (PFAF: Grieve, 1971)	*Gall*: tannins (PFAF: Grieve, 1971)
Fagaceae	*Quercus serrata* Thunb.	*Gall*: Astringent, gastrointestinal or digestive disorders, blood purification & clotting (PFAF: Grieve, 1971), External parasites (PFAF: Grieve, 1971)	*Gall*: tannins (PFAF: Grieve, 1971)
Fagaceae	*Quercus variabilis* Blume.	*Gall*: Astringent, gastrointestinal or digestive disorders, blood purification & clotting (PFAF: Grieve, 1971), External parasites (PFAF: Grieve, 1971)	*Gall*: tannins (PFAF: Grieve, 1971)
Gramineae	*Echinochloa crus-galli* (L.) Beauv	*Grass*: Tonic & stimulant, burns/lesions & skin disorders, blood purification & clotting, antioxidant & cancer-related, inflammation & inflammatory disorders (PFAF: Duke, 1983), *Shoot*: Burns/lesions & skin disorders, blood purification & clotting (PFAF: Duke & Ayensu, 1985; Chopra *et al.*, 1986)	
Gramineae	*Miscanthus sinensis* Anderss.	*Young Stem*: Antidote, blood purification & clotting, metabolic and circulatory disorders, inflammation & inflammatory disorders, febrifuge, renal disorders & calculi (PFAF: Duke and Ayensu, 1985), Renal disorders & calculi, febrifuge (PFAF: Fogarty, 1977)	

Family	Species	Ethnomedicinal uses	Chemical constituents
Gramineae	*Oryza sativa* L.	**Bran**: Antioxidant & cancer-related (Higashi-Okai *et al.*, 2004; Han *et al.*, 2004), Cold/flu & respiratory diseases, immuno-modulation & antiallergic (Lee *et al.*, 2006b, 2006c), Antioxidant & cancer-related (Chen *et al.*, 2006)	**Bran**: 2-arylbenzofuran, 2-(3, 4-dihydroxyphenyl)-4, 6-dihydroxybenzofuran-3-carboxylic acid methyl ester, oryzafuran (Han *et al.*, 2004), anthocyanins (Chen *et al.*, 2006)
Gramineae	*Phyllostachys bambusoides* Sieb. et Zucc.	**Shoot**: Blood purification & clotting, renal disorders & calculi (PFAF: Duke and Ayensu, 1985), Metabolic and circulatory disorders (Cho *et al.*, 2009a), Neurological & stress-related disorders (Hong *et al.*, 2010), **Leaf**: Febrifuge (PFAF: Duke and Ayensu, 1985)	**Shoot**: pyrolyzates (Hong *et al.*, 2010)
Gramineae	*Setaria viridis* P. Beauv.	**Plant**: Inflammation & inflammatory disorders (PFAF: Chopra *et al.*, 1986), **Aerial Parts**: Inflammation & inflammatory disorders (Kim *et al.*, 1997 cited in Kwon *et al.*, 2002), Antioxidant & cancer-related (Kwon *et al.*, 2002), **Seed**: Renal disorders & calculi, burns/lesions & skin disorders, febrifuge, tonic & stimulant (PFAF: Duke and Ayensu, 1985)	**Aerial Parts**: orientin 2'-O-xyloside, tricin-7-O-B-D-glucoside (Kwon *et al.*, 2002)
Gramineae	*Sporobolus indicus* R. Br.	**Plant**: Metabolic and circulatory disorders, inflammation & inflammatory disorders, antibacterial, sex & pregnancy (PFAF: Duke and Ayensu, 1985)	
Gramineae	*Zea mays* L.	**Plant**: Antioxidant & cancer-related, metabolic and circulatory disorders (PFAF: Duke and Ayensu, 1985), **Root**: Renal disorders & calculi (PFAF: Duke and Ayensu, 1985), **Leaf**: Renal disorders & calculi (PFAF: Duke and Ayensu, 1985), **Corn Silk**: Oral hygiene & disease, renal disorders & calculi, tonic & stimulant, metabolic and circulatory disorders (PFAF: Grieve, 1971; Launert, 1981; Duke and Ayensu, 1985; Yeung, 1985), Metabolic and circulatory disorders (PFAF: Launert, 1981; Duke and Ayensu, 1985), Inflammation & inflammatory disorders, renal disorders & calculi, antibacterial (PFAF: Foster and Duke, 2000), Gastrointestinal or digestive disorders (Launert, 1981), Renal disorders & calculi (Suzuki *et al.*, 2005; Velazquez *et al.*, 2005), **Seed**: Renal disorders & calculi, tonic & stimulant, burns/lesions & skin disorders, inflammation &	**Seed**: allantoin (PFAF: Foster and Duke, 2000), **Corn Cob**: anthocyanins (Gutierrez *et al.*, 2009)

Table 17.2. (cont.)

Family	Latin name	Biological activity	Chemical constituents
		inflammatory disorders (PFAF: Grieve, 1971), Antioxidant & cancer-related, antifungal (PFAF: Duke and Ayensu, 1985), Anthelmintic (Kozan et al., 2006), **Corn Cob**: Blood purification & clotting: nose, sex & pregnancy (PFAF: Duke and Ayensu, 1985), Antioxidant & cancer-related (Gutierrez et al., 2009)	
Hamamelidaceae	*Malus sieboldii* Rehder	**Leaf**, **Stem**: Antioxidant & cancer-related (Kim et al., 2010b)	**Leaf**, **Stem**: catechin, chlorogenic acid, quercetin (Kim et al., 2010b)
Illiciaceae	*Illicium religiosum* Seib. et Zucc.	**Plant**: Renal disorders & calculi, oral hygiene & disease, tonic & stimulant (PFAF: Hedrick, 1972), Antioxidant & cancer-related, inflammation & inflammatory disorders (Kim et al., 2009a), **Leaf**: Antibacterial (PFAF: Chopra et al., 1986), **Fruit**: Gastrointestinal & digestive disorders, tonic & stimulant (PFAF: Chopra et al., 1986; Lust, 2001), **Seed**: Antibacterial (PFAF: Chopra et al., 1986)	
Iridacea	*Iris japonica* Thunb.	**Plant**: Inflammation & inflammatory disorders, cold/flu & respiratory diseases, burns/lesions & skin disorders (PFAF: Duke and Ayensu, 1985)	
Lamiaceae	*Rabdosia inflexa* (Thunb.) Hara	**Leaf**: Antioxidant & cancer-related (Fujita et al., 1982)	**Leaf**: kaurene diterpenoids: inflexin, inflexinol (Fujita et al., 1982)
Lardizabalaceae	*Akebia quinata* Decne.	**Stem**: General analgesic, antifungal, inflammation & inflammatory disorders, astringent, antiperspirant, renal disorders & calculi, febrifuge, sex & pregnancy, gastrointestinal or digestive disorders, tonic & stimulant, burns/lesions & skin disorders (PFAF: Stuart, 1911; Kariyone, 1971; Duke and Ayensu, 1985; Bown, 1995), Antibacterial, antifungal, renal disorders & calculi, sex & pregnancy (PFAF: Bown, 1995), Renal disorders & calculi, inflammation & inflammatory disorders, sex & pregnancy (Yen, 1992), Renal disorders & calculi, inflammation & inflammatory disorders (Ou, 1999), inflammation & inflammatory disorders, general analgesic (Choi et al., 2005), Inflammation & inflammatory disorders (Kim	**Stem**: aristolochic acid, oleanolic acid, hederagenin (Yen, 1992), saponins, sapogenins (Choi et al., 2005)

Family	Species	Traditional uses	Constituents
		et al., 1997), **Fruit**: Inflammation & inflammatory disorders, blood purification & clotting, renal disorders & calculi, febrifuge, gastrointestinal or digestive disorders, tonic & stimulant, antioxidant & cancer-related (PFAF: Duke and Ayensu, 1985), General analgesic, renal disorders & calculi, muscle-related (Kimura and Kimura, 1991), **Root**: Febrifuge (PFAF: Duke and Ayensu, 1985), renal disorders & calculi, inflammation & inflammatory disorders, muscle-related (Kimura and Kimura, 1991), **Bark**: Renal disorders & calculi, general analgesic, burns/lesions & skin disorders, inflammation & inflammatory disorders, sedative, sex & pregnancy (Kimura and Kimura, 1991)	**Stem**: aristolochic acid, oleanolic acid, hederagenin (Yen, 1992)
Lardizabalaceae	*Akebia trifoliata* Koidz.	**Bark**: Renal disorders & calculi, general analgesic, burns/lesions & skin disorders, inflammation & inflammatory disorders, sedative, sex & pregnancy (Kimura and Kimura, 1991), **Root**: Renal disorders & calculi, inflammation & inflammatory disorders, muscle-related (Kimura and Kimura, 1991), **Fruit**: General analgesic, renal disorders & calculi, muscle-related (Kimura and Kimura, 1991), **Stem**: General analgesic, antibacterial, antifungal, inflammation & inflammatory disorders, antioxidant & cancer-related, tonic & stimulant, cardiotonic, renal disorders & calculi, sex & pregnancy (PFAF: Fogarty, 1977; Yeung, 1985; Bown, 1995), Renal disorders & calculi, inflammation & inflammatory disorders, sex & pregnancy (Yen, 1992)	
Lardizabalaceae	*Stauntonia hexaphylla* (Thunb.) Decaisne.	**Root**: Inflammation & inflammatory disorders, renal disorders & calculi (PFAF: Duke and Ayensu, 1985), General analgesic, cardiotonic, renal disorders & calculi (Kimura and Kimura, 1991), **Fruit**: Inflammation & inflammatory disorders, renal disorders & calculi (PFAF: Duke and Ayensu, 1985), renal disorders & calculi (PFAF: Hedrick, 1972), **Stem**: Inflammation & inflammatory disorders, renal disorders & calculi (PFAF: Duke and Ayensu, 1985), General analgesic, cardiotonic, renal disorders & calculi (Kimura and Kimura, 1991)	

Table 17.2. (cont.)

Family	Latin name	Biological activity	Chemical constituents
Lauraceae	*Actinodaphne lancifolia* (Sieb. et Zucc.) Meisn.	**Stem**: Antioxidant & cancer-related (Kim et al., 2004c)	**Stem**: lactonic & lignan compounds (Kim et al., 2004c)
Lauraceae	*Cinnamomum camphora* Sieb.	**Leaf**: General analgesic, muscle-related, oral hygiene & disease, tonic & stimulant, metabolic and circulatory disorders, cold/flu & respiratory diseases (PFAF: Stuart, 1911; Uphof, 1968; Schery, 1972; Fogarty, 1977; Bown, 1995), Gastrointestinal or digestive disorders, anthelmintic, inflammation & inflammatory disorders, muscle-related, sedative, cardiotonic, gastrointestinal & digestive disorders, antiperspirant, tonic & stimulant, cold/flu & respiratory diseases, burns/lesions & skin disorders, neurological & stress-related disorders (PFAF: Grieve, 1971; Duke and Ayensu, 1985; Chopra et al., 1986; Bown, 1995), Inflammation & inflammatory disorders, external parasites, general analgesic, burns/lesions & skin disorders, oral hygiene & disease, gastrointestinal or digestive disorders, cold/flu & respiratory diseases, neurological & stress-related disorders (Ou, 1999), Antifungal (Dutta et al., 2007), antioxidant & cancer-related, inflammation & inflammatory disorders (Lee et al., 2006a), **Seed**: External parasites (Zhou et al., 2000; Liu et al., 2006a), **Bark/Wood**: antioxidant & cancer-related (Ling and Liu, 1996), General analgesic, muscle-related, oral hygiene & disease, tonic & stimulant, metabolic and circulatory disorders, cold/flu & respiratory diseases (PFAF: Stuart, 1911; Uphof, 1968; Schery, 1972; Fogarty, 1977; Bown, 1995), Gastrointestinal or digestive disorders, anthelmintic, inflammation & inflammatory disorders, muscle-related, sedative, cardiotonic, antiperspirant, tonic & stimulant, cold/flu & respiratory diseases, burns/lesions & skin disorders, neurological & stress-related disorders (PFAF: Grieve, 1971; Duke and Ayensu, 1985; Chopra et al., 1986; Bown, 1995),	Camphor

Family	Species		
		Inflammation & inflammatory disorders, external parasites, general analgesic, burns/lesions & skin disorders, oral hygiene & disease, gastrointestinal or digestive disorders, cold/flu & respiratory diseases, neurological & stress-related disorders (Ou, 1999), External parasites (Hashimoto et al., 1997; Hiramatsu and Miyazaki, 2001), *Camphor*: Gastrointestinal or digestive disorders, inflammation & inflammatory disorders, liver diseases, neurological & stress-related disorders, burns/lesions & skin disorders (Kimura and Kimura, 1991), External parasites (Yang et al., 2004; Kim et al., 2007b)	*Bark*: lanostane triterpenes (Sharma et al., 1994), *Leaf*: isosericenine, caryophyllene oxide, alpha-cadinol (Furuno et al., 1994)
Lauraceae	*Neolistea sericea* Koidz.	*Bark*: External parasites (Sharma et al., 1994), *Leaf*: External parasites (Furuno et al., 1994)	
Leguminosae	*Albizzia julibrissin* Durazz.	*Flower*: Gastrointestinal or digestive disorders, sedative, tonic & stimulant (PFAF: Yeung, 1985; Duke and Ayensu, 1985; Bown, 1995), Sedative, cold/flu & respiratory diseases, neurological & stress-related disorders (PFAF: Yeung, 1985; Bown, 1995), sedative (Kang et al., 2000), *Stem Bark*: Neurological & stress-related disorders, metabolic and circulatory disorders, burns/lesions & skin disorders, general analgesic (Kimura and Kimura, 1991; Ou, 1999), General analgesic, anthelmintic, gastrointestinal & digestive disorders, sex & pregnancy, sedative, tonic & stimulant (PFAF: Duke and Ayensu, 1985; Stuart, 1911; Yeung, 1985), Burns/lesions & skin disorders, sedative, inflammation & inflammatory disorders (PFAF: Bown, 1995), Antioxidant & cancer-related (Ikeda et al., 1997; Jung et al., 2003; Zou et al., 2006; Fuchs et al., 2009; Hua et al., 2009)	*Flower*: flavonol glycosides: quercitrin, isoquercitrin (Kang et al., 2000), *Stem Bark*: triterpenoid saponins: julibrosides (Ikeda et al., 1997; Zou et al., 2006; Fuchs et al., 2009; Hua et al., 2009)
Leguminosae	*Desmodium oxyphyllum* DC.	*Plant*: Sex & pregnancy (Yoo et al., 2005)	*Plant*: phytoestrogens (Yoo et al., 2005)
Leguminosae	*Glycine max* (L.) Merrill	*Seed*: Blood purification & clotting: acidosis, metabolic and circulatory disorders, muscle-related (PFAF: Duke, 1983), Antiperspirant, gastrointestinal or digestive disorders, febrifuge, cold/flu & respiratory diseases, general analgesic, sedative (PFAF: Yeung,	*Seed*: unsaturated fatty acid, lecithin (PFAF: Duke, 1983), soy protein (Yen, 1992)

Table 17.2. (cont.)

Family	Latin name	Biological activity	Chemical constituents
		1985), Antidote, gastrointestinal or digestive disorders, cardiotonic, renal disorders & calculi; liver diseases (PFAF: Duke and Ayensu, 1985), Febrifuge, cold/flu & respiratory diseases, general analgesic, neurological & stress-related disorders (Yen, 1992), Cold/flu & respiratory diseases, febrifuge (Ou, 1999), Burns/lesions & skin disorders (Kimura and Kimura, 1991), *Seed Pods*: Oral hygiene & disease, burns/lesions & skin disorders (PFAF: Duke and Ayensu, 1985), *Seed Sprouts*: Gastrointestinal or digestive disorders (PFAF: Duke & Ayensu, 1985), muscle-related (PFAF: WHO, 1998), Inflammation & inflammatory disorders, renal disorders & calculi, antiperspirant, cold/flu & respiratory diseases (PFAF: Yeung, 1985), *Leaf*: Antidote (PFAF: Duke and Ayensu, 1985), *Flower*: Eye-related (PFAF: Duke and Ayensu, 1985), *Stem*: Inflammation and inflammatory disorders, antifungal (PFAF: Duke and Ayensu, 1985), *Root*: Astringent (Chopra *et al.*, 1986; Duke, 1983)	
Leguminosae	*Lespedeza cyrtobotrya* Miq.	*Aerial Parts*: Burns/lesions & skin disorders, antioxidant & cancer-related (Mori-Hongo *et al.*, 2009), *Leaf*: Antioxidant & cancer-related (Mori-Hongo *et al.*, 2009)	*Aerials, Leaf*: lipophilic flavonoids (Mori-Hongo *et al.*, 2009)
Leguminosae	*Lespedeza pilosa* Sieb. et Zucc.	*Plant*: Inflammation & inflammatory disorders, gastrointestinal or digestive disorders, sedative (PFAF: Flora of China, http://www.efloras.org/flora_page.aspx?flora_id=2)	
Leguminosae	*Pueraria lobata* (Willd.) Ohwi	*Root*: General analgesic, gastrointestinal or digestive disorders, burns/lesions & skin disorders, metabolic and circulatory disorders, ear-related (Yen, 1992), Muscle-related, cold/flu & respiratory diseases, febrifuge, general analgesic, burns/lesions & skin disorders, gastrointestinal or digestive disorders, eye-related, ear-related, cardiotonic, metabolic and circulatory disorders, neurological & stress-related disorders, sex & pregnancy (Ou, 1999), Antiperspirant, febrifuge, general analgesic (Kimura and Kimura, 1991), Antidote,	*Root*: flavonoids, triterpenoids, saccharides (Yen, 1992), daidzein (Ou, 1999), puerarin (PFAF: Yeung, 1985), daidzin, daidzein (PFAF: Bown, 1995), *Flower*: daidzin, daidzein (PFAF: Bown, 1995)

		gastrointestinal or digestive disorders, febrifuge, muscle-related, antiperspirant, metabolic and circulatory disorders (PFAF: Kariyone, 1971; Duke and Ayensu, 1985; Yeung, 1985; Bown, 1995; Foster and Duke, 2000), general analgesic (PFAF: Duke and Ayensu, 1985), antiviral (PFAF: Chevallier, 2001), metabolic and circulatory disorders (PFAF: Yeung, 1985), *Leaf*: Astringent, metabolic and circulatory disorders (PFAF: Duke and Ayensu, 1985), *Flower*: Antidote, gastrointestinal or digestive disorders, febrifuge, muscle-related, antiperspirant, metabolic and circulatory disorders (PFAF: Kariyone, 1971; Duke and Ayensu, 1985; Yeung, 1985; Bown, 1995; Foster and Duke, 2000), general analgesic (PFAF: Duke and Ayensu, 1985), *Seed*: Gastrointestinal or digestive disorders (PFAF: Duke and Ayensu, 1985; Foster and Duke, 2000), *Stem*: Sex & pregnancy, inflammation & inflammatory disorders, burns/lesions & skin disorders, oral hygiene & disease (PFAF: Duke and Ayensu, 1985; Foster and Duke, 2000)	
Leguminosae	*Trifolium repens* L.	*Plant*: Tonic & stimulant, burns/lesions & skin disorders, blood purification & clotting, antibacterial, inflammation & inflammatory disorders (PFAF: Duke and Ayensu, 1985), Cold/flu & respiratory diseases, febrifuge, sex & pregnancy (PFAF: Moerman, 1998), *Aerial Shoots*: anthelmintic (Tangpu *et al.*, 2004), Antioxidant & cancer-related (Lellau and Liebezeit, 2003), *Leaf*: Inflammation & inflammatory disorders (PFAF: Duke and Ayensu, 1985), *Flower*: Eye-related (PFAF: Moerman, 1998)	
Leguminosae	*Wisteria floribunda* DC.	*Gall*: Antioxidant & cancer-related (Heo *et al.*, 2005)	
Liliaceae	*Lilium japonicum* Thunb.	*Bulb*: Cold/flu & respiratory diseases, tonic & stimulant, sedative, febrifuge (PFAF: Kariyone, 1971; Yeung, 1985)	
Liliaceae	*Ophiopogon japonicus* Ker Gawler.	*Tuber*: Cold/flu & respiratory diseases, blood purification & clotting, febrifuge, oral hygiene & disease, gastrointestinal or digestive disorders (Yen, 1992), Cold/flu & respiratory diseases,	*Tuber*: steroids, saponins, flavonoids, saccharides (Yen, 1992)

Table 17.2. (cont.)

Family	Latin name	Biological activity	Chemical constituents
Liliaceae	*Smilax china* L.	gastrointestinal or digestive disorders, febrifuge, oral hygiene & disease, metabolic and circulatory disorders, cardiotonic, sedative (Ou, 1999), Cold/flu & respiratory diseases, inflammation & inflammatory disorders, febrifuge, cardiotonic, tonic & stimulant (Kimura and Kimura, 1991), *Plant*: Neurological & stress-related disorders, antioxidant & cancer-related (Ichikawa *et al.*, 2003) *Root*: Antidote, inflammation & inflammatory disorders, renal disorders & calculi, sex & pregnancy (Kimura and Kimura, 1991), Alterative, antibacterial, gastrointestinal & digestive disorders, blood purification & clotting, antiperspirant, renal disorders & calculi (PFAF: Stuart, 1911; Grieve, 1971; Fogarty, 1977; Duke and Ayensu, 1985), Antibacterial, renal disorders & calculi, burns/lesions & skin disorders, tonic & stimulant, inflammation & inflammatory disorders, liver diseases, gastrointestinal or digestive disorders (PFAF: Grieve, 1971; Bown, 1995), General analgesic, inflammation & inflammatory disorders, renal disorders & calculi, burns/lesions & skin disorders, gastrointestinal or digestive disorders, antibacterial (Ou, 1999), Inflammation & inflammatory disorders, antioxidant & cancer-related (Shao *et al.*, 2007), Inflammation & inflammatory disorders, general analgesic (Shu *et al.*, 2006), Antioxidant & cancer-related (Lee *et al.*, 2001)	*Root*: steroidal saponins (Shao *et al.*, 2007)
Magnoliaceae	*Magnolia kobus* DC. var. *borealis* Sarg.	*Plant*: General analgesic (PFAF: Kariyone, 1971), *Bud*: General analgesic, oral hygiene & disease, cold/flu & respiratory diseases (Kimura and Kimura, 1991), *Bark*: Antioxidant & cancer-related (Park *et al.*, 2010), Inflammation & inflammatory disorders (Kang *et al.*, 2008b), Neurological & stress-related disorders (Yang *et al.*, 2006)	*Bark*: germacranolide sesquiterpenes (Park *et al.*, 2010)
Magnoliaceae	*Magnolia obovata* Thunb.	*Plant*: Anthelmintic, gastrointestinal or digestive disorders, blood purification & clotting, sex & pregnancy, cold/flu & respiratory	*Bark*: sesquiterpenoids, monoterpenoids, phenols, alkaloids, tannoids (Yen, 1992),

Family	Species	Uses	Phytochemicals / References
		diseases, eye-related, sedative, tonic & stimulant (PFAF: Stuart, 1911; Kariyone, 1971), **Bark**: Gastrointestinal or digestive disorders, cold/flu & respiratory diseases (Yen, 1992), Antioxidant & cancer-related (Schuhly et al., 2001), Antioxidant & cancer-related, gastrointestinal or digestive disorders, antibacterial (Cho et al., 2008), Antioxidant & cancer-related (Lee et al., 2008a, 2009c; Youn et al., 2008a), Neurological & stress-related disorders, inflammation & inflammatory disorders (Oh et al., 2005; Seo et al., 2007), Antifungal (Bang et al., 2000), Antibacterial (Chang et al., 1998), Gastrointestinal or digestive disorders (Kawai et al., 1994), Inflammation & inflammatory disorders, general analgesic (Wang et al., 1992), **Bud**: Gastrointestinal or digestive disorders, renal disorders & calculi, cold/flu & respiratory diseases (Kimura and Kimura, 1991)	magnolol, honokiol (Cho et al., 2008), lignans (Youn et al., 2008a), obovatol (Lee et al., 2008a, 2009c; Ock et al., 2010)
Moraceae	*Broussonetia kazinoki* Sieb.	**Root**: Burns/lesions & skin disorders (Baek et al., 2009), Antioxidant & cancer-related (Ko et al., 1999), **Stem Bark**: Metabolic and circulatory disorders (Cha et al., 2008), burns/lesions & skin disorders (Baek et al., 2009)	**Stem Bark**: 1, 3-diphenylpropanes, alkaloids, flavonoids (Baek et al., 2009). **Root**: 1, 3-diphenylpropanes (Baek et al., 2009), prenylflavonoids (Ko et al., 1999). **Leaf**: flavonoids, phenols, α-tocopherol (Konyahioglu et al., 2005), triterpenoids (Saeed and Sabir 2002), **Latex**: ficin (de Amorin et al., 1999)
Moraceae	*Ficus carica* L.	**Plant**: Antioxidant & cancer-related (PFAF: Duke and Ayensu, 1985), **Leaf**: Inflammation & inflammatory disorders, gastrointestinal or digestive disorders (PFAF: Duke and Ayensu, 1985), **Stem**: Burns/lesions & skin disorders, inflammation & inflammatory disorders (PFAF: Grieve, 1971; Duke and Ayensu, 1985), General analgesic, burns/lesions & skin disorders (PFAF: Chiej, 1984), Antioxidant & cancer-related, inflammation & inflammatory disorders (Lansky et al., 2008), **Fruit**: Gastrointestinal or digestive disorders, oral hygiene & disease, inflammation & inflammatory disorders (PFAF: Duke and Ayensu, 1985; Grieve, 1971), Tonic & stimulant, sex & pregnancy (PFAF: Duke and Ayensu, 1985), Burns/lesions & skin disorders, oral hygiene & disease, inflammation & inflammatory disorders (PFAF: Grieve, 1971), Gastrointestinal or digestive	

Table 17.2. (cont.)

Family	Latin name	Biological activity	Chemical constituents
		disorders (PFAF: Chevallier, 2001), Antioxidant & cancer-related, inflammation & inflammatory disorders (Lansky et al., 2008), Inflammation & inflammatory disorders, gastrointestinal or digestive disorders (Gilani et al., 2008), **Latex**: Burns/lesions & skin disorders, inflammation & inflammatory disorders (PFAF: Duke and Ayensu, 1985; Grieve, 1971), General analgesic, burns/lesions & skin disorders (PFAF: Chiej, 1984), Antioxidant & cancer-related, inflammation & inflammatory disorders (Lansky et al., 2008), Antioxidant & cancer-related (Rubnov et al., 2001), Anthelmintic (de Amorin et al., 1999)	
Moraceae	*Morus bombycis* Koidz.	**Bark**: Renal disorders & calculi, cold/flu & respiratory diseases, inflammation & inflammatory disorders (PFAF: Fogarty, 1977), Cold/flu & respiratory diseases, blood purification & clotting, inflammation & inflammatory disorders, gastrointestinal or digestive disorders (Yen, 1992), Inflammation & inflammatory disorders, renal disorders & calculi, cold/flu & respiratory diseases (Kimura and Kimura, 1991), Metabolic and circulatory disorders, antibacterial, inflammation & inflammatory disorders, cold/flu & respiratory diseases, renal disorders & calculi (Daigo et al., 1986; Venkatesh and Chauhan, 2008), Metabolic and circulatory disorders (Heo et al., 2007), **Leaf**: Febrifuge, general analgesic, eye-related, inflammation & inflammatory disorders (Yen, 1992), Febrifuge, cold/flu & respiratory diseases, antibacterial (Kimura and Kimura, 1991), metabolic and circulatory disorders, antiviral (Venkatesh and Chauhan, 2008), Metabolic and circulatory disorders (Nojima et al., 1998)	**Bark**: flavonoids, benzofurans, coumarins (Yen, 1992), quinones: kwanons G, H, phytoalexins: moracin A-Z, albanins A-H, mulberrofuran I (Venkatesh and Chauhan 2008), 2, 5-dihydroxy-4, 3'-di (beta-D-glucopyranosyloxy)-trans-stilbene (DGTS) (Heo et al., 2007), **Leaf**: flavonoids, steroids (Yen, 1992), N-methyl-1-deoxynojinimycin (Venkatesh and Chauhan, 2008), fagomine (Nojima et al., 1998)
Myricaceae	*Myrica rubra* Seib et Zucc.	**Plant**: Astringeñ, gastrointestinal & digestive disorders, metabolic and circulatory disorders (PFAF: Stuart, 1911), Cardiotonic, gastrointestinal or digestive disorders, antibacterial (PFAF: Duke	**Bark**: diaryloheptanoids (Morikawa, 2007)

Family	Species	Medicinal uses	Chemical constituents
		and Ayensu, 1985), **Bark**: Anthelmintic, gastrointestinal or digestive disorders, antidote (Kimura and Kimura, 1991), Antidote, burns/lesions & skin disorders (PFAF: Duke and Ayensu, 1985), Astringent, external parasites, antidote, gastrointestinal or digestive disorders (Japanese Medicinal Plant Research Center Online Database, http://wwwts9.nibio.go.jp, http://wwwts9.nibio.go.jp), inflammation & inflammatory disorders (Morikawa, 2007), Antioxidant & cancer-related (Kuo et al., 2004a, 2004b), Antiviral (Cheng et al., 2003), liver diseases (Ohta et al., 1992), **Fruit**: Cold/flu & respiratory diseases, gastrointestinal or digestive disorders (PFAF: Duke and Ayensu, 1985), **Seed**: Antiperspirant (PFAF: Duke and Ayensu, 1985), **Leaf**: Antioxidant & cancer-related (Yang et al., 2003)	**Leaf**: iridoid glycosides (Sung et al., 2005), **Fruit**: nuezhnide, alpha-mannitol, oleanolic acid, anthocyan (Yen, 1992)
Oleaceae	*Ligustrum japonicum* Thunb.	**Leaf**: Inflammation & inflammatory disorders, burns/lesions & skin disorders, antibacterial, gastrointestinal or digestive disorders (Kimura and Kimura, 1991), Neurological & stress-related disorders (Sung et al., 2005), **Fruit**: Neurological & stress-related disorders, muscle-related, inflammation & inflammatory disorders, hair-related (Yen, 1992), Tonic & stimulant (PFAF: Duke and Ayensu, 1985), **Flower**: Cold/flu & respiratory diseases, burns/lesions & skin disorders (PFAF: Duke and Ayensu, 1985)	
Oleaceae	*Osmanthus fragrans* var. *auranticus* Makino.		
Papaveraceae	*Corydalis incisa* (Thunb.) Pers.	**Plant**: Inflammation & inflammatory disorders (PFAF: Duke and Ayensu, 1985), **Aerial Parts**: Antioxidant & cancer-related, inflammation & inflammatory disorders, febrifuge, burns/lesions & skin disorders, gastrointestinal or digestive disorders, liver diseases, general analgesic (Choi et al., 2007), Antifungal (Ma et al., 1999)	**Aerial Parts**: isoquinoline alkaloids (Choi et al., 2007)
Papaveraceae	*Macleaya cordata* (Willd.) R. Br.	**Leaf, Stem**: Inflammation & inflammatory disorders, gastrointestinal or digestive disorders, general analgesic, renal disorders & calculi (PFAF: Duke and Ayensu, 1985), Burns/lesions & skin disorders, antifungal (PFAF: Duke and Ayensu, 1985), Inflammation & inflammatory disorders, antidote, external parasites (Kimura and Kimura, 1991), antifungal, antibacterial (Liu et al., 2009)	**Leaf, Stem**: isoquinoline alkaloids (Liu et al., 2009)

Table 17.2. (cont.)

Family	Latin name	Biological activity	Chemical constituents
Pinaceae	*Pinus densiflora* Seib et Zucc.	***Resin***: Burns/lesions & skin disorders, inflammation & inflammatory disorders (Kimura and Kimura, 1991), Renal disorders & calculi, inflammation & inflammatory disorders, cold/flu & respiratory diseases, burns/lesions & skin disorders, gastrointestinal or digestive disorders, anthelmintic, antibacterial (PFAF: Grieve, 1971)	***Resin***: turpentine
Pinaceae	*Pinus thunbergii* Parl.	***Resin***: Burns/lesions & skin disorders, inflammation & inflammatory disorders (Kimura and Kimura, 1991), Renal disorders & calculi, inflammation & inflammatory disorders, cold/flu & respiratory diseases, burns/lesions & skin disorders, gastrointestinal or digestive disorders, anthelmintic, antibacterial (PFAF: Grieve, 1971)	***Resin***: turpentine
Polygonaceae	*Polygonum cuspidata* Sieb et Zucc.	***Rhyzome***: Liver diseases, renal disorders & calculi, burns/lesions & skin disorders, antidote, febrifuge, gastrointestinal or digestive disorders, cold/flu & respiratory diseases, metabolic and circulatory disorders, inflammation & inflammatory disorders, antibacterial, antiviral, general analgesic (Ou, 1999), Gastrointestinal or digestive disorders, renal disorders & calculi, sex & pregnancy, cold/flu & respiratory diseases, sedative, burns/lesions & skin disorders (Kimura and Kimura, 1991)	
Polygonaceae	*Polygonum filiforme* Thunb.	***Plant***: Astringent, tonic & stimulant, cold/flu & respiratory diseases (PFAF: Stuart, 1911; Chopra *et al.*, 1986), ***Leaf***: Cold/flu & respiratory diseases (PFAF: Moerman, 1998)	
Polygonaceae	*Rumex acetosa* L. (*Polygonum acetosa syn.*)	***Plant***: Gastrointestinal or digestive disorders, blood purification & clotting (PFAF: Grieve, 1971; Chiej, 1984; Duke and Ayensu, 1985; Lust, 2001), Muscle-related, burns/lesions & skin disorders (PFAF: Launert, 1981), Antiviral (Gescher *et al.*, 2009), antibacterial, oral hygiene & disease (Anke *et al.*, 2007; Loehr *et al.*, 2009), Antioxidant & cancer-related (Ito *et al.*, 1980), ***Leaf***: Astringent, gastrointestinal or digestive disorders, renal disorders & calculi, febrifuge (PFAF: Grieve, 1971; Chiej *et al.*, 1984; Bown, 1995;	***Plant***: proanthocyanidins (Anke *et al.*, 2007)

Family	Species	Ethnomedicinal uses	Chemical constituents
		Lust, 2001), febrifuge, metabolic and circulatory disorders, burns/lesions & skin disorders, antifungal (PFAF: Grieve, 1971), **Root**: Blood purification & clotting (PFAF: Grieve, 1971), **Stem**: Astringent, renal disorders & calculi, blood purification & clotting (PFAF: Grieve, 1971; Chiej, 1984; Duke and Ayensu, 1985; Lust, 2001), Liver diseases, renal disorders & calculi, blood purification & clotting (PFAF: Grieve, 1971), bone-related (PFAF: Manandhar, 2002)	
Rosaceae	*Duchesnea indica*	***Plant***: Metabolic and circulatory disorders, antidote, febrifuge (PFAF: Fogarty, 1977), ***Flower***: Metabolic and circulatory disorders (PFAF: Duke and Ayensu, 1985), ***Leaf***: Burns/lesions & skin disorders, antifungal, gastrointestinal or digestive disorders (PFAF: Fogarty, 1977), ***Fruit***: Burns/lesions & skin disorders, antifungal (PFAF: Duke and Ayensu, 1985)	
Rosaceae	*Eriobotrya japonica* (Thunb.) Lindl.	***Leaf***: General analgesic, anti-bacterial, gastrointestinal or digestive disorders, cold/flu & respiratory diseases, antiviral, astringent, renal disorders & calculi (PFAF: Fogarty, 1977; Chiej, 1984; Duke and Ayensu, 1985; Yeung, 1985; WHO, 1998), Oral hygiene & disease, cold/flu & respiratory diseases, astringent, febrifuge (PFAF: Chiej, 1984; Bown, 1995), Cold/flu & respiratory diseases, blood purification & clotting, gastrointestinal or digestive disorders (Yen, 1992), cold/flu & respiratory diseases, gastrointestinal or digestive disorders, antibacterial, antiviral (Ou, 1999), Renal disorders & calculi, cold/flu & respiratory diseases, gastrointestinal or digestive disorders, tonic & stimulant, burns/lesions & skin disorders (Kimura and Kimura, 1991), Immuno-modulation & antiallergic (Kim et al., 2009d), Metabolic and circulatory disorders (Qa'dan et al., 2009), ***Shoot***: Astringent, oral hygiene & disease, cold/flu & respiratory diseases, febrifuge (PFAF: Chiej, 1984; Bown, 1995), ***Seed***: Antioxidant & cancer-related (Yoshioka et al., 2010), ***Fruit***: Astringent, cold/flu & respiratory diseases, sedative, gastrointestinal	***Leaf***: saponins: ursolic acid, oleanic acid, cyanophore glycosides: amygdalin (Yen, 1992), Cinchonain Ib (Qa'dan et al., 2009)

Table 17.2. (cont.)

Family	Latin name	Biological activity	Chemical constituents
		or digestive disorders (PFAF: Chiej, 1984; Chopra et al., 1986), *Flower*: Cold/flu & respiratory diseases (PFAF: Duke and Ayensu, 1985; Chopra et al., 1986)	
Rosaceae	*Kerria japonica* DC.	*Flower*: Cold/flu & respiratory diseases, sex & pregnancy (PFAF: Duke and Ayensu, 1985; Stuart, 1911). *Wood*: Inflammation & inflammatory disorders (Wu et al., 2008)	*Wood*: flavonoids (Wu et al., 2008)
Rosaceae	*Photinia glabra* Maxim.	*Plant*: Burns/lesions & skin disorders, gastrointestinal or digestive disorders, anthelmintic, liver diseases (PFAF: Stuart, 1911; Duke and Ayensu, 1985)	
Rosaceae	*Prunus grayana* Maxim.	*Leaf*: Antioxidant & cancer-related (Osawa et al., 1991)	*Leaf*: wax: tocopherol derivatives: prunasols A and B (Osawa et al., 1991)
Rosaceae	*Prunus jamasakura* Sieb.	*Fruit*: Cold/flu & respiratory diseases, gastrointestinal or digestive disorders, tonic & stimulant (PFAF: Bown, 1995). *Bark*: Antioxidant & cancer-related (Miyazawa et al., 2003), External parasites (Yoshioka et al., 1990), Cold/flu & respiratory diseases (Kimura and Kimura, 1991)	*Fruit*: amygdalin, prunasin, hydrocyanic acid (PFAF: Bown, 1995), *Bark*: sakuranetin (Miyazawa et al., 2003), flavonoids (Yoshioka et al., 1990)
Rosaceae	*Prunus mume* Siebold & Zucc.	*Stem Bark*: Metabolic and circulatory disorders, liver diseases (Xia et al., 2010), *Fruit*: Gastrointestinal or digestive disorders, blood purification & clotting, anthelmintic (Yen, 1992), Gastrointestinal or digestive disorders, cold/flu & respiratory diseases, anthelmintic, general analgesic, blood purification & clotting, antibacterial (Ou, 1999), Antibacterial, febrifuge, muscle-related, astringent, gastrointestinal & digestive disorders, oral hygiene & disease, anthelmintic (PFAF: Stuart, 1911; Kariyone, 1971; Fogarty, 1977; Yeung, 1985; Bown, 1995), Antibacterial (PFAF: WHO, 1998), cold/flu & respiratory diseases, gastrointestinal or digestive disorders, Anthelmintic (PFAF: Stuart, 1911; Bown, 1995; WHO, 1998), Gastrointestinal or digestive disorders, metabolic and circulatory disorders, cold/flu & respiratory diseases (PFAF: Chevallier, 2001),	*Fruit*: organic acids, triterpenoid, steroid (Yen, 1992), amygdalin and prunasin (PFAF: Bown, 1995), trimethyl citrate and hexanedioic acid (Bae and Lee, 2003), mumefural, HMF, citric acid, malic acid, furfuryl alcohol (Chuda et al., 1999), *Flower*: chlorogenic acid isomers (Shi et al., 2009)

Rosaceae		

burns/lesions & skin disorders, antifungal (PFAF: Bown, 1995), muscle-related, gastrointestinal & digestive disorders, febrifuge (PFAF: Duke and Ayensu, 1985), Cold/flu & respiratory diseases, gastrointestinal or digestive disorders, tonic & stimulant (PFAF: Bown, 1995), Cold/flu & respiratory diseases, gastrointestinal or digestive disorders, burns/lesions & skin disorders (Jung et al., 2010), Metabolic and circulatory disorders (Chuda et al., 1999), metabolic and circulatory disorders (Park et al., 2009), Muscle-related, tonic & stimulant (Kim et al., 2008b; Yamada et al., 2008), antioxidant & cancer-related (Jeong et al., 2006; Adachi et al., 2007; Mori et al., 2007), Antibacterial (Bae and Lee, 2003), *Flower*: Antioxidant & cancer-related (Matsuda et al., 2003; Shi et al., 2009), *Seed*: Bone-related (Youn et al., 2008b)

Seed: laetrile: vitamin B17 (PFAF: Duke and Ayensu, 1985), cyanogenic glycosides: amygdalin, prunasin (Fukuda et al., 2003)

Prunus persica (L.) Batsch

Fruit: Liver diseases (Lee et al., 2009a), sex & pregnancy (Kim et al., 2008a), *Leaf*: Astringent, renal disorders & calculi, cold/flu & respiratory diseases, febrifuge, gastrointestinal or digestive disorders, anthelmintic, sedative (PFAF: Duke and Ayensu, 1985; Lust, 2001), Gastrointestinal or digestive disorders, cold/flu & respiratory diseases (PFAF: Bown, 1995), Gastrointestinal or digestive disorders, sex & pregnancy, renal disorders & calculi, burns/lesions & skin disorders (PFAF: Lust, 2001), Burns/lesions & skin disorders (Abbasi et al., 2010), gastrointestinal or digestive disorders (Gilani et al., 2000), *Flower*: Renal disorders & calculi, sedative, anthelmintic (PFAF: Grieve, 1971; Duke and Ayensu, 1985; Yeung, 1985; Lust, 2001), Gastrointestinal or digestive disorders (PFAF: Bown, 1995), burns/lesions & skin disorders, antioxidant & cancer-related (Kim et al., 2000), *Stem*: Astringent, sedative (PFAF: Grieve, 1971; Duke and Ayensu, 1985; Yeung, 1985; Lust, 2001), *Seed*: Cold/flu & respiratory diseases, burns/lesions & skin disorders, gastrointestinal or digestive disorders, sedative (PFAF: Grieve, 1971; Duke and Ayensu, 1985; Yeung, 1985; Lust, 2001), Gastrointestinal or digestive

Table 17.2. (cont.)

Family	Latin name	Biological activity	Chemical constituents
		disorders, cold/flu & respiratory diseases, sex & pregnancy, tonic & stimulant (PFAF: Bown, 1995), antioxidant & cancer-related (Fukuda *et al.*, 2003), **Bark**: Renal disorders & calculi, cold/flu & respiratory diseases, sedative (PFAF: Grieve, 1971), gastrointestinal or digestive disorders, cold/flu & respiratory diseases (PFAF: Bown, 1995), **Root Bark**: Inflammation & inflammatory disorders, liver diseases (PFAF: Duke and Ayensu, 1985)	
Rosaceae	*Prunus x yedoensis* Matsum.	**Fruit**: Cold/flu & respiratory diseases, gastrointestinal or digestive disorders, tonic & stimulant (PFAF: Bown, 1995), **Leaf**: antifungal (Ito and Kumazawa, 1995), **Bark**: Cold/flu & respiratory diseases (Kimura and Kimura, 1991), Burns/lesions & skin disorders, inflammation & inflammatory disorders (Kang *et al.*, 2008a)	**Fruit**: amygdalin, prunasin (PFAF: Bown, 1995)
Rosaceae	*Pyrus pyrifolia* (Burm.) Nak. var. culta	**Plant, Fruit**: Burns/lesions & skin disorders, astringent, febrifuge, neurological & stress-related disorders, cold/flu & respiratory diseases (PFAF: Stuart, 1911)	
Rosaceae	*Rubus corchorifolus* L. f.	**Fruit**: Sex & pregnancy, renal disorders & calculi, eye-related (Yen, 1992)	**Fruit**: organic acids, saccharides (Yen, 1992)
Rosaceae	*Rubus crataegifolius* Bunge	**Root**: Antioxidant & cancer-related (Lee *et al.*, 2000), immuno-modulation & antiallergic (Ni *et al.*, 2009)	**Root**: α-glucan (Ni *et al.*, 2009)
Rosaceae	*Rubus hirsutus* Thunb.	**Fruit**: Sex & pregnancy (PFAF: Duke and Ayensu, 1985), **Leaf**: Eye-related (PFAF: Duke and Ayensu, 1985)	
Rosaceae	*Rubus parvifolius* L.	**Stem**: Neurological & stress-related disorders (Xu *et al.*, 2008b), **Leaf**: Astringent (PFAF: Lassak and McCarthy, 1983), Burns/lesions & skin disorders, metabolic and circulatory disorders (PFAF: Duke and Ayensu, 1985), Neurological & stress-related disorders (Xu *et al.*, 2008b), **Root**: Astringent (PFAF: Lassak and McCarthy, 1983), Burns/lesions & skin disorders, metabolic and circulatory disorders (PFAF: Duke and Ayensu, 1985)	

Rubiaceae	*Galium spurium* L.	**Plant**: Blood purification & clotting, sex & pregnancy (PFAF: Bown, 1995)	**Plant**: asperuloside (PFAF: Bown, 1995)
Rubiaceae	*Gardinia jasmioides* var. *frandiflora* Nakai.	**Fruit**: Febrifuge, liver diseases, eye-related, blood purification & clotting, cold/flu & respiratory diseases, burns/lesions & skin disorders, muscle-related (Yen, 1992), Liver diseases, febrifuge, sedative, neurological & stress-related disorders, renal disorders & calculi, liver diseases, burns/lesions & skin disorders, blood purification & clotting, general analgesic, inflammation & inflammatory disorders, metabolic and circulatory disorders (Ou, 1999), Febrifuge, blood purification & clotting, inflammation & inflammatory disorders, sedative, renal disorders & calculi, gastrointestinal or digestive disorders, sedative (Kimura and Kimura, 1991)	**Fruit**: iridoids, monoterpenoids (Yen, 1992)
Rubiaceae	*Paederia scandens* (Lour) Merrill	**Plant**: General analgesic, gastrointestinal or digestive disorders, anthelmintic, inflammation & inflammatory disorders, sex & pregnancy, neurological & stress-related disorders, general analgesic, burns/lesions & skin disorders (PFAF: Duke and Ayensu, 1985; Fogarty, 1977), Antibacterial, gastrointestinal or digestive disorders, Antioxidant & cancer-related (Wang and Huang, 2005), *Leaf*: Gastrointestinal or digestive disorders, general analgesic, cold/ flu & respiratory diseases, metabolic and circulatory disorders, neurological & stress-related disorders, liver diseases, burns/lesions & skin disorders, antidote (Ou, 1999), Oral hygiene & disease, inflammation & inflammatory disorders, general analgesic, burns/ lesions & skin disorders, renal disorders & calculi, gastrointestinal or digestive disorders, antioxidant & cancer-related, antiviral (Kapadia et al., 1996)	
Rutaceae	*Orixa japonica* Thunb.	**Plant**: Febrifuge (PFAF: Duke and Ayensu, 1985), antibacterial, inflammation & inflammatory disorders (Kim et al., 2008c), Metabolic and circulatory disorders (Sharma et al., 2005), *Leaf*: External parasites (Funayama et al., 2001; Park and Shin, 2005), external parasites, muscle-related (Funayama et al., 2001), *Stem*: Antioxidant & cancer-related (Ito et al., 2004)	*Leaf*: quinoline alkaloids: japonine, eduline (Funayama et al., 2001)

Table 17.2. (cont.)

Family	Latin name	Biological activity	Chemical constituents
Rutaceae	*Poncirus trifoliata* Rafin.	*Fruit*: Inflammation & inflammatory disorders; immuno-modulation & antiallergic (Lee et al., 1996; Park et al., 2005; Zhou et al., 2007a), Gastrointestinal or digestive disorders (Lee et al., 2009b; Choi et al., 2010), Antioxidant & cancer-related, liver diseases (Hong et al., 2008), Antibacterial (Kim et al., 1999). *Bark*: Antiviral (PFAF: Duke and Ayensu, 1985)	*Fruit*: 21-methylmelianodiols (Zhou et al., 2007a; Hong et al., 2008), neohesperidin, poncirin (Lee et al., 2009b), ponciretin (Kim et al., 1999)
Rutaceae	*Zanthoxylum ailanthoides* (Sieb. et Zucc.)	Resin in *Bark & Roots*: Cold/flu & respiratory diseases, gastrointestinal & digestive disorders, tonic & stimulant (PFAF: Stuart, 1911; Sargent, 1965), *Root Bark*: Antiviral (Cheng et al., 2005a), *Leaf*: Antioxidant & cancer-related (Chung et al., 2006)	*Root Bark*: sesquiterpenoids (Cheng et al., 2005a)
Rutaceae	*Zanthoxylum piperitum* DC.	*Fruit*: General analgesic, gastrointestinal or digestive disorders, cold/flu & respiratory diseases, oral hygiene & disease, anthelmintic, burns/lesions & skin disorders (Yen, 1992), Gastrointestinal or digestive disorders, general analgesic, cold/flu & respiratory diseases, external parasites (Kimura and Kimura, 1991), *Resin*: Tonic & stimulant (PFAF: Sargent, 1965)	*Fruit*: monoterpenoids, alkanes, arenes, alkaloids, coumarins, phenylpropanoids (Yen, 1992)
Solanaceae	*Physalis alkekengi* L.	*Plant*: General analgesic, febrifuge, cold/flu & respiratory diseases (PFAF: Duke and Ayensu, 1985; Launert, 1981; Usher, 1974; Fogarty, 1974; Stuart, 1911), Renal disorders & calculi, burns/lesions & skin disorders (PFAF: Chopra et al., 1986), Antifungal, cold/flu & respiratory diseases, inflammation & inflammatory disorders, general analgesic, febrifuge (Helvaci et al., 2010), *Aerial Parts*: Antibacterial, antioxidant & cancer-related (Helvaci et al., 2010), *Fruit*: Gastrointestinal or digestive disorders, renal disorders & calculi (PFAF: Grieve, 1971; Launert, 1981; Duke and Ayensu, 1985), Febrifuge, renal disorders & calculi (PFAF: Grieve, 1971; Bown, 1995), Renal disorders & calculi (PFAF: Launert, 1981), sex & pregnancy (Vessal et al., 1991), *Leaf*: Febrifuge, tonic & stimulant (PFAF: Grieve, 1971), burns/lesions & skin disorders, inflammation & inflammatory disorders (PFAF: Bown, 1995), *Stem*: Febrifuge, tonic & stimulant (PFAF: Grieve 1971), *Seed*: Sex & pregnancy (PFAF: Duke and Ayensu, 1985)	*Aerial Parts*: physalins (Helvaci et al., 2010)

Solanaceae	*Tubocapsicum anomalum* (Franch. et Savat.) Makino	*Leaf, Stem, Root*: Antioxidant & cancer-related (Chang *et al.*, 2007; Hsieh *et al.*, 2007)	*Leaf, Stem, Root*: withanolides (Chang *et al.*, 2007; Hsieh *et al.*, 2007)
Sterculiaceae	*Firmiana platanifolia* (L.f.) Marsili	*Leaf*: Inflammation & inflammatory disorders, burns/lesions & skin disorders (PFAF: Duke and Ayensu, 1985), *Root*: Inflammation & inflammatory disorders (PFAF: Duke and Ayensu, 1985), *Seed*: Gastrointestinal or digestive disorders (Kimura and Kimura, 1991), Inflammation & inflammatory disorders, cold/flu & respiratory diseases, febrifuge (PFAF: Duke and Ayensu, 1985), *Stem*: Neurological & stress-related disorders (Son *et al.*, 2005)	*Stem*: neolignans (Son *et al.*, 2005)
Styracaceae	*Styrax japonica* Sieb.et Zucc.	*Plant*: Antifungal (Choi *et al.*, 2006), *Leaf*: Antibacterial (Kim and Shin, 2004a), *Stem Bark*: Burns/lesions & skin disorders (Kim *et al.*, 2004b), Blood purification & clotting, metabolic and circulatory disorders, inflammation & inflammatory disorders (Min *et al.*, 2004a), Antioxidant & cancer-related (Min *et al.*, 2004b), antioxidant & cancer-related (Kim *et al.*, 2004b), *Stem*: Burns/lesions & skin disorders (Moon *et al.*, 2005, 2006), *Fruit*: Soap (Kimura and Kimura, 1991)	*Leaf*: 2-hexenal, n-hexenal, germacrene B, nerol, 3-hexen-1-ol, trans-2-heptenal (Kim and Shin, 2004a), *Bark*: norlignans, terpenes (Min *et al.*, 2004a, b), *Stem*: triterpenoids (Moon *et al.*, 2005)
Symplocaceae	*Symplocos prunifolia* Sieb. et Zucc.	*Leaf*: Gastrointestinal or digestive disorders (PFAF: Duke and Ayensu, 1985)	
Taxaceae	*Torreya nucifera* Sieb. et. Zucc.	Gastrointestinal & digestive disorders, cold/flu & respiratory diseases (PFAF: Duke and Ayensu, 1985), *Seed*: Anthelmintic (PFAF: Duke and Ayensu, 1985)	
Taxodiaceae	*Cryptomeria japonica* D. Don.	*Resin*: Blood purification & clotting, sex & pregnancy, antibacterial (PFAF: Duke and Ayensu, 1985), *Bark*: External parasites (Cheng *et al.*, 2003), antifungal, antibacterial (Yu *et al.*, 1997), antifungal (Cheng *et al.*, 2005b), *Leaf*: External parasites (Cheng *et al.*, 2003; Wang *et al.*, 2006), Antibacterial (Yu *et al.*, 1997; Cha *et al.*, 2007), Antifungal (Ogata and Ogiyama, 2000; Cheng *et al.*, 2005b), metabolic and circulatory disorders (Suzuki and Aoki, 1994)	*Bark*: monoterpene hydrocarbons: 3-carene & limonene, sesquiterpene hydrocarbons: cadala-1, 3, 5-triene, valencene & γ-muurolene (Cheng *et al.*, 2005b), *Leaf*: 3-carene, p-cymene, limonene, β-myrcene, γ-terpinene, α-terpinene, 4-terpineol (Wang *et al.*,

Table 17.2. (cont.)

Family	Latin name	Biological activity	Chemical constituents
			2006), volatile oils: isophyllodecene, ferruginol, phenolics, acids (Yu *et al.*, 1997), *α*-pinene, sabinene, terpinen-4-ol, *α*-terpineol, elemol, 10 (15)-cadinen-4-ol (Cha *et al.*, 2007), sesquiterpene hydrocarbons, diterpene hydrocarbons (Cheng *et al.*, 2005b)
Theaceae	*Camellia japonica* L.	*Flower*: Gastrointestinal or digestive disorders, blood purification & clotting (Yoshikawa *et al.*, 2007), antibacterial (Kim *et al.*, 2001b), *Fruit*: Antiviral (Park *et al.*, 2002), bone-related (Hatano *et al.*, 1995), *Leaf*: Antioxidant & cancer-related (Onodera *et al.*, 2006), Gastrointestinal or digestive disorders, anthelmintic (Harder *et al.*, 2003), Bone-related (Hatano *et al.*, 1995), antiviral (Park *et al.*, 2002), Antifungal (Nagata *et al.*, 1985), *Seed*: Burns/lesions & skin disorders (Kimura and Kimura, 1991), antiviral (Akihisa *et al.*, 2004), Inflammation & inflammatory disorders (Akihisa *et al.*, 1997)	*Flower*: triterpene oligoglycosides (Yoshikawa *et al.*, 2007), *Fruit*: camelliatannins (Park *et al.*, 2002), *Leaf*: flavonol glycosides (Onodera *et al.*, 2006), camelliatannins (Hatano *et al.*, 1995; Park *et al.*, 2002), triterpenoid saponins (Nagata *et al.*, 1985), *Seed*: triterpenoids (Akihisa *et al.*, 2004)
Theaceae	*Camellia sinensis* O. Kuntze.	*Leaf*: Cardiotonic, tonic & stimulant, renal disorders & calculi, cold/flu & respiratory diseases, astringent (PFAF: Grieve, 1971; Karriyone, 1971; Duke, 1983; Duke and Ayensu, 1985; Chopra *et al.*, 1986), Gastrointestinal or digestive disorders, antiviral (PFAF: Duke and Ayensu, 1985; Bown, 1995), Burns/lesions & skin disorders, inflammation & inflammatory disorders, eye-related (PFAF: Duke and Ayensu, 1985; Bown, 1995; Moerman, 1998), *Tea*: Cardiotonic, oral hygiene & disease (PFAF: Chevallier, 2001), Antiprotozoan, antibacterial, gastrointestinal or digestive disorders, antiviral, metabolic and circulatory disorders, sedative, neurological & stress-related disorders (PFAF: Bown, 1995)	*Leaf*: tannins (Duke, 1983)
Theaceae	*Cleyera japonica* Thunb.	*Leaf*: Antioxidant & cancer-related (Hou *et al.*, 2003b)	

Ulmaceae	*Celtis sinensis* Planch.	**Root Bark**: Antioxidant & cancer-related, inflammation & inflammatory disorders (Kim et al., 2005b), Inflammation & inflammatory disorders, cold/flu & respiratory diseases, gastrointestinal or digestive disorders (PFAF: Duke and Ayensu, 1985)	**Bark**: triterpenoids, amide compounds, lignan glycosides (Kim et al., 2005b)
Umbelliferae	*Daucus carota* Linn.	**Plant**: Anthelmintic, gastrointestinal & digestive disorders, renal disorders & calculi, sex & pregnancy, eye-related, tonic & stimulant (PFAF: Grieve, 1971; Launert, 1981; Bown, 1995), **Leaf**: Renal disorders & calculi, sex & pregnancy (PFAF: Chevallier, 2001), **Flower**: Metabolic and circulatory disorders (PFAF: Weiner, 1972), **Root**: Anthelmintic (PFAF: Weiner, 1972; Foster and Duke, 2000; Chevallier, 2001), Sex & pregnancy (PFAF: Weiner, 1972), renal disorders & calculi (PFAF: Foster and Duke, 2000), Cardiotonic (Muralidharan et al., 2008), Renal disorders & calculi (Chaterjee and Prakesh cited in Muralidharan et al., 2008), Liver diseases (Bishayee et al., 1995), Anthelmintic (Urban et al., 2008), **Seed**: Renal disorders & calculi (PFAF: Weiner, 1972; Duke and Ayensu, 1985), Gastrointestinal & digestive disorders, sex & pregnancy, anthelmintic (PFAF: Duke and Ayensu, 1985; Grieve, 1971), Gastrointestinal & digestive disorders, sex & pregnancy (PFAF: Bown, 1995), Sex & pregnancy: contraceptive (PFAF: Foster and Duke, 2000), Sex & pregnancy: abortifacient (PFAF: Chevallier, 2001), Burns/lesions & skin disorders: antiwrinkle (PFAF: Bown, 1995), General analgesic, inflammation & inflammatory disorders (Prochezhian and Ansari, 2000 cited in Muralidharan et al., 2008), Tonic & stimulant, burns/lesions & skin disorders, tonic & stimulant, renal disorders & calculi, liver diseases, anthelmintic, muscle-related (Lawless, 1999 cited in Sokovic et al., 2009), Neurological & stress-related disorders, metabolic and circulatory disorders: cholesterol (Vasudevan and Parle, 2006), **Fruit**: Antibacterial, antifungal (Sokovic et al., 2009)	**Leaf**: porphyrins (PFAF: Chevallier, 2001)
Urticaceae	*Boehmeria longispica* Steud.	**Leaf**: Febrifuge (PFAF: http://www.efloras.org/flora_page. aspx?flora_id=2), antifungal (Inagaki et al., 2008)	**Leaf**: Antibacterial, antifungal (Sokovic et al., 2009)

Table 17.2. (cont.)

Family	Latin name	Biological activity	Chemical constituents
Urticaceae	*Boehmeria nipononivea* Koidz.	*Leaf*: Febrifuge (PFAF: Flora of China, http://www.efloras.org/flora_page.aspx?flora_id=2), antifungal (Inagaki *et al.*, 2008)	
Verbenaceae	*Callicarpa japonica* Thunb.	*Leaf*: External parasites (Cantrell *et al.*, 2005, 2006), antibacterial (Kim and Shin, 2004b), blood purification & clotting, metabolic and circulatory disorders, antibacterial, cold/flu & respiratory diseases (Bae, 2000 cited in Kim and Shin, 2004), antiviral: herpes, sex & pregnancy (Hayashi *et al.*, 1997), antiviral (Tsuchiya *et al.*, 1985), *Shoot*: Metabolic and circulatory disorders, antibacterial, cold/flu & respiratory diseases (Bae, 2000 cited in Kim and Shin, 2004b)	*Leaf*: terpenoids: callicarenal, spathulenol, intermedeol (Cantrell *et al.*, 2005), flavonoids (Tsuchiya *et al.*, 1985), 5, 6, 7-trimethoxyflavone (Kim and Shin, 2004b), *Shoot*: 5, 6, 7-trimethoxyflavone (Kim and Shin, 2004b)
Verbenaceae	*Clerodendron trichotomum* Thunb.	*Leaf*: Inflammation & inflammatory disorders, general analgesic, sedative, burns/lesions & skin disorders, metabolic and circulatory disorders, antiprotozoan: malaria, cold/flu & respiratory diseases (Ou, 1999), Inflammation & inflammatory disorders (Park and Kim, 2007), Antioxidant & cancer-related, inflammation & inflammatory disorders, metabolic and circulatory disorders, sedative, general analgesic (Chae *et al.*, 2004, 2005, 2006), Antioxidant & cancer-related (Lee *et al.*, 2004), Inflammation & inflammatory disorders (Choi *et al.*, 2004), Antiviral: HIV (Kim *et al.*, 2001a), antioxidant & cancer-related (Nagao *et al.*, 2001), Metabolic and circulatory disorders (Lu *et al.*, 1994), Metabolic and circulatory disorders, renal disorders & calculi (Wang and Liao, 1990), *Stem*: Metabolic and circulatory disorders: hypertension (Kang *et al.*, 2003)	*Leaf*: clerodendronin A & B (Ou, 1999), phenylpropanoid glycosides: trichotomoside (Chae *et al.*, 2006), *Stem*: phenylpropanoid glycosides (Kang *et al.*, 2003)
Vitaceae	*Ampelopsis brevipedunculata* Trautv.	*Leaf*: Inflammation & inflammatory disorders, blood purification & clotting, febrifuge (PFAF: Fogarty, 1977). Antibacterial (Kundakovic *et al.*, 2008a, 2008b), *Stem*: Inflammation & inflammatory disorders, blood purification & clotting, renal disorders & calculi, liver diseases (Wu *et al.*, 2004), Antiviral (Sun *et al.*, 1986 cited in Wu *et al.*, 2004), *Root*: Inflammation & inflammatory disorders, blood purification & clotting, renal disorders & calculi, liver diseases (Wu *et al.*, 2004), *Fruit*: Inflammation & inflammatory disorders, blood purification & clotting, febrifuge (PFAF: Fogarty, 1977)	*Stem*, *Root*: triterpenes: lupeol, quercetin, flavonoids: kaempferol, and resveratrol (Wu *et al.*, 2004), *Fruit*: ionone, phenylpropanoid glycosides, hydroquinone glucosides, oligostilbenes (ampelopsins), triterpenes (Kundakovic *et al.*, 2008a, 2008b)

Family	Plant	Medicinal activities	Compounds
Vitaceae	*Cayratia japonica* (Thunb.) Gagnep.	*Plant*: Antioxidant & cancer-related (Han et al., 2007), *Root*: Antioxidant & cancer-related (Lee and Houghton, 2005)	*Plant*: flavonoids (Han et al., 2007)
Vitaceae	*Parthenocissus tricuspidata* (Siebold & Zucc.) Planch.	*Leaf*: Antiprotozoan: malaria (Son et al., 2007), antioxidant & cancer-related (Kim et al., 2005c). Antiprotozoan: malaria (Kim et al., 2005c; Son et al., 2007; Tanaka et al., 2008 cited in Kundakovic et al., 2008b)	*Leaf*: stilbene glycosides (Son et al., 2007), *Stem*: stilbene glycosides (Kundakovic et al., 2008a, 2008b)
Vitaceae	*Vitis flexuosa* Thunb.	*Fruit*: Tonic & stimulant (PFAF: Duke and Ayensu, 1985), *Sap*: Tonic & stimulant (PFAF: Duke and Ayensu, 1985), *Root*: Muscle-related (PFAF: Duke and Ayensu, 1985)	
Zingiberaceae	*Zingiber mioga* (Thunb.) Roscoe	*Flower Bud*: Antibacterial (Ohara et al., 2008), inflammation & inflammatory disorders (Abe et al., 2006), Antioxidant & cancer-related, inflammation & inflammatory disorders (Kim et al., 2005d), antibacterial, antifungal (Abe et al., 2004), antioxidant & cancer-related (Miyoshi et al., 2003), neurological & stress-related disorders (Suzuki et al., 1979)	*Flower Bud*: labdane-type trialdehyde (Abe et al., 2006), aframodial, galanal B, [6]-gingerol, galanolactone (Kim et al., 2005d), diterpene dialdehydes: galanal A, galanal B, miogadial (Abe et al., 2004)

17.3   The future

The list in Table 17.2 was prepared to facilitate future work on this topic at Arashiyama and to stimulate similar work at other study sites across Japan. We attempted to draw attention to the fact that there is an abundance of information already available concerning the bioactivity of various plants in this country, and that there is considerable diversity in the potential pharmacological activity being obtained through the ingestion of these plants in the natural diet of Japanese macaques. We use this work to emphasise the point that plant food species lists can be used not only to describe the ecological diversity of a habitat and the nutritional importance of certain plants to the survival of a species, but also can provide a window into understanding the greater potential of plants for their healing and disease-preventing effects. Plants that may not be eaten as frequently as others can still play an important role in the survival and long-term viability of a species.

Acknowledgements

M. A. Huffman is indebted to the late Nobuo Asaba, former director of the Iwatayama Monkey Park between 1975 and 2001, who made life and research on the mountain immensely enjoyable and memorable. His great love and respect for the monkeys, his boundless generosity and warmth and his dedication to supporting research on the mountain, made work possible. We thank Drs Charmalie A. D. Nahallage and Sachi Sri Kanta for their contribution to the development of the Japanese macaque diet and antiparasitic item database. Their efforts were instrumental in getting this database started. A. J. J. MacIntosh thanks F. Kanou, T. Kaneko and especially K. MacIntosh for their help with deciphering and translating Japanese texts detailing macaque dietary items and ethnomedicinal uses of various plant items. A. J. J. MacIntosh was supported by the Japan Ministry of Education, Culture, Sports, Science and Technology (MEXT) by way of a Monbukagakusho scholarship for doctoral research.

References

Abbasi, A. M., Khan, M. A., Ahmad, M. *et al.* (2009). Medicinal plants used for the treatment of jaundice and hepatitis based on socio-economic documentation. *African Journal of Biotechnology*, **8**, 1643–1650.
    (2010). Ethnopharmacological application of medicinal plants to cure skin diseases and in folk cosmetics among the tribal communities of North-West Frontier Province, Pakistan. *Journal of Ethnopharmacology*, **128**, 322–335.

Abe, M., Ozawa, Y., Uda, Y. *et al.* (2004). Antimicrobial activities of diterpene dialde-hydes, constituents from myoga (*Zingiber mioga* Roscoe), and their quantitative analysis. *Bioscience Biotechnology and Biochemistry*, **68**, 1601–1604.

(2006). A novel labdane-type trialdehyde from Myoga (*Zingiber mioga* Roscoe) that potently inhibits human platelet aggregation and human 5-lipoxygenase. *Bioscience Biotechnology and Biochemistry*, **70**, 2494–2500.

Adachi, M., Suzuki, Y., Mizuta, T. *et al.* (2007). The '*Prunus mume* Sieb. et Zucc' (Ume) is a rich natural source of novel anti-cancer substance. *International Journal of Food Properties*, **10**, 375–384.

Agra, M. F., Baracho, G. S., Nurit, K., Basilio, I. and Coelho, V. P. M. (2007). Medicinal and poisonous diversity of the flora of '*Cariri Paraibano*', Brazil. *Journal of Ethnopharmacology*, **111**, 383–395.

Ahn, E. M., Bang, M. H., Song, M. C. *et al.* (2006). Cytotoxic and ACAT-inhibitory sesquiterpene lactones from the root of *Ixeris dentata* forma *albiflora*. *Archives of Pharmacal Research*, **29**, 937–941.

Ahn, Y. J., Lee, H. S., Oh, H. S., Kim, H. T. and Lee, Y. H. (2005). Antifungal activ-ity and mode of action of *Galla rhois*-derived phenolics against phytopathogenic fungi. *Pesticide Biochemistry and Physiology*, **81**, 105–112.

Akihisa, T., Yasukawa, K., Kimura, Y. *et al.* (1997). Triterpene alcohols from camellia and sasanqua oils and their anti-inflammatory effects. *Chemical and Pharmaceutical Bulletin*, **45**, 2016–2023.

Akihisa, T., Tokuda, H., Ukiya, M. *et al.* (2004). 3-epicabraleabydroxylactone and other triterpenoids from camellia oil and their inhibitory effects on Epstein-Barr virus activation. *Chemical and Pharmaceutical Bulletin*, **52**, 153–156.

Alpinar, K., Ozyurek, M., Kolak, U. *et al.* (2009). Antioxidant capacities of some food plants wildly grown in Ayvalik of Turkey. *Food Science and Technology Research*, **15**, 59–64.

Ambrosio, S. R., Furtado, N., de Oliveira, D. C. R. *et al.* (2008). Antimicrobial activ-ity of kaurane diterpenes against oral pathogens. *Zeitschrift fur Naturforschung Section C – A Journal of Biosciences*, **63** (5–6), 326–330.

Anke, J., Petereit, F. and Hensel, A. (2007). Isolation, structure elucidation, cytotoxicity and anti-adhesive properties of proanthocyanidins from *Rumex acetosa* L. *Planta Medica*, **73**, 326.

Baba, M., Jin, Y. R., Mizuno, A. *et al.* (2002). Studies on cancer chemoprevention by traditional folk medicines XXIV. Inhibitory effect of a coumarin derivative, 7-iso-pentenyloxycoumarin, against tumor-promotion. *Biological and Pharmaceutical Bulletin*, **25**, 244–246.

Bae, J. H. and Lee, S. M. (2003). Identification of antimicrobial substances from *Prunus mume* on the growth of food-borne pathogens. *Food Science and Biotechnology*, **12**, 128–132.

Baek, Y. H., Huh, J. E., Lee, J. D., Choi, D. Y. and Park, D. S. (2006). Effect of *Aralia cordata* extracts on cartilage protection and apoptosis inhibition. *Biological and Pharmaceutical Bulletin*, **29**, 1423–1430.

Baek, Y. S., Ryu, Y. B., Curtis-Long, M. J. *et al.* (2009). Tyrosinase inhibitory effects of 1,3-diphenylpropanes from *Broussonetia kazinoki*. *Bioorganic and Medicinal Chemistry*, **17**, 35–41.

Bang, K. H., Kim, Y. K., Min, B. S. *et al.* (2000). Antifungal activity of magnolol and honokiol. *Archives of Pharmacal Research*, **23**, 46–49.

Ben Salah-Abbes, J., Abbes, S., Abdel-Wahhab, M. A. and Oueslati, R. (2009). *Raphanus sativus* extract protects against Zearalenone induced reproductive toxicity, oxidative stress and mutagenic alterations in male Balb/c mice. *Toxicon*, **53**, 525–533.

Bishayee, A., Sarkar, A. and Chatterjee, M. (1995). Hepatoprotective activity of carrot (*Daucus carota* L) against carbon-tetrachloride intoxication in mouse-liver. *Journal of Ethnopharmacology*, **47**, 69–74.

Bown, D. (1995). *Encyclopedia of Herbs and Their Uses*. London: Dorling Kindersley.

Byeon, S. E., Choi, W. S., Hong, E. K. *et al.* (2009). Inhibitory effect of saponin fraction from *Codonopsis lanceolata* on immune cell-mediated inflammatory responses. *Archives of Pharmacal Research*, **32**, 813–822.

Cantrell, C. L., Klun, J. A., Bryson, C. T., Kobaisy, M. and Duke, S. O. (2005). Isolation and identification of mosquito bite deterrent terpenoids from leaves of American (*Callicarpa americana*) and Japanese (*Callicarpa japonica*) beautyberry. *Journal of Agricultural and Food Chemistry*, **53**, 5948–5953.

Cantrell, C. L., Klun, J. A., Bryson, C. T. and Duke, S. O. (2006). Isolation and identification of mosquito bite-deterrent constituents from leaves of *Callicarpa americana* and *Callicarpa japonica*. *Abstracts of Papers of the American Chemical Society*, **231**.

Cha, J. D., Jeong, M. R., Jeong, S. I. *et al.* (2007). Chemical composition and antimicrobial activity of the essential oil of *Cryptomeria japonica*. *Phytotherapy Research*, **21**, 295–299.

Cha, J. Y., Kim, Y. T., Kim, H. S. and Cho, Y. S. (2008). Antihyperglycemic effect of stem bark powder from paper mulberry (*Broussonetia kazinoki* sieb.) in type 2 diabetic Otsuka Long-Evans Tokushima fatty rats. *Journal of Medicinal Food*, **11**, 499–505.

Chae, S., Kim, J. S., Kang, K. A. *et al.* (2004). Antioxidant activity of jionoside D from *Clerodendron trichotomum*. *Biological and Pharmaceutical Bulletin*, **27**, 1504–1508.

(2005). Antioxidant activity of isoacteoside from *Clerodendron trichotomum*. *Journal of Toxicology and Environmental Health – Part a – Current Issues*, **68**, 389–400.

Chae, S., Kang, K. A., Kim, J. S., Hyun, J. W. and Kang, S. S. (2006). Trichotomoside: a new antioxidative phenylpropanoid glycoside from *Clerodendron trichotomum*. *Chemistry and Biodiversity*, **3**, 41–48.

Chang, B. S., Lee, Y. M., Ku, Y., Bae, K. H. and Chung, C. P. (1998). Antimicrobial activity of magnolol and honokiol against periodontopathic microorganisms. *Planta Medica*, **64**, 367–369.

Chang, F. R., Hsieh, Y. C., Chang, Y. F. *et al.* (2010). Inhibition of the Epstein-Barr virus lytic cycle by moronic acid. *Antiviral Research*, **85**, 490–495.

Chang, H. C., Chang, F. R., Wang, Y. C. *et al.* (2007). A bioactive withanolide Tubocapsanolide A inhibits proliferation of human lung cancer cells via repressing Skp2 expression. *Molecular Cancer Therapeutics*, **6**, 1572–1578.

Chen, J. H., Tsai, S. J. and Chen, H. I. (1999). Welsh onion (*Allium fistulosum* L) extracts alter vascular responses in rat aortae. *Journal of Cardiovascular Pharmacology*, **33**, 515–520.

Chen, P. N., Kuo, W. H., Chiang, C. L. *et al.* (2006). Black rice anthocyanins inhibit cancer cells invasion via repressions of MMPs and u-PA expression. *Chemico-Biological Interactions*, **163**, 218–229.

Chen, Y. H., Chen, H. Y., Hsu, C. L. and Yen, G. C. (2007). Induction of apoptosis by the *Lactuca indica* L. in human leukemia cell line and its active components. *Journal of Agricultural and Food Chemistry*, **55**, 1743–1749.

Cheng, M. J., Lee, K. H., Tsai, I. L. and Chen, I. S. (2005a). Two new sesquiterpenoids and anti-HIV principles from the root bark of *Zanthoxylum ailanthoides*. *Bioorganic and Medicinal Chemistry*, **13**, 5915–5920.

Cheng, S. S., Chang, H. T., Chang, S. T., Tsai, K. H. and Chen, W. J. (2003). Bioactivity of selected plant essential oils against the yellow fever mosquito *Aedes aegypti* larvae. *Bioresource Technology*, **89**, 99–102.

Cheng, S. S., Lin, H. Y. and Chang, S. T. (2005b). Chemical composition and antifungal activity of essential oils from different tissues of Japanese cedar (*Cryptomeria japonica*). *Journal of Agricultural and Food Chemistry*, **53**, 614–619.

Chevallier, A. (2001). *Encyclopedia of Medicinal Plants*. St. Leonards: Dorling Kindersley.

Chiej, R. (1984). *Encyclopedia of Medicinal Plants*. Edinburgh: MacDonald.

Cho, H., Cho, K. A., Jia, S., Cho, S. J. and Choi, D. (2009). Influence of bamboo oil supplementation on blood lipid concentration in serum. *Journal of Industrial and Engineering Chemistry*, **15**, 281–284.

Cho, J. H., Lee, J. Y., Sim, S. S., Whang, W. K. and Kim, C. J. (2010). Inhibitory effects of diterpene acids from root of *Aralia cordata* on IgE-mediated asthma in guinea pigs. *Pulmonary Pharmacology and Therapeutics*, **23**, 190–199.

Cho, S. O., Ban, J. Y., Kim, J. Y. *et al.* (2009). *Aralia cordata* protects against amyloid beta protein (25–35)-induced neurotoxicity in cultured neurons and has anti-dementia activities in mice. *Journal of Pharmacological Sciences*, **111**, 22–32.

Cho, S. Y., Lee, J. H., Bae, K. H., Kim, Y. S. and Jeong, C. S. (2008). Anti-gastric effects of magnolol and honokiol from the stem bark of *Magnolia obovata*. *Biomolecules and Therapeutics*, **16**, 270–276.

Choi, G. J., Kim, J. C., Jang, K. S. *et al.* (2006). In vivo antifungal activities of 67 plant fruit extracts against six plant pathogenic fungi. *Journal of Microbiology and Biotechnology*, **16**, 491–495.

Choi, J., Jung, H. J., Lee, K. T. and Park, H. J. (2005). Antinociceptive and anti-inflammatory effects of the saponin and sapogenins obtained from the stem of *Akebia quinata*. *Journal of Medicinal Food*, **8**, 78–85.

Choi, J. H., Whang, W. K. and Kim, H. J. (2004). Studies on the anti-inflammatory effects of *Clerodendron trichotomum* Thunberg leaves. *Archives of Pharmacal Research*, **27**, 189–193.

Choi, K. H., Il Jeong, S., Hwang, B. S. *et al.* (2010). Hexane extract of *Poncirus trifoliata* (L.) Raf. stimulates the motility of rat distal colon. *Journal of Ethnopharmacology*, **127**, 718–724.

Choi, S. U., Baek, N. I., Kim, S. H. *et al.* (2007). Cytotoxic isoquinoline alkaloids from the aerial parts of *Corydalis incisa*. *Archives of Pharmacal Research*, **30**, 151–154.

Chon, S. U., Heo, B. G., Park, Y. S., Kim, D. K. and Gorinstein, S. (2009). Total phenolics level, antioxidant activities and cytotoxicity of young sprouts of some traditional Korean salad plants. *Plant Foods for Human Nutrition*, **64**, 25–31.

Chopra, R. N., Nayar, S. L. and Chopra, I. C. (1986). *Glossary of Indian Medicinal Plants*. New Delhi: Council of Scientific and Industrial Research.

Chuda, Y., Ono, H., Ohnishi-Kameyama, M. *et al.* (1999). Mumefural, citric acid derivative improving blood fluidity from fruit-juice concentrate of Japanese apricot (*Prunus mume* Sieb. et Zucc). *Journal of Agricultural and Food Chemistry*, **47**, 828–831.

Chung, H. S., Jeong, H. J., Han, M. J. *et al.* (2002). Nitric oxide and tumor necrosis factor-alpha production by *Ixeris dentata* in mouse peritoneal macrophages. *Journal of Ethnopharmacology*, **82**, 217–222.

Chung, Y. C., Chien, C. T., Teng, K. Y. and Chou, S. T. (2006). Antioxidative and mutagenic properties of *Zanthoxylum ailanthoides* Sieb & Zucc. *Food Chemistry*, **97**, 418–425.

Chung, Y. S., Choi, Y. H., Lee, S. J. *et al.* (2005). Water extract of *Aralia elata* prevents cataractogenesis in vitro and in vivo. *Journal of Ethnopharmacology*, **101**, 49–54.

Daigo, K., Inamori, Y. and Takemoto, T. (1986). Studies on the constituents of the water extract of the root of mulberry tree (*Morus bombycis* Koidz). *Chemical and Pharmaceutical Bulletin*, **34**, 2243–2246.

Dalvi, R. R. (1985). Suguinarine: its potential as a liver toxic alkaloid present in the seeds of *Argemone mexicana*. *Experientia*, **41**, 77–78.

Dan, N. V. and Nhu, D. T. (1989). *Medicinal Plants in Vietnam*. Manila: World Health Organization.

de Amorin, A., Borba, H. R., Carauta, J. P. P., Lopes, D. and Kaplan, M. A. C. (1999). Anthelmintic activity of the latex of *Ficus* species. *Journal of Ethnopharmacology*, **64**, 255–258.

Ding, Y., Nguyen, H. T., Choi, E. M., Bae, K. and Kim, Y. H. (2009a). Rhusonoside A, a new megastigmane glycoside from *Rhus sylvestris*, increases the function of osteoblastic MC3T3-E1 cells. *Planta Medica*, **75**, 158–162.

Ding, Y., Nguyen, H. T., Kim, S. I., Kim, H. W. and Kim, Y. H. (2009b). The regulation of inflammatory cytokine secretion in macrophage cell line by the chemical constituents of *Rhus sylvestris*. *Bioorganic and Medicinal Chemistry Letters*, **19**, 3607–3610.

Doo, H. K., Shim, Y. J., Kim, E. H. *et al.* (2003). Anti-hyperglycemic activity of aqueous extract from the gall of *Rhus chinensis* and its protective effect on beta-cell injury. *Diabetes*, **52**, A549.

Duke, J. A. (1983). *Handbook of Energy Crops*. URL: http://www.hort.purdue.edu/newcrop

Duke, J. A. and Ayensu, E. S. (1985). *Medicinal Plants of China*. Algonac, MI: Reference Publications, Inc.

Dutta, B. K., Karmakar, S., Naglot, A., Aich, J. C. and Begam, M. (2007). Anticandidial activity of some essential oils of a mega biodiversity hotspot in India. *Mycoses*, **50**, 121–124.

El, S. N. and Karakaya, S. (2004). Radical scavenging and iron-chelating activities of some greens used as traditional dishes in Mediterranean diet. *International Journal of Food Sciences and Nutrition*, **55**, 67–74.

Evanno, L., Jossang, A., Nguyen-Pouplin, J. *et al.* (2008). Further studies of the nord-iterpene (+)-harringtonolide isolated from *Cephalotaxus harringtonia var. drupacea*: absolute configuration, cytotoxic and antifungal activities. *Planta Medica*, **74**, 870–872.

Fashing, P. J., Dierenfeld, E. S. and Mowry, C. B. (2007). Influence of plants and soil chemistry on food selection, ranging patterns, and biomass of *Colobus guereza* in Kakamega Forest, Kenya. *International Journal of Primatology*, **28**, 673–703.

Felton, A. M., Felton, A., Lindenmayer, D. B. and Foley, W. J. (2009). Nutritional goals of wild primates. *Functional Ecology*, **23**, 70–78.

Fogarty, J. E. (1977). *A Barefoot Doctor's Manual*. Philadelphia, PA: Running Press.

Forbey, J. S., Harvey, A. L., Huffman, M. A. *et al.* (2009). Exploitation of secondary metabolites by animals: a response to homeostatic challenges. *Integrative and Comparative Biology*, **49**, 314–328.

Foster, S. and Duke, J. A. (2000). *A Field Guide to Medicinal Plants*, 2nd edn. Boston, MA: Houghton Mifflin Company.

Freeland, W. J. and Janzen, D. H. (1974). Strategies in herbivory by mammals: the role of plant secondary compounds. *American Naturalist*, **108**, 269–289.

Fuchs, H., Bachran, D., Panjideh, H. *et al.* (2009). Saponins as tool for improved targeted tumor therapies. *Current Drug Targets*, **10**, 140–151.

Fujita, T., Takeda, Y., Yuasa, E. *et al.* (1982). Structure of inflexinol, a new cyto-toxic diterpene from *Rabdosia inflexa*. *Phytochemistry*, **21**, 903–905.

Fukuda, T., Ito, H., Mukainaka, T. *et al.* (2003). Anti-tumor promoting effect of glycosides from *Prunus persica* seeds. *Biological and Pharmaceutical Bulletin*, **26**, 271–273.

Funayama, S., Tanaka, R., Kumekawa, Y. *et al.* (2001). Rat small intestine muscle relaxation alkaloids from *Orixa japonica* leaves. *Biological and Pharmaceutical Bulletin*, **24**, 100–102.

Furumai, T., Yamakawa, T., Yoshida, R. and Igarashi, Y. (2003). Clethramycin, a new inhibitor of pollen tube growth with antifungal activity from *Streptomyces hygroscopicus* TP-A0623 I. Screening, taxonomy, fermentation, isolation and biological properties. *Journal of Antibiotics*, **56**, 700–704.

Furuno, T., Terada, Y., Yano, S., Uehara, T. and Jodai, S. (1994). Activities of leaf oils and their components from Lauraceae trees against house dust mites. *Mokuzai Gakkaishi*, **40**, 78–87.

Games, P. D., dos Santos, I. S., Mello, E. O. *et al.* (2008). Isolation, characterization and cloning of a cDNA encoding a new antifungal defensin from *Phaseolus vulgaris* L. seeds. *Peptides*, **29**, 2090–2100.

Gescher, K., Bicker, J., Hafezi, W., Kuhn, J. and Hensel, A. (2009). Potent antiviral effect of a polyphenol-enriched *Rumex acetosa* L. extract against herpes simplex virus type 1 via interaction with viral envelope proteins. *Planta Medica*, **75**, 884.

Ghayur, M. N. and Gilani, A. H. (2005). Gastrointestinal stimulatory and uterotonic activities of dietary radish leaves extract are mediated through multiple pathways. *Phytotherapy Research*, **19**, 750–755.

    (2006). Radish seed extract mediates its cardiovascular inhibitory effects via muscarinic receptor activation. *Fundamental and Clinical Pharmacology*, **20**, 57–63.

Gilani, A. H. and Ghayur, A. N. (2004). Pharmacological basis for the gut stimulatory activity of *Raphanus sativus* leaves. *Journal of Ethnopharmacology*, **95**, 169–172.

Gilani, A. H., Aziz, N., Ali, S. M. and Saeed, M. (2000). Pharmacological basis for the use of peach leaves in constipation. *Journal of Ethnopharmacology*, **73**, 87–93.

Gilani, A. H., Mehmood, M. H., Janbaz, K. H., Khan, A. U. and Saeed, S. A. (2008). Ethnopharmacological studies on antispasmodic and antiplatelet activities of *Ficus carica*. *Journal of Ethnopharmacology*, **119**, 1–5.

Glander, K. E. (1982). The impact of plant secondary compounds on primate feeding behavior. *American Journal of Physical Anthropology*, **25** (Suppl. 3), S1–S18.

Grieve, M. (1971). *A Modern Herbal*. New York, NY: Dover Publications.

Gu, Q., Wang, R. R., Zhang, X. M. *et al.* (2007). A new benzofuranone and anti-HIV constituents from the stems of *Rhus chinensis*. *Planta Medica*, **73**, 279–282.

Gutierrez, A. G., Acevedo, J. A., Ballarte, L. N. *et al.* (2009). Anthocyanins, total phenolic compounds and antioxidant activity of purple corn (*Zea mays* L) cobs: extraction method. *Boletin Latinoamericano Y Del Caribe De Plantas Medicinales Y Aromaticas*, **8**, 509–518.

Ha, T. J., Jang, D. S., Lee, J. R. *et al.* (2003). Cytotoxic effects of sesquiterpene lactones from the flowers of *Hemisteptia lyrata* B. *Archives of Pharmacal Research*, **26**, 925–928.

Han, S. B., Lee, C. W., Yoon, Y. D. *et al.* (2005). Prevention of arthritic inflammation using an oriental herbal combination BDX-1 isolated from *Achyranthes bidentata* and *Atractylodes japonica*. *Archives of Pharmacal Research*, **28**, 902–908.

Han, S. J., Ryu, S. N. and Kang, S. S. (2004). A new 2-arylbenzofuran with antioxidant activity from the black colored rice (*Oryza sativa* L.) bran. *Chemical and Pharmaceutical Bulletin*, **52**, 1365–1366.

Han, X. H., Hong, S. S., Hwang, J. S. *et al.* (2007). Monoamine oxidase inhibitory components from *Cayratia japonica*. *Archives of Pharmacal Research*, **30**, 13–17.

Hanya, G., Kiyono, M., Takafumi, H., Tsujino, R. and Agetsuma, N. (2007). Mature leaf selection of Japanese macaques: effects of availability and chemical content. *Journal of Zoology*, **273**, 140–147.

Haraguchi, H., Kataoka, S., Okamoto, S., Hanafi, M. and Shibata, K. (1999). Antimicrobial triterpenes from *Ilex integra* and the mechanism of antifungal action. *Phytotherapy Research*, **13**, 151–156.

Harder, A., Schmitt-Wrede, H. P., Krucken, J. *et al.* (2003). Cyclooctadepsipeptides – an anthelmintically active class of compounds exhibiting a novel mode of action. *International Journal of Antimicrobial Agents*, **22**, 318–331.

Hashimoto, K., Ohtani, Y. and Sameshima, K. (1997). The termiticidal activity and its transverse distribution in camphor (*Cinnamomum camphora*) wood. *Mokuzai Gakkaishi*, **43**, 566–573.

Hatano, T., Han, L., Taniguchi, S. *et al.* (1995). Camelliatannin-D, a new inhibitor of bone-resorption, from *Camellia japonica*. *Chemical and Pharmaceutical Bulletin*, **43**, 2033–2035.

Hayashi, K., Hayashi, T., Otsuka, H. and Takeda, Y. (1997). Antiviral activity of 5,6,7-trimethoxyflavone and its potentiation of the antiherpes activity of acyclovir. *Journal of Antimicrobial Chemotherapy*, **39**, 821–824.

Hayashi, Y. (1941). General overview of the plants of Arashiyama. *Yagai Hakubutsu*, **3**, 21–28. (in Japanese).

Hedrick, U. P. (1972). *Sturtevant's Edible Plants of the World*. New York, NY: Dover Publications.

Helvaci, S., Kokdil, G., Kawai, M. *et al.* (2010). Antimicrobial activity of the extracts and physalin D from *Physalis alkekengi* and evaluation of antioxidant potential of physalin D. *Pharmaceutical Biology*, **48**, 142–150.

Heo, J. C., Park, J. Y., Lee, J. M. *et al.* (2005). *Wisteria floribunda* gall extract inhibits cell migration in mouse B16F1 melanoma cells by regulating CD44 expression and GTP-RhoA activity. *Journal of Ethnopharmacology*, **102**, 10–14.

Heo, S. I., Jin, Y. S., Jung, M. J. and Wang, M. H. (2007). Antidiabetic properties of 2,5-dihydroxy-4,3'-di(beta-D-glucopyranosyloxy)-trans-stilbene from mulberry (*Morus bombycis* Koidzumi) root in streptozotocin-induced diabetic rats. *Journal of Medicinal Food*, **10**, 602–607.

Higashi-Okai, K., Kanbara, K., Amano, K. *et al.* (2004). Potent antioxidative and anti-genotoxic activity in aqueous extract of Japanese rice bran: association with peroxidase activity. *Phytotherapy Research*, **18**, 628–633.

Hiramatsu, Y. and Miyazaki, Y. (2001). Effect of volatile matter from wood chips on the activity of house dust mites and on the sensory evaluation of humans. *Journal of Wood Science*, **47**, 13–17.

Hong, E. J., Na, K. J., Choi, I. G., Choi, K. C. and Jeung, E. B. (2004). Antibacterial and antifungal effects of essential oils from coniferous trees. *Biological and Pharmaceutical Bulletin*, **27**, 863–866.

Hong, E. J., Jung, E. M., Lee, G. S. *et al.* (2010). Protective effects of the pyrolyzates derived from bamboo against neuronal damage and hematoaggregation. *Journal of Ethnopharmacology*, **128**, 594–599.

Hong, J. Y., Min, H. Y., Xu, G. H. *et al.* (2008). Growth inhibition and G1 cell cycle arrest mediated by 25-methoxyhispidol A, a novel triterpenoid, isolated from the fruit of *Poncirus trifoliata* in human hepatocellular carcinoma cells. *Planta Medica*, **74**, 151–155.

Hou, C. C., Lin, S. J., Cheng, J. T. and Hsu, F. L. (2003a). Antidiabetic dimeric guianolides and a lignan glycoside from *Lactuca indica*. *Journal of Natural Products*, **66**, 625–629.

Hou, W. C., Lin, R. D., Cheng, K. T. *et al.* (2003b). Free radical-scavenging activity of Taiwanese native plants. *Phytomedicine*, **10**, 170–175.

Hsieh, C. C., Kuo, Y. H., Kuo, C. C. *et al.* (2010). Chamaecypanone C, a novel skeleton microtubule inhibitor, with anticancer activity by trigger caspase 8-Fas/FasL dependent apoptotic pathway in human cancer cells. *Biochemical Pharmacology*, **79**, 1261–1271.

Hsieh, P. W., Huang, Z. Y., Chen, J. H. *et al.* (2007). Cytotoxic withanolides from *Tubocapsicum anomalum. Journal of Natural Products*, **70**, 747–753.

Hua, H., Feng, L., Zhang, X. P., Zhang, L. F. and Jin, J. (2009). Anti-angiogenic activity of juliboside J(8), a natural product isolated from *Albizia julibrissin. Phytomedicine*, **16**, 703–711.

Huffman, M. A. (1984). Plant foods and foraging behavior of the Arashiyama Japanese macaques (in Japanese). In *Arashiyama Japanese Monkeys: Arashiyama Natural History Research Station Report, Volume 3*, ed. N. Asaba. Osaka: Osaka Seihan Printers, pp. 55–65.

Huffman, M. (1991). History of the Arashiyama Japanese macaques in Kyoto, Japan. In *The Monkeys of Arashiyama: Thirty-five Years of Study in Japan and the West*, eds. L. M. Fedigan and P. J. Asquith. Albany, NY: State University of New York Press, pp. 21–53.

Huffman, M. A. (2007). Primate self-medication. In *Primates in Perspective*, eds. C. Campbell, A. Fuentes, K. MacKinnon, M. Panger and S. K. Bearder. Oxford: University of Oxford Press, pp. 677–690.

Hung, J. Y., Yang, C. J., Tsai, Y. M., Huang, H. W. and Huang, M. S. (2008). Antiproliferative activity of aucubin is through cell cycle arrest and apoptosis in human non-small cell lung cancer A549 cells. *Clinical and Experimental Pharmacology and Physiology*, **35**, 995–1001.

Ichikawa, H., Wang, X. J. and Konishi, T. (2003). Role of component herbs in antioxidant activity of Shengmai San: a traditional Chinese medicine formula preventing cerebral oxidative damage in rat. *American Journal of Chinese Medicine*, **31**, 509–521.

Ikeda, T., Fujiwara, S., Araki; K. *et al.* (1997). Cytotoxic glycosides from *Albizia julibrissin. Journal of Natural Products*, **60**, 102–107.

Inagaki, H., Yamaguchi, A., Kato, K. *et al.* (2008). Screening of weed extracts for anti-fungal properties against *Colletotrichum lagenarium*, the causal agent of anthracnose in cucumber. *Weed Biology and Management*, **8**, 276–283.

Ito, C., Itoigawa, M., Furukawa, A. *et al.* (2004). Quinolone alkaloids with nitric oxide production inhibitory activity from *Orixa japonica. Journal of Natural Products*, **67**, 1800–1803.

Ito, H., Hidaka, H., Shimura, K. and Sakakibara, J. (1980). Anti-tumor effect and biological activity of *Rumex acetosa* L. *Japanese Journal of Pharmacology*, **30**, 111.

Ito, T. and Kumazawa, K. (1995). Precursors of antifungal substances from cherry leaves (*Prunus yedoensis* Matsumura). *Bioscience Biotechnology and Biochemistry*, **59**, 1944–1945.

Itoh, T., Oyama, M., Takimoto, N. *et al.* (2009). Inhibitory effects of sesquiterpene lactones isolated from *Eupatorium chinense* L. on IgE-mediated degranulation in rat basophilic leukemia RBL-2H3 cells and passive cutaneous anaphylaxis reaction in mice. *Bioorganic and Medicinal Chemistry*, **17**, 3189–3197.

Iwamoto, T. (1982). Food and nutritional condition of free ranging Japanese monkeys on Koshima islet during winter. *Primates*, **23**, 153–170.

Jaman, M. F., Takemoto, H. and Huffman, M. (2010). The foraging behavior of Japanese macaques (*Macaca fuscata*) in a forested enclosure: effects of nutrient composition, energy and its seasonal variation on the consumption of natural plant foods. *Current Zoology*, **56**, 198–208.

Jang, D. S., Yang, M. S., Ha, T. J. and Park, K. H. (1999). Hemistepsins with cytotoxic activity from *Hemisteptia lyrata*. *Planta Medica*, **65**, 765–766.

Jang, D. S., Yoo, N. H., Lee, Y. M. *et al.* (2008). Constituents of the flowers of *Erigeron annuus* with inhibitory activity on the formation of advanced glycation end products (AGEs) and aldose reductase. *Archives of Pharmacal Research*, **31**, 900–904.

Jang, D. S., Yoo, N. H., Kim, N. H. *et al.* (2010). 3,5-Di-O-caffeoyl-epi-quinic acid from the leaves and stems of *Erigeron annuus* inhibits protein glycation, aldose reductase, and cataractogenesis. *Biological and Pharmaceutical Bulletin*, **33**, 329–333.

Jeong, J. T., Moon, J. H., Park, K. H. and Shin, C. S. (2006). Isolation and characterization of a new compound from *Prunus mume* fruit that inhibits cancer cells. *Journal of Agricultural and Food Chemistry*, **54**, 2123–2128.

Jin, J. Y., Kim, J. M., Yang, H. K. *et al.* (2006). Antihypertensive and vasodilator effects of ethanolic extract of *Aralia elata* in the spontaneously hypertensive rats. *Acta Pharmacologica Sinica*, **27**, 131.

Jun, W., Hur, J. M., Yang, E. J. *et al.* (2007). Protective properties of quercetin and luteolin from *Petasites japonicus* leaves against A beta (25–35)-induced neurotoxicity in B103 cells. *Planta Medica*, **73**, P472.

Jung, B. G., Cho, S. J., Koh, H. B., Han, D. U. and Lee, B. J. (2010). Fermented Maesil (*Prunus mume*) with probiotics inhibits development of atopic dermatitis-like skin lesions in NC/Nga mice. *Veterinary Dermatology*, **21**, 184–191.

Jung, M. J., Chung, H. Y., Kang, S. S. *et al.* (2003). Antioxidant activity from the stem bark of *Albizzia julibrissin*. *Archives of Pharmacal Research*, **26**, 458–462.

Jung, S. M., Choi, S. I., Park, S. M. and Heo, T. R. (2008). Synergistic antimicrobial effect of *Achyranthes japonica* Nakai extracts and *Bifidobacterium supernatants* against *Clostridium difficile*. *Food Science and Biotechnology*, **17**, 402–407.

Kang, D. G., Lee, Y. S., Kim, H. J., Lee, Y. M. and Lee, H. S. (2003). Angiotensin converting enzyme inhibitory phenylpropanoid glycosides from *Clerodendron trichotomum*. *Journal of Ethnopharmacology*, **89**, 151–154.

Kang, G. J., Lee, H. J., Yoon, W. J. *et al.* (2008a). *Prunus yeddensis* inhibits the inflammatory chemokines, MDC and TARC, by regulating the STAT1-signaling pathway in IFN-gamma-stimulated HaCaT human keratinocytes. *Biomolecules and Therapeutics*, **16**, 394–402.

Kang, J. S., Lee, K. H., Han, M. H. *et al.* (2008b). Antiinflammatory activity of methanol extract isolated from stem bark of *Magnolia kobus*. *Phytotherapy Research*, **22**, 883–888.

Kang, K. A., Lee, K. H., Zhang, R. *et al.* (2007). Protective effects of *Castanopsis cuspidata* through activation of ERK and NF-kappa B on oxidative cell death induced by hydrogen peroxide. *Journal of Toxicology and Environmental Health – Part a – Current Issues*, **70**, 1319–1328.

Kang, T. H., Jeong, S. J., Kim, N. Y., Higuchi, R. and Kim, Y. C. (2000). Sedative activity of two flavonol glycosides isolated from the flowers of *Albizzia julibrissin* Durazz. *Journal of Ethnopharmacology*, **71**, 321–323.

Kapadia, G. J., Sharma, S. C., Tokuda, H., Nishino, H. and Ueda, S. (1996). Inhibitory effect of iridoids on Epstein-Barr virus activation by a short-term in vitro assay for anti-tumor promoters. *Cancer Letters*, **102**, 223–226.

Kariyone, T. (1971). *Atlas of Medicinal Plants*. Osaka: Takeda Chemical Industries.

Kawai, T., Kinoshita, K., Koyama, K. and Takahashi, K. (1994). Antiemetic principles of *Magnolia obovata* bark and *Zingiber officinale* rhyzome. *Planta Medica*, **60**, 17–20.

Kim, D. H., Bae, E. A. and Han, M. J. (1999). Anti-*Helicobacter pylori* activity of the metabolites of poncirin from *Poncirus trifoliata* by human intestinal bacteria. *Biological and Pharmaceutical Bulletin*, **22**, 422–424.

Kim, D. H., Jung, S. J., Chung, I. S. *et al.* (2005a). Ergosterol peroxide from flowers of *Erigeron annuus* L. as an anti-atherosclerosis agent. *Archives of Pharmacal Research*, **28**, 541–545.

Kim, D. K., Lim, J. P., Kim, J. W., Park, H. W. and Eun, J. S. (2005b). Antitumor and antiinflammatory constituents from *Celtis sinensis*. *Archives of Pharmacal Research*, **28**, 39–43.

Kim, D. Y., Kang, S. H. and Ghil, S. H. (2010a). *Cirsium japonicum* extract induces apoptosis and anti-proliferation in the human breast cancer cell line MCF-7. *Molecular Medicine Reports*, **3**, 427–432.

Kim, H. J., Saleem, M., Seo, S. H., Jin, C. and Lee, Y. S. (2005c). Two new antioxidant stilbene dimers, parthenostilbenins A and B from *Parthenocissus tricuspidata*. *Planta Medica*, **71**, 973–976.

Kim, H. J., Woo, E. R., Shin, C. G. *et al.* (2001a). HIV-1 integrase inhibitory phenylpropanoid glycosides from *Clerodendron trichotomum*. *Archives of Pharmacal Research*, **24**, 286–291.

Kim, H. W., Murakami, A., Abe, M. *et al.* (2005d). Suppressive effects of mioga ginger and ginger constituents on reactive oxygen and nitrogen species generation, and the expression of inducible pro-inflammatory genes in macrophages. *Antioxidants and Redox Signaling*, **7**, 1621–1629.

Kim, I. G., Kang, S. C., Kim, K. C., Choung, E. S. and Zee, O. P. (2008a). Screening of estrogenic and antiestrogenic activities from medicinal plants. *Environmental Toxicology and Pharmacology*, **25**, 75–82.

Kim, J. C., Choi, G. J., Lee, S. W. *et al.* (2004a). Screening extracts of *Achyranthes japonica* and *Rumex crispus* for activity against various plant pathogenic fungi and control of powdery mildew. *Pest Management Science*, **60**, 803–808.

Kim, J. S., Shim, S. H., Sungwook, C. *et al.* (2005e). Saponins and other constituents from the leaves of *Aralia elata*. *Chemical and Pharmaceutical Bulletin*, **53**, 696–700.

Kim, J. Y., Kim, S. S., Oh, T. H. *et al.* (2009a). Chemical composition, antioxidant, antielastase, and anti-inflammatory activities of *Illicium anisatum* essential oil. *Acta Pharmaceutica*, **59**, 289–300.

Kim, K. H., Kim, Y. H. and Lee, K. R. (2007a). Isolation of quinic acid derivatives and flavonoids from the aerial parts of *Lactuca indica* L. and their hepatoprotective activity in vitro. *Bioorganic and Medicinal Chemistry Letters*, **17**, 6739–6743.

Kim, K. Y., Davidson, P. M. and Chung, H. J. (2001b). Antibacterial activity in extracts of *Camellia japonica* L. petals and its application to a model food system. *Journal of Food Protection*, **64**, 1255–1260.

Kim, M. B., Park, J. S. and Lim, S. B. (2010b). Antioxidant activity and cell toxicity of pressurised liquid extracts from 20 selected plant species in Jeju, Korea. *Food Chemistry*, **122**, 546–552.

Kim, M. H., Lee, J., Yoo, D. S. *et al.* (2009b). Protective effect of stress-induced liver damage by saponin fraction from *Codonopsis lanceolata*. *Archives of Pharmacal Research*, **32**, 1441–1446.

Kim, M. R., Lee, H. H., Hahm, K. S., Moon, Y. H. and Woo, E. R. (2004b). Pentacyclic triterpenoids and their cytotoxicity from the stem bark of *Styrax japonica* S. et Z. *Archives of Pharmacal Research*, **27**, 283–286.

Kim, O. S., Kim, Y. S., Jang, D. S., Yoo, N. H. and Kim, J. S. (2009c). Cytoprotection against hydrogen peroxide-induced cell death in cultured mouse mesangial cells by erigeroflavanone, a novel compound from the flowers of *Erigeron annuus*. *Chemico-Biological Interactions*, **180**, 414–420.

Kim, S., Park, S. H., Lee, H. N. and Park, T. (2008b). *Prunus mume* extract ameliorates exercise-induced fatigue in trained rats. *Journal of Medicinal Food*, **11**, 460–468.

Kim, S. H., Kwon, Y. E., Park, W. H., Jeon, H. and Shin, T. Y. (2009d). Effect of leaves of *Eriobotrya japonica* on anaphylactic allergic reaction and production of tumor necrosis factor-alpha. *Immunopharmacology and Immunotoxicology*, **31**, 314–319.

Kim, S. I., Na, Y. E., Yi, J. H., Kim, B. S. and Ahn, Y. J. (2007b). Contact and fumigant toxicity of oriental medicinal plant extracts against *Dermanyssus gallinae* (Acari: Dermanyssidae). *Veterinary Parasitology*, **145**, 377–382.

Kim, S. S., Kim, J. Y., Lee, N. H. and Hyun, C. G. (2008c). Antibacterial and anti-inflammatory effects of Jeju medicinal plants against acne-inducing bacteria. *Journal of General and Applied Microbiology*, **54**, 101–106.

Kim, S. Y., Son, K. H., Chang, H. W., Kang, S. S. and Kim, H. P. (1997). Inhibitory effects of plant extracts on adjuvant-induced arthritis. *Archives of Pharmacal Research*, **20**, 313–317.

Kim, T. D., Lee, J. Y., Cho, B. J., Park, T. W. and Kim, C. J. (2010c). The analgesic and anti-inflammatory effects of 7-oxosandaracopimaric acid isolated from the roots of *Aralia cordata*. *Archives of Pharmacal Research*, **33**, 509–514.

Kim, Y., Min, H. Y., Park, H. J. *et al.* (2004c). Suppressive effect of inducible nitric oxide synthase (iNOS) expression by the methanol extract of *Actinodaphne lancifolia*. *Phytotherapy Research*, **18**, 853–856.

Kim, Y. H., Yang, H. E., Kim, J. H. *et al.* (2000). Protection of the flowers of *Prunus persica* extract from ultraviolet B-induced damage of normal human keratinocytes. *Archives of Pharmacal Research*, **23**, 396–400.

Kim, Y. S. and Shin, D. H. (2004a). Volatile components and antibacterial effects of simultaneous steam distillation and solvent extracts from the leaves of *Styrax japonica* S. et Z. *Food Science and Biotechnology*, **13**, 561–565.

(2004b). Volatile constituents from the leaves of *Callicarpa japonica* thunb. and their antibacterial activities. *Journal of Agricultural and Food Chemistry*, **52**, 781–787.

Kimura, K. and Kimura, T. (1991). *Medicinal Plants of Japan*. Osaka: Hoikusha Publishing Co., Ltd.

Ko, H. H., Yen, M. H., Wu, R. R., Won, S. J. and Lin, C. N. (1999). Cytotoxic isoprenylated flavans of *Broussonetia kazinoki*. *Journal of Natural Products*, **62**, 164–166.

Koyama, S., Yamaguchi, Y., Tanaka, S. and Motoyoshiya, J. (1997). A new substance (Yoshixol) with an interesting antibiotic mechanism from wood oil of Japanese traditional tree (Kiso-Hinoki), *Chamaecyparis obtusa*. *General Pharmacology*, **28**, 797–804.

Kozan, E., Kupeli, E. and Yesilada, E. (2006). Evaluation of some plants used in Turkish folk medicine against parasitic infections for their in vivo anthelmintic activity. *Journal of Ethnopharmacology*, **108**, 211–216.

Kumar, K. H., Charles, M., Kumar, R. and Vasanta, M. (2008). Diuretic activity of *Raphanus sativus*. *Indian Journal of Pharmacology*, **40**, 610.

Kundakovic, T., Stanojkovic, T., Grubin, J. *et al.* (2008a). Antioxidant and cytotoxic activity of *Ampelopsis brevipedunculata* and *Parthenocissus tricuspidata*. *Planta Medica*, **74**, 954.

Kundakovic, T., Stanojkovic, T., Milenkovic, M. *et al.* (2008b). Cytotoxic, antioxidant, and antimicrobial activities of *Ampelopsis brevipedunculata* and *Parthenocissus tricuspidata* (Vitacaea). *Archives of Biological Sciences*, **60**, 641–647.

Kuo, P. L., Hsu, Y. L., Lin, T. C. and Lin, C. C. (2004a). Prodelphinidin B-2 3,3′-di-O-gallate from *Myrica rubra* inhibits proliferation of A549 carcinoma cells via blocking cell cycle progression and inducing apoptosis. *European Journal of Pharmacology*, **501**, 41–48.

Kuo, P. L., Hsu, Y. L., Lin, T. C., Lin, L. T. and Lin, C. C. (2004b). Induction of apoptosis in human breast adenocarcinoma MCF-7 cells by prodelphinidin B-2 3,3′-di-O-gallate from *Myrica rubra* via Fas-mediated pathway. *Journal of Pharmacy and Pharmacology*, **56**, 1399–1406.

Kuo, Y. C., Kuo, Y. H., Lin, Y. L. and Tsai, W. J. (2006). Yatein from *Chamaecyparis obtusa* suppresses herpes simplex virus type 1 replication in HeLa cells by interruption of the immediate-early gene expression. *Antiviral Research*, **70**, 112–120.

Kuroyanagi, M., Ikeda, R., Gao, H. Y. *et al.* (2008). Neurite outgrowth-promoting active constituents of the Japanese cypress (*Chamaecyparis obtusa*). *Chemical and Pharmaceutical Bulletin*, **56**, 60–63.

Kwon, H. Y., Rhyu, M. R. and Lee, Y. (2008). The effects of *Cirsium japonicum* on lipid profile in ovariectomized rats. *Biomolecules and Therapeutics*, **16**, 293–298.

Kwon, Y. S., Kim, E. Y., Kim, W. J., Kim, W. K. and Kim, C. M. (2002). Antioxidant constituents from *Setaria viridis*. *Archives of Pharmacal Research*, **25**, 300–305.

Lansky, E. P., Paavilainen, H. M., Pawlus, A. D. and Newman, R. A. (2008). *Ficus spp.* (fig): Ethnobotany and potential as anticancer and anti-inflammatory agents. *Journal of Ethnopharmacology*, **119**, 195–213.

Lassak, E. V. and McCarthy, T. (1983). *Australian Medicinal Plants*. Sydney: Methuen.

Launert, E. (1981). *Edible and Medicinal Plants of Britain and Northern Europe*. London: Hamlyn.

Lee, C. C. and Houghton, P. (2005). Cytotoxicity of plants from Malaysia and Thailand used traditionally to treat cancer. *Journal of Ethnopharmacology*, **100**, 237–243.

Lee, C. K., Park, K. K., Hwang, J. K., Lee, S. K. and Chung, W. Y. (2009a). Extract of *Prunus persica* flesh (PPFE) improves chemotherapeutic efficacy and protects against nephrotoxicity in cisplatin-treated mice. *Phytotherapy Research*, **23**, 999–1005.

Lee, E. B., Kim, O. J., Kang, S. S. and Jeong, C. (2005a). Araloside A, an antiulcer constituent from the root bark of *Aralia elata*. *Biological and Pharmaceutical Bulletin*, **28**, 523–526.

Lee, H. J., Hyun, E. A., Yoon, W. J. *et al.* (2006a). In vitro anti-inflammatory and anti-oxidative effects of *Cinnamomum camphora* extracts. *Journal of Ethnopharmacology*, **103**, 208–216.

Lee, H. J., Seo, S. M., Lee, O. K. *et al.* (2008a). Lignans from the bark of *Magnolia kobus*. *Helvetica Chimica Acta*, **91**, 2361–2366.

Lee, I., Kim, H., Youn, U. *et al.* (2008b). Absolute configuration of a diterpene with an acyclic 1,2-diol moiety and cytotoxicity of its analogues from the aerial parts of *Aralia cordata*. *Bulletin of the Korean Chemical Society*, **29**, 1839–1842.

Lee, J. H., Ham, Y. A., Choi, S. H. *et al.* (2000). Activity of crude extract of *Rubus crataegifolius* roots as a potent apoptosis inducer and DNA topoisomerase I inhibitor. *Archives of Pharmacal Research*, **23**, 338–343.

Lee, J. H., Choi, S. I., Lee, Y. S. and Kim, G. H. (2008c). Antioxidant and anti-inflammatory activities of ethanol extract from leaves of *Cirsium japonicum*. *Food Science and Biotechnology*, **17**, 38–45.

Lee, J. H., Lee, S. H., Kim, Y. S. and Jeong, C. S. (2009b). Protective effects of neohesperidin and poncirin isolated from the fruits of *Poncirus trifoliata* on potential gastric disease. *Phytotherapy Research*, **23**, 1748–1753.

Lee, K. T., Lee, K. W., Lee, Y. and Park, H. J. (2004). Induction of differentiation by acteoside isolated from *Clerodendron trichotomum* in human promyelocytic HL-60 leukemia cells via cell cycle arrest. *Cancer Epidemiology Biomarkers and Prevention*, **13**, 1912S.

Lee, K. W., Jung, H. J., Park, H. J. *et al.* (2005b). B-D-xylopyranosyl-(1 -> 3)-beta-D-glucuronopyranosyl echinocystic acid isolated from the roots of *Codonopsis lanceolata* induces caspase-dependent apoptosis in human acute promyelocytic leukemia HL-60 cells. *Biological and Pharmaceutical Bulletin*, **28**, 854–859.

Lee, S. E., Ju, E. M. and Kim, J. H. (2001). Free radical scavenging and antioxidant enzyme fortifying activities of extracts from *Smilax china* root. *Experimental and Molecular Medicine*, **33**, 263–268.

Lee, S. H., Choi, S. M., Sohn, Y. S., Kang, K. K. and Yoo, M. (2006b). Effect of *Oryza sativa* extract on the progression of airway inflammation and remodeling in an experimental animal model of asthma. *Planta Medica*, **72**, 405–410.

Lee, S. H., Sohn, Y. S., Kang, K. K., Kwon, J. W. and Yoo, M. (2006c). Inhibitory effect of DA-9201, an extract of *Oryza sativa* L., on airway inflammation and bronchial hyperresponsiveness in mouse asthma model. *Biological and Pharmaceutical Bulletin*, **29**, 1148–1153.

Lee, S. Y., Cho, J. S., Yuk, D. Y. *et al.* (2009c). Obovatol enhances docetaxel-induced prostate and colon cancer cell death through inactivation of nuclear transcription factor-kappa B. *Journal of Pharmacological Sciences*, **111**, 124–136.

Lee, Y. M., Kim, D. K., Kim, S. H., Shin, T. Y. and Kim, H. M. (1996). Antianaphylactic activity of *Poncirus trifoliata* fruit extract. *Journal of Ethnopharmacology*, **54**, 77–84.

Lellau, T. F. and Liebezeit, G. (2003). Cytotoxic and antitumor activities of ethanolic extracts of salt marsh plants from the Lower Saxonian Wadden Sea, Southern North Sea. *Pharmaceutical Biology*, **41**, 293–300.

Li, Z. Y., Gu, Y. C., Irwin, D. *et al.* (2009). Further *Daphniphyllum* alkaloids with insecticidal activity from the bark of *Daphniphyllum macropodum* MIQ. *Chemistry and Biodiversity*, **6**, 1744–1750.

Liao, Z. Y., Chen, X. L. and Wu, M. J. (2010). Antidiabetic effect of flavones from *Cirsium japonicum* DC in diabetic rats. *Archives of Pharmacal Research*, **33**, 353–362.

Liaw, C. C., Kuo, Y. H., Hwang, T. L. and Shen, Y. C. (2008). Eupatozansins A-C, sesquiterpene lactones from *Eupatorium chinense* var. *tozanense*. *Helvetica Chimica Acta*, **91**, 2115–2121.

Lima, J. M., Silva, C. A., Rosa, M. B. *et al.* (2009). Phytochemical prospecting of *Sonchus oleraceus* and its toxicity to *Artemia salina*. *Planta Daninha*, **27**, 7–11.

Ling, J. and Liu, W. Y. (1996). Cytotoxicity of two new ribosome-inactivating proteins, cinnamomin and camphorin, to carcinoma cells. *Cell Biochemistry and Function*, **14**, 157–161.

Liu, C. H., Mishra, A. K., Tan, R. X. *et al.* (2006a). Repellent and insecticidal activities of essential oils from *Artemisia princeps* and *Cinnamomum camphora* and their effect on seed germination of wheat and broad bean. *Bioresource Technology*, **97**, 1969–1973.

Liu, H., Wang, J. H., Zhao, J. L. *et al.* (2009). Isoquinoline alkaloids from *Macleaya cordata* active against plant microbial pathogens. *Natural Product Communications*, **4**, 1557–1560.

Liu, S. J., Luo, X., Li, D. X. *et al.* (2006b). Tumor inhibition and improved immunity in mice treated with flavone from *Cirsium japonicum* DC. *International Immunopharmacology*, **6**, 1387–1393.

Liu, S. J., Zhang, J., Li, D. *et al.* (2007). Anticancer activity and quantitative analysis of flavone of *Cirsium japonicum* DC. *Natural Product Research*, **21**, 915–922.

Loehr, G., Beikler, T., Bicker, J. and Hensel, A. (2009). Antiadhesive effects of natural extracts out of *Myrothamnus flabellifolia* Welw. and *Rumex acetosa* L. against *Porphyromonas gingivalis*. *Planta Medica*, **75**, 897–898.

Lu, G. W., Miura, K., Yukimura, T. and Yamamoto, K. (1994). Effects of extract from *Clerodendron trichotomum* on blood pressure and renal function in rats and dogs. *Journal of Ethnopharmacology*, **42**, 77–82.

Lust, J. (2001). *The Herb Book*. New York: Beneficial Books.

Ma, W. G., Fukushi, Y. and Tahara, S. (1999). Fungitoxic alkaloids from Hokkaido *Corydalis* species. *Fitoterapia*, **70**, 258–265.

MacIntosh, A. J. J. and Huffman, M. A. (2010). Toward understanding the role of diet in host-parasite interactions: the case for Japanese macaques. In *The Japanese Macaques*, eds. N. Nakagawa, M. Nakamichi and H. Sugiura. Tokyo: Springer, pp. 323–344.

MacIntosh, A. J. J., Hernandez, A. D. and Huffman, M. A. (2010). Host age, sex, and reproductive seasonality affect nematode parasitism in wild Japanese macaques. *Primates*, **51**, 353–364.

Manandhar, N. P. (2002). *Plants and People of Nepal*. Portland, OR: Timber Press.

Matsuda, H., Morikawa, T., Ishiwada, T. *et al.* (2003). Medicinal flowers. VIII. Radical scavenging constituents from the flowers of *Prunus mume*: Structure of prunose III. *Chemical and Pharmaceutical Bulletin*, **51**, 440–443.

Matthews, V. (1994). *The New Plantsman*, vol. 1. Royal Horticultural Society.

McCutcheon, A. R., Ellis, S. M., Hancock, R. E. W. and Towers, G. H. N. (1994). Antifungal screening of medicinal plants of British Columbian native peoples. *Journal of Ethnopharmacology*, **44**, 157–169.

McKey, D. (1974). Adaptive patterns in alkaloid physiology. *American Naturalist*, **108**, 305–320.

Min, B. S., Oh, S. R., Ahn, K. S. *et al.* (2004a). Anti-complement activity of norlignans and terpenes from the stem bark of *Styrax japonica*. *Planta Medica*, **70**, 1210–1215.

Min, B. S., Cui, S., Lee, H. K., Sok, D. E., and Kim, M. R. (2005). A new furofuran lignan with antioxidant and antiseizure activities from the leaves of *Petasites japonicus*. *Archives of Pharmacal Research*, **28**, 1023–1026.

Min, L. S., Na, M. K., Oh, S. R. *et al.* (2004b). New furofuran and butyrolactone lignans with antioxidant activity from the stem bark of *Styrax japonica*. *Journal of Natural Products*, **67**, 1980–1984.

Minakami, K. and Jitouzono, T. (2002). Detoxifying effects of *Achyranthes japonica* extract against Habu venom. *Japanese Journal of Pharmacology*, **88**, 288.

Miyazawa, M., Kinoshita, H. and Okuno, Y. (2003). Antimutagenic activity of sakuranetin from *Prunus jamasakura*. *Journal of Food Science*, **68**, 52–56.

Miyoshi, N., Nakamura, Y., Ueda, Y. *et al.* (2003). Dietary ginger constituents, galanals A and B, are potent apoptosis inducers in human T lymphoma Jurkat cells. *Cancer Letters*, **199**, 113–119.

Moerman, D. (1998). *Native American Ethnobotany*. Portland, OR: Timber Press.

Moon, H. I., Kim, M. R., Woo, E. R. and Chung, J. H. (2005). Triterpenoid from *Styrax japonica* SIEB. et ZUCC, and its effects on the expression of matrix metalloproteinases-1 and type 1 procollagen caused by ultraviolet irradiated cultured primary human skin fibroblasts. *Biological and Pharmaceutical Bulletin*, **28**, 2003–2006.

Moon, H. I., Lee, J. and Chung, J. H. (2006). The effect of erythrodiol-3-acetate on the expressions of matrix metalloproteinase-1 and type-1 procollagen caused by ultraviolet irradiated cultured primary old aged human skin fibroblasts. *Phytomedicine*, **13**, 707–711.

Mori, S., Sawada, T., Okada, T. *et al.* (2007). New anti-proliferative agent, MK615, from Japanese apricot '*Prunus mume*' induces striking autophagy in colon cancer cells in vitro. *World Journal of Gastroenterology*, **13**, 6512–6517.

Mori-Hongo, M., Yamaguchi, H., Warashina, T. and Miyase, T. (2009). Melanin synthesis inhibitors from *Lespedeza cyrtobotrya*. *Journal of Natural Products*, **72**, 63–71.

Morikawa, T. (2007). Search for bioactive constituents from several medicinal foods: hepatoprotective, antidiabetic, and antiallergic activities. *Journal of Natural Medicines*, **61**, 112–126.

Muralidharan, P., Balamurugan, G. and Kumar, P. (2008). Inotropic and cardioprotective effects of *Daucus carota* Linn. on isoproterenol-induced myocardial infarction. *Bangladesh Journal of Pharmacology*, **3**, 74–79.

Murata, G. and Hazama, N. (1968). Flora of Arashiyama, Kyoto, and plant food of Japanese monkeys. *Iwatayama Shizenshi Kenkyujo Chosa Kenkyu Hokoku*, **2**, 1–59. (in Japanese)

Nagao, T., Abe, F. and Okabe, H. (2001). Antiproliferative constituents in the plants 7. Leaves of *Clerodendron bungei* and leaves and bark of *C. trichotomum*. *Biological and Pharmaceutical Bulletin*, **24**, 1338–1341.

Nagata, T., Tsushida, T., Hamaya, E. *et al.* (1985). Camellidins, antifungal saponins isolated from *Camellia japonica*. *Agricultural and Biological Chemistry*, **49**, 1181–1186.

Nakagawa, N., Iwamoto, T., Yokota, T. and Soumah, A. G. (1996). Inter-regional and inter-seasonal variations of food quality in Japanese macaques: constraints of digestive volume and feeding time. In *Evolution and Ecology of Macaque Societies*, eds. J. E. Fa and D. G. Lindburg. Cambridge: Cambridge University Press, pp. 207–234.

Ni, W. H., Zhang, X., Bi, H. T. *et al.* (2009). Preparation of a glucan from the roots of *Rubus crataegifolius* Bge. and its immunological activity. *Carbohydrate Research*, **344**, 2512–2518.

Nojima, H., Kimura, I., Chen, F. J. *et al.* (1998). Antihyperglycemic effects of N-containing sugars from *Xanthocercis zambesiaca, Morus bombycis, Aglaonema treubii* and *Castanospermum australe* in streptozotocin-diabetic mice. *Journal of Natural Products*, **61**, 397–400.

Ock, J., Han, H. S., Hong, S. H. *et al.* (2010). Obovatol attenuates microglia-mediated neuroinflammation by modulating redox regulation. *British Journal of Pharmacology*, **159**, 1646–1662.

Ogata, H. and Ogiyama, K. (2000). Chemical compositions and antipathogenic activities of constituent fatty acids from neutral wax in foliage of *Cryptomeria japonica* D. Don. *Mokuzai Gakkaishi*, **46**, 54–62.

Oh, K., Seo, J., Lee, S. *et al.* (2005). Anxiolytic effects of CBPharm-001 isolated from *Magnolia obovata* and its possible molecular mechanisms. *Behavioural Pharmacology*, **16** (Suppl.), S40.

Ohara, A., Saito, F. and Matsuhisa, T. (2008). Screening of antibacterial activities of edible plants against *Streptococcus mutans*. *Food Science and Technology Research*, **14**, 190–193.

Ohta, S., Sakurai, N., Kamogawa, A. *et al.* (1992). Protective effects of the bark of *Myrica rubra* Sieb et Zucc on experimental liver injuries. *Yakugaku Zasshi – Journal of the Pharmaceutical Society of Japan*, **112**, 244–252.

Onodera, K., Hanashiro, K. and Yasumoto, T. (2006). Camellianoside, a novel antioxidant glycoside from the leaves of *Camellia japonica*. *Bioscience Biotechnology and Biochemistry*, **70**, 1995–1998.

Ooi, L. S. M., Wang, H., Luk, C. W. and Ooi, V. E. C. (2004). Anticancer and antiviral activities of *Youngia japonica* (L.) DC (Asteraceae, Compositae). *Journal of Ethnopharmacology*, **94**, 117–122.

Osawa, T., Kumazawa, S. and Kawakishi, S. (1991). Prunasol A and prunasol B, novel antioxidative tocopherol derivatives isolated from the leaf wax of *Prunus grayana*. *Agricultural and Biological Chemistry*, **55**, 1727–1731.

Ou, M. (1999). *Regular Chinese Medicinal Handbook*. Taipei: Warmth Publications, Inc.

Park, E. A., Chae, H. J., Woo, J. H. and Chyun, J. H. (2009). Effect of dietary supplementation of maesil (*Prunus mume*) concentrates on the improvement of diabetes in rats. *Annals of Nutrition and Metabolism*, **55**, 487.

Park, E. K., Sung, A. H., Trinh, H. T. *et al.* (2008). Lactic acid bacterial fermentation increases the antiallergic effects of *Ixeris dentata*. *Journal of Microbiology and Biotechnology*, **18**, 308–313.

Park, H. W., Lee, J. H., Choi, S. U. *et al.* (2010). Cytotoxic germacranolide sesquiterpenes from the bark of *Magnolia kobus*. *Archives of Pharmacal Research*, **33**, 71–74.

Park, I. K. and Shin, S. C. (2005). Fumigant activity of plant essential oils and components from garlic (*Allium sativum*) and clove bud (*Eugenia caryophyllata*) oils against the Japanese termite (*Reticulitermes speratus kolbe*). *Journal of Agricultural and Food Chemistry*, **53**, 4388–4392.

Park, J. C., Hur, J. M., Park, J. G. *et al.* (2002). Inhibitory effects of Korean medicinal plants and camelliatannin H from *Camellia japonica* on human immunodeficiency virus type 1 protease. *Phytotherapy Research*, **16**, 422–426.

Park, M. A. and Kim, H. J. (2007). Anti-inflammatory constituents isolated from *Clerodendron trichotomum* tunberg leaves (CTL) inhibit pro-inflammatory gene expression in LPS-stimulated RAW 264.7 macrophages by suppressing NF-kappa B activation. *Archives of Pharmacal Research*, **30**, 755–760.

Park, S. H., Park, E. K. and Kim, D. H. (2005). Passive cutaneous anaphylaxis-inhibitory activity of flavanones from *Citrus unshiu* and *Poncirus trifoliata*. *Planta Medica*, **71**, 24–27.

Pitts, B. J. R. and Meyerson, L. R. (1981). Inhibition of Na, K-ATPase activity and ouabian binding by sanguinarine. *Drug Development Research*, **1**, 43–49.

Pop, A., Cranganu, D. and Constantin, N. (1998). *Raphanus sativus* seeds lectins-epiphysal hormones supramolecular complexes with antitumor activity. *Naunyn-Schmiedebergs Archives of Pharmacology*, **358**, 5256.

Pyun, M. S. and Shin, S. (2006). Antifungal effects of the volatile oils from *Allium* plants against *Trichophyton* species and synergism of the oils with ketoconazole. *Phytomedicine*, **13**, 394–400.

Qa'dan, F., Verspohl, E. J., Nahrstedt, A., Petereit, F. and Matalka, K. Z. (2009). Cinchonain Ib isolated from *Eriobotrya japonica* induces insulin secretion in vitro and in vivo. *Journal of Ethnopharmacology*, **124**, 224–227.

Reid, B. E. (1977). *Famine Foods of the Chiu-Huang Pen-ts'ao*. Taipei: Southern Materials Centre.

Rubnov, S., Kashman, Y., Rabinowitz, R., Schlesinger, M. and Mechoulam, R. (2001). Suppressors of cancer cell proliferation from fig (*Ficus carica*) resin: isolation and structure elucidation. *Journal of Natural Products*, **64**, 993–996.

Saad, E. E. (2009). Cytotoxic and antibacterial constituents from the roots of *Sonchus oleraceus* L. growing in Egypt. *Pharmacognosy Magazine*, **5**, 324–328.

Saito, S., Ebashi, J., Sumita, S. *et al.* (1993). Comparison of cytoprotective effects of saponins isolated from leaves of *Aralia elata* Seem (Araliaceae) with synthesized bisdesmosides of oleanoic acid and hederagenin on carbon tetrachloride-induced hepatic-injury. *Chemical and Pharmaceutical Bulletin*, **41**, 1395–1401.

Saleem, M., Kim, H. J., Jin, C. and Lee, Y. S. (2004). Antioxidant caffeic acid derivatives from leaves of *Parthenocissus tricuspidata*. *Archives of Pharmacal Research*, **27**, 300–304.

Sargent, C. S. (1965). *Manual of the Trees of N. America*. New York, NY: Dover Publications.

Schaffer, S., Schmitt-Schillig, S., Muller, W. E. and Echert, G. P. (2005). Antioxidant properties of Mediterranean food plant extracts: geographical differences. *Journal of Physiology and Pharmacology*, **56** (Suppl.), S115–S124.

Schery, R. W. (1972). *Plants for Man*, 2nd edn. Englewood Cliffs, NJ: Prentice Hall.

Schuhly, W., Khan, I. and Fischer, N. H. (2001). The ethnomedicinal uses of Magnoliaceae from the southeastern United States as leads in drug discovery. *Pharmaceutical Biology*, **39**, 63–69.

Seo, J. J., Lee, S. H., Lee, Y. S. *et al.* (2007). Anxiolytic-like effects of obovatol isolated from *Magnolia obovata* – involvement of GABA/benzodiazepine receptors complex. *Progress in Neuro-Psychopharmacology and Biological Psychiatry*, **31**, 1363–1369.

Serit, M., Okubo, T., Su, R. H. *et al.* (1991). Antibacterial compounds from oak, *Quercus acuta* Thunb. *Agricultural and Biological Chemistry*, **55**, 19–23.

Shao, B., Guo, H. Z., Cui, Y. J. *et al.* (2007). Steroidal saponins from *Smilax china* and their anti-inflammatory activities. *Phytochemistry*, **68**, 623–630.

Sharma, M. C., Ohira, T. and Yatagai, M. (1994). Lanostane triterpenes from the bark of *Neolitsea sericea*. *Phytochemistry*, **37**, 201–203.

Sharma, N., Sharma, V. K. and Seo, S. Y. (2005). Screening of some medicinal plants for anti-lipase activity. *Journal of Ethnopharmacology*, **97**, 453–456.

Shi, J. Y., Gong, J. Y., Liu, J., Wu, X. Q. and Zhang, Y. (2009). Antioxidant capacity of extract from edible flowers of *Prunus mume* in China and its active components. *Lwt – Food Science and Technology*, **42**, 477–482.

Shibano, M., Kakutani, K., Taniguchi, M., Yasuda, M. and Baba, K. (2008). Antioxidant constituents in the dayflower (*Commelina communis* L.) and their alpha-glucosidase-inhibitory activity. *Journal of Natural Medicines*, **62**, 349–353.

Shim, Y. J., Do, H. K., Ah, S. Y. *et al.* (2003). Inhibitory effect of aqueous extract from the gall of *Rhus chinensis* on alpha-glucosidase activity and postprandial blood glucose. *Journal of Ethnopharmacology*, **85**, 283–287.

Shimizu, M. and Tomoo, T. (1994). Antiinflammatory constituents of topically applied crude drugs. 5. Constituents and antiinflammatory effect of aoki, *Aucuba japonica* Thunb. *Biological and Pharmaceutical Bulletin*, **17**, 665–667.

Shu, X. S., Gao, Z. H. and Yang, X. L. (2006). Anti-inflammatory and anti-nociceptive activities of *Smilax china* L. aqueous extract. *Journal of Ethnopharmacology*, **103**, 327–332.

Sokovic, M., Stojkovic, D., Glamoclija, J. *et al.* (2009). Susceptibility of pathogenic bacteria and fungi to essential oils of wild *Daucus carota*. *Pharmaceutical Biology*, **47**, 38–43.

Son, I. H., Chung, I. M., Lee, S. J. and Moon, H. I. (2007). Antiplasmodial activity of novel stilbene derivatives isolated from *Parthenocissus tricuspidata* from South Korea. *Parasitology Research*, **101**, 237–241.

Son, Y. K., Lee, M. H. and Han, Y. N. (2005). A new antipsychotic effective neolignan from *Firmiana simplex*. *Archives of Pharmacal Research*, **28**, 34–38.

Stajner, D., Milic, N., Canadanovic-Brunet, J. *et al.* (2006). Exploring *Allium* species as a source of potential medicinal agents. *Phytotherapy Research*, **20**, 581–584.

Stajner, D., Igic, R., Popovic, B. M. and Malencic, D. (2008). Comparative study of antioxidant properties of wild growing and cultivated *Allium* species. *Phytotherapy Research*, **22**, 113–117.

Stuart, G. A. (1911). *Chinese Materia Medica*. Shanghai: Presbyterian Mission Press.

Sung, S. H., Kim, E. S., Lee, K. Y., Lee, M. K. and Kim, Y. C. (2005). A new neuroprotective compound of *Ligustrum japonicum* leaves. *Planta Medica*, **72**, 62–64.

Suzuki, M. and Aoki, T. (1994). Effects of volatile compounds from leaf oil on blood pressure after exercising. *Mokuzai Gakkaishi*, **40**, 1243–1250.

Suzuki, R., Okada, Y. and Okuyama, T. (2005). The favorable effect of style of *Zea mays* L. on streptozotocin induced diabetic nephropathy. *Biological and Pharmaceutical Bulletin*, **28**, 919–920.

Suzuki, Y., Taguchi, K., Hagiwara, Y. and Kajiyama, K. (1979). Pharmacological studies on *Zingiber mioga*. 2. Effects on the central nervous system. *Folia Pharmacologica Japonica*, **75**, 731–746.

Takasaki, M., Konoshima, T., Tokuda, H. *et al.* (1999). Anti-carcinogenic activity of *Taraxacum* plant. II. *Biological and Pharmaceutical Bulletin*, **22**, 606–610.

Tanaka, S., Koyama, S., Haniu, H., Yamaguchi, Y. and Motoyoshiya, J. (1999). In vitro and in vivo antitumor activity of Yoshixol(TR) against murine L1210 leukemic cells. *General Pharmacology*, **33**, 179–186.

Tangpu, V., Temjenmongla, K. and Yadav, A. K. (2004). Anticestodal activity of *Trifolium repens* extract. *Pharmaceutical Biology*, **42**, 656–658.

Taniguchi, H., Kobayashi-Hattori, K., Tenmyo, C. *et al.* (2006). Effect of Japanese radish (*Raphanus sativus*) sprout (Kaiware-daikon) on carbohydrate and lipid metabolisms in normal and streptozotocin-induced diabetic rats. *Phytotherapy Research*, **20**, 274–278.

Tobinaga, S., Takeuchi, N., Kasama, T. *et al.* (1983). Anti-histaminic and anti-allergic principles of *Petasites japonicus*. *Chemical and Pharmaceutical Bulletin*, **31**, 745–748.

Tomatsu, M., Kondo, T., Yoshikawa, T. *et al.* (2004). An apoptotic inducer, aralin, is a novel type II ribosome-inactivating protein from *Aralia elata*. *Biological Chemistry*, **385**, 819–827.

Tsuchiya, Y., Shimizu, M., Hiyama, Y. *et al.* (1985). Antiviral activity of natural occurring flavonoids in vitro. *Chemical and Pharmaceutical Bulletin*, **33**, 3881–3886.

Uphof, J. C. (1968). *Dictionary of Economic Plants*. Wurzburg: Cramer.

Urban, J., Kokoska, L., Langrova, I. and Matejkova, J. (2008). In vitro anthelmintic effects of medicinal plants used in Czech Republic. *Pharmaceutical Biology*, **46**, 808–813.

Usher, G. (1974). *A Dictionary of Plants Used by Man*. London: Constable and Company.

Vargas, S., Perez, R. M., Perez, S., Zavala, M. A. and Perez, C. (1999). Antiurolithiatic activity of *Raphanus sativus* aqueous extract on rats. *Journal of Ethnopharmacology*, **68**, 335–338.

Vasudevan, M. and Parle, M. (2006). Pharmacological evidence for the potential of *Daucus carota* in the management of cognitive dysfunctions. *Biological and Pharmaceutical Bulletin*, **29**, 1154–1161.

Velazquez, D. V. O., Xavier, H. S., Batista, J. E. M. and de Castro-Chaves, C. (2005). *Zea mays* L. extracts modify glomerular function and potassium urinary excretion in conscious rats. *Phytomedicine*, **12**, 363–369.

Venkatesh, K. P. and Chauhan, S. (2008). Mulberry: life enhancer. *Journal of Medicinal Plants Research*, **2**, 271–278.

Venkateswarlu, S., Panchagnula, G. K. and Subbaraju, G. V. (2004). Synthesis and anti-oxidative activity of 3',4',6,7-tetrahydroxyaurone, a metabolite of *Bidens frondosa*. *Bioscience Biotechnology and Biochemistry*, **68**, 2183–2185.

Vessal, M., Mehrani, H. A. and Omrani, G. H. (1991). Effects of an aqueous extract of *Physalis alkekengi* fruit on estrus cycle, reproduction and uterine creatine-kinase BB-isozyme in rats. *Journal of Ethnopharmacology*, **34**, 69–78.

Vilela, F. C., Padilha, M. D., dos Santos-e-Silva, L. *et al.* (2009a). Evaluation of the antinociceptive activity of extracts of *Sonchus oleraceus* L. in mice. *Journal of Ethnopharmacology*, **124**, 306–310.

Vilela, F. C., Soncini, R. and Giusti-Paiva, A. (2009b). Anxiolytic-like effect of *Sonchus oleraceus* L. in mice. *Journal of Ethnopharmacology*, **124**, 325–327.

Vilela, F. C., Bitencourt, A. D., Cabral, L. D. M. *et al.* (2010a). Anti-inflammatory and antipyretic effects of *Sonchus oleraceus* in rats. *Journal of Ethnopharmacology*, **127**, 737–741.

Vilela, F. C., Padilha, M. D., Alves-da-Silva, G., Soncini, R. and Giusti-Paiva, A. (2010b). Antidepressant-like activity of *Sonchus oleraceus* in mouse models of immobility tests. *Journal of Medicinal Food*, **13**, 219–222.

Wakibara, J. V., Huffman, M. A., Wink, M. *et al.* (2001). The adaptive significance of geophagy for Japanese macaques (*Macaca fuscata*) at Arashiyama, Japan. *International Journal of Primatology*, **22**, 495–520.

Wang, G. J. and Liao, C. (1990). The antihypertensive and diuretic effects of *Clerodendron trichotomum* water crude extract. *European Journal of Pharmacology*, **183**, 1825–1825.

Wang, J. P., Hsu, M. F., Raung, S. L. *et al.* (1992). Antiinflammatory and analgesic effects of magnolol. *Naunyn-Schmiedebergs Archives of Pharmacology*, **346**, 707–712.

Wang, R. R., Gu, Q., Wang, Y. H. *et al.* (2008). Anti-HIV-1 activities of compounds isolated from the medicinal plant *Rhus chinensis*. *Journal of Ethnopharmacology*, **117**, 249–256.

Wang, S. Y., Chang, H. N., Lin, K. T. *et al.* (2003). Antioxidant properties and phytochemical characteristics of extracts from *Lactuca indica*. *Journal of Agricultural and Food Chemistry*, **51**, 1506–1512.

Wang, S. Y., Lai, W. C., Chu, F. H. *et al.* (2006). Essential oil from the leaves of *Cryptomeria japonica* acts as a silverfish (*Lepisma saccharina*) repellent and insecticide. *Journal of Wood Science*, **52**, 522–526.

Wang, Y. C. and Huang, T. L. (2005). Screening of anti-*Helicobacter pylori* herbs deriving from Taiwanese folk medicinal plants. *FEMS Immunology and Medical Microbiology*, **43**, 295–300.

Weiner, M. A. (1972). *Earth Medicine, Earth Food*. New York, NY: MacMillan.

World Health Organization (WHO) (1998). *Medicinal Plants in the Republic of Korea: Information on 150 Commonly Used Medicinal Plants*. Geneva: World Health Organization.

Worman, C. O. and Chapman, C. A. (2005). Seasonal variation in the quality of a tropical ripe fruit and the response of three frugivores. *Journal of Tropical Ecology*, **21**, 689–697.

Wu, J., Feng, J. Q. and Zhao, W. M. (2008). A new lignan and anti-inflammatory flavonoids from *Kerria japonica*. *Journal of Asian Natural Products Research*, **10**, 435–438.

Wu, M. J., Yen, J. H., Wang, L. S. and Weng, C. Y. (2004). Antioxidant activity of porcelainberry (*Ampelopsis brevipedunculata* (Maxim.) Trautv.). *American Journal of Chinese Medicine*, **32**, 681–693.

Xia, D. Z., Wu, X. Q., Yang, Q., Gong, J. Y. and Zhang, Y. (2010). Anti-obesity and hypolipidemic effects of a functional formula containing *Prumus mume* in mice fed high-fat diet. *African Journal of Biotechnology*, **9**, 2463–2467.

Xu, B. J., Zheng, Y. N. and Sung, C. K. (2005). Natural medicines for alcoholism treatment: a review. *Drug and Alcohol Review*, **24**, 525–536.

Xu, L. P., Wang, H. and Yuan, Z. (2008a). Triterpenoid saponins with anti-inflammatory activity from *Codonopsis lanceolata*. *Planta Medica*, **74**, 1412–1415.

Xu, L. R., Wang, J. S., Li, H. Z. *et al.* (2008b). Effects of *Rubus parvifolius* L. on neuronal apoptosis and expression of apoptosis-related proteins in a rat model of focal cerebral ischemic-reperfusion injury. *Neural Regeneration Research*, **3**, 742–746.

Yagi-Chaves, S. N., Liu, G., Yamashita, K. *et al.* (2006). Effect of five triterpenoid compounds isolated from root bark of *Aralia elata* on stimulus-induced superoxide generation, tyrosyl or serine/threonine phosphorylation and translocation of P47(phox), p67(phox), and rac to cell membrane in human neutrophils. *Archives of Biochemistry and Biophysics*, **446**, 84–90.

Yamada, K., Ina, H., Matsumoto, K. and Miyazaki, T. (2008). Effects of the pickled fruit and the constituents of the fruit of *Prunus mume* for relieving tension in man and rats. *Planta Medica*, **74**, 1181.

Yang, J. K., Choi, M. S., Seo, W. T. *et al.* (2007). Chemical composition and antimicrobial activity of *Chamaecyparis obtusa* leaf essential oil. *Fitoterapia*, **78**, 149–152.

Yang, L. L., Chang, C. C., Chen, L. G. and Wang, C. C. (2003). Antitumor principle constituents of *Myrica rubra* var. acuminata. *Journal of Agricultural and Food Chemistry*, **51**, 2974–2979.

Yang, Y. C., Lee, E. H., Lee, H. S., Lee, D. K. and Ahn, Y. J. (2004). Repellency of aromatic medicinal plant extracts and a steam distillate to *Aedes aegypti*. *Journal of the American Mosquito Control Association*, **20**, 146–149.

Yang, Y. J., Park, J. I., Lee, H. J. *et al.* (2006). Effects of (+)-eudesmin from the stem bark of *Magnolia kobus* DC. var. borealis Sarg. on neurite outgrowth in PC12 cells. *Archives of Pharmacal Research*, **29**, 1114–1118.

Yaniv, Z. and Bachrach, U. (2005). *Handbook of Medicinal Plants*. Binghamton, NY: Haworth Printers Inc.

Yen, K. Y. (1992). *The Illustrated Chinese Materia Medica*. Taipei: SMC Publishing.

Yeo, H. S., Kim, D. W. and Jun, C. Y. (2007). Neuroprotective effect of *Cirsium japonicum* and silibinin on lipopolysaccharide-induced inflammation in BV2 microglial cells. *Journal of Korean Oriental Internal Medicine*, **28**, 166–175.

Yeung, H. C. (1985). *Handbook of Chinese Herbs and Formulas*. Los Angeles, CA: H. C. Yeung.

Yi, J. M., Hong, S. H., Lee, H. J. *et al.* (2002). *Ixeris dentata* green sap inhibits both compound 48/80-induced anaphylaxis-like response and IgE-mediated anaphylactic response in murine model. *Biological and Pharmaceutical Bulletin*, **25**, 5–9.

Yin, J., Heo, S. I., Jung, M. J. and Wang, M. H. (2008). Antioxidant activity of fractions from 70% methanolic extract of *Sonchus oleraceus* L. *Food Science and Biotechnology*, **17**, 1299–1304.

Yoo, H. H., Kim, T., Ahn, S. *et al.* (2005). Evaluation of the estrogenic activity of leguminosae plants. *Biological and Pharmaceutical Bulletin*, **28**, 538–540.

Yoo, N. H., Jang, D. S., Yoo, J. L. *et al.* (2008). Erigeroflavanone, a flavanone derivative from the flowers of *Erigeron annuus* with protein glycation and aldose reductase inhibitory activity. *Journal of Natural Products*, **71**, 713–715.

Yoshikawa, M., Harada, E., Matsuda, H. *et al.* (1993). Elatoside-A and Elatoside-B, potent inhibitors of ethanol absorption in rats from the bark of *Aralia elata* Seem – the structure-activity-relationships of oleanolic acid oligoglycosides. *Chemical and Pharmaceutical Bulletin*, **41**, 2069–2071.

Yoshikawa, M., Yoshizumi, S., Ueno, T. *et al.* (1995). Medicinal foodstuffs. 1. Hypoglycemic constituents from a garnish foodstuff taranome, the young shoot of *Aralia elata* Seem – Elatoside-G, Elatoside-H, Elatoside-I, Elatoside-J, and Elatoside-K. *Chemical and Pharmaceutical Bulletin*, **43**, 1878–1882.

Yoshikawa, M., Morikawa, T., Asao, Y. *et al.* (2007). Medicinal flowers. XV. The structures of noroleanane- and oleanane-type triterpene oligoglycosides with gastroprotective and platelet aggregation activities from flower buds of *Camellia japonica*. *Chemical and Pharmaceutical Bulletin*, **55**, 606–612.

Yoshioka, A., Etoh, H., Yagi, A., Sakata, K. and Ina, K. (1990). Isolation of flavonoids and cerebrosides from the bark of *Prunus jamasakura* as repellents against the blue mussel, *Mytilus edulis*. *Agricultural and Biological Chemistry*, **54**, 3355–3356.

Yoshioka, S., Hamada, A., Jobu, K. *et al.* (2010). Effects of *Eriobotrya japonica* seed extract on oxidative stress in rats with non-alcoholic steatohepatitis. *Journal of Pharmacy and Pharmacology*, **62**, 241–246.

Yoshizawa, Y., Kawaii, S., Urashima, M. *et al.* (2000). Differentiation-inducing effects of small fruit juices on HL-60 leukemic cells. *Journal of Agricultural and Food Chemistry*, **48**, 3177–3182.

Youn, U., Chen, Q. C., Lee, I. S. *et al.* (2008a). Two new lignans from the stem bark of *Magnolia obovata* and their cytotoxic activity. *Chemical and Pharmaceutical Bulletin*, **56**, 115–117.

Youn, Y. N., Lim, E., Lee, N. *et al.* (2008b). Screening of Korean medicinal plants for possible osteoclastogenesis effects in vitro. *Genes and Nutrition*, **2**, 375–380.

Yu, J. Q., Komada, H., Yokoyama, H. *et al.* (1997). Sugi (*Cryptomeria japonica* D. Don) bark, a potential growth substrate for soilless culture with bioactivity against some soilborne diseases. *Journal of Horticultural Science*, **72**, 989–996.

Zhang, M., Liu, G., Tang, S. H. *et al.* (2006). Effect of five triterpenoid compounds from the buds of *Aralia elata* on stimulus-induced superoxide generation, tyrosyl phosphorylation and translocation of cytosolic compounds to the cell membrane in human neutrophils. *Planta Medica*, **72**, 1216–1222.

Zhou, H. Y., Shin, E. M., Guo, L. Y. *et al.* (2007a). Anti-inflammatory activity of 21(alpha, beta)-methylmelianodiols, novel compounds from *Poncirus trifoliata* Rafinesque. *European Journal of Pharmacology*, **572**, 239–248.

Zhou, X., Li, X. D., Yuan, J. Z., Tang, Z. H. and Liu, W. Y. (2000). Toxicity of cinnamomin – a new type II ribosome-inactivating protein to bollworm and mosquito. *Insect Biochemistry and Molecular Biology*, **30**, 259–264.

Zhou, X., Zhou, B. L., Li, Z. W. and Dong, C. F. (2007b). Inhibitory and preventive effects of plants extracts against *Verticillium dahliae*. *Allelopathy Journal*, **20**, 145–155.

Zou, K., Zhao, Y. Y. and Zhang, R. Y. (2006). A cytotoxic saponin from *Albizia julibrissin*. *Chemical and Pharmaceutical Bulletin*, **54**, 1211–1212.

Part IV

Management and education

18 Birth control in female Japanese macaques at Iwatayama Monkey Park, Arashiyama

KEIKO SHIMIZU

Heterosexual consortship between Cooper-65–75–84–96 (male) and Rakushi-59–79–92 (female), with her infant female (Rakushi-59–79–92–09) at Arashiyama, November 2009 (photo by N. Gunst).

18.1   Introduction

Wild animal populations are primarily kept within the limits of their food supplies and habitat size. However, under certain circumstances such as food provisioning, uncontrolled increases in population sizes in some species may cause social problems to humans, through the destruction of local vegetation and crop damage (Kirkpatrick and Turner, 1991). Although various measures

The Monkeys of Stormy Mountain: 60 Years of Primatological Research on the Japanese Macaques of Arashiyama, eds. Jean-Baptiste Leca, Michael A. Huffman and Paul L. Vasey. Published by Cambridge University Press. © Cambridge University Press 2012.

of population control (e.g. hunting, poisoning, trapping and relocation) have been attempted around the world, these methods have only transient effects. Moreover, these methods are not always ethically acceptable from the standpoint of animal welfare.

The appropriate approach to controlling fertility may vary according to the target species and the intended outcome, which includes consideration of such factors as relative efficacy and safety, economic value, delivery mechanism, non-target and environmental effects and social acceptability (Asa, 2005; Kirkpatrick and Frank, 2005; Barfield *et al.*, 2006). Each approach raises ethical and social acceptability issues (Barr *et al.*, 2002).

Over the past 30 years, managing wildlife populations through fertility control has been the focus of considerable interest (Fagerstone *et al.*, 2002). A number of contraceptive methods have been developed for use in various mammals. These include hormone implants, surgical procedures, chemical sterilants and vaccines (Barr *et al.*, 2002; Asa, 2005; Kirkpatrick and Frank, 2005; Barfield *et al.*, 2006). To evaluate these various contraceptives, the characteristics of the ideal wildlife fertility control agent should first be identified. These characteristics include (1) a high degree of effectiveness, (2) a lack of toxicity and harmful side effects, particularly to pregnant animals, (3) reversibility and flexible duration of action, to preserve the reproductive and genetic integrity of target animals, (4) low cost, (5) minimal or no effect on social organisation and behaviour, (6) remote delivery, preferably with a single administration and (7) inability of the contraceptive agent to be passed from the treated animal to predators, scavengers or humans through the food chain (Kirkpatrick and Turner, 1991; Kirkpatrick and Rutberg, 2001). Despite their great potential interest, fertility control methods have been applied in only a few situations, and their practical experience in macaques is very limited.

In the case of free-ranging Japanese macaques, provisioning was initiated for research, education and tourism more than 50 years ago. Feeding macaques with highly nutritious and large amounts of food caused a rapid increase in population sizes in most monkey parks throughout Japan (Sugiyama and Ohsawa, 1982). Consequently, monkey parks and adjacent agricultural areas face problems associated with monkey overpopulation.

Iwatayama Monkey Park in Arashiyama, like many other monkey parks, has experienced the effects of artificial feeding with increased birth rates, lower infant mortality and younger age at first birth, especially in high-ranking kin group members (Mori, 1979; Sugiyama and Ohsawa, 1982). The expanding monkey population has led to an increasing number of conflicts with local human populations. This chapter describes the effects of hormonal contraceptives previously used on the Japanese macaque group living in the Iwatayama Monkey Park of Arashiyama.

18.2   Contraception in female Japanese macaques

18.2.1   *Medroxyprogesterone acetate*

Injectable medroxyprogesterone acetate (MPA: Promone E, Upjohn Pty. Limited, Rydalmere, NSW) is a long-acting synthetic progesterone used as a contraceptive by more than 10 million women in approximately 80 nations (Coutinho *et al.*, 1966; Coutinho and De Souza, 1968; Prahalada *et al.*, 1985). Subsequently, MPA was applied as contraception in some non-human primate species (black howlers, cotton-top tamarins, orangutans and ruffed lemurs), two carnivores (red foxes and timber wolves), and one brown bear species in North American zoos (Porton *et al.*, 1990). These tests demonstrated the effectiveness of MPA. Based on these findings, we explored the possibility of using MPA for contraception in female Japanese macaques by investigating the effect of a single subcutaneous administration of MPA on their menstrual cyclicity, endocrinology and sexual behaviour.

Suppression of menstrual cyclicity by MPA

Before the application of MPA to free-ranging Japanese macaques, the long-acting efficacy of a single subcutaneous injection of MPA was examined. For this purpose, the cynomolgus macaque (*Macaca fascicularis*) is a more appropriate study species than the Japanese macaque (*Macaca fuscata*) because cynomolgus macaques ovulate all year round, whereas Japanese macaques are seasonal breeders and ovulate only during the breeding season (autumn and winter months) (Nigi, 1975; Nozaki *et al.*, 1992; Shimizu, 2005).

After four weeks of pre-treatment blood sampling, five females were given a single subcutaneous injection of 15 mg/kg MPA in sterile water. Blood samples were obtained from the cubital vein three times a week. Blood sampling continued until the resumption of their menstrual cycles. Plasma endocrine profiles and the incidence of menses throughout the experimental period are shown in Figure 18.1 and Figure 18.2. In these females, during the pre-treatment period, plasma concentrations of oestradiol showed a sharp rise and fall in the late follicular phase. Plasma levels of progesterone rose markedly after the mid-cycle oestradiol and luteinising hormone (LH) peaks, reaching maximal concentrations during the mid-luteal phase, demonstrating that all females exhibited normal ovulatory menstrual cycles during the pre-treatment period.

In control females, that received only saline, no notable changes were observed in endocrine and menstruation profiles after saline treatment (Figure 18.1): they showed continued regular cyclic endocrine changes characteristic

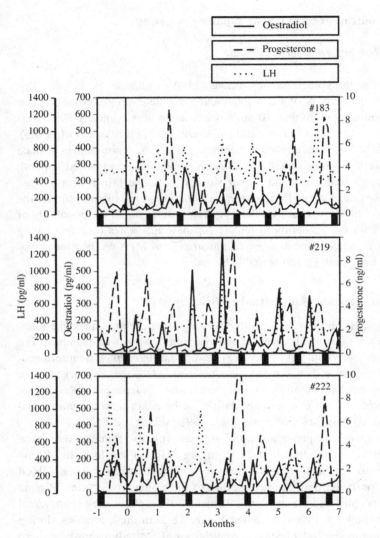

Figure 18.1. Changes in plasma concentrations of oestradiol (solid line), progesterone (broken line) and luteinising hormone (LH) (dotted line) before and after sterile water injection in control monkeys. Black vertical bars at the bottom column reflect the days of menses, and vertical solid lines indicate the day of vehicle injection. Each graph corresponds to a particular study subject whose identification number is mentioned in top right corner.

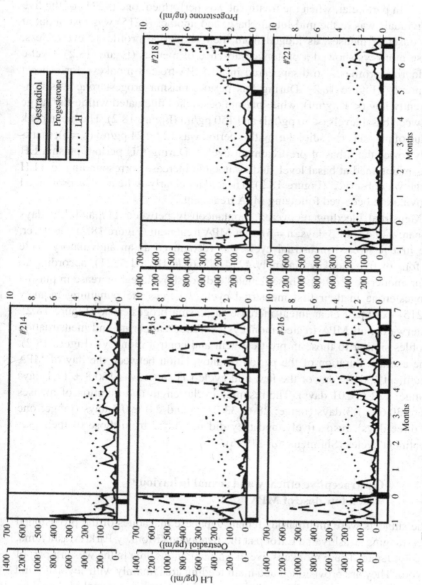

Figure 18.2. Changes in plasma concentrations of oestradiol (solid line), progesterone (broken line) and luteinising hormone (LH) (dotted line) before and after 15 mg/kg MPA injection in experimental monkeys. Black horizontal bars at the bottom column reflect the days of menses, and vertical lines indicate the day of MPA injection.

of ovulatory cycles. By contrast, in all MPA-treated monkeys, menstrual cycles completely disappeared up to at least 3 months after MPA injections (Figure 18.2). In particular, when the treatment was performed, one (#221) of the five individuals was in the mid-luteal phase, and another (#215) was just about at the onset of menses, as judged from their endocrine profiles. Nevertheless, these monkeys showed a prolonged absence of menses (Figure 18.2). Cyclic endocrine profiles also disappeared in all MPA-treated monkeys for at least 4 months (Figure 18.2). During this period, plasma progesterone was consistently low (< 1 ng/ml), while plasma oestradiol fluctuated within the range between basal levels (< 50 pg/ml) and 150 pg/ml (Figure 18.2). The mean peak value of plasma oestradiol during this period was 127 ± 21 pg/ml (n = 5), markedly lower than that of pre-treatment cycles. During this period, plasma LH was maintained at basal level, and no notable increase corresponding to a LH surge was observed (Figure 18.2). These data clearly indicate that prolonged anovulation occurred following MPA treatment.

Menstrual bleeding recovered spontaneously between 117 and 176 days (mean: 149.4 ± 11.3 days; n = 5) after MPA treatment (Figure 18.2). However, the first recovered menstrual cycle was recognised as an anovulatory cycle in four of five monkeys (namely #214, #216, #218 and #221), according to their endocrine profiles. These females did not show luteal increase in plasma progesterone in their first menstrual cycle, whereas the remaining monkey (#215) exhibited clear luteal increase in plasma progesterone (Figure 18.2). Afterwards, all MPA-treated monkeys exhibited endocrine and menstruation profiles similar to those in pre-treatment and control monkeys (Figure 18.2). The average duration of the period of anovulation between the day of MPA treatment and the day of the first mid-cycle LH surge was 160.8 ± 17.1 days (range: 122 to 201 days). The resumed cycle length and duration of menses were 30.6 ± 1.3 days (range: 28 to 33 days) and 2.0 ± 0.3 days (range: one to three days), respectively, and they did not differ from those of their pre-treatment cycles (Shimizu *et al.*, 1996).

Contraceptive efficacy and sexual behaviour with a low dose of MPA

The subjects were multiparous female Japanese macaques of the provisioned free-ranging Arashiyama E troop at the Iwatayama Monkey Park, Arashiyama, Kyoto, Japan. These monkeys live in a mountainous area in the western part of Kyoto. They are provisioned at a main feeding area, mainly with soybeans and wheat, five times a day. The vegetation in their home range includes deciduous and evergreen forest, grass-bush fields and cultivated land. They also feed on these natural food sources (see Huffman and MacIntosh: chapter 17). The

Table 18.1. *Mean age, body weight and social rank of study subjects*

Group		n	Age (yr) mean	S.D.	Body weight (kg) mean	S.D.	Rank mean	S.D.
MPA (7.5 mg/kg)	lactation	10	12.4	5.0	8.4	1.7	13.0	8.0
	non-lactation	5	13.8	5.4	7.1	0.9	29.8	8.3
	total	15	12.8	5.0	8.0	1.6	18.6	11.0
MPA (19.4 mg/kg)	lactation	3	16	1.5	8.3	1.4	32	2.6
	non-lactation	2	14	4.2	6.8	1.7	35	5.7
	total	5	15.4	6.7	7.7	1.5	33.2	3.8
control	lactation	14	12.4	4.4	8.8	1.7	17.0	10.3
	non-lactation	10	16.9	5.1	7.3	1.1	26.3	0.6
	total	24	14.3	5.1	8.4	1.6	20.9	12.0

group members were individually identified and the birth dates of all individuals were collected by the Park staff. In late 1993, the group consisted of 170 individuals, including 39 multiparous females. In this group, about half of the mature females had suckling infants every autumn. These 39 multiparous females were used as study subjects.

Fifteen of the 39 multiparous females were captured randomly and received a single subcutaneous injection of 7.5 ± 0.4 mg/kg of MPA during the period between late September and early October 1993, just before the onset of the breeding season. The remaining 24 females were used as intact controls (Table 18.1). Observations of sexual behaviour such as mounting, copulation and ejaculation were made daily between 08:00 and 10:00 a.m., using *ad libitum* sampling (Altmann, 1974) throughout the entire breeding season, from 1 September 1993 to 28 February 1994. Copulation was defined as one or more series of mounting by sexually mature males on females terminated by ejaculation. The presence of a coagulated semen plug at the opening of genitalia was also regarded as evidence for the occurrence of copulation. Delivery was noted in all females during the ensuing birth season.

All data obtained were expressed as the group mean ± s.d. Differences between two means were evaluated with Student's t-test or the Cochran–Cox test. Fisher's exact probability test was also used on some occasions. All statistical tests with $p < 0.05$ were considered significant (cf. Shimizu *et al.*, 1996).

Copulatory behaviour and fecundity rate

MPA treatment was performed during the transition period between the non-breeding and breeding seasons. During the ensuing breeding season, copulations were observed in 16 of 24 control monkeys (66.7%), and nine of 15 MPA-treated monkeys (60.0%). The remaining eight control and six MPA-treated monkeys did not copulate throughout the breeding season. Thus, no significant difference was observed in the number of copulating animals between the two groups (Table 18.2). However, a significant difference was found between the two groups in the onset of copulations during the breeding season (Figures 18.3 and 18.4). In control monkeys, the first copulations were observed during the period between 19 September 1993 and 20 January 1994 (median: November 14; n = 16) (Figure 18.3), whereas in MPA-treated monkeys, they tended to be concentrated in the middle of the breeding season (range: 10 October 1993 to 1 January 1994; median: 21 December; n = 9) (Figure 18.4).

In the following birth season, 11 of the 24 control and five of the 15 MPA-treated monkeys delivered live offspring, and thus no significant difference was found in fecundity rates between the two groups (Table 18.2). However, a significant difference was found in the delivery dates (F-test, $p < 0.05$; Figures 18.3 and 18.4). In control monkeys, deliveries occurred during the period between 16 April and 24 July (Figure 18.3), whereas in MPA-treated monkeys,

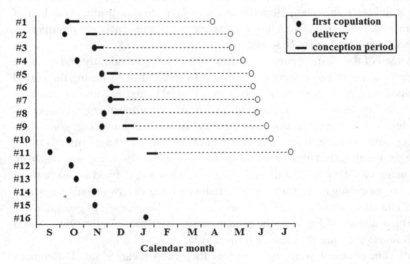

Figure 18.3. Distribution of the first copulation days (•), estimated conception periods (horizontal black bars) and delivery days (○) of control monkeys in the provisioned free-ranging Japanese macaques. The ranges of estimated conception periods (horizontal lines) and gestation periods (dotted line) are superimposed.

Table 18.2. *Comparison of copulatory and fecundity rates in MPA treatment and control Japanese macaques ('+': observed, '−': not observed)*

		Copulation +	Copulation −	Delivery +	Delivery −	Total
MPA (7.5 mg/kg)	non-lactation	5(50%)	5	3(30%)	7	10
	lactation	4(80%)	1	2(40%)	3	5
	total	9(60%)	6	5(33.3%)	10	15
MPA (19.4 mg/kg)	non-lactation	0(0%)	3	0(0%)	3	3
	lactation	0(0%)	2	0(0%)	2	2
	total	0(0%)	NS 5	0(0%)	NS 5	5
control	non-lactation	8(57.1%)	6	5(35.7%)	9	14
	lactation	8(80%)	2	6(60%)	4	10
	total	16(66.7%)	8	11(45.8%)	13	24

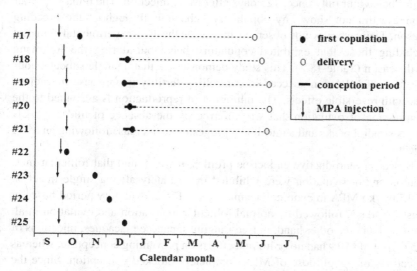

Figure 18.4. Distribution of the first copulation days (●), estimated conception periods (horizontal black bars) and delivery days (○) after MPA treatment in the provisioned free-ranging Japanese macaques. Arrows indicate the MPA injections. The ranges of estimated conception periods (horizontal lines) and gestation periods (dotted line) are superimposed.

they were concentrated within 20 days in the late birth season (range: 4 June to 23 June; Figure 18.4). Mean age, body weight, lactation and dominance rank of both groups are arranged in Table 18.1. The differences in these parameters between the two groups were not statistically significant (Shimizu *et al.*, 1996).

Contraceptive efficacy and sexual behaviour with a high dose of MPA

As described above, a low dose of MPA had no or limited effects on contraception or sexual behaviour. Accordingly, in the subsequent breeding season, during the period between late September and early October 1994, a high dose of MPA (19.4 ± 2.4 mg/kg) was given to five monkeys of the Arashiyama E group. They were multiparous females and three of them were lactating. The physical parameters of these monkeys are shown in Table 18.1. Their sexual behaviour and delivery were also recorded.

Each of the five monkeys received two and half times more MPA than that administered in 1993. Following MPA-injections, one monkey exhibited copulatory behaviour only once, 125 days after MPA injection. The remaining four monkeys did not show any copulatory behaviour throughout the breeding season. Delivery was not observed in any of the five experimental females, including those that exhibited copulatory behaviour during the following birth season (Table 18.2). This study demonstrates that a single subcutaneous injection of MPA can succeed in controlling fertility in Japanese macaques, and with reversible effects. The inhibition of reproduction is attributed to the suppression of ovulation that was inferred by the absence of mid-cycle LH and oestradiol peaks and a mid-luteal progesterone increase following an MPA injection.

Based on reproductive endocrine profiles, it was found that follicular maturation and/or ovulation were inhibited immediately after a single injection of 15 mg/kg MPA in cynomolgus macaques. The anovulatory period lasted at least 120 days, followed by normal follicular maturation and ovulation in all monkeys. On the other hand, in free-ranging Japanese macaques, injections of 7.5 mg/kg of MPA had no or limited effects on preventing conception, whereas injections of a high dose of MPA completely blocked conception. Since the breeding season of Japanese macaques lasts about 4 months (late autumn through mid-winter: Nigi, 1976; Takahata, 1980; Nozaki *et al.*, 1992; Shimizu, 2005), the latter finding suggests that 19.4 mg/kg of MPA was effective for about 5 or more months.

Thus, although the differences in housing conditions and species may influence the efficacy of MPA, these results suggest that the effect of MPA on the suppression of ovulatory functions is dose-dependent, and that effective fertility control in macaques can be achieved by choosing an appropriate amount of MPA. The present study further demonstrated that MPA treatment was effective at any phase of the ovarian cycle. This finding provides another advantage for applying MPA to wild animals, since it is generally difficult to outwardly detect their ovarian stage.

According to an estimation of the conception days based on a mean duration of 173 ± 7 days for the gestation in Japanese macaques (Nigi, 1976; Shimizu, 1988; Nozaki *et al.*, 1990), it was evaluated that five females of the MPA-treated group conceived during 76 and 92 days after MPA injection (mean: 83.4 ± 3.2 days). The high incidence of pregnancies at this particular time was statistically significant. This result suggests that the onset of ovulation was somewhat delayed by MPA treatment, which might result in synchronised ovulations among troop members.

In the present study, plasma levels of oestradiol fluctuated within a limited range during the MPA-treatment in cynomolgus macaques. These results suggest that ovarian follicular growth is not completely inhibited by MPA-treatment. However, since peak levels of plasma oestradiol during the MPA-treatment were significantly lower than those in mid-cycle oestradiol surges, follicular maturation seems to be blocked by MPA treatment. Regarding the sites of action of MPA, both possibilities of direct action on the ovary and indirect action through the hypothalamus should be considered. Further studies are needed to elucidate these possibilities.

This study revealed that following MPA treatment, the first recovered cycle was anovulatory in most animals. To account for this result, the following arguments could be considered. The effect of MPA was gradually reduced with the elapse of time, and follicular development resumed gradually. Secretion of oestradiol also increased gradually, which stimulated endometrium development. However, since ovarian functions were not fully recovered at that time, the developing follicles became atretic, and the stimulation of endometrium by ovarian hormones also declined, which resulted in the onset of menses. In conclusion, this study indicated that a single subcutaneous injection of MPA was a valuable and effective strategy for controlling fertility in wild monkey populations. This hormone-induced sterility was reversible; the females eventually returned to normal cycles and experienced normal pregnancy (Shimizu *et al.*, 1996).

18.2.2   *Chrolmadinon acetate (CMA)*

Long-acting Chrolmadinon acetate (CMA: Prostal, ASKA Pharmaceutical Co., Ltd., Tokyo, Japan) is a synthetic progesterone used as a contraceptive in companion animals. I investigated the efficacy of oral and subdermal administration of CMA on menstrual cyclicity and hormonal profiles as a way to explore the possibility of using CMA for contraception in female Japanese macaques.

Subdermal implant of CMA

Before the application to free-ranging Japanese macaques, the long-acting efficacy of the subdermal CMA implants was examined. Four sexually mature, normally cycling, multiparous female Japanese macaques were used. They were housed individually in an air-conditioned room with controlled temperature (20 ± 5 °C) and lighting (lights on: 6:00 a.m. to 6:00 p.m.) at the Primate Research Institute, Kyoto University, Japan. After 4 weeks of pretreatment blood sampling, they were administered an implant containing 150 mg of CMA subcutaneously. Blood samples were obtained three times a week. Plasma concentrations of reproductive hormones were determined by specific radioimmunoassays.

Changes in plasma levels of oestradiol and progesterone before and after CMA implants in experimental monkeys are shown in Figures 18.5a–d. After administration of the CMA implant, menstrual bleeding completely disappeared and cyclic endocrine profiles were not observed (Figures 18.5a, b and d). After the implants, plasma CMA levels increased immediately, then gradually decreased (Figure 18.6). These hormonal profiles clearly indicate that prolonged anovulation occurred for at least 3 years following the CMA implant. On the other hand, one implanted female (Mff 615) showed monthly menstrual bleeding and a typical hormonal pattern of ovulatory cycle in the year following the CMA implant (Figure 18.5c). Moreover, Mff 615 became pregnant in the breeding season of the third year and delivered a normal newborn (Figure 18.5c). This suggests that the implant of Mff 615 dropped within a year after treatment (Figure 18.5c). This evidence demonstrated that the effect of the CMA implant was reversible and after its removal, the ovulatory menstrual returned and normal pregnancy could occur in Japanese macaques.

Oral administration of CMA

As the reversibility of CMA effects was confirmed, CMA was administered orally to sexually mature, normal cycling, multiparous female Japanese macaques at Iwatayama Monkey Park, Arashiyama. Females were orally administered 12.5 mg of CMA once a week throughout the breeding season from 1994 to 2008. The fecundity rate for 15 years is shown in Figure 18.7. It was very low (average 6.5%) in CMA-treated macaques, compared with that of non-treated macaques (Figure 18.7). Thus, oral administration of CMA also proved to be effective in controlling the fertility of Japanese macaques.

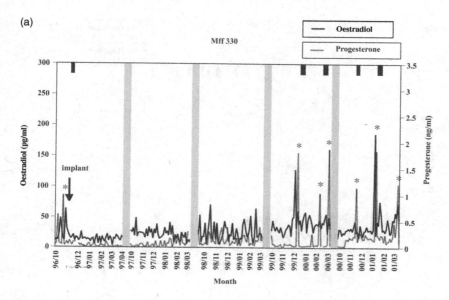

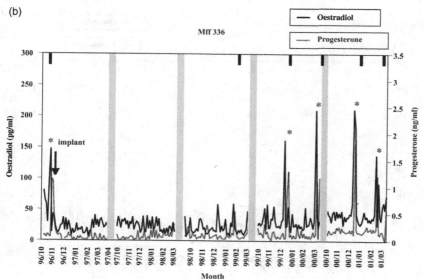

Figure 18.5. Changes in plasma levels of oestradiol and progesterone before and after CMA implant in experimental macaques. Black vertical bars at the top column reflect the days of menses and grey vertical lines indicate non-breeding seasons. Arrow represents CMA implant date. Asterisks in each column represent estimated ovulation. (a), (b) and (d) Menstrual bleedings and cyclic endocrine profiles completely disappeared after CMA implant. Hormonal profiles clearly indicate that a prolonged anovulation occurred following CMA-implant. (c) Mff 615 showed monthly menses and typical ovulatory hormonal profiles were observed at the second year. This female became pregnant during the breeding season of the third year.

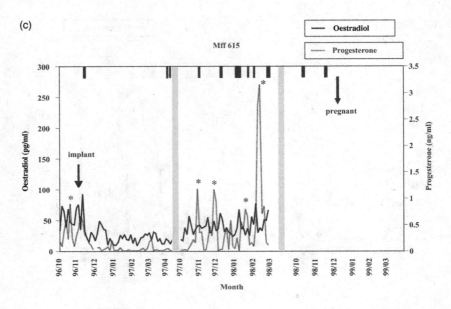

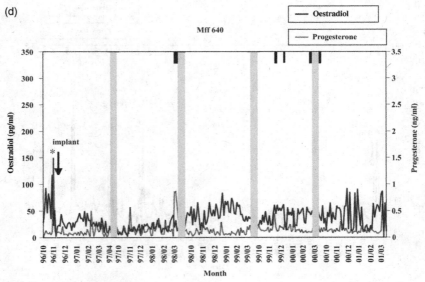

Figure 18.5. (*cont.*)

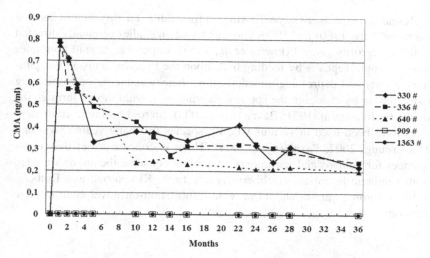

Figure 18.6. Changes in plasma CMA levels after administration of the implant. After administration of the implant, plasma CMA levels increased immediately, then gradually decreased. On the other hand, CMA was not detected in control monkeys (#909 and #1363) throughout the experimental periods.

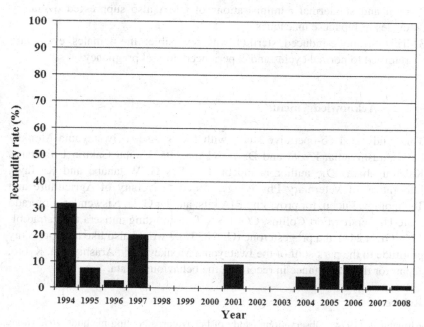

Figure 18.7. Fecundity rate (%) in CMA treated macaques. The average value (6.5%) was very low.

Research on the use of steroid hormones for wildlife fertility control became common in the 1960s and 1970s and was based on studies originally directed at human fertility control (Pincus *et al.*, 1958). In general, steroid hormones work as contraceptives by feeding back upon the hypothalamus and/or pituitary glands and depressing gonadotropins, thereby reducing or eliminating ovulation, or by changing the speed with which the ovum moves through the oviducts (Johansson, 1971; Bakry *et al.*, 2008). Steroid hormone contraceptives have been used in various wildlife species for many years (Sivin and Moo-Young, 2002; Pazol *et al.*, 2004). There is concern about the consequences for humans and scavengers who might consume the animals treated with synthetic hormones (Deliberto *et al.*, 1998; Kirkpatrick and Rutberg, 2001). Ongoing studies should provide information on novel contraceptive methods.

18.3   Conclusion

1. A single subcutaneous injection of MPA suppressed ovulatory cycles in dose-dependent and reversible manners in macaques.
2. Oral and subdermal administrations of CMA also suppressed ovulatory cycles in Japanese macaques.
3. This hormone-induced sterility was reversible; the females eventually returned to normal cycles and experienced normal pregnancy.

Acknowledgements

This study is a co-operative study with Mr S. Asaba, Iwatayama Monkey Park, Arashiyama, Kyoto and Dr Y. Takenoshita, Chubu Gakuin University, Kakamigahara. The author is indebted to Drs G. Watanabe and K. Taya, Laboratory of Veterinary Physiology, Tokyo University of Agriculture and Technology, Fuchu, for providing RIA kits; and Dr G. D. Niswender, Colorado State University, Fort Collins, CO, USA, for providing antisera to oestradiol-17β (GDN 244) and progesterone (GDN 337). I would also like to express my gratitude to the park staff of the Iwatayama Monkey Park, Arashiyama, Kyoto, Japan for their assistance in recording the behavioural data.

References

Altmann, J. (1974). Observational study of behaviour: sampling methods. *Behaviour*, **49**, 227–267.

Asa, C. S. (2005). Types of contraception. In *Wildlife Contraception: Issues, Methods, and Applications*, eds. C. S. Asa and I. J. Porton. Baltimore, MD: The Johns Hopkins University Press, pp. 29–52.

Bakry, S., Merhi, Z. O., Sclise, T. J. *et al.* (2008). Depot-medroxyprogesterone acetate: an update. *Archives of Gynecology and Obstetrics*, **278**, 1–12.

Barfield, J. P., Nieschlag, E. and Cooper, T. G. (2006). Fertility control in wildlife: humans as a model. *Contraception*, **73**, 6–22.

Barr, J. J. F., Lurz, P. W. W., Shirley, M. D. F. and Rushton, S. P. (2002). Evaluation of immunocontraception as a publicly acceptable form of vertebrate pest species control: the introduced grey squirrel in Britain as an example. *Environmental Management*, **30**, 342–351.

Coutinho, E. M., De Souza, J. C. and Csapo, A. I. (1966). Reversible sterility induced by medroxypreogesterone injections. *Fertility and Sterility*, **17**, 261–266.

Coutinho, E. M. and De Souza, J. C. (1968). Conception control by monthly injections of medroxyprogesterone suspension and a long-acting oestrogen. *Journal of Reproduction and Fertility*, **15**, 209–214.

Deliberto, T. J., Gese, E. M., Knowlton, F. F. *et al.* (1998). Fertility control in coyotes: Is it a potential management tool? In *Proceeding of 18th Vertebrate Pest Conference*, eds. R. O. Baker and A. C. Crabb. Davis, CA: University of California, pp. 144–149.

Fagerstone, K. A., Coffey, M. A., Curtis, P. D. *et al.* (2002). Wildlife fertility control. *Wildlife Society Technical Review*, **2**, 2–29.

Johansson, E. D. B. (1971). Depression of the progesterone levels in women treated with synthetic gestagens after ovulation. *Acta Endocrinologica*, **68**, 779–792.

Kirkpatrick, J. F. and Frank, K. M. (2005). Contraception in free-ranging wildlife. In *Wildlife Contraception: Issues, Methods, and Applications*, eds. C. S. Asa and I. J. Porton. Baltimore, MD: The Johns Hopkins University Press, pp. 195–221.

Kirkpatrick, J. F. and Rutberg, A. T. (2001). Fertility control in animals. In *The State of the Animals*, eds. A. N. Rowan and D. J. Washington. Salem, DC: Humane Society of the US, pp. 183–198.

Kirkpatrick, J. F. and Turner, Jr, J. F. (1991). Reversible contraception in nondomestic animals. *Journal of Zoo and Wildlife Medicine*, **22**, 392–408.

Mori, A. (1979). An experiment on the relation between the feeding speed and the caloric intake through leaf eating in Japanese monkeys. *Primates*, **20**, 185–195.

Nigi, H. (1975). Menstrual cycle and some other related aspects of Japanese monkeys (*Macaca fuscata*). *Primates*, **16**, 207–216.

    (1976). Some aspects related to conception of the Japanese monkey (*Macaca fuscata*). *Primates*, **17**, 81–87.

Nozaki, M., Watanabe, G., Taya, K. *et al.* (1990). Changes in circulating inhibin levels during pregnancy and early lactation in the Japanese monkey. *Biology of Reproduction*, **43**, 444–449.

Nozaki, M., Mori, Y. and Oshima, K. (1992). Environmental and internal factors affecting seasonal breeding of Japanese monkeys (*Macaca fuscata*). In *Topics in Primatology, Volume 3*, eds. S. Matano, R. H. Tuttle, H. Ishida and H. Goodman. Tokyo: University of Tokyo Press, pp. 301–317.

452   *K. Shimizu*

Pazol, K., Wilson, M. E. and Wallen, K. (2004). Medroxyprogesterone acetate antagonises the effects of estrogen treatment on social and sexual behavior in female macaques. *Journal of Clinical Endocrinology and Metabolism*, **89**, 2998–3006.

Pincus, G., Rock, J., Garcia, C. R. *et al.* (1958). Fertility control with oral medication. *American Journal of Obstetrics and Gynecology*, **75**, 1333–1346.

Porton, I., Asa, C. and Baker, A. (1990). Survey results on the use of birth control methods in primates and carnivores in North American zoos. *AAZPA Annual Conference Proceedings*, **34**, 489–497.

Prahalada, S., Carroad, E., Cukierski, M. and Hendrickx, A. G. (1985). Embryotoxicity of a single dose of medroxyprogesterone acetate (MPA) and maternal serum MPA concentrations in cynomolgus monkey (*Macaca fascicularis*). *Teratology*, **32**, 421–432.

Shimizu, K. (1988). Ultrasonic assessment of pregnancy and fetal development in three species of macaque monkey. *Journal of Medical Primatology*, **17**, 247–256.

(2005). Studies on reproductive endocrinology in non-human primates: application of non-invasive methods. *Journal of Reproduction and Development*, **51**, 1–13.

Shimizu, K., Takenoshita, Y., Mitsunaga, F. and Nozaki, M. (1996). Suppression of ovarian function and successful contraception in macaque monkeys following a single injection of medroxyprogesterone acetate. *Journal of Reproduction and Development*, **42**, 147–155.

Sivin, I. and Moo-Young, A. (2002). Recent developments in contraceptive implants at the Population Council. *Contraception*, **65**, 113–119.

Sugiyama, Y. and Ohsawa, Y. (1982). Population dynamics of Japanese monkeys with special reference to the effect of artificial feeding. *Folia Primatologica*, **39**, 238–263.

Takahata, Y. (1980). The reproductive biology of a free-ranging troop of Japanese monkeys. *Primates*, **21**, 303–329.

19 Importance of the Arashiyama-Kyoto Japanese macaques in science and environmental education

YUJI TAKENOSHITA AND
YUKIYO MAEKAWA

Female Japanese macaque grooming her infant at Arashiyama (photo Y. Takenoshita).

The Iwatayama Monkey Park, Arashiyama, is a field zoo (Hirose, 1984). Thus it is supposed to perform functions expected from zoological parks in general, i.e. research, education, conservation and recreation. In this chapter, we highlight the use of Japanese macaques and the field site of Arashiyama in

The Monkeys of Stormy Mountain: 60 Years of Primatological Research on the Japanese Macaques of Arashiyama, eds. Jean-Baptiste Leca, Michael A. Huffman and Paul L. Vasey. Published by Cambridge University Press. © Cambridge University Press 2012.

education to show that the scientific value of the monkeys of Arashiyama goes beyond research.

19.1   Iwatayama Monkey Park, Arashiyama, as a field site for education

In the first place, even without special educational programmes, the Iwatayama Monkey Park, Arashiyama, in itself provides general visitors with opportunities to learn and think about human–wildlife relationships. For example, as opposed to ordinary zoological gardens, humans visiting Iwatayama Monkey Park are to be 'in the cage'. It means that while Arashiyama-Kyoto macaques are free-ranging, those who want to keep apart from the monkeys have to enter a resting house situated in the centre of the provisioning site. This perspective may allow visitors to break with the stereotyped view of human–animal relations in which humans are free agents and animals are to be controlled. On the other hand, if visitors want to stay outside with the monkeys, they are required to respect three fundamental regulations, based on a respectful attitude towards the macaques: (1) do not stare at, (2) do not touch and (3) do not feed the monkeys. As long as these basic rules are respected, a peaceful coexistence between humans and the Japanese macaques of Arashiyama is possible. However, if these regulations are violated, human observers may be shown aggression by the monkeys. Therefore by respecting the regulations, the visitors of Iwatayama Monkey Park may acquire a respectful attitude towards wildlife in general, sometimes unconsciously.

The Arashiyama field site has a long history of collaboration between researchers, naturalists and educators in research as well as educational activities. In the early stages of the site exploitation, both researchers and local educators conducted surveys of Japanese macaques either independently or in collaboration. For example, Mr Eiji Ohta, a primary school teacher, conducted the first survey on the monkeys of Arashiyama in 1948, prior to the survey conducted by Itani and Kawamura in 1951 (Asaba, 1984; Huffman, 1991; Ohta: chapter 2). After the success of provisioning the Arashiyama monkeys, Iwatayama Shizen Yuenchi (i.e. Nature Park, hereafter ISY) was founded by those people in 1957. ISY was aimed to be a field museum that promoted tourism and culture development in the region (Asaba, 1984). In 1973, Arashiyama Shizen Kenkyukai (i.e. Nature Study Group, hereafter ASK) was established in order to support research and education activities in Arashiyama. The ASK was organised by the researchers, local naturalists and educators. The ASK, in collaboration with ISY, conducted various educational activities such as nature observation walks, fieldwork guidance and lectures (Ohta, 1974), along with research activities.

In 1979, Arashiyama Shizen Tomonokai (i.e. Nature Club, hereafter AST) took over these educational activities (Suzuki and Asaba, 1984).

Today, although ASK and AST have ceased, educational activities in Arashiyama still continue. The majority of these activities are conducted and/or supported by Mr Shinsuke Asaba, director of the Iwatayama Monkey Park, Arashiyama (established from ISY) and the park staff. The park regularly receives and actively supports school excursions and fieldwork practice in science courses for various schools (from preschool to university undergraduate levels). It also conducts its own fieldwork courses during school holiday periods.

In contrast to the vigorous educational activities conducted by the park, commitment to educational activities by Arashiyama researchers has decreased lately. This tendency goes against the general trend in science education throughout Japan. Recently, the collaboration between scientists and science educators has been considered one of the solutions to the problems of Japanese science education (Ogawa, 2006), and the number of actual collaboration cases is increasing. At Takasakiyama Monkey Park (Oita prefecture), one of the oldest study sites of provisioned Japanese macaques, Dr Kurita, a primatologist who belongs to the local education board, developed a learning course for primary school children (Kurita, 2007).

Thus, commitment of primate researchers to science education should be reinforced in Arashiyama. It does not mean that we should go back to 'the good old days' but we should seek a new type of collaboration between researchers and educators that meets contemporary needs of science and environmental education. In this chapter, we first review the current situation of science education in Japan. We then discuss the positive role of primatologists as collaborative educators and the implication of the Iwatayama Monkey Park, Arashiyama as a field site for science education. Finally, we introduce our own educational practices as an example of a contemporary attempt to find a new type of collaboration between educators and researchers.

19.2   Problems in Japanese science education

19.2.1 *'Achievement crisis' in Japanese education*

For the last two decades, there has been a nationwide debate on the children's 'achievement crisis' in the Japanese society (Tsuneyoshi, 2004). The decline in students' academic ability in maths- and science-related subjects is emphasised by many critics in pedagogy (Nishimura *et al.*, 1999). It is often argued that the reforms of curriculum guidelines by the Japanese Ministry of Education

(MEXT) are the major reasons for this crisis (Nishimura *et al.*, 1999). Since the 1980s, the so-called 'relaxed education' concept was introduced into the curriculum guideline. School days were reduced from 6 days to 5 days per week, and accordingly, learning contents were reduced by about 30% from those included in the curriculum guideline during the 1970s.

On the other hand, there are researchers in pedagogy who advocate that there is no evidence for a decline in Japanese schoolchildren's academic ability over the last 20 years (Tsuneyoshi, 2004). The results of major international investigations of education, such as the 'Program for International Student Assessment' (PISA) conducted by the Organisation for Economic Cooperation and Development (OECD) and 'Trends in International Mathematics and Science Study' (TIMSS) by the International Association for Evaluation of Education Achievements (IEA) indicate that education achievements of Japanese students have been kept at a very high level of performance, compared with other countries participating in these assessments, even after the introduction of the 'relaxed education system' (Martin *et al.*, 2007).

It might not be fruitful to discuss whether a students' achievement crisis has actually occurred, nor to discuss how student academic ability has declined quantitatively. Rather, it is important to pay attention to the learning content, i.e. what Japanese students learn and what they do not learn, as well as when and how they learn each subject (Tsuneyoshi, 2004).

19.2.2   *Poverty of life sciences content in the Japanese science curriculum and achievement*

When we examine the science education curriculum in Japan, we find that it is poor in life sciences content. IEA's TIMSS is an international mathematics and science assessment conducted every 4 years. TIMSS evaluates students' achievements in mathematics and science for 4th and 8th grade students. It also assesses factors associating scientific achievements and the maths/science curriculum and instruction (Martin *et al.*, 2007).

The fourth cycle of TIMSS conducted in 2007 revealed that Japanese students' science achievement was top ranking among the participant countries both at the 4th and 8th grade level. However, although Japanese students' achievement in science is high in general, their performance in natural history-related content is relatively low. At the 4th grade, Japanese students' performance is relatively poor in life sciences and earth sciences content. At the 8th grade, performance in earth science content was relatively lower than in biology, physics and chemistry.

The relatively low achievement in natural history subjects can be explained by the characteristics of the Japanese science curriculum and instruction, which seem to overemphasise physics and chemistry, whereas biology and earth sciences are by and large disregarded. First, compared with the former, less time is devoted to life sciences and biology in Japan. Second, fewer contents of life sciences and biology are implemented in the Japanese science curriculum. Third, within life sciences and biology contents, the topics related to ecology, behaviour, biodiversity and nature conservation are poor or almost absent in the Japanese science curriculum. For example, the concept of evolution of life is not integrated into the compulsory education curriculum.

In response to criticisms that the relaxed education system may have caused a crisis in Japanese students' academic ability, MEXT revised the curriculum guidelines in 2008. The time devoted to science in compulsory education increased by 23%, and many topics removed from the 'relaxed education' curriculum during the past two decades were restored in the new guideline. However, the tendency to overemphasise physics remains, and still, evolutionary concepts are not taught. The poor life sciences contents in the Japanese educational program are problematic not only because this lack of education may reduce the ability of Japanese people to contribute to wildlife conservation issues, including the maintenance of biodiversity, which is a global task, but also because it may result in Japanese children's lack of interest in and dislike of science.

19.2.3   Children's disinterest in and dislike of science in Japan

The aim of science education is not only to improve children's scientific knowledge, but also to improve their attitude towards science, i.e. to encourage students to love and appreciate the importance of science. In this regard, disinterest, dislike and a lack of self-confidence in science among students have been pointed out to be some of the major problems in Japanese science education.

Japanese students seem less interested in science than students from other countries. In addition, the importance of science is less understood by students in Japan than in other countries. TIMSS 2007 indicates that the percentage of Japanese children who dislike science is much larger than in other countries. This also indicates that the percentage of Japanese students who think science is important is much smaller than the international average. The results of prior TIMSS assessments conducted in 1995 and 2003 show the same trend, suggesting that disinterest and dislike in science by Japanese students might be regarded as a rather long-term tendency.

This tendency may be related to the overemphasis on physics in the science curriculum. Kato (2009) conducted a questionnaire-based investigation of Japanese students from the 3rd to 10th grade in order to examine their attitude towards science. It revealed that the percentage of students who disliked science increased as the grade advanced. This trend was true for all areas of science covered, but differed between grades and among areas. While the percentage of students who disliked physics and chemistry drastically increased from the 7th to 10th grade, the percentage of students who disliked biology increased only slightly. When the students were asked if they liked or disliked each field of science separately (i.e. physics, chemistry, biology and earth sciences), the percentage of students who 'liked' biology (41%) was much larger than that who said they liked physics (18%), chemistry (20%) or earth sciences (24%). On the contrary, the percentage of students who 'disliked' biology (11%) was much smaller than that who said they disliked the other fields (32%, 36% and 24%, respectively).

An overemphasis on physics in itself is not an undesired education policy, since students with a better knowledge of physics and chemistry may be better equipped to contribute to Japan's high-tech industry. In addition, a physics-oriented curriculum might be a product of Japanese culture. During the post-war era, Japan produced two Nobel Prize winners in physics, Hideki Yukawa and Shin'ichiro Tomonaga. Such particular events may have increased the Japanese public interest in physics and consequently, led to a greater emphasis on physics in the overall science curriculum. If, as we assume, an overemphasis on physics resulted in students being disinterested and/or disliking science, it should be noted that a reconsideration of the science curriculum is currently required.

Biological topics, especially those related to ecology, animal behaviour and the relations between organisms and their environment are more likely to trigger students' interest as they are more familiar and visible biological phenomena. It might be easier or more interesting for students to learn about such empirical topics rather than physical laws that can only be understood through laboratory experiments. It might also be easier to understand the connection between such topics and the students' daily life. Devoting more time to develop an attractive learning programme in biology for school children may help reduce the growing disinterest and dislike of science by Japanese students. Consequently, this may contribute to achieve the major objective of the Japanese science curriculum, i.e. to develop a 'scientific way of thinking'. Moreover, learning biology, especially the interactions between biology and ecology, is essential for fostering an attitude for the conservation of biodiversity.

However, there are difficulties for school science teachers in developing education programmes in biology. TIMSS 2007 showed that the percentage of Japanese school science teachers who feel very well prepared for science lessons is lower than the international average. While this is a general trend in all fields of science, it was more noticeable in the case of biology. Furthermore, among biology contents, this trend is particularly marked for topics related to biodiversity, adaptation, ecological systems and interactions between humans and their environment (Martin *et al.*, 2007).

19.3   Teaching primatology in science education

One of the solutions to develop a good biology education programme is collaboration between school teachers and biologists. Ogawa (2006) pointed out the prospects of collaboration between science educators and science practitioners or researchers. The combination of teaching and science expertise may provide a more effective and attractive science programme to schoolchildren.

Primatologists can make important contributions to the development of a biology programme. First, because the Primate order is a taxon we humans belong to, we share many characteristics, and non-human primates (monkeys and apes) are good candidates to learn about the continuity between humankind and other living organisms. Second, primates show a high within-taxon diversity in many life-related aspects. There are nocturnal and diurnal species, arboreal and terrestrial species, frugivores, herbivores, folivores, insectivores, omnivores and grazers, solitary, monogamous, polygamous, polyandrous and multi-male multi-female species, etc. From all these characteristics, one can learn about the nature of biodiversity and the evolution of diversity from a common ancestor. Third, because primates are social animals, they provide various opportunities to observe different types of social behaviours and interactions. Obviously, observing social interactions is more attractive to schoolchildren than observing the behaviour of solitary animals. Fourth, because almost all living primate species are, to some extent, endangered, they represent a good taxon to learn about the relations between human activities and biodiversity, and the need for wildlife conservation.

There are two additional advantages for primatologists to contribute to biology education, especially in Japan. First, Japan is a non-human primate ranging country. Japanese macaques are endemic and quite popular animals in Japan. Using Japanese macaques as a learning resource provides schoolchildren not only with the opportunity to learn about the monkeys themselves, but also to learn about nature in the country where they live. Second, primatology has long been a popular science among Japanese people (Tachibana, 1991). This is not only due to the presence of Japanese macaques, but also because most Japanese people are aware that Japanese primatologists have made a unique and major contribution to the field of primatology in the world (Asquith, 1991; de Waal, 2001). Learning about primatology from active primatologists in

Japan might provide schoolchildren with an opportunity to closely connect to up-to-date science. Such an experience may improve the attitude of schoolchildren not only towards primatology but also to science in general, and may help reduce the students' disinterest and dislike for science.

19.4   Benefits of Arashiyama as a field site for science education

We chose Arashiyama as the location to conduct our field course owing to the following advantages:

1. Easy access and safety. The site is located at the edge of Kyoto City, and the monkeys are always present at the park. The Iwatayama Monkey Park, Arashiyama, maintains the facilities as a tourist site. There are numerous trails providing close access to the monkeys in the park, making it easy for the students, who have no previous field experience, to safely follow and observe the monkeys.
2. Free-ranging monkeys. There are no boundaries that separate students from the monkeys. When observing a monkey, students are able to go into the middle of the troop and follow their focal animals whenever and wherever they move. Tracking free-ranging monkeys provides students with a more impressive experience than the one consisting of observing captive animals in a zoo from outside the enclosure.
3. The monkeys are well habituated. Compared with wild populations of Japanese macaques, it is easier to get close to them. This allows even inexperienced students to observe monkey behaviour and morphology up close and in detail.
4. All monkeys are identified, and individual profiles, complete information on the maternal kin relationships, dominance ranks and long-term records of the demographic data are available (Huffman, 1991).
5. Logistical support by park staff. From the early stages of field research at Arashiyama, monkeys were provisioned, not only for research but also for tourism and education (Hirose, 1984). The Iwatayama Monkey Park can be considered a sort of field zoo, and it provides four essential functions required by zoological gardens, namely research, education, conservation and recreation. Therefore, the park staff is always ready to provide support to educational activities. All individual profiles of the Arashiyama-Kyoto troop members are maintained by the park staff, and are open to visitors for any purpose.
6. For decades, the study site has been used for research purposes by both Japanese and foreign researchers. Sharing the same material, i.e. monkeys,

with active researchers might be a great motivator for students to study not only biology but science in general.

19.5   Primate field course in Arashiyama

We have been involved in the development and practice of primate-based fieldwork courses for high-school students at Arashiyama since 2007. The courses are prepared for students of Nanzan Girls High School (NGH). NGH is a private high school in Nagoya, founded in 1948. This school introduced a system integrating junior and senior high school together to encompass a curriculum from the 7th through to the 12th grade.

We have been conducting field courses at Arashiyama since 2007. These courses are organised as extra-curricular science activities. The targets of the course are students of the 8th–12th grades who wish to participate. Since 2008, the programme has been conducted under the auspices of the Science Partnership Project (SPP), supported by the Japan Science and Technology Agency (JST). The SPP supports activities conducted by universities, science centres and the boards of education to enhance elementary through high school children's interest in science, technology and mathematics, encouraging children to develop an inquiring mind (Japan Science and Technology Agency, 2010). Below, we present overviews of actual activities of the courses conducted during the first 3 years.

19.5.1   *The 2007 field course*

The first course of this programme was conducted on 13 October 2007. As it was our first attempt, the aim of the course was still vague and our prospectus was not well developed. Seventeen students (eleven 10th grade students and six 11th grade students) participated. The educators consisted of a primatologist instructor (YT) and five high-school science teachers, including YM. Neither pre- nor post-field course lectures were conducted. The field course lasted two hours (2:00–4:00 p.m.).

For all the students, it was their first experience observing free-ranging animals. During the first 30 minutes, the instructor conducted a general introduction to Japanese macaque behaviour and society, and explained the function/ meaning of several social behaviours and interactions, i.e. grooming, spatial/ social proximity, facial expressions such as grimace and types of vocalisations. How to individually identify monkeys using such characteristics as scars, moles etc. was explained with examples.

The main topic of the course was to investigate mother–infant relations in Japanese macaques. We chose this topic partly because it was an issue familiar to students; therefore, it might lead students to contrast what they observe in monkeys to what they experience in their daily life, and also because it is generally enjoyable for people to watch mother–infant pairs.

After the general introduction, students were divided into four groups of four or five members. Each group was assigned a mother and infant pair to observe. Then each group was divided into two subgroups, one had to follow the mother and the other one had to follow the infant. After 10 minutes learning to individually identify their study subjects, each subgroup was required to follow their focal animal for a total of 30 minutes over a 1-hour period.

Students were required to record the geographical position of focal animals on a map every 5 minutes to measure the mother–infant distances for later analysis. In addition, they were required to record *ad libitum* any social behaviour that was of interest to them.

All the subgroups following the mothers succeeded in collecting 30 minutes of observations within an hour, although they lost their focal animal several times in the process. In contrast, the subgroups following the infants had many difficulties. They lost their focal animal more often because of the difficulty in identifying each infant. When the focal infants joined other infants, students became confused and mistook other infants for their focal subject. Logically, an observer should be able to follow a focal animal by keeping a close watch over it, even if individual recognition is inaccurate. However, for inexperienced high-school students, it turned out to be a difficult task.

We also found that it was difficult for students to locate both the geographical location of the focal animal and themselves on a map. Consequently, we failed to obtain enough data for the intended analysis. Nonetheless, students who participated in the course seemed to have enjoyed the fieldwork experience. Many students stated that the course was stimulating and that their attitude towards primatology was more or less positive. However, when we evaluated the course from the viewpoint of effectiveness as a science education programme, there remained a lot of areas for improvement. In particular, we became aware of the need to identify the students' aptitude for fieldwork and how to design the field course in better ways.

19.5.2   The 2008 field course

The second course was conducted on 11 October 2008. The participants consisted of 23 students (twenty 10th grade students, two 11th grade students, one 12th grade student). The educators consisted of a primatologist instructor (YT) and two school science teachers, including YM. The time schedule was the same as in 2007 (2:00 to 4:00 p.m.). This time, we specified the primary purpose of the course, i.e. to learn that nonmaterial characteristics of animals such as behaviour and social interactions can be measured, quantified and statistically analysed, much like morphological and physiological characteristics.

Accordingly, we organised the fieldwork course to practise the most common methods to record and analyse animal behaviour, i.e. focal animal following, scan sampling, one-zero score sampling and *ad libitum* sampling method (Altmann, 1974; Martin and Bateson, 1986). In addition, we organised a lecture afterwards to demonstrate how to summarise, process and visualise the data. The main topic of the course was again mother–infant relations in the Japanese macaque. The students were divided into 11 pairs. Each pair was assigned a mother (a female with dependent infants), or an infant. As in 2007, each student pair was required to follow their focal animal for a total of 30 minutes within a 1-hour time period, after an initial 10-minute period spent becoming familiar with the physical traits of their study subjects.

Instead of recording the geographical position of the focal animal, we chose two common parameters to assess affiliative relations between monkeys, grooming and proximity. The students used the scan sampling method to record mother–infant proximity. They were then required to record in their field notebook whether the infant of the focal animal was within 1- or 3-metres proximity, every 3 minutes. If students chose to follow an infant, they were required to record if its mother was within 1- or 3-metres proximity. At the same time, students were required to count the total number of other monkeys within 1 or 3 metres of the focal animal.

The students were then required to use the one-zero score sampling method to record frequency of grooming between mothers and infants. They had to record if the mother groomed the infant or if the infant groomed the mother during the 3-minute intervals between scan sampling for proximity. It has been argued that one-zero score sampling is not a valid method to measure the frequency and/or duration of a behaviour (Kraemer, 1979; Martin and Bateson, 1986). However, based on the 2007 course, we learned that it was not possible for students without previous field experience to record the beginning and ending time of each grooming bout. One-zero score sampling was easier and more conventional (Rhine and Linville, 1980) for our purposes. All the student pairs successfully completed the 30 minutes of focal animal follows within the time limit. Moreover, all were successful in recording all the required data.

The lecture that followed was conducted at NGH on 24 October 2008, two weeks after the field programme. First, the students wrote down the data recorded in their field notebooks at Arashiyama into check-sheets. The check-sheets were paper-based because not all students had yet learned to use spreadsheet software. The primatologist instructor (YT) assembled the check-sheets, entered data into a statistical software package, and displayed basic data analysis and how to make graphs. The frequency of mother–infant grooming was compared between individuals to test possible differences in intensity of mother–infant bonds according to the infant's sex, age, as well as mother's age and rank. The results were illustrated in box plot format on a screen. We acknowledge that this analysis was no more than a demonstration since the sample size and total duration of observation were too small to run statistical tests. Nonetheless, it was enough to achieve the aim of our course. The students learned that one can provide quantitative behavioural data analyses, and graph the results to visually examine subtle differences in the social relations of mother–infant pairs that would be hard to detect otherwise.

19.5.3   The 2009 field course

In our latest field course, conducted in 2009, the aims of the course were revised and elaborated as follows:

1. To learn that non-material characteristics of animals such as behaviour and social interactions can be measured, quantified and statistically analysed, much like morphological and physiological characteristics, through the practice of common sampling methods used in primate behavioural studies.
2. To learn that sociality is not a unique characteristic of humans, through the observation of various social interactions among Japanese macaques.
3. To learn that each individual exhibits its own characteristics and/or personality.
4. To use an evolutionary perspective to understand our human society, through the observation of social interactions in Japanese macaques, and contrast them with those occurring in our daily life.

In 2009, we conducted a lecture prior to the field course. The field course was conducted for 2 consecutive days. The proximate topic of the field course was always the mother–infant relationship. Eighteen students (two 7th grade, seven 10th grade and nine 11th grade students) participated in the programme.

Pre-field course lecture

A pre-field course lecture was conducted at NGH on 30 October 2009. The lecture consisted of a review on primate behavioural studies and the method-ologies of behavioural sampling (Altmann, 1974; Martin and Bateson, 1986).

Several examples of social behaviour, facial expressions and vocalisations were shown using videos of monkeys recorded at Arashiyama. We also introduced some of the prior studies conducted in Arashiyama by Japanese and foreign researchers.

Field course: Day 1

A 2-day field course was conducted on 21 and 22 November 2009. The educators consisted of two primatologist instructors, including YT, and two science teachers, including YM. The course was conducted from 1:00 to 4:00 p.m. on the first day, and from 9:00 a.m. to 1:00 p.m. on the second day.

The programme for the first day was almost the same as in 2008. The students were divided into pairs and each was assigned a mother with a dependant infant, that they had to follow for a total of 30 minutes within a 1-hour period. During focal animal follows, the students were required to record in their field notebooks if the infant of the focal animal was within 1- or 3-metres proximity as well as the number of monkeys within 1 or 3 metres from the focal animal, every 3 minutes. They also had to record if the mother groomed the infant or if the infant groomed the mother during the 3-minute intervals between scan sampling for proximity. In addition, the students were required to record if their focal animal fed on natural or provisioned food during each of the 3-minute intervals between scan samples for proximity. Finally, we instructed them to record all other behaviours and interactions that they estimated worth noting, as notes. As in 2008, all students successfully completed focal animal follows for a total of 30 minutes within the time limit. Moreover, all pairs successfully recorded the required data.

We conducted a workshop in the evening of the first day. First, we practised data entry and analysis. Each pair of students entered the sampling data into a spreadsheet software by themselves, under the guidance of the instructors. Next, the instructors assembled all the data and demonstrated different types of analyses like in the post-field course lecture in 2008.

Finally, we conducted a discussion session. Students were free to talk about their impressions of the monkeys, the field exercise and what can be derived from the data they collected. Many students talked about their sense of affinity with their focal animal and their own interpretation of the personality of their focal animals. The significance of the students' responses is discussed later.

Field course: Day 2

The programme of the second day was the same as the first day, except that the sampling time was prolonged to 1 hour of focal sampling time within a 2-hour period. Although the sampling time was twice as long as the first day, all the student pairs completed their tasks. We asked the students to analyse the data they collected on the second day later at home.

At the end of the second day, we conducted another discussion session. Most of the students stated that their impression towards their focal animal had changed from that of the first day, and consequently, they understood that more observation was required to know the animal, and that long-term study was important for the behavioural study of primates.

19.6   Changes in students' attitudes towards science through the field course

Subsequent to the field course in 2009, we presented a questionnaire to the students to learn about the changes in their attitude towards science. The results are summarised below.

- Q. Did you acquire a liking of school science after this programme?
    - ◦ A1. I liked science, and this programme helped me like it even more: 11 students.
    - ◦ A2. I liked science, and I like it as much as before participating in this programme: 5 students.
    - ◦ A3. I disliked science, but this programme helped me like it: 1 student.
    - ◦ A4. I disliked science, and this programme did not change anything at all: 1 student.
    - ◦ A5. I disliked science, and this programme made me dislike it even more: 0 student.

- Q. Did your interest in science and technology increase after this programme?
    - ◦ A1. I have been interested in science and technology, and after this programme, I am even more interested than before: 11 students.
    - ◦ A2. I have been interested in science and technology, and after this programme, I am interested in them as much as before: 3 students.
    - ◦ A3. I was not interested in science and technology, but I became interested in them after this programme: 4 students.
    - ◦ A4. I am not interested in science and technology even after this programme: 0 student.
    - ◦ A5. I am not interested in science and technology, and after this programme, I am less interested than before: 0 student.

- Q. Do you wish to get a job that is related to science and technology (e.g. scientist, engineer, etc.)?
  - ○ A1. I wish to do so more than before participating in this programme: 5 students.
  - ○ A2. I wished to do so even before participating this programme: 3 students.
  - ○ A3. I did not wish to do so, but after this programme, I do wish to: 1 student.
  - ○ A4. I did not wish to do so, and after participating this programme, still do not: 7 students.
  - ○ A5. I have no wish to do so, and this program reinforced that: 0 student.

The answers seem to indicate that our programme had a positive influence on the attitudes of students towards science. This suggests to us that observing the behaviour of free-ranging Japanese macaques in Arashiyama is a good way to help change the attitude of students towards science. From this exercise, we conclude that such a programme is an effective countermeasure against the current trend of students' disinterest and dislike of science in Japan.

However, we should note that it is not clear that our practice is effective to improve the attitude towards science of every student, since our programme provided the opportunity only to those students already motivated to voluntarily participate. Therefore the majority of the participants already had a more or less positive attitude towards science. Whether the programme that we developed can have a positive influence even on those students who are not already interested in science is the subject for future research.

19.7 Educational value of individual identification and focal animal following

The fact that the students acquired a sense of affinity towards their focal individuals indicates the educational value of focal animal following in developing the students' sense of wildlife conservation. For primate researchers, focal animal following is one of the common methods used to systematically collect behavioural data for statistical analysis (Altmann, 1974; Martin and Bateson, 1986). On the other hand, it is suggested that identifying an individual monkey and following it for some time raises a student's emotional empathy to the focal animal. A feeling of empathy towards wildlife provides a fundamental basis for the attitude towards the protection of nature and conservation of biodiversity.

19.8   Invitation to researchers to get involved in science education

Through the practice of our field course programme at Arashiyama, we rediscovered the importance of collaboration between researchers and teachers in developing a good science course. A good science course requires both a specialised knowledge of the study subject and an understanding of the academic ability of students who participate in the programme. Researchers are particularly qualified to provide the former, while the teachers are more familiar with the latter.

As noted in other chapters of this book, Arashiyama monkeys are used not only for behavioural studies (e.g. chapters 10–16), but also for demographic (chapter 5), ecological (chapter 17), physiological (chapters 7 and 18) and reproductive studies (chapters 6, 8 and 9). Researchers working on all these subjects are able to contribute from their own fields of expertise. Therefore, we invite more researchers to get involved in the educational activities at Arashiyama, and encourage those from other countries who read this book, especially those who are from a non-human primate ranging country, to create their own programmes in their respective countries.

Acknowledgements

We thank Mr Shinsuke Asaba, current director of Iwatayama Monkey Park, Arashiyama and staff of the park for their hospitality and for facilitating the field course. We thank Dr Eiji Inoue for his help in developing the programme and his participation in the field course. We thank Ms Yasuyo Senda and the teachers of Nanzan Girls High School for their help in conducting the field course. We thank the Japan Science and Technology Agency for their financial support. Finally, we thank all the students of NGH who participated in the course.

References

Altmann, J. (1974). Observational study of behavior: sampling methods. *Behaviour*, **49**, 227–267.

Asaba, N. (1984). The last thirty years of the Iwatayama Monkey Park and perspectives for the future. In *The Japanese Monkeys of Arashiyama. Iwatayama Shizenshi Kenkyujo Houkoku, Vol. 3*, eds. H. Suzuki and N. Asaba. Kyoto: Arashiyama Shizenshi Kenkyujo, pp. 15–20. (in Japanese)

Asquith, P. (1991). Primate research groups in Japan: orientations and East-West differences. In *The Monkeys of Arashiyama: Thirty Five Years of Research in Japan and the West*, eds. L. Fedigan and P. Asquith. Albany, NY: State University of New York, pp. 81–98.

de Waal, F. (2001). *The Ape and the Sushi Master: Cultural Reflections by a Primatologist*. New York, NY: Basic Books.

Hirose, S. (1984). Arashiyama is a natural history science museum. In *The Japanese Monkeys of Arashiyama. Iwatayama Shizenshi Kenkyujo Houkoku, Volume 3*, eds. H. Suzuki and R. Asaba. Kyoto: Arashiyama Shizenshi Kenkyujo, pp. 80–81. (in Japanese)

Huffman, M. (1991). History of the Arashiyama Japanese macaques in Kyoto, Japan. In *The Monkeys of Arashiyama: Thirty Five Years of Research in Japan and the West*, eds. L. Fedigan and P. Asquith. Albany, NY: State University of New York, pp. 21–53.

Japan Science and Technology Agency (2010). Science Partnership Project (SPP): JST Projects for Promoting Public Understanding of Science and Technology.

Kato, J. (2009). Science education and students' dislike of science subjects in secondary school. *Annals of Kobe Shoin Women's University*, **50**, 65–80. (in Japanese)

Kraemer, H. C. (1979). One-zero sampling in the study of primate behavior. *Primates*, **20**, 237–244.

Kurita, H. (2007). Preliminary study on the knowledge of junior high school students about Japanese macaques: Recommendations for science education at Monkey Parks. *Primate Research*, **23**, 17–23. (in Japanese with English abstract)

Martin, M. O., Mullis, I. V. S. and Foy, P. (2007). TIMSS 2007 International Science Report.

Martin, P. and Bateson, P. (1986). *Measuring Behaviour: An Introductory Guide*. Cambridge: Cambridge University Press.

Nishimura, K., Tose, N. and Okabe, T. (1999). *Bunsu ga dekinai daigakusei [College students that cannot solve problems of fractions]*. Tokyo: Keizai Shinposha. (in Japanese)

Ogawa, M. (2006). Possibility of a new type of science education research through authentic collaboration among societal sectors, practitioners and researchers. *Journal of Science Education in Japan*, **30**, 121–131. (in Japanese)

Ohta, E. (1974). The activities of Arashiyama Shizen Kenkyukai 1974. *Arashiyama Shizen Kenkyukai Kaiho*, **2**, 16–20. (in Japanese)

Rhine, R. J. and Linville, A. K. (1980). Properties of one-zero scores in observational studies of primate social behavior: the effect of assumptions on empirical analyses. *Primates*, **21**, 111–122.

Suzuki, H. and Asaba, R. (1984). *The Japanese Monkeys of Arashiyama. Iwatayama Shizenshi Kenkyujo Houkoku, Vol. 3*. Kyoto: Arashiyama Shizenshi Kenkyujo. (in Japanese)

Tachibana, T. (1991). *The Frontiers of Primatology*. Tokyo: Heibonsha. (in Japanese)

Tsuneyoshi, R. (2004). The new Japanese educational reforms and the achievement 'crisis' debate. *Educational Policy*, **18**, 364–394.

Appendix: Bibliography of publications on the Arashiyama macaques

(updated from Fedigan, L. M. and Asquith, P. J. (1991). *The Monkeys of Arashiyama: Thirty-five Years of Research in Japan and the West*. Albany, NY: State University of New York Press)

Ando, A. (1988). Grooming relationships of the transplanted Arashiyama A troop of Japanese macaques (*Macaca fuscata*). In *Research Reports of the Arashiyama West and East Groups of Japanese Monkeys*. Osaka: Laboratory of Ethological Studies, Faculty of Human Sciences, Osaka University, pp. 41–49.

Anonymous (1972). Arashiyama West: Japanese monkeys find new home in Texas. *Primate Record*, **3**, 3–6.

  (1974). Despite difficulties Japanese troop adapting to Texas. *Primate Record*, **5**, 14.

  (1984). Macaques in Texas. *National Geographic World*, **106**, 11–14.

  (1994). Hunting ban lifted in Arashiyama, Kyoto. *Japan Primate Newsletter*, **4**, 5–6.

Asaba, N. (1984). The last thirty years of the Iwatayama Monkey Park and perspectives for the future. In *The Japanese Monkeys of Arashiyama. Iwatayama Shizenshi Kenkyujo Houkoku, Vol. 3*, eds. H. Suzuki and N. Asaba. Kyoto: Arashiyama Shizenshi Kenkyujo, pp. 15–20. (in Japanese)

Asaba, N. and Suzuki, H. [eds]. (1984). Birth lists and family trees for the Arashiyama monkeys. In *Bulletin of Iwatayama Institute of Natural History Iwatayama Shizenshi Kenkyujo Hokoku. Vol. 3*. Kyoto: Arashiyama Shizen Kenkyujo, pp. 83–114. (in Japanese) [note: there is no primary author(s) for this article]

Asquith, P. (1989). Provisioning and the study of free-ranging primates: history, effects and prospects. *Yearbook of Physical Anthropology*, **32**, 129–158.

  (1991). Primate research groups in Japan: Orientations and East-West differences. In *The Monkeys of Arashiyama: Thirty-five Years of Research in Japan and the West*, eds. L. Fedigan and P. Asquith. Albany, NY: State University of New York Press, pp. 81–98.

Baxter, J. and Fedigan, L. M. (1979). Grooming and consort partner selection in a troop of Japanese monkeys (*Macaca fuscata*). *Archives of Sexual Behavior*, **8**, 445–458.

Bélisle, P. (2002). *Apparentement et co-alimentation chez le macaque japonais*. PhD dissertation, University of Montréal.

Bélisle, P. and Chapais, B. (2001). Tolerated co-feeding in relation to degree of kinship in Japanese macaques. *Behaviour*, **138**, 487–509.

Blount, B. G. (1985). 'Girney' vocalizations among Japanese macaques females: context and function. *Primates*, **26**, 424–435.

470

Bramblett, C. A. (1980). Arashiyama 'A' Japanese troop to remain intact: support solicited. *Laboratory Primate Newsletter*, **19**, 15–16.

(1994). *Patterns of Primate Behavior*, 2nd edn. Prospect Heights, IL: Waveland Press.

Bramblett, C. A., Noyes, M. J. S., Bramblett, S. and Clark, L. G. (1981). Behavior and adaptation of free ranging Japanese macaques in Texas to relocation to similar habitat. *American Journal of Primatology*, **1**, 31. (Abstract)

Bramblett, S. S., Bramblett, C. A. and Noyes, M. J. S. (1981). Successful relocation of Arashiyama West Japanese monkeys. *International Primate Protection League Newsletter*, **8**, 9–11.

Brandon-Jones, C. (1995). '*The Monkeys of Arashiyama: Thirty-five Years of Research in Japan and the West*, eds. L. Fedigan and P. Asquith. Albany, NY: State University of New York, 353 pp.' [Book review]. *Primate Eye*, **55**, 37–38.

Bullard, J. (1984). Mount prompting behaviors of female Japanese macaques (*Macaca fuscata*). M.A. thesis, Department of Anthropology, University of Alberta, Edmonton, Canada.

Casey, D. E. and Clark, T. W. (1976). Some spacing relations among the central males of a transplanted troop of Japanese macaques (Arashiyama West). *Primates*, **17**, 433–450.

Chalmers, A. (2008). Life histories and hormones: variations by habitat in three populations of *Macaca fuscata*. Master's thesis. Kyoto University, Faculty of Science, Laboratory of Human Evolution Studies.

Chalmers, A. and Shimizu, K. (2008). Evolving life history traits: the influence of environment and nutrition on three populations of *Macaca fuscata*. Paper presented before the *22nd International Primatological Society Meetings*. *Primate Eye*, **96**, #857. (Abstract)

Chapais, B. (1985). An experimental analysis of a mother-daughter rank reversal in Japanese macaques *(Macaca fuscata)*. *Primates*, **26**, 407–423.

(1988). Experimental matrilineal inheritance of rank in female Japanese macaques. *Animal Behaviour*, **36**, 1025–1037.

(1988). Rank maintenance in female Japanese macaques: experimental evidence for social dependency. *Behaviour*, **104**, 41–59.

(1991). Matrilineal dominance in Japanese macaques: The contribution of an experimental approach. In *The Monkeys of Arashiyama: Thirty-five Years of Research in Japan and the West*, eds. L. M. Fedigan and P. J. Asquith. Albany, NY: State University of New York Press, pp. 251–273.

(1992). The role of alliances in the social inheritance rank among female primates. In *Coalitions and Alliances in Humans and Other Animals*, eds. A. Harcourt and F. M. B. de Waal. Oxford: Oxford University Press, pp. 29–60.

(2001). Primate nepotism: what is the explanatory value of kin selection? *International Journal of Primatology*, **22**, 203–229.

(2006). Kinship, competence and cooperation in primates. In *Cooperation in Primates and Humans: Mechanism and Evolution*, eds. P. M. Kappeler and C. P. van Schaik. Berlin: Springer-Verlag, pp. 47–66.

Chapais, B. and Bélisle, P. (2004). Constraints on kin selection in primate groups. In *Kinship and Behavior in Primates*, eds. B. Chapais and C. M. Berman. New York: Oxford University Press, pp. 365–386.

Chapais, B. and Gauthier, C. (1993). Early agonistic experience and the onset of matrilineal rank acquisition in Japanese macaques. In *Juvenile Primates: Life-History, Development, and Behavior*, eds. M. E. Pereira and L. A. Fairbanks. New York, NY: Oxford University Press, pp. 246–258.

Chapais, B. and Larose, F. (1988). Experimental rank reversals among peers in *Macaca fuscata*: rank is maintained after removal of kin support. *American Journal of Primatology*, **16**, 31–42.

Chapais, B. and Mignault, C. (1991). Homosexual incest avoidance among females in captive Japanese macaques. *American Journal of Primatology*, **23**, 177–183.

Chapais, B. and St-Pierre, C.-E. G. (1997). Kinship bonds are not necessary for maintaining matrilineal rank in captive Japanese macaques. *International Journal of Primatology*, **18**, 375–385.

Chapais, B., Girard, M. and Primi, G. (1991). Non-kin alliances, and the stability of matrilineal dominance relations in Japanese macaques. *Animal Behaviour*, **41**, 481–491.

Chapais, B., Prud'homme, J. and Teijeiro, S. (1994). Dominance competition among siblings in Japanese macaques: Constraints on nepotism. *Animal Behaviour*, **48**, 1335–1347.

Chapais, B., Gauthier, C., Prud'homme, J. and Vasey, P. (1997). Relatedness threshold for nepotism in Japanese macaques. *Animal Behaviour*, **53**, 1089–1101.

Chapais, B., Savard, L. and Gauthier, C. (2001). Kin selection and the distribution of altruism in relation to degree of kinship in Japanese macaques *(Macaca fuscata)*. *Behavioral Ecology and Sociobiology*, **49**, 493–502.

Clark, T. W. (1978). Agonistic behavior in a transplanted troop of Japanese macaques: Arashiyama West. *Primates*, **19**, 141–151.

(1979). Food adaptations of a transplanted Japanese macaque troop (Arashiyama West) in the U.S. *Primates*, **20**, 399–410.

(1981). A successful case of adoption of a two-month old infant Japanese monkey by his two-year old sister. *American Journal of Primatology*, **1**, 318. (Abstract)

Clark, T. W. and Huckabee, J. W. (1977). Elemental hair analysis of Japanese macaque transplanted to the United States. *Primates*, **18**, 299–303.

Clark, T. W. and Mano, T. (1975). Transplantation and adaptation of a troop of Japanese monkeys to a Texas brushland habitat. In *Contemporary Primatology: Proceedings of the 5th International Congress of Primatology*, eds. S. Kondo, M. Kawai and A. Ehara. Basel: S. Karger, pp. 358–361.

Collinge, N. E. (1985). Variation in weaning-related behaviors of mothers and infants in semi-free ranging Japanese macaques *(Macaca fuscata)*. M.A. thesis, Department of Anthropology, University of Alberta.

(1987). Weaning variability in semi-free ranging Japanese macaques *(Macaca fuscata)*. *Folia Primatologica*, **48**, 137–150.

(1991). Variability in aspects of the mother-infant relationship in Japanese macaques during weaning. In *The Monkeys of Arashiyama: Thirty-five Years of Research in Japan and the West*, eds. L. M. Fedigan and P. J. Asquith. Albany, NY: State University of New York Press, pp. 157–174.

Demmet, M. (1976). Feeding ecology of Japanese macaques (*Macaca fuscata*) in southern Texas. M. Sc. Thesis, Department of Zoology, University of Wisconsin, Madison.

Ehardt, C. L. (1980). The structure of social relationships in infant Japanese macaques (*Macaca fuscata*). *Dissertation Abstract International*, **A41**, 3173 (1981).

(1982). Patterned interactions and ontogeny of relationships in infant Japanese macaques. *American Journal of Physical Anthropology*, **57**, 184–185. (Abstract)

(1983). Birth season associations of adult Japanese macaque females without infants. *American Journal of Physical Anthropology*, **60**, 191. (Abstract)

(1984). Birth season associations of adult male Japanese macaques. *American Journal of Physical Anthropology*, **63**, 155. (Abstract)

(1987). Birth-season interactions of adult female Japanese macaques (*Macaca fuscata*) without newborn infants. *International Journal of Primatology*, **8**, 245–259.

Ehardt, C. (1991). Birth season contingencies and the affiliative behavior of adult male Japanese macaques. In *The Monkeys of Arashiyama: Thirty-five Years of Research in Japan and the West*, eds. L. M. Fedigan and P. J. Asquith. Albany, NY: State University of New York Press, pp. 227–248.

Ehardt, C. L. and Blount, B. G. (1981). Patterns of visual behavior among infant Japanese macaques. *American Journal of Primatology*, **1**, 318. (Abstract)

(1984). Mother-infant visual interaction in Japanese macaques. *Developmental Psychobiology*, **17**, 391–405.

Fedigan, L. M. (1973). The classification of predators by Japanese macaques (*Macaca fuscata*) in the mesquite chaparral habitat of south Texas. *American Journal of Physical Anthropology*, **40**, 135. (Abstract)

(1974). Wania 6672: A video tape showing the social development of an abnormal infant in a semi-free-ranging troop of Japanese macaques (*Macaca fuscata*) at La Moca, Texas. *American Journal of Physical Anthropology*, **40**, 135. (Abstract)

(1975). Role behaviors in the Arashiyama West troop of Japanese monkeys (*Macaca fuscata*). *Dissertation Abstract International*, **A35**, 4765–4766.

(1976). A study of roles in the Arashiyama West troop of Japanese monkeys (*Macaca fuscata*). A Monograph in the Series: *Contributions to Primatology, Vol. 9*. Basel: S. Karger.

(1981). Demographic characteristics of the Arashiyama West group of Japanese macaques. *American Journal of Primatology*, **1**, 316. (Abstract)

(1982). *Primate Paradigms: Sex Roles and Social Bonds*. Montreal: Eden Press.

(1991). History of the Arashiyama West Japanese macaques in Texas. In *The Monkeys of Arashiyama: Thirty-five Years of Research in Japan and the West*, eds. L. M. Fedigan and P. J. Asquith. Albany, NY: State University of New York Press, pp. 54–73.

Fedigan, L. (1991). Life span and reproduction in Japanese macaque females. In *The Monkeys of Arashiyama: Thirty-five Years of Research in Japan and the West*, eds. L. M. Fedigan and P. J. Asquith. Albany, NY: State University of New York Press, pp. 140–154.

Fedigan, L. M. and Asquith, P. J. (1990). The monkeys of Arashiyama: 35 years of study in the East and the West. *Proceedings of the 13th Congress of the International Primatological Society*, **11**, 107. (Abstract)

(1991). *The Monkeys of Arashiyama: Thirty-five Years of Research in Japan and the West*. Albany, NY: State University of New York Press.

(1992). The history of Arashiyama research as a microcosm of larger trends in primatology. In *Topics in Primatology, Vol. 2, Behavior, Ecology and Conservation*, eds. N. Itoigawa, Y. Sugiyama, G. P. Sackett and R. K. R. Thompson. Tokyo: University of Tokyo Press, pp. 67–77.

Fedigan, L. M. and Fedigan, L. (1976). The social development of a handicapped infant in a free-living troop of Japanese monkeys. In *Primate Biosocial Development*, eds. F. E. Poirier and S. Chevalier -Skolnikoff. New York, NY: Garland Press, pp. 205–222.

Fedigan, L.M. and Gouzoules, H. (1978). The consort relationship in a troop of Japanese monkeys. In *Recent Advances in Primatology, Vol. 1: Behaviour*, eds. D. J. Chivers and J. Herbert. London: Academic Press, pp. 493–495.

Fedigan, L. M. and Griffin, L. (1996). Determinants of reproductive seasonality in Japanese macaques. In *Evolution and Ecology of Macaque Societies*, eds. J. E. Fa and D. G. Lindburg. Cambridge: Cambridge University Press, pp. 368–387.

Fedigan, L. M. and Pavelka, M. S. M. (1994). The physical anthropology of menopause. In *Strength in Diversity: A Reader in Physical Anthropology*, eds. A. Herring and L. Chan. Toronto: Canadian Scholars Press, pp. 103–126.

(2001). Is there adaptive value to reproductive termination in Japanese macaques? A test of maternal investment hypotheses. *International Journal of Primatology*, **22**, 109–125.

(2007). Reproductive cessation in female primates: comparisons of Japanese macaques and humans. In *Primates in Perspective*, eds. C. Campbell, A. Fuentes, K. MacKinnon, M. Panger and S. Bearder. Oxford: Oxford University Press, pp. 437–447.

Fedigan, L. and Pavelka, M. (2010). Menopause: interspecific comparisons of reproductive termination in female primates. In *Primates in Perspective*, eds. C. J. Campbell, A. Fuentes, K. C. MacKinnon, S. K. Bearder and R. M. Stumpf. New York, NY: Oxford University Press, pp. 488–499.

Fedigan, L. M. and Zohar, S. (1997). Sex differences in mortality of Japanese macaques: 21 years of data from the Arashiyama West population. *American Journal of Physical Anthropology*, **102**, 161–175.

Fedigan, L. M., Gouzoules, H. and Gouzoules, S. (1981). Selected features of the demography of Japanese monkeys. *Canadian Review of Physical Anthropology*, **3**, 76. (Abstract)

(1983). Population dynamics of Arashiyama West Japanese monkeys. *International Journal of Primatology*, **3**, 307–321.

Fedigan, L. M., Fedigan, L., Gouzoules, S., Gouzoules, H. and Koyama, N. (1986). Lifetime reproductive success in female Japanese macaques. *Folia Primatologica*, **47**, 143–157.

Fujimoto, M. (2004). Grooming interactions of adult female Japanese monkeys: comparison between mating and non-mating seasons. *Reichorui Kenkyu/Primate Research*, **20** (suppl. 1), 25. (Abstract, in Japanese)

(2004). Social grooming among adult female Japanese monkeys at Arashiyama E-group: comparison between mating and non-mating seasons. Master's thesis, Doshisha University, Japan. (in Japanese)

(2010). Seasonal changes in partners and interactions of social grooming among free-ranging adult female Japanese monkeys (*Macaca fuscata*) of Arashiyama E-group, Kyoto. The 23rd Congress of the International Primatological Society, Kyoto, Japan. *Reichorui Kenkyu/Primate Research*, **26** (suppl. 1), Abstract #138.

(2010). A study about evolutionary basis of short-distance communication in the primates: social grooming of Japanese macaques and peering behavior of wild chimpanzees. Ph.D. thesis, the University of Shiga Prefecture, Japan. (in Japanese)

Fujimoto, M. and Takeshita, H. (2004). Sequential analysis of grooming interactions in female Japanese monkeys (*Macaca fuscata*). Paper presented before *The 2nd International Workshop for Young Psychologists on Evolution and Development of Cognition*, Kyoto, Japan. Program and Abstracts: p. 36.

(2004). Analysis of behavioral sequences and social interactions of grooming among adult female Japanese macaques. Paper presented before *The 23rd Annual Congress of the Japan Ethological Society*, Proceeding abstract #55. (in Japanese)

(2007). Analysis of behavioral sequences in grooming interactions between adult females in the Arashiyama-E group of Japanese monkeys (*Macaca fuscata*). *Japanese Journal of Animal Psychology*, **57**, 61–71. (in Japanese with English summary)

Furuya, Y. (1969). On the fission of troops of Japanese monkeys. *Primates*, **10**, 47–69.

Gotoh, S. (2000). Regional differences in the infection of wild Japanese macaques by gastrointestinal helminth parasites. *Primates*, **41**, 291–298.

Gouzoules, H. (1977). Arashiyama West: a status report. *International Primate Protection League Newsletter*, **4**, 2.

(1980). A description of genealogical rank changes in a troop of Japanese monkeys (*Macaca fuscata*). *Primates*, **21**, 262–267.

(1981). Biosocial determinants of behavioral variability in infant Japanese monkeys (*Macaca fuscata*). *Dissertation Abstract International*, **B41**, 4415.

(1992). 'The Monkeys of Arashiyama: Thirty-five Years of Research in Japan and the West, eds. L. Fedigan and P. Asquith. Albany, NY: State University of New York, 353 pp.' [Book review]. *International Journal of Primatology*, **13**, 691–693.

Gouzoules, H. and Goy, R. W. (1983). Physiological and social influences on mounting behavior of troop-living female monkeys (*Macaca fuscata*). *American Journal of Primatology*, **5**, 39–49.

Gouzoules, H., Fedigan, L. M. and Fedigan, L. (1975). Responses of a transplanted troop of Japanese macaques (*Macaca fuscata*) to bobcat (*Lynx rufus*) predation. *Primates*, **16**, 335–349.

Gouzoules, H., Gouzoules, S. and Fedigan, L. M. (1981). Japanese monkey group translocation: effects on seasonal breeding. *International Journal of Primatology*, **2**, 323–334.

Gouzoules, H., Gouzoules, S. and Fedigan, L. (1981). Dominance and reproduction among female Japanese monkeys (*Macaca fuscata*). *International Journal of Primatology*, **1**, 317. (Abstract)

Gouzoules, H., Gouzoules, S. and Fedigan, L. M. (1982). Behavioral dominance and reproductive success in female Japanese monkeys (*Macaca fuscata*). *Animal Behaviour*, **30**, 1138–1150.

Gouzoules, H., Fedigan, L. M., Gouzoules, S. and Fedigan, L. (1984). Comments on reproductive senescence among female Japanese macaques. *Journal of Mammalogy*, **65**, 341–342.

Gouzoules, S. (née Manly). (1981). Social relationships of adult female Japanese monkeys (*Macaca fuscata*). Ph.D. dissertation, Department of Anthropology, University of Chicago.

Green, S. (1975). Dialects in Japanese monkeys: Vocal learning and cultural transmission of locale specific vocal behavior. *Zeitschrift für Tierpsychologie*, **38**, 304–331.

(1975). Variations of vocal pattern with social situation in the Japanese monkey (*Macaca fuscata*): a field study. In *Primate Behavior, Vol. 4*, ed. L. A. Rosenblum. London: Academic Press, pp. 1–102.

(1977). Comparative aspects of vocal signals including speech. In *Recognition of Complex Acoustic Signals*, ed. T. H. Bullock. Berlin: Dahlem Konferezen, pp. 209–237.

Grewal, B. S. (1980). Social relationships between adult central males and kinship groups of Japanese monkeys at Arashiyama with some aspects of troop organization. *Primates*, **21**, 161–180.

(1980). Changes in relationships of nulliparous and parous females of Japanese monkeys at Arashiyama with some aspects of troop organization. *Primates*, **21**, 330–339.

(1981). Self wrist biting in Arashiyama B troop of Japanese monkeys (*Macaca fuscata fuscata*). *Primates*, **22**, 277–280.

(1984). Paternal care and partial breakage of matrilineal organization in Arashiyama-B troop of Japanese monkeys, *Macaca fuscata*. In *Current Primate Researches*, eds. M. L. Roonwal, S. M. Mohonot and N. S. Rathore. Jodhpur: Jodhpur University, pp. 295–300.

Griffin, L. (1990). The fission of the Arashiyama troop of Japanese monkeys living in south Texas, Spring of 1989. *Proceedings of the 13th Congress of the International Primatological Society*, **11**, 9. (Abstract)

Hanamura, S. (2005). Male socio-spatial distribution and male-female social relationships in a provisioned troop of Japanese macaques. *The Society for Ecological Anthropology News Letter*, **11**, 23–25. (Abstract, in Japanese)

(2005). Male spatial distribution in the society of Japanese macaques. *Reichorui Kenkyu/Primate Research*, **21**, suppl. S2–3. (Abstract, in Japanese)

(2008). Male socio-spatial distribution and male-female interactions resulting in male escape in a provisioned troop of Japanese macaques (*Macaca fuscata*) at

Arashiyama. Paper presented before the *22nd International Primatological Society Meetings. Primate Eye*, **96**, #243. (Abstract)

Hauser, M. and Tyrell, G. (1984). Old age and its behavioral manifestations: a study on two species of macaque. *Folia Primatologica*, **43**, 24–35.

Hayama, S. (1965). Morphological studies of *Macaca fuscata*. II. The sequence of epiphyseal union by roentgenographic estimation. *Primates*, **6**, 249–269.

Hayashi, Y. (1941). General overview of the plants of Arashiyama. *Yagai Hakubutsu*, **3**, 21–28. (in Japanese).

Hazama, N. (1957). *The Study of Monkeys*. Tokyo: Koseisha. (in Japanese)

(1962). On the weight measurement of wild Japanese monkeys at Arashiyama. *Iwatayama Monkey Park Bulletin*, **1**, 1–54. (in Japanese)

(1964). Weighing wild Japanese monkeys in Arashiyama. *Primates*, **3–4**, 81–104.

(1965). Wild Japanese monkeys at Mt. Hieizan. In *Monkey Sociological Studies*, eds. S. Kawamura and J. Itani. Tokyo: Chukoronsha, pp. 375–401.

(1972). *The Man who Became a Monkey*. Tokyo: Raicho-sha. (in Japanese)

Hirose, S. (1960). Biographies of leader males, 3 – Lincoln of the Arashiyama troop. *Yaen*, **6**, 20. (in Japanese)

(1984). At the beginning of provisioning. In *The Japanese Monkeys of Arashiyama. Iwatayama Shizenshi Kenkyūjo Hōkoku, Vol. 3*, eds. H. Suzuki and R. Asaba. Kyoto: Arashiyama Shizenshi Kenkyūjo, pp. 21–24. (in Japanese)

(1984). Arashiyama is a natural history science museum. In *The Japanese Monkeys of Arashiyama. Iwatayama Shizenshi Kenkyūjo Hōkoku, Vol. 3*, eds. H. Suzuki and R. Asaba. Kyoto: Arashiyama Shizenshi Kenkyūjo, pp. 80–81. (in Japanese)

Huffman, M. A. (1981). Preferential mating and partner selection in female Japanese macaques (*Macaca fuscata*) at Arashiyama. *American Journal of Primatology*, **1**, 316–317 (Abstract)

(1984). Stone-play of *Macaca fuscata* in Arashiyama B troop: transmission of a non-adaptive behavior. *Journal of Human Evolution*, **13**, 725–735.

(1984). Plant foods and foraging behavior of the Arashiyama Japanese macaques. In *Arashiyama Japanese Monkeys: Arashiyama Natural History Research Station Report, Volume 3*, ed. N. Asaba. Osaka: Osaka Seihan Printers, pp. 55–65. (in Japanese)

(1984). Mount-sequence intrusion and male-male competition in *Macaca fuscata*. *International Journal of Primatology*, **5**, 349 (Abstract).

(1985). Monkeys of Arashiyama. *Kaleidoscope Kyoto*, **11**, 10–14.

(1986). The misunderstood monkey: primates, potatoes and cultural evolution. *Whole Earth Review*, **52**, 26–29.

(1986). Mate choice and partner preference in female Japanese macaques. *Primate Report*, **14**, 110. (Abstract)

(1987). Consort intrusion and female mate choice in Japanese macaques (*Macaca fuscata*). *Ethology*, **75**, 221–234.

(1987). Mate choice and partner preference in female Japanese monkeys. *Reichorui Kenkyu/Primate Research*, **3**, 150. (Abstract).

(1990). Influences of female partner preference on male reproductive success in Japanese macaques. *American Journal of Primatology*, **20**, 199. (Abstract)

(1991). History of Arashiyama Japanese Macaques in Kyoto, Japan. In *The Monkeys of Arashiyama: Thirty-five Years of Research in Japan and the West*, eds. L. M. Fedigan and P. J. Asquith. Albany, NY: State University of New York Press, pp. 21–53.

(1991). Mate selection and partner preferences in female Japanese macaques. In *The Monkeys of Arashiyama: Thirty-five Years of Research in Japan and the West*, eds. L. M. Fedigan and P. J. Asquith. Albany, NY: State University of New York Press, pp. 101–122.

(1991). Consort relationship duration, conception, and social relationships in female Japanese macaques. In *Primatology Today*, eds. A. Ehara, T. Kimura, O. Takenaka and M. Iwamoto. Amsterdam: Elsevier Science Publishers B. V. (Biomedical Division), pp. 199–202.

(1991). Females have the right to choose their mates. In *Kimono Watching, A Guide to Animal Watching in Kyoto Prefecture*, ed. M. Kawamichi. Kyoto: Kyoto Shimbunsha, pp. 146–148.

(1991). A cultural play behavior innovated by young macaques. In *Kimono Watching, A Guide to Animal Watching in Kyoto Prefecture*, ed. M. Kawamichi. Kyoto: Kyoto Shimbunsha, pp. 148–149.

(1991). Stone handling in monkeys. In *The World of Animals: Live Strategies '10 – learning, remembering, applying'*, Weekly Encyclopedia. Tokyo: Asahi News, **10**, 1–317.

(1991). Stone handling: play culture innovated by the young. In *Primate Natural History*, eds. T. Nishida, K. Izawa and T. Kano. Tokyo: Heibon-sha, pp. 491–504.

(1992). Influences of female partner preferences on potential reproductive outcome in Japanese macaques. *Folia Primatologica*, **59**, 77–89.

(1996). Acquisition of innovative cultural behaviors in non-human primates: a case study of stone handling, a socially transmitted behavior in Japanese macaques. In *Social Learning in Animals: The Roots of Culture*, eds. B. G. Galef and C. Heyes. Orlando, FL: Academic Press, pp. 267–289.

Huffman, M. A. and Hirata, S. (2003). Biological and ecological foundations of primate behavioral tradition. In *The Biology of Tradition: Models and Evidence*, eds. D. M. Fragaszy and S. Perry. Cambridge: Cambridge University Press, pp. 267–296.

Huffman, M. A. and Quiatt, D. (1986). Stone handling by Japanese macaques (*Macaca fuscata*): implications for tool use of stones. *Primates*, **27**, 413–423.

Huffman, M. A., Nahallage, C. A. D. and Leca, J.-B. (2008). Cultured monkeys, social learning cast in stones. *Current Directions in Psychological Science*, **17**, 410–414.

Huffman, M. A., Leca, J.-B. and Nahallage, C. A. D. (2010). Cultured Japanese macaques – a multidisciplinary approach to stone handling behavior and its implications for the evolution of behavioral tradition in nonhuman primates. In *The Japanese Macaques*, eds. N. Nakagawa, M. Nakamichi and H. Sugiura. Tokyo: Springer, pp. 191–219.

Inoue, E. (2004). Which males sire many infants in a provisioned group of Japanese macaques? Paper presented before the *International Symposium supported by Kyoto University 21COE Biodiversity Program, African Great Apes: Evolution, Diversity and Conservation, Conference Program*, p. 18.

(2006). Which males do females choose? Paternity analyses of Japanese macaques using hair samples. In *Animals from the Windows of Genes*, eds. M. Murayama, K. Watanabe and A. Takenaka. Kyoto: Kyoto University Press, pp. 147–166. (in Japanese)

(2008). Paternity and female mate choice in Japanese macaques. Paper presented before the *20th Annual Meeting of the Human Behavior and Evolution Society, Conference Program*, p. 67.

Inoue, E. and Takenaka, O. (2003). Paternity analyses of Japanese macaques: which males attain high reproductive success? Paper presented before the *19th congress of Primate Society of Japan, Conference Program*, p. 37. (in Japanese)

(2004). Male mating success in a provisioned group of Japanese macaque (*Macaca fuscata*). *Reichorui Kenkyu/Primate Research*, **20** (Suppl.), 34–35. (in Japanese)

(2008). The effect of male tenure and female mate choice on paternity in free-ranging Japanese macaques. *American Journal of Primatology*, **70**, 62–68.

Inoue, M. (1988). Soil-eating of Japanese macaques (*Macaca fuscata*) at Arashiyama, Kyoto. *Reichorui Kenkyu/Primate Research*, **3**, 103–111. (in Japanese)

(1995). Application of paternity discrimination by DNA polymorphism to the analysis of the social behavior of primates. *Human Evolution*, **10**, 53–62.

Inoue, M., Mitsunaga, F., Ohasawa, H. *et al.* (1992). Paternity testing in captive Japanese macaques (*Macaca fuscata*) using DNA fingerprinting. In *Paternity in Primates: Genetic Tests and Theories*, eds. R. D. Martin, A. F. Dixson and E. J. Wickings. Basel: S. Karger, pp. 131–140.

Ishimoto, G., Toyomasu, T. and Uemura, K. (1965). Electrophoretic study of serum proteins of Japanese macaques. *Primates*, **8**, 271–282.

Iwamoto, M. (1967). Morphological studies of *Macaca fuscata*: VI Somatometry. *Primates*, **12**, 151–174.

Iwasaki, N., Sprague, D. and Takenoshita, Y. (2004). Performance evaluation of GPS telemetry for a middle-small mammal. *Theory and Applications of GIS*, **12**, 107–113. (in Japanese with English abstract)

Iwata, N. (1963). Biography of leader males, 11 – the carefree Yajisan (Y). *Yaen*, **14**, 23. (in Japanese)

Jack, K. (1995). Mating strategies of peripheral male Japanese macaques. M.A. thesis, University of Calgary.

(1997). Mating success among male Japanese macaques at Arashiyama West. Paper presented at the *20th Annual Meeting of the American Society of Primatologists. American Journal of Primatology*, **42**, 118. (Abstract)

Jack, K. M. and Pavelka, M. S. M. (1997). The behavior of peripheral males during the mating season in *Macaca fuscata*. *Primates*, **38**, 369–377.

Johnston, T. (1975). Adaptations of communication patterns in Japanese monkeys to a brushland environment in southern Texas. M.Sc. Thesis, Department of Zoology, University of Wisconsin, Madison.

(1976). Theoretical considerations in the adaptations of animal communication systems. *Journal of Theoretical Biology*, **57**, 43–72.

Joiner, G. N., Russell, L. H., Bush, D. E. *et al.* (1975). A spontaneous neuropathy of free-ranging Japanese macaques. *Laboratory Animal Science*, **25**, 232–237.

Kamada, J. (1988). A cross-sectional study on development of visual behaviors in free-ranging Japanese macaques. In *Research Reports of the Arashiyama West and East Groups of Japanese Monkeys*. Osaka: Laboratory of Ethological Studies, Faculty of Human Sciences, Osaka University, pp. 75–86.

Kawai, M., Azuma, S. and Yoshiba, K. (1967). Ecological studies of reproduction in Japanese monkeys (*Macaca fuscata*). I. Problems of the birth season. *Primates*, **8**, 35–74.

Kawamura, S. (1959). The process of sub-cultural propagation among Japanese macaques. *Primates*, **2**, 43–60.

Kitahara-Frisch, J. (1991). Culture and primatology: East and West. In *The Monkeys of Arashiyama: Thirty-five Years of Research in Japan and the West*, eds. L. M. Fedigan and P. J. Asquith. Albany, NY: State University of New York Press, pp. 74–80.

Kitajima, M., Ohkura, Y., Shotake, T. and Nozawa, K. (1975). Genetic polymorphisms of blood proteins in the troops of Japanese macaques, *Macaca fuscata*: erythrocyte esterase polymorphism in *Macaca fuscata*. *Primates*, **16**, 399–404.

Kondo-Ikemura, K. (1988). The daily activity rhythm of a Japanese monkey troop at Arashiyama West, Texas, in summer season. In *Research Reports of the Arashiyama West and East Groups of Japanese Monkeys*. Osaka: Laboratory of Ethological Studies, Faculty of Human Sciences, Osaka University, pp. 19–28.

Kotera, S. and Suzuki, K. (1961–1962). On supply of Japanese monkeys for research at Japan Monkey Centre. *Primates*, **3**, 47–58.

Koyama, N. (1966). Rank and play of a wild Japanese monkey troop in Arashiyama. M.A. thesis, Osaka City University. (in Japanese)

(1967). On dominance rank and kinship of a wild troop in Arashiyama. *Primates*, **8**, 189–216.

(1970). Changes in dominance rank and division of wild Japanese monkey troop in Arashiyama. *Primates*, **11**, 335–390.

(1974). Division of Arashiyama troop – fissioning mechanism and blood relationships. *Anima*, **11**, 31–36. (in Japanese)

(1977). Social structure of Japanese monkeys. In *Jinruigakukōza. 2. Reichōrui*, ed. J. Itani. Tokyo: Yukazankaku Press, pp. 255–276. (in Japanese)

(1979). Grooming strategy of a female Japanese monkey during the mating season. In *Sociobiological Studies on the Role of Sex in the Japanese Monkey Society*, eds. M. Kawai and S. Azuma. Kyoto: Kyoto University Primate Research Institute, pp. 21–26.

(1980). Demography of Japanese monkeys. In *Following Japanese Wildlife*, ed. N. Asahi. Tokyo: Tokai University Press, pp. 8–34.

(1980). Saru shakai ni okeru ningenmi no aru kannkei. *Hyojyunka to Hinshitsukanri*, **33**, 39–44.

(1981). Mating strategies and spatial structure of the Japanese monkey troop. In *Studies of the Society of Japanese Monkeys Based on Ethology*, ed. S. Kawamura. Kyoto: Kyoto University Primate Research Institute, pp. 19–40.

(1984). Population dynamics of the Arashiyama troop of Japanese monkeys. In *The Japanese Monkeys of Arashiyama. Iwatayama Shizenshi Kenkyūjo Hōkoku. Vol. 3*, eds. N. Asaba and H. Suzuki. Kyoto: Arashiyama Shizenshi Kenkyūjo, pp. 30–38.

(1985). Playmate relationships among individuals of the Japanese monkey troop in Arashiyama. *Primates*, **26**, 390–406.

(1986). Analysis of the birth data in the Japanese monkey troop at Arashiyama. *Reichorui Kenkyu/Primate Research*, **2**, 173. (Abstract)

(1991). Grooming relationships in the Arashiyama group of Japanese monkeys. In *The Monkeys of Arashiyama: Thirty-five years of Research in Japan and the West*, eds. L. M. Fedigan and P. J. Asquith. Albany, NY: State University of New York Press, pp. 211–226.

Koyama, N. and Nishida, K. (1977). The present status and protection of Japanese monkeys at Arashiyama, Kyoto City. *Nihonzaru*, **3**, 94–99. (in Japanese)

Koyama, N. and Norikoshi, K. (1984). The Arashiyama-West Japanese monkeys. In *The Japanese Monkeys of Arashiyama. Iwatayama Shizenshi Kenkyūjo Hōkoku. Vol. 3*, eds. N. Asaba and H. Suzuki. Kyoto: Arashiyama Shizenshi Kenkyūjo, pp. 66–70.

Koyama, N., Norikoshi, K. and Mano, T. (1975). Population dynamics of Japanese monkeys at Arashiyama. In *Contemporary Primatology: Proceedings of the 5th International Congress of Primatology*, eds. S. Kondo, M. Kawai and A. Ehara. Basel: S. Karger, pp. 411–417.

Koyama, N., Norikoshi, K., Mano, T. and Takahata, Y. (1980). Population changes of Japanese monkeys at Arashiyama. In *Demographic Study on the Society of Wild Japanese Monkeys*, ed. Y. Sugiyama. Kyoto: Kyoto University Primate Research Institute, pp. 19–33. (in Japanese)

Koyama, N., Takahata, Y., Huffman, M. A., Norikoshi, K. and Suzuki, H. (1992). Reproductive parameters of female Japanese macaques: thirty years' data from the Arashiyama troops, Japan. *Primates*, **33**, 33–47.

Koyama, T. (1988). A comparative study on gregariousness and level of tension in the Arashiyama West (Texas) and East (Kyoto) troops. In *Research reports of the Arashiyama West and East groups of Japanese monkeys*. Osaka: Laboratory of Ethological Studies, Faculty of Human Science, Osaka University, pp. 29–40.

Koyama, T., Fujii, H. and Yonekawa, F. (1981). Comparative studies of gregariousness and social structure among seven feral groups of Japanese monkeys at Arashiyama (A troop). *Jinruigaku Zasshi*, **80**, 381–383. (in Japanese)

Kubota, K. (1989). An examination of hand preference in Arashiyama-R troop of Japanese monkeys. *Neuroscience Research*, **9**, S29. (Abstract)

(1990). Preferred hand use in the Japanese macaque troop, Arashiyama-R, during visually guided reaching for food pellets. *Primates*, **31**, 393–406.

(1991). Preferred hand use of the Japanese macaques during visually guided reachings. In *Primatology Today*, eds. A. Ehara, T. Kimura, O. Takenaka and

M. Iwamoto. Amsterdam: Elsevier Science Publishers B. V. (Biomedical Division), pp. 269–270.

Kurita, H. (2010). Stochastic variation in sex ratio in infant mortality rates due to small samples in provisioned Japanese macaques (*Macaca fuscata*) populations. *Primates*, **51**, 75–78.

Lair, S., Chapais, B., Higgins, R., Mirkovic, R. and Martineau, D. (1996). Myeloencephalitis associated with a *viridans* group *Steptococcus* in a colony of Japanese macaques (*Macaca fuscata*). *Veterinary Pathology*, **33**, 99–103.

Lance, M. (1985). Monkey business. *Southwest*, **14**, 53–57.

Lanigan, C. M. S. (1991). Short guide to DNA techniques for genealogy construction, with special attention to Japanese macaques. In *The Monkeys of Arashiyama: Thirty-five Years of Research in Japan and the West*, eds. L. M. Fedigan and P. J. Asquith. Albany, NY: State University of New York Press, pp. 291–318.

Larose, F. (1988). Analyse expérimentale de la dominance généalogique entre males juveniles chez *Macaca fuscata*. M.A. thesis, Department of Anthropology, University of Montreal.

Leca, J.-B. (2010). The Japanese macaques of Arashiyama: demographic studies, behavioral research and management efforts. *Congress of the International Primatological Society*, Kyoto, Japan. *Reichorui Kenkyu/Primate Research*, **26** (suppl.), 67. (Abstract)

Leca, J.-B. and Huffman, M. A. (2008). Comparative and longitudinal approaches in cultural primatology: the case of stone handling behavior in Japanese macaques. *Annual Meeting of Primate Society of Japan*, Tokyo, Japan. *Reichorui Kenkyu/Primate Research*, **24** (suppl.), 11. (Abstract)

Leca, J.-B., Gunst, N., Nahallage, C. A. D. and Huffman, M. A. (2004). Stone handling as a behavioral tradition: a comparative study in nine troops of Japanese macaques (*Macaca fuscata*). *Reichorui Kenkyu/Primate Research*, **20** (suppl.), 25. (Abstract)

(2006). Stone handling in Japanese macaques (*Macaca fuscata*): a 10-troop comparative study about a behavioral tradition. *American Journal of Primatology*, **68** (Suppl. 1), 110–111. (Abstract)

Leca, J.-B., Gunst, N. and Huffman, M. A. (2007). Japanese macaque cultures: inter- and intra-troop behavioural variability of stone handling patterns across 10 troops. *Behaviour*, **144**, 251–281.

(2007). Age-related differences in the performance, diffusion, and maintenance of stone handling, a behavioral tradition in Japanese macaques. *Journal of Human Evolution*, **53**, 691–708.

Leca, J.-B., Gunst, N., Nahallage, C. A. D. and Huffman, M. A. (2007). Stone handling as a behavioral tradition in Japanese macaques (*Macaca fuscata*): charting inter- and intra-group diversity and investigating ecological and socio-demographic contexts of transmission. *American Journal of Primatology*, **69** (Suppl. 1), 77. (Abstract)

Leca, J.-B., Gunst, N. and Huffman, M. A. (2008). Food provisioning and stone handling tradition in Japanese macaques: a comparative study of ten troops. *American Journal of Primatology*, **70**, 803–813.

(2008). Of stones and monkeys: testing ecological constraints on stone handling, a behavioral tradition in Japanese macaques. *American Journal of Physical Anthropology*, **135**, 233–244.

Leca, J.-B., Gunst, N., Nahallage, C. A. D. and Huffman, M. A. 2008. Stone handling in Japanese macaques: an exaptive traditional behavior? *Congress of the International Primatological Society*, Edinburgh, Scotland. *Primate Eye*, **96**, #689. (Abstract)

Leca, J.-B., Gunst, N. and Huffman, M. A. (2009). Determinants of behavioral innovations and constraints on their diffusion and maintenance in Japanese macaques. *Annual Meeting of Primate Society of Japan*, Gifu, Japan. *Reichorui Kenkyu/ Primate Research*, **25** (suppl.), 14–15. (Abstract)

(2010). Principles and levels of laterality in unimanual and bimanual stone handling patterns by Japanese macaques. *Journal of Human Evolution*, **58**, 155–165.

(2010). Indirect social influence in the maintenance of the stone handling tradition in Japanese macaques (*Macaca fuscata*). *Animal Behaviour*, **79**, 117–126.

(2010). The first case of dental flossing by a Japanese macaque (*Macaca fuscata*): implications for the determinants of behavioral innovation and the constraints on social transmission. *Primates*, **51**, 13–22.

(2010). Thirty years of stone handling tradition in Japanese macaques: implications for cumulative culture and stone-tool use evolution. *Congress of the International Primatological Society*, Kyoto, Japan. *Reichorui Kenkyu/Primate Research*, **26** (suppl.), 71. (Abstract)

(2011). Complexity in object manipulation by Japanese macaques (*Macaca fuscata*): a cross-sectional analysis of manual coordination in stone handling patterns. *Journal of Comparative Psychology*, **125**, 61–71.

Lester, B. (1996). Arashiyama West Institute/Texas snow monkey sanctuary. *OWM Tag Newsletter*, **3**, 12.

Lucotte, G., Fedigan, L. M. and Fedigan, L. (1984). Effets d'echantillonage sur la variation electrophoretique chez *Macaca fuscata*. *Biochemical Systematics and Ecology*, **12**, 349–351.

Mahaney, W. C., Hancock, R. G. V. and Inoue, M. (1993). Geochemistry and clay mineralogy of soils eaten by Japanese macaques. *Primates*, **34**, 85–91.

Malinov, M. R. and de Larmoy, C. W. (1967). The electrocardiogram of *Macaca fuscata*. *Folia Primatologica*, **7**, 284–291.

Manly, S. (1977). A study of the relationship between spatial proximity and genetic relatedness in female Japanese monkeys (*Macaca fuscata*). M.A. thesis, Department of Anthropology, University of Chicago.

Mano, T. (1972). Inter-individual spatial structure and inter-troop relations of Arashiyama A and B troop. *Kyoto University Primate Research Institute*, **2**, 27–30. (in Japanese)

Mano, T., Norikoshi, K. and Koyama, N. (1972). Report on tattoo numbers, individual names, age and body weight of Japanese monkeys at Arashiyama (A troop). *Journal of Anthropological Society of Nippon*, **80**, 381–383. (in Japanese)

Masataka, N. (1983). Psycholinguistic analysis of alarm calls of Japanese monkeys (*Macaca fuscata fuscata*). *American Journal of Primatology*, **5**, 111–125.

(1988). Temporal and sequalae analysis of food calling behavior in Japanese macaques. In *Research reports of the Arashiyama West and East groups of Japanese monkeys*. Osaka: Laboratory of Ethological Studies, Faculty of Human Science, Osaka University, pp. 51–55.

McDonald, M. S. (1983). The courtship behavior of female Japanese monkeys. M.A. thesis, Department of Anthropology, McMaster University, Hamilton.

Mehlman, P. and Chapais, B. (1988). Differential effects of kinship, oestrus and dominance on female allogrooming in a captive group of *Macaca fuscata*. *Primates*, **29**, 195–217.

Miyadi, D. (1966). *The Story of Monkeys*. Tokyo: Iwanami-Shoten. (in Japanese)

Murata, G. and Hazama, N. (1968). Flora of Arashiyama, Kyoto, and plant food of Japanese monkeys. In *Iwatayama Shizenshi Kenkyūjo Chosa Kenkyu Hōkoku*, **2**, 1–59. (in Japanese)

Nakajima, K. (1958). A natural troop of Japanese monkeys at Arashiyama. *Yaen*, **1**, 18–22. (in Japanese)

(1984). The origin of names given to monkeys of the Arashiyama troop. In *The Japanese Monkeys of Arashiyama. Iwatayama Shizenshi Kenkyujo Houkoku, Volume 3*, eds. H. Suzuki and N. Asaba. Kyoto: Arashiyama Shizenshi Kenkyujo, pp. 25–29. (in Japanese)

Nakamichi, M. (1984). Behavioral characteristics of old female Japanese monkeys in a free-ranging group. *Primates*, **25**, 192–203.

(1984). Aging and behavior in females of the Arashiyama A group (Texas) and the Arashiyama B group (Kyoto). In *The Japanese Monkeys of Arashiyama. Iwatayama Shizenshi Kenkyujo Houkoku, Vol. 3*, eds. H. Suzuki and N. Asaba. Kyoto: Arashiyama Shizenshi Kenkyujo, pp. 71–79. (in Japanese)

(1988). Aging and behavioral changes of female Japanese macaques. In *Research reports of the Arashiyama West and East groups of Japanese monkeys*. Osaka: Laboratory of Ethological Studies, Faculty of Human Science, Osaka University, pp. 87–97.

(1991). Behavior of old females: Comparisons of Japanese monkeys in the Arashiyama East and West groups. In *The Monkeys of Arashiyama: Thirty-five Years of Research in Japan and the West*, eds. L. M. Fedigan and P. J. Asquith. Albany, NY: State University of New York Press, pp. 175–193.

Nass, G. G. (1976). Eruption and attrition in the dentition of a troop of *Macaca fuscata fuscata*. M. Sc. Thesis, Department of Zoology, University of Wisconsin, Madison.

(1977). Inter-group variation in the dental eruption sequence in *Macaca fuscata fuscata*. *Folia Primatologica*, **28**, 306–314.

(1981). Sex differences in tooth wear of *Macaca fuscata*, the Arashiyama-A troop in Texas. *Primates*, **22**, 266–276.

Negayama, K. (1988). Early mother-infant relationships of the Arashiyama A and B troops in summer. In *Research Reports of the Arashiyama West and East Groups of Japanese Monkeys*. Osaka: Laboratory of Ethological Studies, Faculty of Human Science, Osaka University, pp. 67–74.

Negayama, K., Ando, A., Hara, A. *et al.* (1988). Basic behavioral profiles among adult of the Arashiyama A troop and the Arashiyama B troop. In *Research Reports of the Arashiyama West and East Groups of Japanese Monkeys.* Osaka: Laboratory of Ethological Studies, Faculty of Human Science, Osaka University, pp. 1–17.

Nigi, H., Tanaka, T. and Noguchi, Y. (1967). Hematological analyses of the Japanese monkey (*Macaca fuscata*). *Primates,* **8**, 107–120.

Nishida, K. (1973). An analysis of spatial relations and social structure in a troop of free-ranging Japanese monkeys at Arashiyama. M.A. thesis, Osaka City University. (in Japanese)

Nishida, T. (1966). A sociological study of solitary male monkeys. *Primates,* **7**, 141–204.

Nishie, H. (2001). Stone handling of Japanese macaques in Arashiyama. *Reichorui Kenkyu/Primate Research,* **17**, 178. (Abstract)

(2002). The proximate factors of stone handling behaviour of Japanese macaques (*Macaca fuscata*) in Arashiyama E troop. *Reichorui Kenkyu/Primate Research,* **18**, 225–232. (in Japanese with English summary)

Noguchi, Y., Tawara, I., Kondo, K., Nigi, H. and Tanaka, T. (1969). Electrocardiographic studies in the Japanese monkey (*Macaca fuscata*) with special reference to the effect of anaesthesia with barbiturates. *Primates,* **10**, 273–283.

Norikoshi, K. (1971). Tests to determine the responsiveness of the free-ranging Japanese monkeys in food getting situations. *Primates,* **12**, 113–124.

(1972). Comparison of particular behaviors between troops observed under the peanut getting situation. *Kyoto University Primate Research Institute,* **2**, 39–41. (in Japanese)

(1973). Sexual relationships in the Arashiyama B troop. *Kyoto University Primate Research Institute,* **3**, 40–41. (in Japanese)

(1973). Lifetime influences of mother-son relationships on males. *Kyoto University Primate Research Institute,* **3**, 63–67. (in Japanese)

(1974). The development of peer-mate relationships in Japanese macaque infants. *Primates,* **15**, 39–46.

(1974). Changes in troop membership of Japanese monkey males and the social structure of wild monkey troops in Arashiyama. In *Life-History of Male Japanese Monkeys: Advances in Field Studies of Japanese Monkeys,* eds. I. K. Wada, S. Azuma and Y. Sugiyama. Inuyama: Kyoto University Primate Research Institute, pp. 35–40. (in Japanese)

(1974). Mother-son relationships in Japanese monkeys. *Kikan Jinruigaku,* **5**, 39–62. (in Japanese)

(1975). Solitary monkeys. *Shizen,* **30**, 35–45. (in Japanese)

(1975). Social structure of Japanese monkeys. *Rinshō Kagaku,* **11**, 239–244. (in Japanese)

(1976). Japanese monkeys. In *Hominizeishon Kenkyuukai,* ed. Scientific American: Special volume on animal sociology. Tokyo: Nihon Keizai Shinbunnsha, pp. 52–61. (in Japanese)

(1977). Group transfer and social structure among male Japanese monkeys at Arashiyama. In *Keishiysu, Shinka, Reichōrui*, eds. T. Kato, S. Nakao and T. Umesao. Tokyo: Chukoron-Sha, pp. 335–370. (in Japanese)

Norikoshi, K. and Koyama, N. (1975). Group shifting and social organization among Japanese monkeys. In *Symposium of the 5th Congress of the International Primatological Society*, eds. S. Kondo, M. Kawai, A. Ehara and S. Kawamura. Tokyo: Japan Science Press, pp. 43–61.

Noyes, N. J. S. (1981). Pregnancy, foetal sex, and aggression in Japanese monkeys. *American Journal of Primatology*, **1**, 317. (Abstract)

Nozaki, J., Mitsunaga, F. and Shimizu, K. (1995). Reproductive senescence in female Japanese monkeys (*Macaca fuscata*): age and season-related changes in hypothalamic-pituitary-ovarian functions and fecundity rates. *Biology of Reproduction*, **52**, 1250–1257.

Nozawa, K. (1972). Population genetics of Japanese monkeys: I. Estimation of the effective troop size. *Primates*, **15**, 39–46.

Nozawa, K., Shotake, T., Ohkura, Y., Kitajima, M. and Tanabe, Y. (1975). Genetic variations within and between troops of *Macaca fuscata fuscata*. In *Contemporary Primatology: Proceedings of the 5th International Congress of Primatology*, eds. S. Kondo, M. Kawai and A. Ehara. Basel: S. Karger, pp. 75–89.

Ogawa, H. (1996). Social interactions among a mother, an adopted infant, and a real infant of Japanese macaques in the Arashiyama F troop. *Reichorui Kenkyu/Primate Research*, **12**, 1–10. (in Japanese with English abstract)

Ogihara, N., Usui, H., Hirasaki, E., Hamada, Y. and Nakatsukasa, M. (2005). Kinematic analysis of bipedal locomotion of a Japanese macaque that lost its forearms due to congenital malformation. *Primates*, **46**, 11–19.

Oguri, H. (2000). Ten-years-later changes of hand preference in reaching for food pellets in the Arashiyama group of Japanese macaques. *Reichorui Kenkyu/Primate Research*, **16**, 270. (Abstract, in Japanese)

Ohta, E. (1974). The activities of Arashiyama Shizen Kenkyukai. *Arashiyama Shizenkenkyūkai Kaihō (Reports of the Arashiyama Nature Research Society)*, **2**, 16–20. (in Japanese)

(1975). Maboroshi no saru wo motomete. *Arashiyama Shizenkenkyūkai Kaihō (Reports of the Arashiyama Nature Research Society)*, **3**, 13–17.

(1976). In Search of the Phantom Monkeys. Part II. *Arashiyama Shizenkenkyūkai Kaihō (Reports of the Arashiyama Nature Research Society)*, **4**, 3–8. (in Japanese)

O'Neill, A. C. (2000). The relationship between ovarian hormones and the behavior of Japanese macaque females (*Macaca fuscata*) during the mating season. Master's thesis, University of Alberta. *Master's Abstracts*, 2001, **39**(6): 1494.

O'Neill, A. C., Fedigan, L. M. and Ziegler, T. E. (2004). Ovarian cycle phase and same sex mating behaviour in Japanese macaque females. *American Journal of Primatology*, **63**, 25–31.

(2004). Relationship between ovarian cycle phase and sexual behavior in female Japanese macaques (*Macaca fuscata*). *American Journal of Physical Anthropology*, **125**, 352–362.

(2006). Hormonal correlates of post-conceptive mating in female Japanese macaques. *Laboratory Primate Newsletter*, **45**, 1–4.

Paterson, J. D. (1993). Behavioural thermoregulation and adaptation in the Arashiyama 'A' troop. *Proceedings of 14th Congress of the International Primatological Society*, **14**, 276. (Abstract)

(1992). Behavioural thermoregulation as a component in adaptation to a new climatic regime: evidence from the Arashiyama 'A' troop. *American Journal of Primatology*, **27**, 50.

(1994). Behavioral thermoregulation and structural adaptation in the Arashiyama 'A' troop. In *Current Primatology, Vol. 3, Behavioural Neuroscience, Physiology, and Reproduction*, eds. J. R. Anderson, J.-J. Roeder, B. Thierry and N. Herrenschmidt. Strasbourg: University Louis Pasteur, pp. 227–236.

(1994). Postural and positional thermoregulatory behavior in the Arashiyama A troop. *Proceedings of the 15th Congress of the International Primatological Society*, **15**, 87.

(1996). Coming to America: acclimation in macaque body structures and Bergmann's rule. *International Journal of Primatology*, **17**, 585–611.

Pavelka (McDonald), M. S. (1985). Courtship behavior of female Japanese monkeys. *Canadian Review of Physical Anthropology*, **4**, 67–75.

Pavelka, M. S. M. (1988). Aging in female Japanese monkeys: Primatological contributions to gerontology. Ph.D. dissertation, Department of Anthropology, University of Alberta, Edmonton.

(1990). Do old females monkeys have a specific social role? *Primates*, **31**, 363–373.

(1991). Sociability in old female Japanese monkeys: human vs. non-human primate aging. *American Anthropologist*, **93**, 588–598.

(1993). *Monkeys of the Mesquite: The Social Life of the South Texas Snow Monkeys.* Dubuque, IA: Kendall/Hunt Publishing Co.

(1994). The nonhuman primate perspective: old age, kinship and social partners in a monkey society. *Journal of Cross-cultural Gerontology*, **9**, 219–229. (Reprinted in 1998 in *Women Among Women: Anthropological Perspectives on Female Age Hierarchies*, eds. J. Dickerson-Putnam and J. K. Brown. Chicago, IL: University of Illinois Press, pp 89–99.)

(1996). The social life of female Japanese macaques. In *The Evolving Female: A Life History Perspective*, eds. M. E. Morbeck, A. Galloway and A. Zihlman. Princeton, NJ: Princeton University Press, pp. 76–85.

(1999). Primate gerontology. In *The Nonhuman Primates*, eds. N. A. Fuentes and P. Dolhinow. Mountain View, CA: Mayfield Publishing Company, pp. 220–225.

(1999). Is there adaptive value to reproductive termination in Japanese macaques? A test of the grandmother hypothesis. *American Association for Physical Anthropology*, Columbus, OH. *American Journal of Physical Anthropology, Annual Meeting Issue.* Suppl. **28**, 126. (Abstract)

(2001). The availability and adaptive value of reproductive and post-reproductive mothers and grandmothers in Japanese macaques. *American Society of Primatology*, Savannah, GA. *American Journal of Primatology*, **54**, 26. (Abstract)

Pavelka, M. S. M. and Fedigan, L. M. (1991). Menopause: a comparative life history perspective. *Yearbook of Physical Anthropology*, **34**, 13–34.

(1999). Reproductive termination in female Japanese monkeys: a comparative life history perspective. *American Journal of Physical Anthropology*, **109**, 455–464.

Pavelka, M. S. M., Gillespie, M. W. and Griffin, L. (1991). Interacting effect of age and rank on the sociability of adult female Japanese monkeys. In *The Monkeys of Arashiyama: Thirty-five Years of Research in Japan and the West*, eds. L. M. Fedigan and P. J. Asquith. Albany, NY: State University of New York Press, pp. 194–207.

Pavelka, M. S. M., Fedigan, L. M. and Zohar, S. (2002). Availability and adaptive value of reproductive and post-reproductive Japanese macaque mothers and grandmothers. *Animal Behaviour*, **67**, 407–414.

Platt, M. M. and Thompson, R. L. (1985). Mirror responses in a Japanese macaque troop (Arashiyama West). *Primates*, **26**, 300–314.

Platt, M. M., Thompson, R. L. and Boatright, S. L. (1991). Monkeys and mirrors: questions of methodology. In *The Monkeys of Arashiyama: Thirty-five Years of Research in Japan and the West*, eds. L. M. Fedigan and P. J. Asquith. Albany, NY: State University of New York Press, pp. 274–290.

Quiatt, D. and Huffman, M. A. (1993). On home bases, nesting sites, activity centers, and new analytic perspectives. *Current Anthropology*, **34**, 68–70.

Quick, L. B. (1977). The social environment of behavior of Japanese macaques. *American Journal of Physical Anthropology*, **47**, 156. (Abstract)

Rasmussen, D. R. (1992). '*The Monkeys of Arashiyama: Thirty-five Years of Research in Japan and the West*, eds. L. Fedigan and P. Asquith. Albany, NY: State University of New York.' [Book review]. *Primates*, **33**, 148–150.

Reinhart, C. J., Pellis, V. C., Thierry, B. *et al.* (2010). Targets and tactics of play fighting: competitive versus cooperative styles of play in Japanese and Tonkean macaques. *International Journal of Comparative Psychology*, **23**, 166–200.

Russon, A. E., Vasey, P. L. and Gauthier, C. (2002). Seeing with the mind's eye: Playing blind in orangutans and Japanese macaques. In *Pretending and Imagination in Animals and Humans*, ed. R. W. Mitchell. Cambridge: Cambridge University Press, pp. 241–254.

Sade, D. S. (1993). '*The Monkeys of Arashiyama: Thirty-five Years of Research in Japan and the West*, eds. L. Fedigan and P. Asquith. Albany, NY: State University of New York.' [Book review]. *American Anthropologist*, **95**, 167–168.

Shimada, M. (2006). Study on social object play among young Japanese macaques. Ph.D. thesis, Graduate School of Kyoto University, Department of Science. (in Japanese)

(2006). Social object play among young Japanese macaques (*Macaca fuscata*) in Arashiyama, Japan. *Primates*, **47**, 342–349.

(2009). An ethnography of play of Japanese macaques: juveniles of Kinkazan, Arashiyama, Koshima, and Shigakogen. In *Anthropology of Children*, ed. N. Kamei. Kyoto: Showado, pp. 81–133. (in Japanese)

(2010). Social object play among juvenile Japanese macaques. In *The Japanese Macaques*, eds. N. Nakagawa, M. Nakamichi and H. Sugiura. Tokyo: Springer, pp. 375–385.

(2010). Comparing play patterns in juvenile Japanese macaques at Arashiyama and other field sites in Japan. The 23rd Congress of the International Primatological Society, Kyoto, Japan. *Reichorui Kenkyu/Primate Research*, **26** (suppl. 1), 71. (Abstract)

Shimizu, K., Takenoshita, Y., Mitsunaga, F., and Nozaki, M. (1996). Suppression of ovarian function and successful contraception in macaque monkeys following a single injection of medroxyprogesterone acetate. *Journal of Reproduction and Development*, **42**, 147–155.

Shotake, T. (1974). Genetic polymorphisms of blood proteins in the troops of Japanese macaques (*Macaca fuscata*). II. Erythrocyte lactate dehydrogenase polymorphism in *Macaca fuscata*. *Primates*, **15**, 297–393.

Shotake, T. and Nozawa, K. (1974). Genetic polymorphisms of blood proteins in the troops of Japanese macaques (*Macaca fuscata*). I. Cytoplasmic malate dehydrogenase polymorphism in *Macaca fuscata* and other nonhuman primates. *Primates*, **15**, 219–226.

Shotake, T. and Ohkura, Y. (1975). Genetic polymorphisms of blood proteins in the troops of Japanese macaques (*Macaca fuscata*). III. Erythrocyte carbonic anhydrase polymorphism in *Macaca fuscata*. *Primates*, **16**, 17–22.

Shotake, T., Ohkura, Y. and Ishimoto, G. (1977). Genetic polymorphisms of the blood proteins in the troops of Japanese macaques (*Macaca fuscata*). V. Erythrocyte phosphoexose isomerase polymorphism. *Primates*, **18**, 285–290.

Shotake, T., Ohkura, Y. and Nozawa, K. (1977). Genetic polymorphisms of the blood proteins in the troops of Japanese macaques (*Macaca fuscata*). VI. Serum transferrrine polymorphism. *Primates*, **18**, 291–297.

Sicotte, P. (1987). Effet de la composition du groupe sur la cohésion du lignage alpha chez *Macaca fuscata* [Effect of group composition on the cohesion of the alpha lineage in *Macaca fuscata*] Mémoire de maîtrise, Université de Montréal. (in French)

Smith, P. A. (1978). The social nexus of Japanese monkeys: Behavior of adult males during the mating season. M.A. thesis, Department of Anthropology, University of Chicago, Chicago.

Sprague, D. S. (1992). Life history and male inter-troop mobility among Japanese macaques (*Macaca fuscata*). *International Journal of Primatology*, **13**, 437–451.

(1992). 'The Monkeys of Arashiyama: Thirty-five Years of Research in Japan and the West, eds. L. Fedigan and P. Asquith. Albany: State University of New York, 353 pp.' [Book review]. *American Journal of Physical Anthropology*, **87**, 239–241.

(1998). Age, dominance rank, natal status, and tenure among male macaques. *American Journal of Physical Anthropology*, **105**, 511–521.

Stephenson, G. R. (1973). Biology of communication and population structure. In *Behavioural Regulators of Behaviour in Primates*, ed. C. R. Carpenter. Lewisburg, PA: Bucknell University Press, pp. 34–55.

(1973). Testing for group-specific communication patterns in Japanese macaques. *Dissertation Abstract International*, **B34**, 4759–4760.

(1973). Testing for group-specific communication patterns in Japanese macaques. In *Precultural Primate Behavior*, ed. E. W. Menzel. Basel: S. Karger, pp. 51–75.

(1975). Social structure of mating activity in Japanese macaques. In *Contemporary Primatology, Symposium of the 5th Congress of the International Primatological Society*, eds. S. Kondo, M. Kawai, A. Ehara and S. Kawamura. Tokyo: Japan Science Press, pp. 65–115.

Suzuki, H. and Asaba, R. (1984). *The Japanese Monkeys of Arashiyama. Iwatayama Shizenshi Kenkyujo Houkoku, Vol. 3*. Kyoto: Arashiyama Shizenshi Kenkyujo. (in Japanese)

Suzuki, H., Mastuura, N. and Kawamishi, T. (1984). Effects of mother kinship and parity on behavioural development of the Japanese macaque. *Journal of Mammalogy*, **44**, 19–30. (in Japanese)

Suzuki, H., Asaba, N., Huffman, M. A. and Kawamishi, T. (1987). On the division of Arashiyama A troop of Japanese monkeys. *Reichorui Kenkyu/Primate Research*, **3**, 77. (Abstract, in Japanese)

Suzuki, H., Asaba, N., Huffman, M. A. and Koyama, N. (1987). On the division of Arashiyama B troop of Japanese monkeys. *Reichorui Kenkyu/Primate Research*, **3**, 177. (Abstract, in Japanese)

Takahata, Y. (1980). The reproductive biology of a free ranging troop of Japanese monkeys. *Primates*, **21**, 303–329.

(1982). The socio-sexual behavior of Japanese monkeys. *Ethology*, **59**, 89–108.

(1982). Social relations between adult males and females of Japanese monkeys in the Arashiyama B troop. *Primates*, **23**, 1–23.

(1982). The socio-sexual behavior of Japanese monkeys. *Zeitschrift für Tierpsychologie*, **59**, 89–108.

(1988). Dominance rank order of adult female Japanese monkeys of the Arashiyama B troop. *Reichorui Kenkyu/Primate Research*, **4**, 19–32. (in Japanese with English summary)

(1991). Diachronic changes in the dominance relations of adult female Japanese monkeys of the Arashiyama B group. In *The Monkeys of Arashiyama: Thirty-five Years of Research in Japan and the West*, eds. L. M. Fedigan and P. J. Asquith. Albany, NY: State University of New York Press, pp. 123–139.

Takahata, Y. and Huffman, M. A. (1994). Diachronic changes in the mating relations of Japanese macaques of the Arashiyama B troop. *Reichorui Kenkyu/Primate Research*, **10**, 150. (in Japanese)

Takahata, Y., Koyama, N., Huffman, M., Norikoshi, K. and Suzuki, H. (1995). Are daughters more costly to produce for Japanese macaque mothers? – Sex of the offspring and subsequent interbirth intervals. *Primates*, **35**, 571–574.

Takahata, Y., Koyama, N. and Suzuki, S. (1995). Do the old aged females experience a long post-reproductive life span? – The cases of Japanese macaques and chimpanzees. *Primates*, **36**, 169–180.

Takahata, Y., Suzuki, S., Agetsuma, N. *et al.* (1998). Reproduction of wild Japanese macaque females of Yakushima and Kinkazan Islands: a preliminary report. *Primates*, **39**, 339–349.

Takahata, Y., Huffman, M. A., Suzuki, S., Koyama, N. and Yamagiwa, J. (1999). Why dominants do not consistently attain high mating and reproductive success: A review of longitudinal Japanese macaque studies. *Primates*, **40**, 143–158.

Takahata, Y., Huffman, M. A. and Bardi, M. (2002). Long-term trends in matrilineal inbreeding among the Japanese macaques of Arashiyama B troop. *International Journal of Primatology*, **23**, 399–410.

Takaragawa, N. (1984). Behavioral development of infants in the Arashiyama troop. In *The Japanese Monkeys of Arashiyama: Iwatayama Shizenshi Kenkyūjo Hōkoku, Vol. 3*, eds. H. Suzuki and N. Asaba. Kyoto: Arashiyama Shizenshi Kenkyūjo, pp. 39–54. (in Japanese)

Takenoshita, Y. (1993). Changes in the alpha male and grooming partners of the alpha female in a provisioned troop of Japanese monkeys. *Monkey*, **250–251**, 5–11. (in Japanese)

(1994). Sexual behavior of young female Japanese monkeys of Arashiyama E troop. *Reichorui Kenkyu/Primate Research*, **10**, 124. (in Japanese)

(1998). Male homosexual behaviour accompanied by ejaculation in a free-ranging troop of Japanese macaques (*Macaca fuscata*). *Folia Primatologica*, **69**, 364–367.

(2009). Sex among Japanese macaques. In *Anthropology of Sex*, eds. K. Okuno, W. Shino and Y. Takenoshita. Yokohama: Shunpusha, pp. 321. (in Japanese)

Takenoshita, Y., Sprague, D. and Iwasaki, N. (2005). Factors affecting success rate and accuracy of GPS collar positioning for free-ranging Japanese macaques. *Reichorui Kenkyu/Primate Research*, **21**, 107–119. (in Japanese with English abstract)

Tanaka, T. (1964). Report on the general survey of the troop in Iwatayama Monkey Park. *Yaen*, **18**, 28–30. (in Japanese)

Tanaka, T. and Nigi, H. (1967). Clinical examinations of the Japanese monkey (*Macaca fuscata*). *Primates*, **8**, 91–108.

Tashiro, Y. (1994). Social interactions of aged female Japanese macaques in Arashiyama-E troop. *Reichorui Kenkyu/Primate Research*, **10**, 125. (in Japanese)

Tokuda, K. (1961–62). A study on the sexual behavior in the Japanese monkey troop. *Primates*, **3**, 1–40.

Uehara, S. (1975). The importance of the temperate forest elements among woody food plants utilized by Japanese monkeys and its possible historical meaning for the establishment of the monkeys' range. In *Contemporary Primatology: Proceedings of the 5th International Congress of Primatology*, eds. S. Kondo, M. Kawai and A. Ehara. Basel: S. Karger, pp. 392–400.

Urban, O. (1984). Sexual behavior of male Japanese monkeys. M.A. thesis, Department of Anthropology, McMaster University, Hamilton.

VanderLaan, D. P., Pellis, S. M. and Vasey, P. L. (2010). Mounting and social play in juvenile male Japanese macaques. *Reichorui Kenkyu/Primate Research*, **26**, 70. (Abstract)

Vasey, P. L. (1995). Homosexual behavior in primates: a review of evidence and theory. *International Journal of Primatology*, **16**, 173–204.

(1996). Interventions and alliance formation between female Japanese macaques, *Macaca fuscata*, during homosexual consortships. *Animal Behaviour*, **52**, 539–551.

(1998). Female choice and inter-sexual competition for female sexual partners in Japanese macaques. *Behaviour*, **135**, 579–597.

(2000). Kama Sutra primates. *Equinox*, **110**, 32–33.

(2002). Sexual partner preference in female Japanese macaques. *Archives of Sexual Behavior*, **31**, 45–56.

(2004). Pre- and post-conflict interactions between female Japanese macaques during homosexual consortships. *International Journal of Comparative Psychology*, **17**, 351–359.

(2004). Sex differences in sexual partner acquisition, retention and harassment during female homosexual consortships in Japanese macaques. *American Journal of Primatology*, **64**, 397–409.

(2004). Homosexual behavior. In *Macaque Societies: A Model for the Study of Social Organization*, eds. B. Thierry, M. Singh and W. Kaumanns. Cambridge: Cambridge University Press, pp. 151–154.

(2004). Same-sex sexual behavior and sexual partner preference in female Japanese macaques: behavioral and neuroanatomical research. *Hormones and Behavior*, **46**, 136. (Abstract)

(2006). The pursuit of pleasure: an evolutionary history of female homosexual behaviour in Japanese macaques. In *Homosexual Behaviour in Animals: An Evolutionary Perspective*, eds. V. Sommer, and P. L. Vasey, 1st edn. New York, NY: Cambridge University Press, pp. 191–219.

(2007). Function and phylogeny: the evolution of same-sex sexual behaviour in primates. *Journal of Psychology and Human Sexuality*, **18**, 215–244. (Reprinted in *Handbook of the Evolution of Human Sexuality*, ed. M. R. Kauth. New York, NY: Haworth Press, pp. 215–244).

2010. Are female Japanese macaques that engage in homosexual behaviour masculinized? Behavioural and neuroanatomical evidence. *Reichorui Kenkyu/Primate Research*, **26**, 69. (Abstract)

Vasey, P. L. and Duckworth, N. (2006). Sexual reward via vulvar, perineal and anal stimulation: a proximate mechanism for female homosexual mounting in Japanese macaques. *Archives of Sexual Behavior*, **35**, 523–532.

(2008). Female-male mounting in Japanese macaques: the proximate role of sexual reward. *Behavioural Processes*, **77**, 405–407.

Vasey, P. L. and Gauthier, C. (2000). Skewed sex ratios and female homosexual activity in Japanese macaques: an experimental analysis. *Primates*, **41**, 17–25.

Vasey, P. L. and Jiskoot, H. (2010). The biogeography and evolution of female homosexual behavior in Japanese macaques. *Archives of Sexual Behavior*, **39**, 1439–1441.

Vasey, P. L. and Pfaus, J. G. (2005). A sexually dimorphic hypothalamic nucleus in a macaque species with frequent female-female mounting and same-sex sexual partner preference. *Behavioural Brain Research*, **157**, 265–272.

Vasey, P. L. and Reinhart, C. J. (2009). Female homosexual behavior in a new group of Japanese macaques: evolutionary implications. *Laboratory Primate Newsletter*, **48**, 8–10.

Vasey, P. L. and Sommer, V. (2006). Homosexual behaviour in animals: topics, hypotheses, and research trajectories. In *Homosexual Behaviour in Animals: An Evolutionary Perspective*, eds. V. Sommer and P. L. Vasey. Cambridge: Cambridge University Press, pp. 3–42.

Vasey, P. L., Chapais, B. and Gauthier, C. (1998). Mounting interactions between female Japanese macaques: testing the influence of dominance and aggression. *Ethology*, **104**, 387–398.

Vasey, P. L., Foroud, A., Duckworth, N. and Kovacovsky, S. D. (2006). Male-female and female-female mounting in Japanese macaques: a comparative study of posture and movement. *Archives of Sexual Behavior*, **35**, 117–129.

Vasey, P. L., Rains, D., VanderLaan, D. P., Duckworth, N. and Kovacovsky, S. D. (2008). Courtship behavior during heterosexual and homosexual consortships in Japanese macaques. *Behavioural Processes*, **78**, 401–407.

Vasey, P. L., VanderLaan, D. P., Rains, D., Duckworth, N. and Kovacovsky, S. D. (2008). Inter-mount social interactions during heterosexual and homosexual consortships in Japanese macaques. *Ethology*, **114**, 564–574.

Wakibara, J. V. (1999). Food habits and geophagy among provisioned Japanese monkeys (*Macaca fuscata*) at Arashiyama. *Reichorui Kenkyu/Primate Research*, **15**, 411. (Abstract)

Wakibara, J. V., Huffman, M. A., Wink, M. *et al.* (2001). The adaptive significance of geophagy for Japanese macaques (*Macaca fuscata*) at Arashiyama, Japan. *International Journal of Primatology*, **22**, 495–520.

Wolfe, L. D. (1976). The sexual behavior of the Arashiyama West troop of Japanese macaques (*Macaca fuscata*). *Dissertation Abstract International*, **A37**, 37–47.

(1978). Age and sexual behavior of Japanese macaques. *Archives of Sexual Behavior*, **7**, 55–68.

(1979). Behavioral patterns of estrous females of the Arashiyama West troop of Japanese macaques. *Primates*, **20**, 525–534.

(1979). Sexual maturation among members of a transported troop of Japanese macaques (*Macaca fuscata*). *Primates*, **20**, 411–418.

(1981). Display behavior of three troops of Japanese monkeys (*Macaca fuscata*). *Primates*, **22**, 24–32.

(1981). A case of male adoption in a troop of Japanese monkeys (*Macaca fuscata*). In *Primate Behavior and Sociobiology*, eds. A. B. Chiarelli and R. S. Corruccini. New York, NY: Springer-Verlag. pp. 156–160.

(1981). The reproductive history of a hybrid female (*Macaca mulatta* + *Macaca fuscata*). *Primates*, **22**, 131–134.

(1981). Arashiyama East and West: a comparative study of reproduction. *American Journal of Primatology*, **1**, 316. (Abstract)

(1984). Female rank and reproductive success among Arashiyama B Japanese macaques (*Macaca fuscata*). *International Journal of Primatology*, **5**, 133–143.

(1984). Japanese macaque female sexual behavior: a comparison of Arashiyama East and West. In *Female Primates: Studies by Women Primatologists*, ed. M. E. Small. New York, NY: Alan R. Liss, pp. 141–157.

(1986). Sexual strategies of female Japanese monkeys. *Human Evolution*, **3**, 267–275.

(1986). Reproductive biology of rhesus and Japanese macaques. *Primates*, **27**, 95–101.

(1991). Human evolution and the sexual behavior of female primates. In *Understanding Behavior: What Primate Studies Tell us About Human Behavior*, eds. J. D. Loy and C. B. Peters. New York, NY: Oxford University Press, pp. 121–151.

Wolfe, L. D. and Gray, P. (1982). Japanese monkeys and popular culture. *Journal of Popular Culture*, **10**, 97–105.

Wolfe, L. D. and Noyes, M. J. S. (1981). Reproductive senescence among female Japanese macaques (*Macaca fuscata fuscata*). *Journal of Mammalogy*, **62**, 698–705.

Yoshida, A. (1988). Alarm and estrous calls of a female Japanese monkey (*Macaca fuscata*): a case study. In *Research Reports of the Arashiyama West and East Groups of Japanese Monkeys*. Osaka: Laboratory of Ethological Studies, Faculty of Human Sciences, Osaka University, pp. 57–65.

Yoshida, T., Matsumuro, M., Miyamoto, S. *et al.* (2001). Monitoring the reproductive status of Japanese monkeys (*Macaca fuscata*) by measurement of the steroid hormones in fecal samples. *Primates*, **42**, 367–373.

Zamma, K. (2002). Grooming site preferences determined by lice infection among Japanese macaques in Arashiyama. *Primates*, **43**, 41–49.

Index